Counting Processes and Survival Analysis

Counting Processes and Survival Analysis

Thomas R. Fleming, Ph.D.
Professor of Biostatistics,
University of Washington and
Fred Hutchinson Cancer Research Center

David P. Harrington, Ph.D.
Professor of Biostatistics,
Harvard University and
Dana Farber Cancer Institute

A Wiley-Interscience Publication
John Wiley & Sons, Inc.
New York / Chichester / Brisbane / Toronto / Singapore

Library of Congress Cataloging-in-Publication Data:

Fleming, Thomas R.
Counting Processes and Survival Analysis / Thomas R. Fleming, David P. Harrington.
p. cm. -- (Wiley series in probability and mathematical statistics. Applied probability and statistics, ISSN 0271-6356)
Includes bibliographical references (p.) and indexes.
1. Point processes. 2. Failure time data analysis.
3. Martingales (Mathematics) I. Harrington David P. II. Title.
III. Series.
QA274.42.F44 1991
519.2'3--dc20 90-13051
CIP

ISBN 0-471-52218-X

Printed in the United States of America

10 9 8 7 6 5 4 3 2 1

To
Joli and Anne

Preface

In recent years, there has been considerable activity in the applied and theoretical statistical literature in methods for analyzing data on events observed over time, and on the study of factors associated with the occurrence rates for those events. This literature is now a rich and important subset of applied statistical work. These useful statistical methods can be cast within a unifying counting process framework, providing an elegant martingale based approach to understanding their properties.

This text explores the martingale approach to the statistical analysis of counting processes, with an emphasis on the application of those methods to censored failure time data. This approach was introduced in the 1970's, and has proved remarkably successful in yielding results about statistical methods for many problems arising in censored data. In 1978 Odd Aalen introduced the multiplicative intensity model for counting processes, and his ideas have led to important advances in statistical methods for counting process data and for the censored data found in many biomedical studies. Martingale methods can be used to obtain simple expressions for the moments of complicated statistics, to calculate and verify asymptotic distributions for test statistics and estimators, to examine the operating characteristics of nonparametric testing methods and semiparametric censored data regression methods, and even to provide a basis for graphical diagnostics in model building with counting process data.

It has been our experience that the interaction between applied problems in censored failure time data and the more theoretical perspective of martingale theory has provided important results for both statistical theory and practice, and that students and researchers benefit from an understanding of both the martingale methods used in counting processes, and the results that become available. Since martingale methods for counting processes are powerful, it is not surprising that it takes some time to establish the background material needed for their use. Until recently, most of the results from the calculus of martingales used in the study of counting process methods have appeared in only the probability research literature, and students of statistics often find this literature a formidable challenge. The recent texts by Karr (1986) and Bremaud (1981) have improved the situation, but those books, along with the more theoretical treatments of Liptser and Shiryayev (1977, 1978) and Elliott (1982), are oriented to engineering applications.

We have tried to give a thorough treatment of both the calculus of martingales needed for the study of counting processes, and of the most important applications of these methods to censored data. We explore both classical problems in asymptotic distribution theory for counting process methods, as well as some newer methods for graphical analyses and diagnostics of censored data. There are already some excellent accounts of the older likelihood approach to failure time data, especially the books by Kalbfleisch and Prentice (1980) and Cox and Oakes (1984), and we do not try to duplicate those efforts here. Gill's research monograph (1980) and Jacobsen's lecture notes (1982) illustrate martingale methods for counting processes, but we have tried to make the theoretical development here more nearly self-contained, and we have given much more emphasis to the regression analysis of censored data.

It is our expectation that students with one year of graduate study in statistics will find the presentation to be at an acceptable level. The prerequisite for this book is a familiarity with a measure theoretic treatment of probability that may be found, for instance, in Chung (1974, Chapters 1-4 and 9). The development of the theory of counting processes, stochastic integrals and martingales is provided, but only to the extent required for applications in survival analysis. In technical parts of the book, such as in Chapter 2, a summary of main results is provided for those who wish to skip the detailed development and proceed directly to applications in later chapters.

Chapter 0 provides motivation for the types of inference and estimation procedures commonly encountered in the analysis of censored failure time data. Appendix A briefly reviews some measure theory concepts, and Chapters 1 and 2 introduce the martingale and counting process framework and indicate how the data analysis methods of Chapter 0 can be reformulated in counting process notation. Chapter 3 considers the small sample moments and large sample consistency of standard test statistics and estimators. Chapter 4 presents censored data regression models and corresponding likelihood methods for inference and estimation, and illustrates the application of these methods. The chapter also introduces regression diagnostics and illustrates the use of martingale based residuals. Convergence in distribution for stochastic processes is introduced in Appendix B. Chapter 5 then discusses the martingale central limit theorems used to derive large sample distribution results for many of the survival analysis methods. Large sample properties of the Kaplan-Meier estimator are presented in Chapter 6, while Chapter 7 considers the large sample null distribution, consistency and efficiency of a class of linear rank statistics. The large sample distribution properties for the regression methods in Chapter 4 are established in Chapter 8.

Data sets are provided in Appendix D and a collection of exercises appear in Appendix E. Exercises have been selected both to provide the reader with practice in the application of martingale methods, as well as to give insight into the martingale calculus itself. Some of the exercises guide the reader through details omitted from the text, and some indicate extensions of results.

We appreciate the help provided by so many people during the development of this book. Elaine Nasco showed tireless dedication while typing the many preliminary drafts with uncommon speed and accuracy. Margaret Sullivan Pepe provided many scientific insights and contributions during the early development of this material, especially that in Chapter 2. Michael Parzen prepared the figures, and his graphical expertise and attention to detail is responsible for their high quality.

We thank all those who generously provided careful review and helpful comments on drafts of this book, including the students and faculty at the University of Washington and Harvard University, Robert Smythe and his students at George Washington University, Robert Wolfe and his students at the University of Michigan, the Wiley reviewers, as well as Peter Sasieni, Jon Wellner, Myrto Lefkopoulou, Luc Watelet, and Janet Andersen. We are indebted to Scott Emerson, who provided considerable help, especially through his development of special routines in the Unix statistical language S for survival analysis methods discussed in Chapters 4 and 6; to Jennifer Thomas and Karen Abbett, who provided support in typing parts of the manuscript; to Herman Callaert, who provided a setting for extended writing during a sabbatical (DPH) in Belgium; and to E. Rolland Dickson at Mayo Clinic, and Howard Jaffe and Alan Izu at Genentech, who provided permission to publish the PBC and CGD data sets.

Our deepest gratitude is reserved for two loving families, whose support and gentle encouragement prevented this project from stopping short of its goal many times. Joli (for TRF) and Anne (for DPH) have given the world a new definition of patience, and have taught us much about the value of balance during a demanding time.

THOMAS R. FLEMING
DAVID P. HARRINGTON

Seattle, Washington
Boston, Massachusetts
September 1990

Contents

CHAPTER 0

The Applied Setting

0.1 INTRODUCTION

This chapter introduces censored failure time data—the central applied focus of this book—and the counting process and martingale methods used with these data. The approach here is a mixture of heuristics and common statistical ideas, and it draws freely from simple results in conditional probability, the analysis of contingency tables, and maximum likelihood inference.

This brief introductory chapter has three goals: to provide the background and description of a data set used in later chapters; to illustrate a few of the nonparametric estimators and test statistics that can be applied to these data; and to show how easily those statistics can be rewritten as stochastic integrals with respect to counting processes and martingales. The data set was collected in a randomized trial in primary biliary cirrhosis of the liver conducted at the Mayo Clinic. It contains survival times and a rich collection of clinical and laboratory measurements that can be studied with censored data regression methods. Censored data from clinical trials are often summarized and analyzed with the Nelson and Kaplan-Meier estimators of the survival function, with logrank significance tests for differences in failure time distributions between two or more groups, and with the proportional hazards regression model. This chapter shows that each of these methods results from natural modifications of well-known techniques in uncensored data, and this discussion should provide sufficient background for readers unfamiliar with censored data and their analysis. Stochastic integrals provide a single, elegant representation for censored data statistics arising from seemingly different problems and lead to a unified approach for studying both small sample and asymptotic properties of these statistics. In fact, the implications of this representation are the main theme of this book. This chapter illustrates the integral representation for the Nelson estimator of the cumulative hazard, the logrank statistic, and proportional hazards regression score statistics.

Integrals with respect to counting processes rely on formulas from Stieltjes integration, and Appendix A summarizes some useful aspects of that technique.

Since this chapter uses only the simplest of those results—writing a finite sum as an integral with respect to a counting measure—a brief reading of Appendix A is more than sufficient preparation for this chapter.

0.2 A DATA SET AND SOME EXAMPLES

Between January, 1974 and May, 1984, the Mayo Clinic conducted a double-blinded randomized trial in primary biliary cirrhosis of the liver (PBC), comparing the drug D-penicillamine (DPCA) with a placebo. There were 424 patients who met the eligibility criteria seen at the Clinic while the trial was open for patient registration. Both the treating physician and the patient agreed to participate in the randomized trial in 312 of the 424 cases. The date of randomization and a large number of clinical, biochemical, serologic, and histologic parameters were recorded for each of the 312 clinical trial patients. The data from the trial were analyzed in 1986 for presentation in the clinical literature. For that analysis, disease and survival status as of July, 1986, were recorded for as many patients as possible. By that date, 125 of the 312 patients had died, with only 11 deaths not attributable to PBC. Eight patients had been lost to follow up, and 19 had undergone liver transplantation. Appendix D contains a subset of the data from the 1986 analysis.

PBC is a rare but fatal chronic liver disease of unknown cause, with a prevalence of about 50-cases-per-million population. The primary pathologic event appears to be the destruction of interlobular bile ducts, which may be mediated by immunologic mechanisms. The data discussed here, and in greater detail in Chapter 4, are important in two respects. First, controlled clinical trials are difficult to complete in rare diseases, and this case series of patients uniformly diagnosed, treated, and followed is the largest existing for PBC. The treatment comparison in this trial is more precise than in similar trials having fewer participants and avoids the bias that may arise in comparing a case series to historical controls. Second, the data present an opportunity to study the natural history of the disease. We will see that, despite the immunosuppressive properties of DPCA, there are no detectable differences between the distributions of survival times for the DPCA and placebo treatment groups. This suggests that these groups can be combined in studying the association between survival time from randomization and clinical and other measurements. In the early to mid 1980s, the rate of successful liver transplant increased substantially, and transplant has become an effective therapy for PBC. The Mayo Clinic data set is therefore one of the last allowing a study of the natural history of PBC in patients who were treated with only supportive care or its equivalent. The PBC data are used in three examples: estimating a survival distribution; testing for differences between two groups; and estimation based on a regression model.

In the notation that follows, I_A is the indicator of the event A, taking value one if A occurs, and zero otherwise. All vectors are considered column vectors, and are denoted in boldface, as in $\mathbf{Z}$; $\mathbf{Z}'$ denotes the transpose of the column vector $\mathbf{Z}$.

Example 0.2.1 Estimating a Failure Time Distribution. For each case, the data in the first four columns in Appendix D give, respectively, the case number, the time in days between the date of randomization and either the date of death, or liver transplant, or the date last known alive as of July, 1986, a binary variable which has value 1 if the time in the second column is the time to death and zero otherwise, and a treatment indicator variable which has value 1 if the case was given DPCA, 2 if the case was given placebo. For patients receiving a liver transplant, denoted by an asterisk in column three, column two contains the time between randomization and the date of transplant, and column three contains a zero.

The binary variable recording death is usually referred to as either a failure indicator or a censoring indicator. The cases with value zero were either known to be alive at the time of the 1986 analysis, could not be located for further follow up, or had been given a liver transplant and consequently removed from study on that date. In these cases, the time between randomization and death is known only to have exceeded the recorded value. Data giving the time to a specific event are commonly called failure time data, and the failure times for cases with censoring indicator equal to zero are said to have been censored on the right. The observed times, censored or uncensored, are called the times at risk or observation times.

Let $F(t) = P\{T \le t\}$, where T is the time between the initiation of therapy and death for a patient treated with DPCA. In uncensored data, an estimator for $F(t)$ would usually be based on a parametric statistical model and appropriate parameter estimates, or on the nonparametric empirical cumulative distribution function. Although many parametric methods have been generalized to censored data, we explore a generalization of the empirical distribution function in this example.

Assume then that T is an arbitrary continuous nonnegative random variable with distribution function $F(t)$ and density function $f(t) = dF(t)/dt$. Heuristic arguments for estimating F can be based on the survivor function $S(t) = P\{T > t\}$ for T and the hazard function

$$\begin{aligned}\lambda(t) &= \lim_{\Delta t \downarrow 0} \frac{1}{\Delta t} P\{t \le T < t + \Delta t | T \ge t\} \\ &= -\left[\frac{d}{dt}\{S(t)\}\right] / S(t) \\ &= f(t)/S(t).\end{aligned}$$

The function $\Lambda(t) = \int_0^t \lambda(u)\,du$ is called the cumulative hazard function for T, and it is easy to show that, for continuous T, $S(t) = \exp\{-\Lambda(t)\}$. Slight changes will be needed in these formulas later when failure time variables have mixed continuous and discrete distributions.

In a sample of n items, let $L \le n$ denote the number of distinct observed failure times. Tied values are common in failure time data since follow-up times are often recorded in days or weeks, and the PBC data have several coincident times. Let $T_1^o < T_2^o < \cdots < T_L^o$ be the ordered observed distinct failure times, and D_k the number of observed failures at T_k^o for $k = 1, \dots, L$. Suppose $0 = t_0 < t_1 < \cdots < t_m = t$ is a partition of the interval $[0, t]$, and let d_ℓ and y_ℓ be, respectively, the number of deaths in $[t_{\ell-1}, t_\ell)$ and the number of cases known not to have failed before time $t_{\ell-1}$. The value y_ℓ is also called the number of cases at risk at $t_{\ell-1}$. For small Δt,

$$\begin{aligned}\Lambda(t+\Delta t) - \Lambda(t) &\approx \lambda(t)\,\Delta t \\ &\approx P\{t \le T < t + \Delta t | T \ge t\}.\end{aligned}$$

Thus, when $y_\ell > 0$, a naive but appealing estimate of $\Lambda(t_\ell) - \Lambda(t_{\ell-1})$ is d_ℓ / y_ℓ, and

$$\hat{\Lambda}(t) = \sum_{\ell : t_\ell \le t} d_\ell / y_\ell$$

should estimate the integrated hazard, if one integrates only over those subintervals $[t_{\ell-1}, t_\ell)$ in which $y_\ell > 0$.

Suppose now that $m \to \infty$ and $\max_{1 \le \ell \le m} |t_\ell - t_{\ell-1}| \to 0$. As the intervals shrink, each interval eventually will contain at most one distinct failure time. In those intervals which contain no failure times, $d_\ell = 0$. In the limit,

$$\hat{\Lambda}(t) = \sum_{k : T_k^o \le t} D_k / \overline{Y}_k,$$

where $\overline{Y}_k$ is the number of cases at risk at T_k^o, $k = 1, \dots, L$.

Nelson (1969) first proposed $\hat{\Lambda}$, and it often is referred to as the Nelson cumulative hazard estimator.

Figure 0.2.1 shows a plot of $\hat{\Lambda}$ for the DPCA group, and Figure 0.2.2 shows $\tilde{S} = \exp\{-\hat{\Lambda}\}$.

When $D_k / \overline{Y}_k \approx 0$,

$$\begin{aligned}\tilde{S}(t) &= \prod_{k : T_k^o \le t} \exp(-D_k / \overline{Y}_k) \\ &\approx \prod_{k : T_k^o \le t} \left(1 - D_k / \overline{Y}_k\right).\end{aligned}$$

The last line above is the Kaplan-Meier (1958) product limit estimator, derived in Chapter 3, and denoted throughout by $\hat{S}$. The Kaplan-Meier estimate for the survival distribution for cases treated with DPCA is also shown in Figure 0.2.2. □

The estimators $\hat{\Lambda}$, $\hat{S}$, and $\tilde{S}$ all have more formal derivations and representations which help explain their properties, and these ideas based on counting processes and martingales are explored in Chapters 3 and 6. The counting process approach uses an integral representation for censored data statistics that has a simple form for $\hat{\Lambda}$. Let T_j be the potential failure time of subject j, and U_j a censoring time,

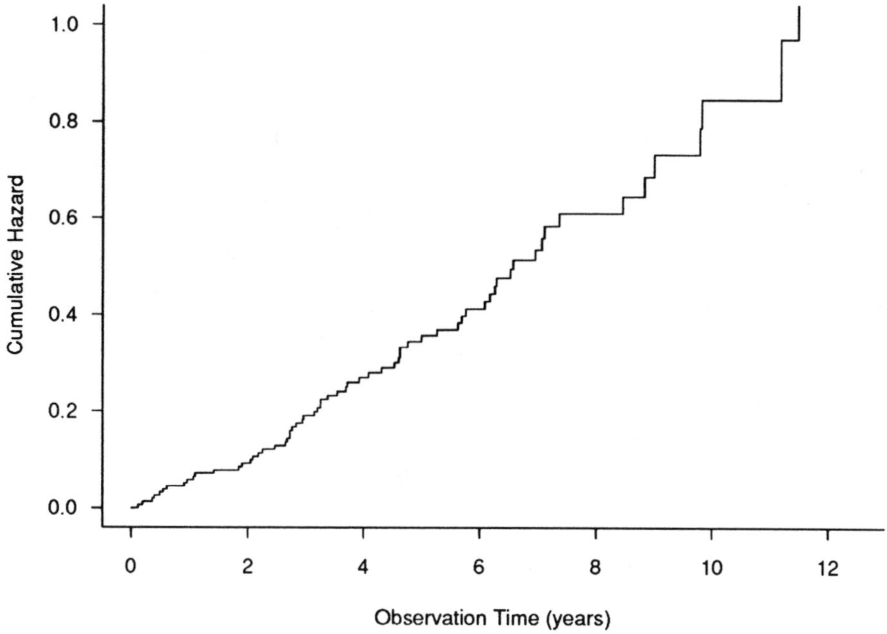

Figure 0.2.1 Nelson cumulative hazard estimate for DPCA group, PBC data.

that is, a time beyond which subject j cannot be observed. The variable T_j will be observed whenever $T_j \leq U_j$. Let $X_j = \min(T_j, U_j)$, and let the processes N_j and Y_j be defined by

$$N_j(t) = I_{\{X_j \leq t, \delta_j = 1\}}$$

and

$$Y_j(t) = I_{\{X_j \geq t\}},$$

where $\delta_j = I_{\{T_j \leq U_j\}}$. If $\overline{N} = \sum_j N_j$ and $\overline{Y} = \sum_j Y_j$, then

$$\hat{\Lambda}(t) = \int_0^t \frac{I_{\{\overline{Y}(u)>0\}}}{\overline{Y}(u)} \, d\overline{N}(u),$$

where we take the integrand to be 0 when both the numerator and denominator vanish. When no statistical model is assumed, information about Λ is available only for $\{u : \overline{Y}(u) > 0\}$, and in fact $\hat{\Lambda}(t)$ really "estimates" the random quantity

$$\Lambda^*(t) = \int_0^t I_{\{\overline{Y}(u)>0\}} \lambda(u) \, du.$$

Simple algebra shows that

$$\hat{\Lambda}(t) - \Lambda^*(t) = \int_0^t \frac{I_{\{\overline{Y}(u)>0\}}}{\overline{Y}(u)} \, d\overline{N}(u) - \int_0^t I_{\{\overline{Y}(u)>0\}} \lambda(u) \, du$$

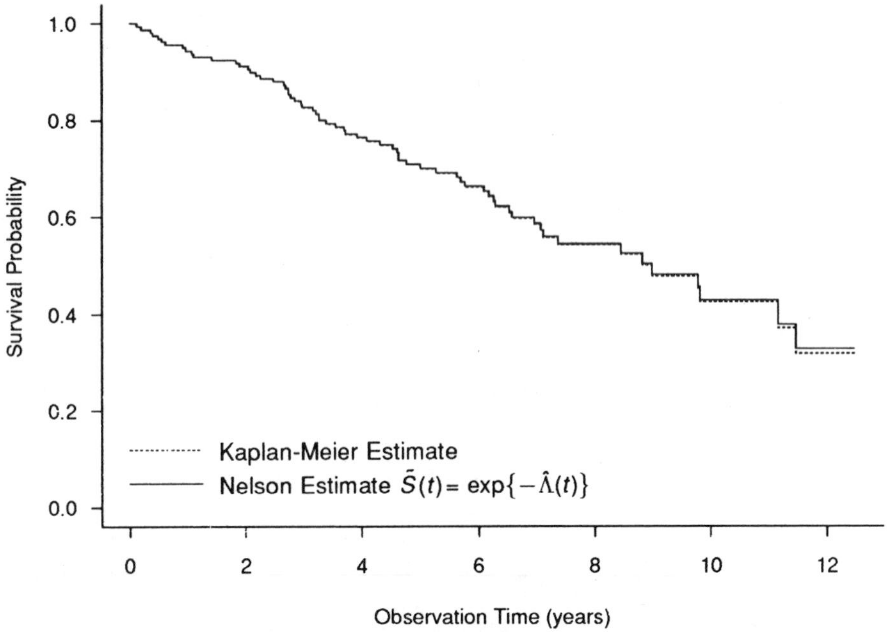

Figure 0.2.2 Estimated survival curves for DPCA group, PBC data.

$$= \int_0^t \frac{I_{\{\overline{Y}(u)>0\}}}{\overline{Y}(u)} \left\{ d\overline{N}(u) - \overline{Y}(u)\lambda(u)\,du \right\},$$

which can also be written as

$$\sum_j \int_0^t \frac{I_{\{\overline{Y}(u)>0\}}}{\overline{Y}(u)}\, dM_j(u), \tag{2.1}$$

with

$$M_j(u) = N_j(u) - \int_0^u Y_j(s)\, d\Lambda(s).$$

The processes M_j play a central role in the later development.

Example 0.2.2 Comparing Two Survival Distributions. Figure 0.2.3 displays the Kaplan-Meier estimates for the distributions of death times for the DPCA and placebo groups. A variety of parametric and nonparametric significance tests can be used to assess observed differences in empirical survival curves. One common nonparametric test is based on the logrank statistic, which has an appealing heuristic derivation and a representation similar to (2.1). We give both here. Chapters 3 and 7 describe more traditional approaches based on likelihood functions and the theory of rank tests.

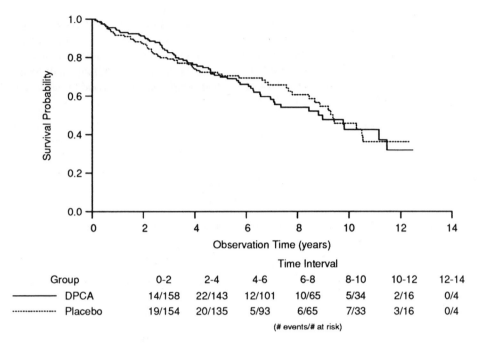

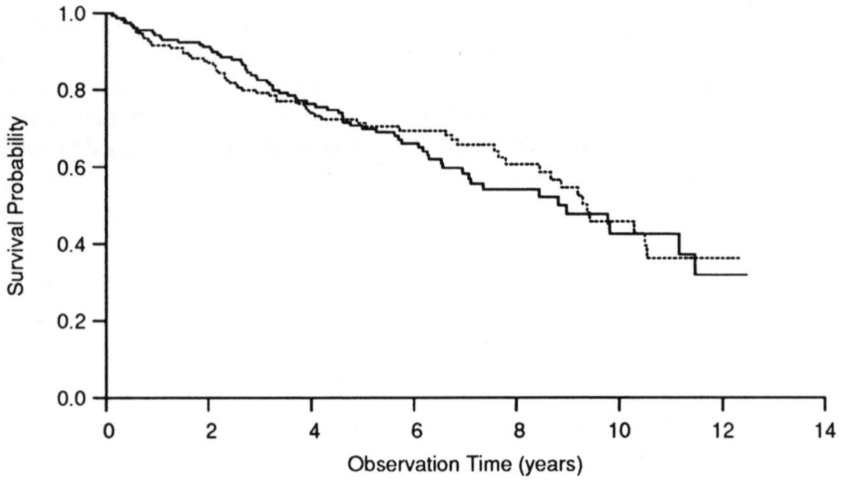

	Time Interval						
Group	0-2	2-4	4-6	6-8	8-10	10-12	12-14
——— DPCA	14/158	22/143	12/101	10/65	5/34	2/16	0/4
········ Placebo	19/154	20/135	5/93	6/65	7/33	3/16	0/4

(# events/# at risk)

Figure 0.2.3 Estimated survival curves for the DPCA and placebo groups, PBC Data. The table below the curves gives the number of failures in each time interval, and the number of cases at risk at the beginning of the interval.

The heuristic development of the logrank statistic uses a conditioning argument based on the numbers at risk of failing just prior to each observed failure time. Let $T_1^o < \ldots < T_L^o$ denote the ordered observed distinct failure times in the sample formed by combining the two groups, and let D_{ik} and $\overline{Y}_{ik}$, $k = 1, \ldots, L$, $i = 1, 2$, denote the number of observed failures and number at risk, respectively, in sample i at time T_k^o. Take D_k and $\overline{Y}_k$ to be the corresponding values in the combined sample. The data at T_k^o can be summarized as in Table 0.2.1.

Table 0.2.1 Numbers of cases failing and not failing at T_k^o from those at risk, by sample.

	Sample		
Failure	1	2	Total
Yes	D_{1k}	D_{2k}	D_k
No	$\overline{Y}_{1k} - D_{1k}$	$\overline{Y}_{2k} - D_{2k}$	$\overline{Y}_k - D_k$
Total	$\overline{Y}_{1k}$	$\overline{Y}_{2k}$	$\overline{Y}_k$

Given $\overline{Y}_{ik}$, the D_{ik} have a binomial distribution with number of trials $\overline{Y}_{ik}$ and, under the hypothesis of a common failure rate λ in the two groups, approximate event probability $\lambda(T_k^o)\Delta t$. Fisher's exact test for equal binomial parameters in this setting is based on conditioning further on D_k and on using the resulting conditional hypergeometric distribution for D_{1k}. This distribution will have conditional mean E_{1k} and variance V_{1k} given by

$$E_{1k} = D_k \frac{\overline{Y}_{1k}}{\overline{Y}_k}$$

and

$$V_{1k} = D_k \frac{\overline{Y}_{1k}(\overline{Y}_{2k})}{\overline{Y}_k^2} \frac{\overline{Y}_k - D_k}{\overline{Y}_k - 1}.$$

Given the margins in each of the L tables at the observed death times,

$$\{D_{11} - E_{11}, \ldots, D_{1L} - E_{1L}\}$$

is a vector of observed-minus-conditionally-expected number of failures across observed failure times, and if we assume these differences are independent,

$$Q = \sum_{k=1}^{L}(D_{1k} - E_{1k})/(\sum_{k=1}^{L} V_{1k})^{1/2}$$

should have approximately a standard normal distribution. The statistic Q is the standardized two-sample logrank statistic and, as we will see later, the normal distribution does in fact often adequately approximate the sampling distribution of Q.

For the PBC data, $Q = .3189$ and $P\{|Q| > .3189\} = .7598$, so these data do not contradict the null hypothesis of no differences between treatment and placebo. □

Some interesting issues arise in this derivation of the logrank statistic. The vector of observed-minus-expected failures does not in fact have independent components (Exercise 0.1) and the central limit theorem usually applied to prove asymptotic normality does not work here. Differences between observed and expected failures are given equal weight, regardless of the size of the risk set (i.e., numbers of cases still under observation) at observed failure times, and such weighting will have implications for the operating characteristics of Q. These more delicate aspects of censored data significance tests are studied with counting process methods in Chapters 3 and 7, and those methods will rely on a representation much like (2.1) for Q and related statistics.

Let (T_{ij}, U_{ij}), $j = 1, \ldots, n_i$, $i = 1, 2$, denote two samples of failure and censoring time pairs, let $X_{ij} = \min(T_{ij}, U_{ij})$, and take $\delta_{ij} = I_{\{T_{ij} \leq U_{ij}\}}$. As in Example 0.2.1, simple counting processes $N_{ij}(t)$ can be defined as

$$N_{ij}(t) = I_{\{X_{ij} \leq t, \delta_{ij} = 1\}},$$

with associated *at-risk* processes

$$Y_{ij}(t) = I_{\{X_{ij} \geq t\}}.$$

If $\overline{N}_i = \sum_j N_{ij}$, $\overline{N} = \sum_i \overline{N}_i$, with similar definitions for $\overline{Y}_i$ and $\overline{Y}$, then under the hypothesis of a common failure rate the numerator of Q may be rewritten as

$$\begin{aligned}
\sum_{k=1}^{L}\{D_{1k} - E_{1k}\} &= \sum_{k=1}^{L} D_{1k} - \sum_{k=1}^{L} \frac{\overline{Y}_{1k}}{\overline{Y}_k} D_k \\
&= \sum_{j=1}^{n_1} \int_0^\infty dN_{1j}(s) - \sum_{i=1}^{2} \sum_{j=1}^{n_i} \int_0^\infty \frac{\overline{Y}_1(s)}{\overline{Y}(s)} dN_{ij}(s) \\
&= \sum_{j=1}^{n_1} \int_0^\infty \frac{\overline{Y}_2(s)}{\overline{Y}(s)} dN_{1j}(s) - \sum_{j=1}^{n_2} \int_0^\infty \frac{\overline{Y}_1(s)}{\overline{Y}(s)} dN_{2j}(s) \\
&= \sum_{j=1}^{n_1} \int_0^\infty \frac{\overline{Y}_1(s)\overline{Y}_2(s)}{\overline{Y}(s)} \{\overline{Y}_1(s)\}^{-1} \{dN_{1j}(s) - Y_{1j}(s)\lambda(s)ds\} \\
&\quad - \sum_{j=1}^{n_2} \int_0^\infty \frac{\overline{Y}_1(s)\overline{Y}_2(s)}{\overline{Y}(s)} \{\overline{Y}_2(s)\}^{-1} \{dN_{2j}(s) - Y_{2j}(s)\lambda(s)ds\} \\
&= \sum_{i=1}^{2} \sum_{j=1}^{n_i} \int_0^\infty (-1)^{i-1} \frac{\overline{Y}_1(s)\overline{Y}_2(s)}{\overline{Y}(s)} \{\overline{Y}_i(s)\}^{-1} dM_{ij}(s) \\
&\equiv \sum_{i,j} \int_0^\infty H_i(s) dM_{ij}(s), \qquad (2.2)
\end{aligned}$$

where $M_{ij}(s) = N_{ij}(s) - \int_0^s Y_{ij}(u)\lambda(u)du$. Although the integrands $H_i = (-1)^{i-1}$ $\overline{Y}_1 \overline{Y}_2 (\overline{Y}\,\overline{Y}_i)^{-1}$ differ in form from that in (2.1), the integrating processes M_{ij} are identical.

The randomization in the Mayo Clinic trial guarantees approximate balance of both recorded and unrecorded covariates for patients in the two treatment groups, and thus makes it unlikely that the treatment comparison is confounded by a higher proportion of poor prognosis patients in one of the two treatment groups. This balance may not be present in nonrandomized or observational studies, and regression models can often be used in an analysis of covariance to adjust for variables that may confound a treatment comparison. Even in a randomized trial, these models can be used to estimate the association between covariates and censored failure times, and to help identify clinical or other features of a disease that place a patient at high risk of early failure. Example 0.2.3 introduces the proportional hazards regression model frequently used with censored data.

Example 0.2.3 A Censored Data Regression Model. In addition to the time at risk, censoring indicator, and treatment variables, Appendix D contains

PBC data on 16 demographic, clinical, biochemical, and histologic measurements made at the time of randomization. Patient age and sex are the demographic variables. Clinical measurements recorded the presence or absence of ascites, hepatomegaly, spiders, and edema. Ascites is an accumulation of fluid in the abdominal cavity; hepatomegaly is a swelling or enlargement of the liver; spiders are vascular lesions formed by the dilation of a small group of blood vessels, generally occurring on the upper chest and arms; and edema is a swelling caused by excess fluid in subcutaneous tissue. Biochemical measurements included the levels of bilirubin, albumin, urine copper, alkaline phosphatase, SGOT, cholesterol and triglycerides. Bilirubin is a liver bile pigment; albumin is a protein found in the blood; urine copper is self-explanatory; alkaline phosphatase and SGOT are enzymes. Serum cholesterol and triglycerides are blood lipoproteins. In addition, the patient's platelet count and prothrombin time were recorded. Prothrombin is a blood coagulation agent, and prothrombin time is the time until a blood sample begins coagulation in a certain laboratory test. Finally, Appendix D contains the histologic stage of the disease, graded 1, 2, 3, or 4, with higher stage disease denoting worse prognosis. Specific coding and units of measurement for all of the variables can be found in that Appendix.

The proportional hazards model (Cox, 1972) is frequently used to estimate the effect of one or more covariates on a failure time distribution. Let $\mathbf{Z}' = (Z_1, \ldots, Z_p)$ denote p measured covariates on a given individual with censored failure time observation $(X = \min(T, U), \delta)$. In the proportional hazards model,

$$\begin{aligned}\lambda(t|\mathbf{Z}) &\equiv \lim_{\Delta t \downarrow 0} \frac{1}{\Delta t} P\{t \le T < t + \Delta t \mid T \ge t, \mathbf{Z}\} \\ &= \lambda_0(t) \exp(\boldsymbol{\beta}'\mathbf{Z}), \end{aligned} \tag{2.3}$$

where $\lambda_0(t)$ is a baseline hazard corresponding to $\mathbf{Z}' = (0, \ldots, 0)$, $\boldsymbol{\beta}' = (\beta_1, \ldots, \beta_p)$ is a vector of regression coefficients, and $\boldsymbol{\beta}'\mathbf{Z}$ is an inner product. Standard likelihood methods cannot be used to estimate $\boldsymbol{\beta}$ when no parametric model is used for λ_0. The components of $\boldsymbol{\beta}$ can be estimated, however, by maximizing a "partial likelihood" function with respect to $\boldsymbol{\beta}$, even when λ_0 is left completely unspecified. This approach is discussed in detail in Chapter 4, and we provide only an outline here.

The partial likelihood function in the proportional hazards model is based on a conditional probability argument. Suppose $(A_1, B_1), (A_2, B_2), \ldots, (A_K, B_K)$ is a collection of pairs of events. Then the likelihood of all $2K$ events is:

$$\begin{aligned} &P\{A_K B_K A_{K-1} B_{K-1} \ldots A_1 B_1\} \\ &\quad = \left[\prod_{k=2}^{K} P\{A_k B_k \mid A_{k-1} B_{k-1} \ldots A_1 B_1\}\right] P\{A_1 B_1\} \\ &\quad = \left[\prod_{k=2}^{K} P\{A_k \mid B_k A_{k-1} B_{k-1} \ldots A_1 B_1\}\right] P\{A_1 \mid B_1\} \end{aligned}$$

$$\times \left[\prod_{k=2}^{K} P\{B_k \mid A_{k-1}B_{k-1} \dots A_1B_1\} \right] P\{B_1\}.$$

When all four terms in the above product depend on a statistical parameter but the last two are neglected, the first two terms form a partial likelihood for that parameter.

For simplicity, consider a censored data regression problem with no tied values in the set of observation times. Again set $X_j = \min(T_j, U_j)$ and $\delta_j = I_{\{X_j = T_j\}}$, so the observed data are $(X_j, \delta_j, \mathbf{Z}_j)$, $j = 1, \dots, n$. As above, let $T_1^o < \dots < T_L^o$ denote the L ordered times of observed failures. Set $T_0^o \equiv 0$ and $T_{L+1}^o \equiv \infty$. Let (k) provide the case label for the patient failing at T_k^o (thus $T_{(k)} = T_k^o$) so the covariates associated with the L failures are $\mathbf{Z}_{(1)}, \dots, \mathbf{Z}_{(L)}$. Let B_k be the event describing the observed times of censoring in the interval $[T_{k-1}^o, T_k^o)$, the case labels associated with the censored times, and the fact that a failure has been observed at T_k^o. If A_k is the event specifying the label, (k), of the case failing at T_k^o, then the observed data are equivalent to the event $B_1A_1 \dots B_LA_LB_{L+1}$ and the likelihood of the data will be

$$P\{B_1A_1 \dots B_LA_LB_{L+1}\}.$$

The baseline hazard λ_0 is unspecified in the proportional hazards model and may in fact be arbitrarily small in the intervals (T_{k-1}^o, T_k^o). In most instances, it is also reasonable to assume that the censoring times do not provide additional information about the failure distribution that would not be available if censoring were not present. (This point is addressed in detail in later chapters.) It seems plausible to assume, then, that the events B_k contain little information about the regression parameter β. A reasonable partial likelihood for β will therefore be

$$\left[\prod_{k=2}^{L} P\{A_k \mid B_kA_{k-1}B_{k-1} \dots A_1B_1\} \right] P\{A_1 \mid B_1\}.$$

For fixed k, since there are no ties in the observation times, it can be established (Chapter 4) that

$$\begin{aligned} P\{A_k \mid B_kA_{k-1}B_{k-1} \dots A_1B_1\} &= \frac{\lambda(T_k^o \mid \mathbf{Z}_{(k)})}{\sum_{j \in R_k} \lambda(T_k^o \mid \mathbf{Z}_j)} \\ &= \frac{\exp(\beta' \mathbf{Z}_{(k)})}{\sum_{j \in R_k} \exp(\beta' \mathbf{Z}_j)}, \end{aligned}$$

where R_k is the set of cases at risk at time T_k^o, that is, $\{j : X_j \geq T_k^o\}$. The partial likelihood for β will thus be

$$L(\beta) = \prod_{k=1}^{L} \frac{\exp(\beta' \mathbf{Z}_{(k)})}{\sum_{j \in R_k} \exp(\beta' \mathbf{Z}_j)}.$$

The log partial likelihood is given by

$$\log L(\beta) = \sum_{k=1}^{L} \left[\beta' \mathbf{Z}_{(k)} - \log \left\{ \sum_{j \in R_k} \exp(\beta' \mathbf{Z}_j) \right\} \right].$$

The efficient score for β, $\mathbf{U}(\beta) = \partial/\partial\beta \log L(\beta)$, is

$$\mathbf{U}(\beta) = \sum_{k=1}^{L} \left\{ \mathbf{Z}_{(k)} - \frac{\sum_{j \in R_k} \mathbf{Z}_j \exp(\beta' \mathbf{Z}_j)}{\sum_{j \in R_k} \exp(\beta' \mathbf{Z}_j)} \right\}. \tag{2.4}$$

Maximum partial likelihood estimates β are found by solving the p simultaneous equations $\mathbf{U}(\beta) = \mathbf{0}$.

When the data contain tied observation times, the argument deriving the partial likelihood becomes more delicate, and is given in detail in Chapter 4. One approximation replaces the partial likelihood with

$$\prod_{k=1}^{L} \frac{\exp(\beta' \mathbf{S}_{(k)})}{\left\{ \sum_{j \in R_k} \exp(\beta' \mathbf{Z}_j) \right\}^{D_k}}$$

where D_k is the number of observed failures at T_k^o, and $\mathbf{S}_{(k)}$ is the sum of the covariates of the subjects failing at T_k^o.

Table 0.2.2 shows the estimated regression coefficients and their standard errors obtained by fitting this model with some of the variables or their transformations in the PBC data set. The variables in this model are age, the presence of edema, the treatment variable, and the natural logarithms of albumin, bilirubin, and prothrombin time. Chapter 4 contains a detailed description of how these variables or their transformations were chosen, so we defer that issue for now. Methods for estimating the standard errors (s.e.) of the regression coefficients and for establishing the asymptotic normality of standardized components of $\hat{\beta}$ are discussed in Chapters 4 and 8.

Table 0.2.2 Regression coefficients and their standard errors from an estimated proportional hazards model in the PBC data.

	$\hat{\beta}$	Standard Error (s.e.)	$\hat{\beta}/\text{s.e.}\hat{\beta}$
Age	0.0347	0.00891	3.89
log (Albumin)	−3.0771	0.71899	−4.28
log (Bilirubin)	0.8840	0.09871	8.96
Edema	0.7859	0.29647	2.65
log (Prothrombin Time)	2.9707	1.01588	2.92
Treatment	0.1360	0.18543	0.73

Table 0.2.2 has several implications. The coefficient for treatment effect is not significantly different from zero, even in a model that adjusts for variables which are clearly associated with failure. As expected, this result is consistent with the value of the logrank test in the unadjusted treatment comparisons. When $\beta_i > 0$, $\lambda(t|\mathbf{Z})$ increases with component Z_i when other covariates are held constant, so the model fit implies that increasing values of age, log bilirubin, log prothrombin time, and the presence of edema are all associated with higher probability of early death. Conversely, increasing values of log albumin levels are associated with lower risk of early death. □

Fitting regression models in censored data is a delicate enterprise with more than a few pitfalls and Chapter 4 discusses some of these complications. For the time being, it is more important to discover the simple "martingale representation" for the score statistic $\mathbf{U}(\beta)$ in Eq. (2.4). As before, let $N_j(t) = I_{\{X_j \le t,\ \delta_j = 1\}}$ and

$$M_j(t) = N_j(t) - \int_0^t I_{\{X_j \ge u\}} \lambda(u|\mathbf{Z}_j) du$$

$$= N_j(t) - \int_0^t I_{\{X_j \ge u\}} \lambda_0(u) e^{\beta' \mathbf{Z}_j} du.$$

Then

$$\mathbf{U}(\beta) = \sum_{j=1}^{n} \int_0^\infty \mathbf{H}_j(u, \beta, \mathbf{Z}_1, \ldots, \mathbf{Z}_n) dN_j(u),$$

where

$$\mathbf{H}_j(u, \beta, \mathbf{Z}_1, \ldots, \mathbf{Z}_n) = \mathbf{Z}_j - \frac{\sum_{i=1}^n \mathbf{Z}_i I_{\{X_i \ge u\}} e^{\beta' \mathbf{Z}_i}}{\sum_{i=1}^n I_{\{X_i \ge u\}} e^{\beta' \mathbf{Z}_i}}.$$

(The integral of the p-dimensional vector $\mathbf{H}_j$ is a simple p-vector whose components are integrals of the components of $\mathbf{H}_j$.) Substitution and some careful algebra show that $\mathbf{U}(\beta)$ can also be written

$$\mathbf{U}(\beta) = \sum_{j=1}^{n} \int_0^\infty \mathbf{H}_j(u, \beta, \mathbf{Z}_1, \ldots, \mathbf{Z}_n) dM_j(u). \quad (2.5)$$

Although the integrands $\mathbf{H}_j$ are more complicated than those in Eqs. (2.1) or (2.2), the form of the score function for β is nevertheless similar to that of the cumulative hazard estimator $\hat{\Lambda}$ and the logrank statistic.

The martingale representation in Eq. (2.5) for $\mathbf{U}(\beta)$ serves several purposes. First, since the representation for the score is so similar to one used for simpler statistics such as $\hat{\Lambda}$ and the logrank statistic, many of the same methods used to study the properties of simple methods of inference for censored data can be used with regression models. This unified approach helps show that many of the identities and formulas used with the proportional hazards regression model are not just great strokes of algebraic luck, but are instead natural consequences of mathematical structure. Second, the counting process representation leads to useful model diagnostics (also discussed in Chapter 4) without which regression models become black boxes hiding potentially embarrassing mistakes in data analysis. Third, and perhaps most important, the representation suggests that these regression models can be used with data more general than censored failure time data. In particular, there is no compelling reason why N_j should only count the occurrence of a single event, and not, say, record the times at which one or more types of events occur repeatedly. The function λ_j then models the rate at which the process N_j jumps. Chapter 4 explores the implication of such a model in a study of the effect of gamma interferon on the rate of infections in children having chronic granulomatous disease, a rare disorder of the immune system.

These examples suggest that methods for analyzing censored data can be studied through the processes $\sum_j \int H_j dM_j$, where $M_j = N_j - \int Y_j \lambda_j$. Chapter 1 begins that study with some necessary background material from the theory of probability and stochastic processes, and with a careful look at the processes $M_j = N_j - \int Y_j \lambda_j$ and $\sum_j \int H_j dM_j$. Chapter 2 completes the study of these simple processes by relaxing some of the regularity conditions used in Chapter 1, and by providing formulas for first and second moments of such processes. In Chapter 3, the tools of the first two chapters are applied to small sample properties of nonparametric estimators of cumulative hazard functions and survival distributions, and to some two-sample significance tests. Chapter 4 examines the proportional hazards regression model and its generalization in depth. An extended treatment of martingale central limit theorems for counting processes is given in Chapter 5. Chapters 6, 7, and 8 apply these limit theorems to asymptotic distribution theory for, respectively, survival distribution estimators, nonparametric significance tests with censored data, and regression models.

Finally, we say a few words here about terminology and notation. The analysis of data recording the time to an event has origins in the construction of population life tables and, more recently, in the study of chronic diseases and in reliability and life-testing in engineering. As a result, censored data are often referred to generically as survival data or failure time data, and we use both terms. Obviously, the event of interest need not be death or failure; the subject could more accurately be called the analysis of censored event time data, or censored waiting time data. The more apropos name obscures the genesis of the subject, however, and for that reason we have chosen to continue using current terminology.

To maintain consistent notation throughout this chapter, the item or subject was indexed by the subscript j. In future chapters requiring double subscripting for sample and subject (i.e., Chapters 3, 6 and 7), by convention i will denote sample, and j will denote subject. In Chapters 2, 4, 5, and 8, subject will be indexed by i.

CHAPTER 1

The Counting Process and Martingale Framework

1.1 INTRODUCTION

Counting process and martingale methods provide direct ways of studying small and large sample properties of estimators and significance tests for right censored failure time data, and provide tools for analyzing event history data more complicated than censored data. The counting process approach relies heavily on results from probability theory and stochastic processes and some preliminary material must be studied to exploit the approach. In this chapter and in the next, the prerequisites for using counting processes and associated martingales are developed in sufficient generality for the rest of the book.

We presume the reader is familiar with the measure theoretic treatment of probability that may be found, for instance, in Chung (1974), Chapters 1 through 4 and Chapter 9, and we freely use the language of that approach. To prevent ambiguity, we restate here a few definitions and results from probability theory and provide a more extended review in Appendix A. We also give a brief outline of the necessary material from stochastic processes. Additional details may be found in Liptser and Shiryayev (1977), in the Appendices in Bremaud (1981), and in the first chapter of Jacod and Shiryayev (1987).

1.2 STOCHASTIC PROCESSES AND STOCHASTIC INTEGRALS

Observational or experimental data gathered over a period of time are often modeled with stochastic processes. In particular, data counting the numbers of events of different types that occur over time can be modeled with counting processes. While our goal is to develop explicit distributional results about methods for analyzing counting process data, the structure of these methods is most clear when studied with some generality. We begin therefore with some definitions for random variables and stochastic processes.

Assume that Ω is a space of outcomes of a random experiment, with each outcome denoted generically by ω, and $\mathcal{F}$ and P are, respectively, a σ-algebra of events and a probability measure on Ω.

Definition 1.2.1. A function Z from Ω to the real line $R = (-\infty, \infty)$ is called a *random variable* or is called *measurable* (relative to $\mathcal{F}$) if

$$\{Z \leq x\} = \{\omega : Z(\omega) \leq x\} \in \mathcal{F},$$

for all x. That is, Z is a measurable mapping from $(\Omega, \mathcal{F}, P)$ to the real line equipped with the Borel σ-algebra $\mathcal{B}$. □

For brevity, we will say that Z is $\mathcal{F}$-measurable, or is an $\mathcal{F}$-measurable random variable.

Definition 1.2.2. A (real-valued) *stochastic process* is a family of random variables $X = \{X(t) : t \in \Gamma\}$ indexed by a set Γ, all defined on the same probability space $(\Omega, \mathcal{F}, P)$. □

The set Γ nearly always indexes time, and is usually either $\{0, 1, 2, ...\}$ (discrete-time processes) or $[0, \infty)$ (continuous-time processes). We will almost always restrict attention to $\Gamma = R^+ = [0, \infty)$ and will denote continuous-time processes by $\{X(t) : t \geq 0\}$, or by just X when there can be no confusion with ordinary random variables. Occasionally, we will refer to "the process $X(t)$" or "the process $X(n)$" (for discrete time) when providing a definition for X at time t or n, rather than use the correct but more cumbersome "the process X whose value at t (or n) is given by $X(t)$ (or $X(n)$)."

For a stochastic process X, the (random) functions $X(\cdot, \omega) : R^+ \rightarrow R, \omega \in \Omega$, are called the sample paths or trajectories of X. A process will be called right- or left-continuous, of bounded variation, increasing, or said to have limits from the left or right if the set of sample paths with the corresponding property has probability one. (The definition of a bounded process given in Definition 1.2.3 is more restrictive, however.)

Two random variables X and Y are called *equivalent* if $P\{X \neq Y\} = 0$, that is, X and Y differ on at most a set of probability zero. The usual notation is that $X = Y$ almost surely (a.s.). A stochastic process X is called a *modification* of another process Y if, for any t, $X(t)$ and $Y(t)$ are equivalent random variables. When X is a modification of Y, it is not difficult to show that, for any positive integer k, $P\{X(t_1) \leq x_1, \ldots, X(t_k) \leq x_k\} = P\{Y(t_1) \leq x_1, \ldots, Y(t_k) \leq x_k\}$, that is, X and Y have the same finite-dimensional distributions. The processes X and Y are *indistinguishable* if $P\{\omega : X(t) = Y(t) \text{ for every } t \geq 0\} = 1$. When X and Y are indistinguishable, the sample paths of X coincide with those of Y for almost all ω. It is possible to show that if both X and Y are left-continuous or both are right-continuous, then X and Y are indistinguishable if and only if they are modifications of one another.

A process X is a mapping from $[0, \infty) \times \Omega$ to the real line, and $X(t, \omega)$ is the value of the random variable $X(t)$ for the outcome $\omega \in \Omega$. If $X(\cdot, \cdot) : R^+ \times \Omega \to R$ is a measurable mapping when R^+ is equipped with the Borel σ-algebra $\mathcal{B}$, then the process X is called *measurable*. In such cases Fubini's theorem implies the sample paths of X are measurable functions from R^+ to R. This will be important when, for fixed ω, X appears in the integrand of integrals over subsets of the positive real line.

It is shown in Appendix A that any process whose paths are right- or left-continuous with probability one is measurable.

There is a large amount of literature on the structure of underlying probability spaces for stochastic processes, but only very little is directly relevant to the applications studied here. Subject to consistency requirements on the joint distributions of $\{X(t_1), \ldots, X(t_k)\}$ for distinct time points $t_1, \ldots, t_k$, the probability space Ω for a process X may be chosen as the space of all paths of X. Jacobsen (1982) illustrates this idea in detail for the counting processes defined later. Integrability and boundedness properties of stochastic processes will often be used later, and those properties frequently referenced are collected in Definition 1.2.3.

Definition 1.2.3. A stochastic process X is

1. *Integrable* if $\sup_{0 \le t < \infty} E|X(t)| < \infty$;
2. *Square integrable* if $\sup_{0 \le t < \infty} E\{X(t)\}^2 < \infty$;
3. *Bounded* if there exists a finite constant Γ such that

$$P\{\sup_{0 \le t < \infty} |X(t)| < \Gamma\} = 1. \qquad \square$$

The following definition will allow the rigorous formulation of the concept of information accruing over time.

Definition 1.2.4.

1. A family of sub-σ-algebras $\{\mathcal{F}_t : t \ge 0\}$ of a σ-algebra $\mathcal{F}$ is called *increasing* if $s \le t$ implies $\mathcal{F}_s \subset \mathcal{F}_t$. (i.e., if for $s \le t$, $A \in \mathcal{F}_s$ implies $A \in \mathcal{F}_t$). An increasing family of sub-σ-algebras is called a *filtration*.
2. When $\{\mathcal{F}_t : t \ge 0\}$ is a filtration, the σ-algebra $\bigcap_{h>0} \mathcal{F}_{t+h}$ is usually denoted by $\mathcal{F}_{t+}$. The corresponding limit from the left, $\mathcal{F}_{t-}$, is the smallest σ-algebra containing all the sets in $\bigcup_{h>0} \mathcal{F}_{t-h}$ and is written $\sigma\left\{\bigcup_{h>0} \mathcal{F}_{t-h}\right\}$ or $\bigvee_{h>0} \mathcal{F}_{t-h}$.
3. A filtration $\{\mathcal{F}_t : t \ge 0\}$ is *right-continuous* if, for any t, $\mathcal{F}_{t+} = \mathcal{F}_t$.
4. A *stochastic basis* is a probability space $(\Omega, \mathcal{F}, P)$ equipped with a right-continuous filtration $\{\mathcal{F}_t : t \ge 0\}$, and is denoted by $(\Omega, \mathcal{F}, \{\mathcal{F}_t : t \ge 0\}, P)$.

5. A stochastic basis is called *complete* if $\mathcal{F}$ contains any subset of a P-null set (so $\mathcal{F}$ is complete) and if each $\mathcal{F}_t$ contains all P-null sets of $\mathcal{F}$. □

Stochastic bases arise frequently in the general theory of stochastic processes developed by the French school of probability (cf. Meyer, 1966, 1976). Right-continuity and completeness arise so often there that this pair of conditions is often referred to as "les conditions habituelles de la théorie générale des processus," or "the usual conditions..." (cf, Dellacherie, 1972).

The most natural filtrations are *histories* of stochastic processes, or families with $\mathcal{F}_t = \sigma\{X(s) : 0 \leq s \leq t\}$, the smallest σ-algebra with respect to which each of the variables $X(s), 0 \leq s \leq t$, is measurable. In this case, $\mathcal{F}_t$ "contains the information" generated by the process X on $[0, t]$.

Definition 1.2.5. A stochastic process $\{X(t) : t \geq 0\}$ is *adapted* to a filtration if, for every $t \geq 0$, $X(t)$ is $\mathcal{F}_t$-measurable. □

Obviously, any process is adapted to its history.

Counting processes and their properties will be central in our development.

Definition 1.2.6. A *counting process* is a stochastic process $\{N(t) : t \geq 0\}$ adapted to a filtration $\{\mathcal{F}_t : t \geq 0\}$ with $N(0) = 0$ and $N(t) < \infty$ a.s., and whose paths are with probability one right-continuous, piecewise constant, and have only jump discontinuities, with jumps of size $+1$. □

The term counting process suggests their most frequent application: $N(t) - N(s)$ will almost always denote the number of events of a certain type occurring in the interval $(s, t]$. The Poisson process is the most prevalent example of a counting process, and some of its properties are explored in Exercise 1.1.

If N is a counting process, f is some (possibly random) function of time, and $0 \leq s < t \leq \infty$, then $\int_s^t f(u)dN(u)$, or more concisely $\int_s^t f dN$, is the Stieltjes integral representation of the sum of the values of f at the jump times of N in the interval $(s, t]$. This notation was used in Chapter 0 for alternative representations of some censored data statistics, and will be used in more general situations where N is replaced by processes that may not be counting processes. Whenever we write $\int_s^t X dY (= \int_{(s,t]} X dY)$, Y will be a right-continuous process of bounded variation on each subinterval of $(s, t]$ and, for almost any ω, $X(\cdot, \omega)$ will be a Borel measurable real-valued function on $(s, t]$. Any bounded variation process Y can be written as the difference $Y_1 - Y_2$ of two nondecreasing processes, and all sample path properties of the integral $\int_s^t X dY$ (considered as a function of t) follow from properties of integrals with respect to nondecreasing functions. In particular, $\int_s^t X dY$ will always be a random variable, and hence $\left\{\int_s^t X dY : s \leq t < \infty\right\}$ will be a stochastic process with index set $[s, \infty)$, where we assume $\int_s^s X dY = 0$. (A proof may be found in Jacobsen, 1982, Proposition 3.2.4). Integrals will often have lower limit zero and, unless otherwise noted, we will henceforth use $\int X dY$ to denote the stochastic process whose value at time t is $\int_0^t X dY$.

When X is left-continuous and Y is right-continuous, the Lebesgue-Stieltjes and Riemann-Stieltjes definitions of $\int_s^t X dY$ coincide. Properties of Lebesgue-Stieltjes integrals are summarized in Appendix A.

When X is sufficiently regular (in a sense defined later) and Y is a special process called a martingale, the process $\int X dY$ will also be a martingale. The integral representation for statistics will allow in these cases the use of moment formulas and central limit theorems for martingales in computing small-sample moments and large-sample distributions of counting process statistics. To develop these ideas, we need a review of the general approach to conditional expectation, the definition of a martingale, and a central example to guide the way.

Example 1.2.1. Let T and U be nonnegative, independent random variables, and assume that the distribution of T has a density. The variable $X = \min(T, U)$, also written $(T \wedge U)$, is a censored observation of the failure time variable T, and $\delta = I_{\{T \le U\}}$ is the indicator variable for the event of an uncensored observation of T. The counting process $\{N(t) : t \ge 0\}$ given at time t by $N(t) = I_{\{X \le t, \delta = 1\}} = \delta I_{\{T \le t\}}$ is basic to the martingale approach to censored data. Let $F(t) = P\{T \le t\}, S(t) = 1 - F(t), \lambda(t) = -d[\log S(t)]/dt$ and $C(u) = P\{U > u\}$. Since T and U are independent,

$$\begin{aligned}\lambda(t) &= \lim_{\Delta t \downarrow 0} \frac{1}{\Delta t} P\{t \le T < t + \Delta t\}[P\{T \ge t\}]^{-1} \\ &= \lim_{\Delta t \downarrow 0} \frac{1}{\Delta t} P\{t \le T < t + \Delta t | T \ge t, U \ge t\},\end{aligned}$$

so

$$P\{t \le T < t + \Delta t | T \ge t, U \ge t\} = \lambda(t)\Delta t + o(\Delta t).$$

If $N(t-) = \lim_{s \uparrow t} N(s)$,

$$\lambda(t)\Delta t \approx P\{N((t + \Delta t)-) - N(t-) = 1 \mid T \ge t, U \ge t\}.$$

Since $N((t + \Delta t)-) - N(t-)$ is a $0, 1$-valued random variable,

$$\lambda(t)\Delta t \approx E\{N((t + \Delta t)-) - N(t-) \mid T \ge t, U \ge t\}.$$

The *hazard function* $\lambda(t)$ thus gives the conditional average rate of change in N over $[t, t + \Delta t)$, given that both the censoring and failure time exceed or equal t, and so indirectly specifies the conditional rate at which N jumps in small intervals.

The process A given by $A(t) = \int_0^t I_{\{X \ge u\}} \lambda(u) du$ is at each fixed t a random variable which should approximate the number of jumps by N over $(0, t]$. In fact,

$$\begin{aligned}
E\{N(t)\} &= P\{X \leq t, \delta = 1\} \\
&= P\{T \leq t, T \leq U\} \\
&= \int_0^t C(u-)dF(u) \\
&= \int_0^t C(u-)S(u)\frac{dF(u)}{S(u)} \\
&= \int_0^t P\{X \geq u\}\lambda(u)du \\
&= E\int_0^t I_{\{X \geq u\}}\lambda(u)du \\
&= E\{A(t)\}.
\end{aligned}$$

Consequently, the centered process $M = N - A$ has mean zero.

Because λ describes the conditional behavior of N, more can be said: it turns out that the conditional mean of $N(t+s) - N(t)$, given all the information in $N(u)$ and $I_{\{X \leq u, \delta=0\}}$ in the interval $[0, t]$, is the conditional mean of $A(t+s) - A(t)$. The corresponding conditional mean of $M(t+s) - M(t)$, given the same information, is zero. □

A precise formulation of conditioning on the path of a stochastic process is needed to establish the claim in the above example.

Suppose we wish to think about a stochastic process X in terms of the conditional distribution of its future behavior given its current evolution, that is, through the conditional distribution of $X(t+s)$ given $\{X(u) : 0 \leq u \leq t\}$. There is no simple analogue of the usual conditional density of a random variable Y, given a second variable Z, when conditioning on, say, the path of a stochastic process, but there is a definition of conditional expectation in the general setting. The general definition can be motivated by properties of conditional expectation in a simple case.

Let (Y, Z) be a pair of continuous random variables with joint density $f(y, z)$. The conditional expectation of Y given Z is

$$E(Y|Z) = \int_{-\infty}^{\infty} yh(y|Z)dy,$$

where $h(y|Z)$ is the conditional density for Y given Z, or

$$h(y|Z) = \frac{f(y, Z)}{\int_{-\infty}^{\infty} f(u, Z)du}.$$

The conditional expectation $E(Y|Z)$ is a function of Z, and for any Borel set $A \subset (-\infty, \infty)$

$$\begin{aligned}
&\int_{z\in A}\int_{y=-\infty}^{\infty} yf(y,z)\,dy\,dz \\
&= \int_{z\in A}\int_{y=-\infty}^{\infty} yh(y|z)\int_{u=-\infty}^{\infty} f(u,z)\,du\,dy\,dz \\
&= \int_{z\in A}\int_{u=-\infty}^{\infty}\left[\int_{y=-\infty}^{\infty} yh(y|z)dy\right] f(u,z)\,du\,dz.
\end{aligned}$$

These last equations imply that for any set $B \in \sigma(Z)$

$$\int_B Y dP = \int_B E(Y|Z)dP.$$

Conditioning on the path up to time t of a process X is conditioning on the σ-algebra $\sigma\{X(u) : 0 \le u \le t\}$, and it is just as easy to define conditional expectation with respect to an arbitrary σ-algebra.

Definition 1.2.7. Suppose Y is a random variable on a probability space $(\Omega, \mathcal{F}, P)$ and let $\mathcal{G}$ be a sub-σ-algebra of $\mathcal{F}$. Let X be a random variable satisfying

1. X is $\mathcal{G}$-measurable; and
2. $\int_B Y dP = \int_B X dP$ for all sets $B \in \mathcal{G}$.

The variable X is called the *conditional expectation* of Y given $\mathcal{G}$, and is denoted by $E(Y|\mathcal{G})$. □

It is a standard result that $E(Y|\mathcal{G})$ exists when $E|Y| < \infty$, and that if W and X are two variables satisfying (1) and (2) above, W and X are equivalent.

If $A \in \mathcal{F}$ is an event, then by $P(A|\mathcal{G})$ we mean $E\{I_A|\mathcal{G}\}$.

A detailed treatment of conditional expectation may be found in Chung (1974, Chapter 9). While the abstract, measure-theoretic definition provides an unambiguous language for treating situations such as conditioning on events of probability zero or on sample paths of a process, the reader will also find that all conditional expectations used in specific cases here have straightforward formulas.

We give one simple example to illustrate a smoothing property of conditional expectation, and then summarize some additional properties.

Example 1.2.2. Suppose $\Omega = [0, 2n)$, n a positive integer, and that $\mathcal{F}$ is the smallest σ-algebra containing the intervals $\{[i, i+1) : i = 0, \ldots, 2n-1\}$. A random variable Y on $(\Omega, \mathcal{F})$ must be constant on the intervals $[i, i+1)$ with value, say, a_i. If $\mathcal{G}$ is the sub-σ-algebra generated by the intervals $\{[2j, 2(j+1)) : j = 0, \ldots, n-1\}$, then $E(Y|\mathcal{G})$ must be constant on the intervals $[2j, 2(j+1))$ in order to satisfy (1). Suppose now that $P\{[i, i+1)\} = 1/2n$ for any $0 \le i \le 2n-1$, so that $P\{[2j, 2(j+1))\} = 1/n$. Then to satisfy (2)

$$E(Y|\mathcal{G}) = (a_{2j} + a_{(2j+1)})/2$$

almost surely on $[2j, 2(j+1))$, for $j = 0, \ldots, n-1$. □

We will use the following properties of conditional expectation as needed and list them here for future reference. Let $(\Omega, \mathcal{F}, P)$ be an arbitrary probability space, X and Y random variables on this space, and $\mathcal{F}_s \subset \mathcal{F}_t \subset \mathcal{F}$ sub-σ-algebras.

1. If $\mathcal{F}_t = \{\emptyset, \Omega\}$, $E(X|\mathcal{F}_t) = EX$ a.s., where $\emptyset$ denotes the empty set.
2. $EE(X|\mathcal{F}_t) = EX$.
3. If $\mathcal{F}_s \subset \mathcal{F}_t$, then $E\{E(X|\mathcal{F}_s)|\mathcal{F}_t\} = E\{E(X|\mathcal{F}_t)|\mathcal{F}_s\} = E(X|\mathcal{F}_s)$ a.s.
4. If $\sigma\{Y\} \subset \mathcal{F}_t$ (i.e., if Y is $\mathcal{F}_t$-measurable) then $E(XY|\mathcal{F}_t) = YE(X|\mathcal{F}_t)$ a.s. Note that this property immediately implies that if Y is $\mathcal{F}_t$-measurable, then $E(Y|\mathcal{F}_t) = Y$ a.s.
5. Let X and Y be independent random variables, and let $\mathcal{G} = \sigma(X)$. Then $E(Y|\mathcal{G}) = E(Y)$ a.s.
6. $E(aX + bY|\mathcal{F}_t) = aE(X|\mathcal{F}_t) + bE(Y|\mathcal{F}_t)$ a.s., for any constants a and b.
7. (Jensen's Inequality). If g is a convex, real-valued function, then $E\{g(X)|\mathcal{F}_t\} \geq g\{E(X|\mathcal{F}_t)\}$.

For two random variables X and Y, we will use $E(Y|X)$ for $E\{Y|\sigma(X)\}$. More generally, if $\mathcal{A}$ is any collection of random variables, $E(Y|\mathcal{A})$ will denote $E\{Y|\sigma(\mathcal{A})\}$.

To illustrate these properties with a simple example, suppose X_1 and X_2 are independent, with $P\{X_i = 0\} = P\{X_i = 1\} = 1/2, i = 1, 2$. Then,

$$E(X_1 + X_2|X_1) = E(X_1|X_1) + E(X_2|X_1) = X_1 + E(X_2|X_1)$$

$$= X_1 + EX_2 = \begin{cases} 1/2 \text{ with probability } 1/2 \\ 3/2 \text{ with probability } 1/2 \end{cases}$$

We complete this section with the definition of a martingale. This seemingly simple definition has surprising implications, and the literature about such processes is extensive. Martingale-based methods have been applied to such diverse areas as sequential analysis, queuing theory, and stochastic differential equations. As mentioned earlier, we will exploit central limit theorems for martingales and formulas for their time dependent first and second moments.

Definition 1.2.8. Let $X = \{X(t) : t \geq 0\}$ be a right-continuous stochastic process with left-hand limits and $\{\mathcal{F}_t : t \geq 0\}$ a filtration, defined on a common probability space. X is called a *martingale* with respect to $\{\mathcal{F}_t : t \geq 0\}$ if

1. X is adapted to $\{\mathcal{F}_t : t \geq 0\}$,
2. $E|X(t)| < \infty$ for all $t < \infty$,
3. $E\{X(t+s)|\mathcal{F}_t\} = X(t)$ a.s. for all $s \geq 0, t \geq 0$.

X is called a *submartingale* if (3) is replaced by

$$E\{X(t+s)|\mathcal{F}_t\} \geq X(t) \quad \text{a.s.}$$

and a *supermartingale* when (3) is replaced by

$$E\{X(t+s)|\mathcal{F}_t\} \leq X(t) \quad \text{a.s.}$$

□

The definition of a martingale does not always include the restriction that almost all paths of the process be right-continuous and have left-hand limits. Since this additional structure always holds in this book and is often required in derivations, there is no important loss of generality in this definition.

Example 1.2.3. A simple and common example of a martingale is a random walk, which is a martingale in discrete time. A discrete-time process $X = \{X(n) : n = 0, 1, \ldots\}$ is a martingale with respect to an increasing sequence of σ-algebras $\{\mathcal{F}_n : n = 0, 1, \ldots\}$ if $X(n)$ is $\mathcal{F}_n$-measurable, $E|X(n)| < \infty$, and $E\{X(n + k)|\mathcal{F}_n\} = X(n)$ a.s. for any $n = 0, 1, \ldots$ and $k = 0, 1, \ldots$.

Let $Y_1, Y_2, \ldots,$ be an independent, identically distributed sequence of random variables with $E(Y_i) = 0$. The process $X(n) = \sum_{i=1}^n Y_i$ is a general random walk, and if each Y_i takes on the two values $\{-1, 1\}$ with equal probability, $X(n)$ is called a simple random walk. Let $\mathcal{F}_n = \sigma\{X(1), ..., X(n)\}$. For any $n \geq 0$, the mapping $\{Y_1, \ldots, Y_n\} \to \{X(1), \ldots, X(n)\}$ is one-to-one, and $\mathcal{F}_n = \sigma\{Y_1, \ldots, Y_n\}$ as well. Thus

$$\begin{aligned} E\{X(n+k)|\mathcal{F}_n\} &= E\left(\sum_{i=1}^{n} Y_i + \sum_{i=n+1}^{n+k} Y_i \middle| Y_1, \ldots, Y_n\right) \\ &= \sum_{i=1}^{n} Y_i + \sum_{i=n+1}^{n+k} E\left(Y_i \middle| Y_1, \ldots, Y_n\right) \\ &= \sum_{i=1}^{n} Y_i \\ &= X(n). \end{aligned}$$

A random walk is a probabilistic model for a fair game in which Y_i represents the amount won on the ith play of the game. $X(n)$ will then be the accumulated winnings (or losses, if $X(n) < 0$) in the first n plays. □

Additional examples are explored in Appendix E.

Two simple properties of a martingale X provide insight into the way martingales are used later. First, since for any $h > 0$, $E\{X(t)|\mathcal{F}_{t-h}\} = X(t-h)$, it is natural to expect that $E\{X(t)|\mathcal{F}_{t-}\} = X(t-)$, and this in fact is true. We state this equality in Proposition 1.2.1. The proof, outlined in Exercise 1.4, relies on convergence theorems for martingales. Second,

$$\begin{aligned} E\{X(t) - X(t-h)|\mathcal{F}_{t-h}\} &= X(t-h) - X(t-h) \text{ a.s.} \\ &= 0 \text{ a.s.} \end{aligned}$$

This last property might be loosely summarized as $E\{dX(t)|\mathcal{F}_{t-}\} = 0$ for a right-continuous martingale. We use this (imprecise) property of X to motivate some arguments later.

Proposition 1.2.1. Let X be a martingale with respect to a filtration $\{\mathcal{F}_t : t \geq 0\}$. Then $E\{X(t)|\mathcal{F}_{t-}\} = X(t-)$ a.s. □

Whenever a martingale M is used, the underlying filtration $\{\mathcal{F}_t : t \geq 0\}$ must be specified. For brevity, we will often say that "M is an $\mathcal{F}_t$-martingale", rather than "M is a martingale with respect to the filtration $\{\mathcal{F}_t : t \geq 0\}$". Similar phrases will be used for other properties that depend on filtrations.

The usual conditions of completeness and right-continuity on the stochastic basis and filtration of an $\mathcal{F}_t$-martingale play somewhat different roles in the general theory of martingales and, consequently, in our development as well. Completeness is a technical requirement needed in the proofs of some theorems in the general theory, such as the Doob-Meyer decomposition for submartingales stated in Section 1.4. Essentially, two processes whose paths differ on only a subset of a P-null set will be indistinguishable when a stochastic basis is complete. Completeness is used, for example, to show that a martingale can be modified so that the set of paths with both right- and left-hand limits at each t has probability 1, and that the modified process is indistinguishable from the original martingale. Any stochastic basis $(\Omega, \mathcal{F}, \{\mathcal{F}_t : t \geq 0\}, P)$ can be completed by replacing $\mathcal{F}$ (respectively, $\mathcal{F}_t$) by its P-completion $\mathcal{F}^P$ (respectively, by $\mathcal{F}_t^P$) which is the smallest σ-algebra generated by $\mathcal{F}$ (respectively, by $\mathcal{F}_t$) and the P-null subsets of $\mathcal{F}$. Any $\mathcal{F}_t$-martingale (or supermartingale) X will be an $\mathcal{F}_t^P$-martingale (respectively, supermartingale), and so properties of X can be established in the completed stochastic basis.

More care must be taken with right-continuity of filtrations. Not all filtrations are right-continuous, nor can right-continuous versions be used with impunity. For example, suppose the filtration $\{\mathcal{F}_t : t \geq 0\}$ is defined to be the history of a stochastic process X, (i.e., $\mathcal{F}_t \equiv \sigma\{X(s) : 0 \leq s \leq t\}$), and A is an $\mathcal{F}$-measurable set such that $0 < P\{A\} < 1$. It is simple to show (Exercise 1.5) that this history fails to be right-continuous at t_0 either for the right-continuous process $X(t) = t + [\min(t, t_0) - t]I_A$, or for the step function process $X(t) = I_{[t>t_0]}I_A$. Right-continuity does play an important role in the general theory, and is used explicitly in the proofs of some of the theorems here, so it must be verified in applications that use these theorems.

Heuristically, for the history of a stochastic process X to be a right-continuous filtration, the new information at t should be in $\mathcal{F}_t$ and there should be no new information infinitesimally after t, for every $t \geq 0$. This structure should hold when X is right-continuous and has sample paths which are step functions. Theorem A.2.6 in Appendix A implies that the history $\{\mathcal{F}_t : t \geq 0\}$ of a counting process N is a right-continuous filtration, and we rely on this result often.

In all subsequent developments, except where specifically indicated, filtrations will be assumed to be right-continuous. In specific applications in this book, filtrations will be the history of counting processes or histories of the form given in Theorem 4.2.3, and they will be right-continuous.

1.3 THE MARTINGALE $M = N - A$

In this section we establish the claim in Example 1.2.1 that, for a censored failure time variable T, the process given by

$$\begin{aligned} M(t) &= I_{\{X \leq t, \delta = 1\}} - \int_0^t I_{\{X \geq u\}} \lambda(u) du \\ &= N(t) - A(t) \end{aligned}$$

is a martingale. We assume first that T has a density function and a hazard function λ, but the approach used here requires only minor modifications under more general assumptions about the distribution of T. Those modifications are outlined later in the section. A heuristic argument provides some insight into the main result of the section.

Example 1.2.1. (continued)
Suppose for the processes defined in Example 1.2.1

$$\mathcal{F}_s = \sigma\{N(u), I_{\{X \leq u, \delta = 0\}} : 0 \leq u \leq s\}.$$

It can be shown (Exercise 1.6) that $\mathcal{F}_{s-}$ is the information in $N(u)$ and $I_{\{X \leq u, \delta=0\}}$ up to, but not including time s, that is,

$$\mathcal{F}_{s-} = \sigma\{N(u), I_{\{X \leq u, \delta = 0\}} : 0 \leq u < s\}.$$

Since $\{X < s\} = \bigcup_{n=1}^{\infty} \{X \leq s - (1/n)\}$, both $\{X < s\}$ and $\{X \geq s\}$ are $\mathcal{F}_{s-}$-measurable events. When T and U are independent, $dN(s)$ is a Bernoulli variable with conditional probability $I_{\{X \geq s\}} \lambda(s) ds$ of being one given $\mathcal{F}_{s-}$, and

$$E\{dN(s) | \mathcal{F}_{s-}\} = I_{\{X \geq s\}} \lambda(s) ds = dA(s).$$

The change in $M = N - A$ over an infinitesimal interval $(s - ds, s]$ is $dM(s) = dN(s) - dA(s)$. Then

$$E\{dM(s) | \mathcal{F}_{s-}\} = 0,$$

because

$$\begin{aligned} E\{dA(s)|\mathcal{F}_{s-}\} &= E\{I_{\{X\geq s\}}\lambda(s)ds|\mathcal{F}_{s-}\} \\ &= I_{\{X\geq s\}}\lambda(s)ds \\ &= dA(s). \end{aligned}$$

In other words, M should be a martingale with respect to $\{\mathcal{F}_s\}$. □

Theorem 1.3.1. Let T be an absolutely continuous failure time random variable and U a censoring time variable with an arbitrary distribution. Set $X = \min(T, U), \delta = I_{\{T\leq U\}}$, and let λ denote the hazard function for T. Define

$$\begin{aligned} N(t) &= I_{\{X\leq t, \delta=1\}}, \\ N^U(t) &= I_{\{X\leq t, \delta=0\}}, \\ \mathcal{F}_t &= \sigma\{N(u), N^U(u):\ 0\leq u\leq t\}. \end{aligned}$$

Then the process M given by

$$M(t) = N(t) - \int_0^t I_{\{X\geq u\}}\lambda(u)du$$

is an $\mathcal{F}_t$-martingale if and only if

$$\lambda(t) = \frac{-\frac{\partial}{\partial u}P\{T\geq u, U\geq t\}|_{u=t}}{P\{T\geq t, U\geq t\}} \quad \text{whenever } P\{X > t\} > 0. \tag{3.1}$$

□

Condition (3.1) has an important interpretation in applications. If we denote the right-hand side of (3.1) by $\lambda^{\#}(t)$, then $\lambda(t)$ is the hazard rate for the failure time variable T, while $\lambda^{\#}(t)$ is

$$\begin{aligned} \frac{-\frac{\partial}{\partial u}P\{T\geq u, U\geq t\}|_{u=t}}{P\{T\geq t, U\geq t\}} &= \lim_{h\downarrow 0}\frac{1}{h}\frac{P\{t\leq T < t+h, U\geq t\}}{P\{T\geq t, U\geq t\}} \\ &= \lim_{h\downarrow 0}\frac{1}{h}P\{t\leq T<t+h|T\geq t, U\geq t\}, \end{aligned} \tag{3.2}$$

i.e., $\lambda^{\#}$ is the hazard for T in the presence of censoring. In competing-risks terminology, λ is the *net* hazard and Eq. (3.2) is the *crude* hazard. Theorem 1.3.1 thus states that M is a martingale if and only if the net and crude hazards are equal.

This condition occurs in the theory of competing risks. Tsiatis (1975) and Peterson (1976) showed that some additional structure such as that imposed by Condition (3.1) is necessary for the distribution of T to be identifiable in inference from observations on (X, δ). Further discussion is provided by Cox (1959), and Cox and Oakes (1984).

Condition (3.1) is slightly weaker than the assumption of statistical independence of T and U. Using Eq. (3.2), Condition (3.1) can be interpreted as

$$P\{s \leq T < s + ds | T \geq s\} = P\{s \leq T < s + ds | T \geq s, U \geq s\}.$$

It is possible to construct interesting examples when T and U are dependent and Condition (3.1) holds (cf., Exercise 1.8). Nevertheless, with a slight abuse in terminology, we will refer to this last condition as *independent censoring*. In situations where T is discrete, independent censoring may take a different form from that of Condition (3.1), but the intent will always be the same.

Proof. It is straightforward to establish (Exercise 1.10) that M is adapted to $\{\mathcal{F}_t : t \geq 0\}$; heuristically, "knowing $\{N(u), N^U(u) : 0 \leq u \leq t\}$" implies "knowing $M(t)$". Rigorously, one need only show that, for any t, $\sigma(M(t)) \subset \mathcal{F}_t \equiv \sigma\{N(u), N^U(u) : 0 \leq u \leq t\}$.

To see that $E|M(t)| < \infty$ for all $t > 0$, note that

$$\begin{aligned} E|M(t)| &\leq EN(t) + E\int_0^t I_{\{X \geq u\}} \lambda(u) du \\ &\leq 1 + \int_0^t P\{X \geq u\} \lambda(u) du \\ &\leq 1 + \int_0^t P\{T \geq u\} \lambda(u) du \\ &= 1 + 1 - S(t) \\ &\leq 2. \end{aligned}$$

Thus, it only remains to show that $E\{M(t+s)|\mathcal{F}_t\} = M(t)$ a.s. for all $s \geq 0, t \geq 0$ if and only if Condition (3.1) holds. Since

$$\begin{aligned} E\{M(t+s)|\mathcal{F}_t\} &= E\left\{N(t+s) - \int_0^{t+s} I_{\{X \geq u\}} \lambda(u) du \mid \mathcal{F}_t\right\} \\ &= N(t) - \int_0^t I_{\{X \geq u\}} \lambda(u) du + E\{N(t+s) - N(t) \mid \mathcal{F}_t\} \\ &\quad - E\left\{\int_t^{t+s} I_{\{X \geq u\}} \lambda(u) du \mid \mathcal{F}_t\right\}, \end{aligned}$$

we must show

$$E\{N(t+s) - N(t)|\mathcal{F}_t\} = E\left\{\int_t^{t+s} I_{\{X \geq u\}} \lambda(u) du \mid \mathcal{F}_t\right\} \text{ a.s.} \tag{3.3}$$

for all $s \geq 0, t \geq 0$ if and only if Condition (3.1) holds. Now,

$$E\{N(t+s) - N(t)|\mathcal{F}_t\} = E\{I_{\{t<X\leq t+s,\delta=1\}}|N(u), N^U(u) : 0 \leq u \leq t\}. \quad (3.4)$$

If either N or N^U has jumped at or before time t, $I_{\{t<X\leq t+s,\delta=1\}} = 0$, so the conditional expectations in Eq. (3.4) must be 0 on the set $\{X \leq t\}$. On the $\mathcal{F}_t$ set $\{N(u) = N^U(u) = 0 \text{ for any } u \in [0,t]\} = \{X > t\}$, the conditional expectations in Eq. (3.4) must be a constant (say k) to satisfy (1) of Definition 1.2.7. To satisfy (2) of that definition,

$$\begin{aligned} kP\{X > t\} &= \int_{\{X>t\}} k dP \\ &= \int_{\{X>t\}} I_{\{t<X\leq t+s,\delta=1\}} dP \\ &= P\{t < X \leq t+s, \delta = 1\}, \end{aligned}$$

so $k = P\{t < X \leq t+s, \delta = 1|X > t\}$. This result, together with Eq. (3.4), implies that

$$\begin{aligned} E\{N(t+s) - N(t)|\mathcal{F}_t\} &= I_{\{X>t\}} P\{t < X \leq t+s, \delta = 1|X > t\} \quad (3.5) \\ &= I_{\{X>t\}} P\{t < X \leq t+s, \delta = 1\}/P\{X > t\}. \end{aligned}$$

On $\{X \leq t\}$, the right-hand side of Eq. (3.3) must be zero. To satisfy (1) and (2) of Definition 1.2.7, $E\{\int_t^{t+s} I_{\{X\geq u\}}\lambda(u)du|\mathcal{F}_t\}$ must be a constant k^* on $\{X > t\}$, and

$$\begin{aligned} k^* P\{X > t\} &= \int_{\{X>t\}} k^* dP \\ &= \int_{\{X>t\}} \int_t^{t+s} I_{\{X\geq u\}}\lambda(u)du\, dP \\ &= E\int_t^{t+s} I_{\{X\geq u\}}\lambda(u)du \\ &= \int_t^{t+s} P\{X \geq u\}\lambda(u)du. \end{aligned}$$

Equation (3.3) is then equivalent to

$$I_{\{X>t\}} \frac{P\{t < X \leq t+s, \delta = 1\}}{P\{X > t\}} = I_{\{X>t\}} \frac{\int_t^{t+s} P\{X \geq u\}\lambda(u)du}{P\{X > t\}} \quad \text{a.s.} \quad (3.6)$$

If $P\{X > t\} = 0$, we have $I_{\{X>t\}} = 0$ a.s. and the above equation is true. If $P\{X > t\} > 0$,

$$\begin{aligned} P\{t < X \le t+s, \delta = 1\} &= P\{t < T \le t+s, T \le U\} \\ &= \int_t^{t+s} \left[-\frac{\partial}{\partial v} P\{T \ge v, U \ge u\}\Big|_{v=u} \right] du \\ &= \int_t^{t+s} P\{X \ge u\} \lambda^{\#}(u) du. \end{aligned}$$

The fact that the equality in Eq. (3.6) must hold on a set of measure one for all nonnegative s and t renders Condition (3.1) necessary (outside a set of Lebesgue measure zero) as well as sufficient. □

The proof of Theorem 1.3.1 implies that

$$M(t) = N(t) - \int_0^t I_{\{X \ge u\}} \lambda^{\#}(u) du$$

is always a martingale, and hence that there is always an increasing process A (in this case $A(t) = \int_0^t I_{\{X \ge u\}} \lambda^{\#}(u) du$) such that $N - A$ is a martingale with respect to the filtration defined in Theorem 1.3.1.

The hazards λ and $\lambda^{\#}$ can be very different under some forms of dependence between T and U, as shown in the following example.

Example 1.3.1. Suppose the pair (T, U) has joint distribution given by

$$P\{T > t, U > s\} = \exp(-\lambda t - \mu s - \theta t s), \quad t \ge 0, s \ge 0,$$

where λ, μ and θ are nonnegative constants and where $\theta \le \lambda\mu$. Then T has an exponential distribution with $\lambda(t) = \lambda$ for $t \ge 0$, while

$$\begin{aligned} \lambda^{\#}(t) &= \frac{\frac{-\partial}{\partial u} P\{T \ge u, U \ge t\}|_{u=t}}{P\{T \ge t, U \ge t\}} \\ &= \lambda + \theta t. \end{aligned}$$

If θ is large, λ and $\lambda^{\#}$ will be very different even for small t. □

In clinical trials in which censoring is caused by the choice of a data analysis time, dependence between T and U is unlikely. In an observational study, however, if censoring is caused by incomplete follow up for reasons associated with failure, Example 1.3.1 shows that the observed rate at which N jumps may give misleading information about λ.

It is not difficult to establish a more general version of Theorem 1.3.1 allowing discontinuities in the distribution of T. This version will be useful later in situations in which samples of right-censored data contain observed failure times with ties. We first need an extension to an earlier definition.

Definition 1.3.1. Let T be a failure time random variable with arbitrary distribution function $F(t) = P\{T \leq t\}$. The cumulative hazard function Λ for T is given by

$$\Lambda(t) = \int_0^t \frac{dF(u)}{1 - F(u-)}. \qquad \square$$

When $F(u) - F(u-) > 0$,

$$\Delta\Lambda(u) = P\{T = u | T \geq u\},$$

which is the natural definition for a hazard function at points of positive probability for T. General hazard functions allow both discrete and continuous failure time distributions, as well as distributions with both discrete and continuous parts, to be handled simultaneously in later chapters. Mixed distributions with discrete and continuous components can arise during a clinical trial when patients are monitored continuously during hospitalization for certain phases of therapy, but only at regular intervals during other times. Certain events may then have an observed time of onset with a mixed distribution.

Theorem 1.3.2. Let T and U be failure and censoring variables, and let $X = \min(T, U)$, $\delta = I_{\{T \leq U\}}$, $N(t) = I_{\{X \leq t, \delta = 1\}}$ and $N^U(t) = I_{\{X \leq t, \delta = 0\}}$. Define $\mathcal{F}_t$ as in Theorem 1.3.1.

The process M given by

$$M(t) = N(t) - \int_0^t I_{\{X \geq u\}} d\Lambda(u)$$

is a martingale with respect to $\mathcal{F}_t$ if and only if

$$\frac{dF(z)}{1 - F(z-)} = \frac{-dP\{T \geq z; U \geq T\}}{P\{T \geq z; U \geq z\}} \tag{3.7}$$

for all z such that $P\{T \geq z, U \geq z\} > 0$. $\square$

Condition (3.7) generalizes (3.1) to the case in which F has discontinuities. In less precise notation it simply says,

$$P\{u \leq T < u + du | T \geq u\} = P\{u \leq T < u + du, U \geq T | T \geq u, U \geq u\},$$

that is, the instantaneous failure probabilities for T with and without observations on U must agree. Exercise 1.7 shows that

$$P\{u \leq T < u{+}du, U \geq T | T \geq u, U \geq u\} = P\{u \leq T < u{+}du | T \geq u, U \geq u\},$$

except possibly when T is a continuous random variable, and $P\{U = u\} > 0$. Condition (3.7) is then another way of stating the condition of independent censoring.

Proof of Theorem 1.3.2. Only a minor change is needed in the proof of Theorem 1.3.1. Indeed, all of the steps leading to (3.6) are still valid. It suffices to observe in this more general case that, when $P\{X > t\} > 0$,

$$\begin{aligned} P\{t < X \leq t+s, \delta = 1\} &= P\{t < T \leq t+s, T \leq U\} \\ &= \int_{z=t}^{t+s} -dP\{T \geq z, U \geq T\} \\ &= \int_{z=t}^{t+s} -P\{X \geq z\} \frac{dP\{T \geq z, U \geq T\}}{P\{T \geq z, U \geq z\}}. \end{aligned}$$

The desired result follows by applying Eq. (3.7) in a manner analogous to the application of (3.1) in the proof of Theorem 1.3.1. □

Any increasing process is a submartingale, and the fact that the increasing process N possesses a "compensator" A such that $N - A$ is a martingale is an instance of a general decomposition theorem for submartingales. The next section sketches the theory behind this more general context, and gives applications to some submartingales associated with $N - A$.

1.4 THE DOOB-MEYER DECOMPOSITION: APPLICATIONS TO QUADRATIC VARIATION

For a submartingale X, it is often possible to find an increasing process A such that $X - A$ is a martingale. Using additional restrictions on X and A specified below, A is unique, and the unique decomposition of X as $M + A$ is called its Doob-Meyer Decomposition. It is useful in statistical problems when A is not only computable, but is closely related to an important parameter, as in Example 1.2.1.

A detailed proof of the Doob-Meyer Decomposition is beyond the scope of this book. Its use, however, requires an understanding of the conditions under which it holds, and those conditions are explored in this section. Many definitions used here are needed in later sections as well, especially in Section 1.5.

We first examine the conditions on A which are sufficient for the existence and uniqueness of the Doob-Meyer Decomposition. The central condition, predictability, arises again when we study integrals with respect to martingale processes.

The idea behind a predictable process is quite simple. A predictable process X is one whose behavior at t is determined by its behavior on $[0, t)$, for any t. In other words, if $\mathcal{F}_s = \sigma\{X(u) : 0 \leq u \leq s\}$, $X(t)$ should be $\mathcal{F}_{t-}$-measurable. The simplest example of a predictable process is an indicator function of certain sets in $R^+ \times \Omega$. If we take a set of the form $(s, t] \times A$, $A \in \mathcal{F}_s$, and define the process $X : [0, \infty) \times \Omega \rightarrow \{0, 1\}$ by

$$X(\cdot,\cdot) = I_{(s,t]\times A}(\cdot,\cdot),$$

then X is either identically zero, or has value one during $(s,t]$ if the event A, observable at time s, has occurred. At any time u, then, the behavior of X at $u+du$ is known. These elementary processes determine the class of all predictable processes.

Definition 1.4.1. Let $(\Omega, \mathcal{F}, P)$ be a probability space with a filtration $\{\mathcal{F}_t : t \geq 0\}$. The σ-algebra on $[0,\infty) \times \Omega$ generated by all sets of the form

$$[0] \times A, \qquad A \in \mathcal{F}_0,$$

and

$$(a,b] \times A, \qquad 0 \leq a < b < \infty, A \in \mathcal{F}_a,$$

is called the *predictable* σ-algebra for the filtration $\{\mathcal{F}_t : t \geq 0\}$. □

Definition 1.4.2.

1. A process X is called *predictable* with respect to a filtration if, as a mapping from $[0,\infty) \times \Omega$ to R, it is measurable with respect to the predictable σ-algebra generated by that filtration. We call X an $\mathcal{F}_t$-predictable process.
2. Let I_B denote the indicator of a set B. A *simple predictable process* X is of the form

$$X = k_0 I_{[0]\times A_0} + \sum_{i=1}^{n} k_i I_{(a_i,b_i]\times A_i},$$

where $A_0 \in \mathcal{F}_0$, $A_i \in \mathcal{F}_{a_i}$, for $i = 1,\ldots,n$, and where $k_0, k_1, \ldots, k_n$ are constants. □

Sets of the form $[0] \times A, A \in \mathcal{F}_0$ or $(a,b] \times A, A \in \mathcal{F}_a$ are called *predictable rectangles*.

This definition leads to the earlier notion of a predictable process, as Proposition 1.4.1 shows. The proof of the proposition is outlined in Exercise 1.12.

Proposition 1.4.1. Let X be an $\mathcal{F}_t$-predictable process. Then, for any $t > 0, X(t)$ is $\mathcal{F}_{t-}$-measurable. □

There are some consequences and alternative characterizations of predictability which provide methods for checking its presence. One appealing characterization is based on the sample path properties of processes.

Lemma 1.4.1. Let $\{\mathcal{F}_t : t \geq 0\}$ be a filtration and X a left-continuous real-valued process adapted to $\{\mathcal{F}_t : t \geq 0\}$. Then X is predictable.

Proof. Let I_A denote the indicator of any interval $A \subset R^+$. We show that a left-continuous process X is a limit of predictable processes, and hence must be predictable. Let

$$X^n(t,\omega) = X(0,\omega)I_{[0]}(t) + \sum_{k=0}^{\infty} X(k/n,\omega)I_{(k/n,(k+1)/n]}(t).$$

Each term in the infinite sum for X^n is predictable since each is the limit of sums of simple predictable processes. Hence X^n is a predictable process. Because X is a.s. left-continuous,

$$X(t) = \lim_{n\to\infty} X^n(t)$$

except on a set of probability zero. □

The practical implications of Lemma 1.4.1 are important: we can insure predictability of a process by either checking for left-continuity or by using its left-continuous version. A stronger version of Lemma 1.4.1 holds. Indeed, while Lemma 1.4.1 says that left-continuous, adapted processes are predictable, Lemma 1.4.2 shows that such processes determine the class of predictable processes.

Lemma 1.4.2. A process X is predictable with respect to $\{\mathcal{F}_t : t \geq 0\}$ if and only if it is measurable with respect to the smallest σ-algebra on $R^+ \times \Omega$ generated by the adapted left-continuous processes.

Proof. Since any adapted left-continuous process is predictable, the σ-algebra generated by the adapted left-continuous processes on $R^+ \times \Omega$ must be contained in the predictable σ-algebra.

To establish inclusion in the other direction, recall that the predictable σ-algebra is generated by the predictable rectangles

$$[0] \times A, \qquad A \in \mathcal{F}_0,$$

$$(a,b] \times A, \qquad 0 \leq a < b < \infty, A \in \mathcal{F}_a,$$

and thus is generated by the processes I_B where B is a predictable rectangle. Since I_B is a left-continuous process, the proof is complete. □

A simple example of a process which is not predictable is a right-continuous martingale having positive probability of a jump at some t_0. By Proposition 1.2.1, $E\{M(t_0)|\mathcal{F}_{t_0-}\} = M(t_0-)$ a.s. Since $P\{M(t_0-) = M(t_0)\} < 1, M(t_0)$ fails to be $\mathcal{F}_{t_0-}$-measurable and Proposition 1.4.1 implies M is not predictable.

Predictable processes arise in two settings in this book. The first is in the unique Doob-Meyer Decomposition of a right-continuous nonnegative submartingale X into the sum of a right-continuous martingale M and an increasing right-continuous predictable process A. The second is in the martingale transform $L \equiv \int H dM$ where H is left-continuous and adapted or, more generally, predictable. The main use of predictability of a process Q is its $\mathcal{F}_{t-}$-measurability, implying

$$E\{Q(t)|\mathcal{F}_{t-}\} = Q(t) \text{ a.s.} \tag{4.1}$$

In the second setting, heuristically,

$$\begin{aligned} E\{dL(t)|\mathcal{F}_{t-}\} &= E\{H(t)dM(t)|\mathcal{F}_{t-}\} \\ &= H(t)E\{dM(t)|\mathcal{F}_{t-}\} \\ &= 0, \end{aligned}$$

where the last equality follows because M is a martingale. Thus predictability of H will be important in establishing that L is a martingale. In the first setting, suppose A is an increasing right-continuous predictable process satisfying $E\{dX(t)|\mathcal{F}_{t-}\} = dA(t)$. Then, because A is predictable, Eq. (4.1) implies $E\{dM(t)|\mathcal{F}_{t-}\} = 0$ so $M \equiv X - A$ should be a martingale.

While Lemma 1.4.2 suggests that many predictable processes are left-continuous, not all predictable processes have this property. In Theorem 1.3.2, the process $M = N - A$ is a martingale, where

$$A(t) = \int_0^t I_{\{X \geq u\}} d\Lambda(u).$$

The process A is right-continuous. However, if Λ has a discontinuity at t, A will not be left-continuous at t on paths where $X \geq t$. Proposition 1.4.2 shows, however, that A is still predictable.

Proposition 1.4.2. Let (T, U) be a failure and censoring time pair, and let $X = \min(T, U)$. If

$$\Lambda(t) = \int_0^t \{1 - F(u-)\}^{-1} dF(u),$$

then the right-continuous process

$$A(t) = \int_0^t I_{\{X \geq u\}} d\Lambda(u)$$

is predictable with respect to the filtration defined in Theorem 1.3.1. □

We will need the following lemma, whose proof is in Elliott (1982, Corollary 6.3.3), to establish the proposition.

Lemma 1.4.3. Let $\{\mathcal{F}_t : t \geq 0\}$ be a filtration on a probability space $(\Omega, \mathcal{F}, P)$. Then the rectangles of the form

$$[0] \times A, \quad A \in \mathcal{F}_0, \quad \text{and} \quad [a, b) \times A, \quad 0 < a < b \leq \infty, \quad A \in \mathcal{F}_{a-}$$

also generate the predictable σ-algebra for $\{\mathcal{F}_t : t \geq 0\}$.

Proof of Proposition 1.4.2. Let

$$A_{mn}(t) = \left\{\Lambda\left(\frac{n+1}{2^m}\right) - \Lambda\left(\frac{n}{2^m}\right)\right\} I_{[\frac{n}{2^m}, \infty)} I_{\{X \geq \frac{n}{2^m}\}}$$

and

$$A_m(t) = \sum_{n=0}^{\infty} A_{mn}(t).$$

(An example of the process A_m is shown in Figure 1.4.1)
The proposition follows if

1. A_{mn} is a predictable process, and
2. $A_m(t) \to A(t)$ a.s., as $m \to \infty$ for each $t \geq 0$.

To establish (1), we need only show that

$$\left[\frac{n}{2m}, \infty\right) \times \left\{X \geq \frac{n}{2m}\right\}$$

is one of the rectangles in Lemma 1.4.3. We must show that $\left\{X \geq \frac{n}{2^m}\right\} \in \mathcal{F}_{(\frac{n}{2^m})-}$.
Recall that

$$\begin{aligned}\mathcal{F}_t &= \sigma\{N(u), N^U(u) : 0 \leq u \leq t\} \\ &= \sigma\left\{I_{\{X \leq u, \delta=1\}}, I_{\{X \leq u, \delta=0\}} : 0 \leq u \leq t\right\}.\end{aligned}$$

Now

$$\left\{X \geq \frac{n}{2m}\right\} = \bigcap_{s < \frac{n}{2^m}} \{X > s\},$$

and since

$$\begin{aligned}\{X > s\} &= \overline{\{X \leq s\}} \\ &= \overline{\{X \leq s, \delta = 0\}} \cap \overline{\{X \leq s, \delta = 1\}} \in \mathcal{F}_s, \\ &\left\{X \geq \frac{n}{2m}\right\} \in \bigvee_{s < \frac{n}{2^m}} \mathcal{F}_s = \mathcal{F}_{(\frac{n}{2^m})-}.\end{aligned}$$

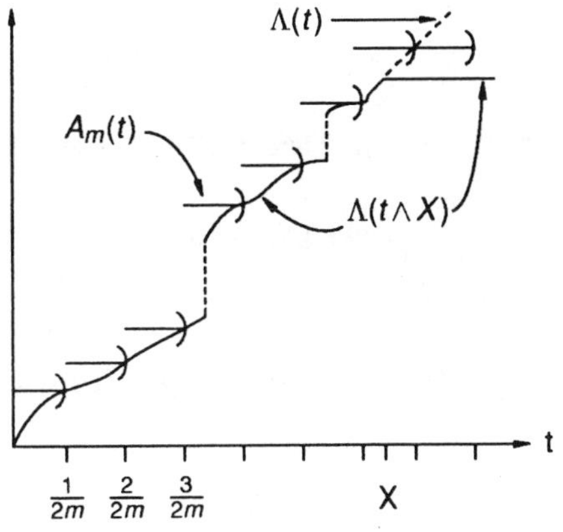

Figure 1.4.1 Approximating $\Lambda(t \wedge X)$ with $A_m(t)$

To establish (2), we note that $A(t) = \Lambda(t \wedge X)$ where $(t \wedge X)$ denotes the minimum of t and X. Since Λ is nondecreasing and right-continuous, it follows that $A_m(t) \downarrow A(t)$ a.s. □

The version of the Doob-Meyer Decomposition we use relies on an important integrability condition.

Definition 1.4.3. A collection of random variables $\{X_t : t \in \tau\}$, where τ is an arbitrary index set, is *uniformly integrable* if

$$\lim_{n\to\infty} \sup_{t\in\tau} E(|X_t| I_{\{|X_t|>n\}}) = 0.$$ □

The following proposition gives conditions often used to check uniform integrability. The proof may be found in Chung (1974, Theorem 4.5.3).

Proposition 1.4.3. The collection $\{X_t : t \in \tau\}$ is uniformly integrable if and only if the following two conditions are satisfied:

a. $\sup_{t\in\tau} E|X_t| < \infty$;
b. For every $\epsilon > 0$, there exists $\delta(\epsilon)$ such that for any set A with $P\{A\} < \delta$,

$$\sup_{t\in\tau} \int_A |X_t| dP < \epsilon.$$ □

A simple example shows that, although tempting, it is not sufficient to check only (a).

Example 1.4.1. Suppose Ω is the unit interval [0, 1], $\mathcal{F}$ the Borel sets on [0, 1], and P Lebesgue measure. For each positive interger t,

$$X_t = tI_{(0,t^{-1})} - tI_{(t^{-1},2t^{-1})}$$

is a random variable on $(\Omega, \mathcal{F}, P)$. Then $E|X_t| = 2$ so, if τ denotes the set of positive integers,

$$\sup_{t\in\tau} E|X_t| < \infty,$$

but

$$\begin{aligned} \sup_{t\in\tau} E(|X_t| I_{\{|X_t|\geq l\}}) &= \sup_{t\in\tau} 2I_{\{t\geq l\}} \\ &= 2 \end{aligned}$$

for all $l \geq 0$. □

There is an additional method for checking uniform integrability.

Proposition 1.4.4. The collection $\{X_t : t \in \tau\}$ is uniformly integrable if and only if there exists an increasing convex function $G : R^+ \to R^+$ such that

$$\lim_{t\to\infty} \frac{G(t)}{t} = \infty$$

and

$$\sup_{t\in\tau} E\{G(|X_t|)\} < \infty. \qquad \square$$

A proof of Proposition 1.4.4, which is due to LaVallée Poussin, may be found in Meyer (1966, Chapter II, T22).

The version of the Doob-Meyer decomposition given here is not the most general, since we state it only for the case of right-continuous, nonnegative submartingales. For right-continuous submartingales of arbitrary sign, either the submartingale must satisfy additional regularity conditions (i.e., must be of the class D or DL as described in Liptser and Shiryayev, 1977, pp. 66–68) or the process M will have somewhat less structure than a martingale (i.e., it will be a local martingale). We will not need the most general decomposition theorem, however.

Theorem 1.4.1 (Doob-Meyer Decomposition). (Meyer [1966, pp 102, 105, and 122] and Liptser and Shiryayev, [1977, Corollary to Theorem 3.8]). Let X be a right-continuous nonnegative submartingale with respect to a stochastic basis $(\Omega, \mathcal{F}, \{\mathcal{F}_t : t \geq 0\}, P)$. Then there exists a right-continuous martingale M and an increasing right-continuous predictable process A such that $EA(t) < \infty$ and

$$X(t) = M(t) + A(t) \text{ a.s.}$$

for any $t \geq 0$. If $A(0) = 0$ a.s., and if $X = M' + A'$ is another such decomposition with $A'(0) = 0$, then for any $t \geq 0$,

$$P\{M'(t) \neq M(t)\} = 0 = P\{A'(t) \neq A(t)\}.$$

If in addition X is bounded, then M is uniformly integrable and A is integrable. $\square$

The decomposition theorem states that for any right-continuous nonnegative submartingale X there is a unique increasing right-continuous predictable process A such that $A(0) = 0$ and $X - A$ is a martingale. Since any adapted nonnegative increasing process with finite expectation is a submartingale (cf. Exercise 1.2), there is a unique process A so that for any counting process N with finite expectation, $N - A$ is a martingale. We summarize this in the next corollary.

Corollary 1.4.1. Let $\{N(t) : t \geq 0\}$ be a counting process adapted to a right-continuous filtration $\{\mathcal{F}_t : t \geq 0\}$ with $EN(t) < \infty$ for any t. Then there exists a unique increasing right-continuous $\mathcal{F}_t$-predictable process A such that $A(0) = 0$ a.s., $EA(t) < \infty$ for any t, and $\{M(t) = N(t) - A(t) : t \geq 0\}$ is a right-continuous $\mathcal{F}_t$-martingale. □

The process A in the Doob-Meyer decomposition is called the *compensator* for the submartingale X. The martingale approach to statistical models for counting processes is useful only in situations in which A is known or can be computed. There are two approaches to obtaining A, and we will use both. For simple models, such as Example 1.2.1 or models built from that example, we will be able to "guess" the form of A, establish that $N - A$ is a martingale, then apply the uniqueness portion of Theorem 1.4.1 to insure that we need look no further. This approach will take us surprisingly far, and indeed will be the foundation for a study of small- and large-sample properties of the Kaplan-Meier estimator for survival functions and of k-sample test statistics for censored failure time data. For some complex models, such as proportional hazards regression models with time-varying covariates, we will assume for a particular compensator A that $N - A$ is a martingale, then specify sufficient conditions which will insure that A is a statistical parameter of interest.

By the (imprecise) property $dA(t) = E\{dN(t)|\mathcal{F}_{t-}\}$, A depends on the filtration $\{\mathcal{F}_t\}$, as the next example shows.

Example 1.4.2. Let T and U be failure and censoring variables, λ the hazard for T, $\lambda^{\#}$ the crude hazard defined by the right-hand side of Eq. (3.1), $X = \min(T, U)$, $Y(t) = I_{\{X \geq t\}}$, and N and N^U as defined in Theorem 1.3.1.

(a) Suppose first that $\mathcal{F}_t = \sigma\{N(s), N^U(s) : 0 \leq s \leq t\}$, and let $\mathcal{F}_t^c = \mathcal{F}_{(t-c)\vee 0}$, where $a \vee b = \max(a, b)$ and c is a nonnegative constant. Let A_c denote the compensator for N based on the filtration $\{\mathcal{F}_t^c : t \geq 0\}$ and A_0 that based on $\{\mathcal{F}_t : t \geq 0\}$. When $0 < c \leq t$, conditioning on $\mathcal{F}_t^c$ is conditioning at time t on only the information available c units of time prior to t, and $\mathcal{F}_t^c \subset \mathcal{F}_t$. Since $E[d\{N(t) - A_c(t)\}|\mathcal{F}_{t-}^c] = 0$,

$$\begin{aligned} dA_c(t) &= E\{N(t) - N(t - dt)|\mathcal{F}_{t-}^c\} \\ &= E\{E\{N(t) - N(t - dt)|\mathcal{F}_{t-}\}|\mathcal{F}_{t-}^c\} \\ &= E\{dA_0(t)|\mathcal{F}_{t-}^c\} \\ &= E\{I_{\{X \geq t\}}\lambda^{\#}(t)dt|\mathcal{F}_{(t-c)-}\} \\ &= E\{Y(t)|\mathcal{F}_{(t-c)-}\}\lambda^{\#}(t)dt \\ &= Y(t - c)P\{Y(t) = 1|Y(t - c) = 1\}\lambda^{\#}(t)dt. \end{aligned}$$

On the set $\{Y(t - c) = 1\} = \{X \geq t - c\}$, $dA_0(t)$ is a binary variable with value 0 when $Y(t) = 0$ and value $\lambda^{\#}(t)dt$ when $Y(t) = 1$. The process $dA_c(t)$ is constant on this set, taking on the value $E\{dA_0(t)|X \geq t - c\}$, and is simply a smoothed version of $dA_0(t)$.

When T and U have the joint distribution

$$P\{T > t, U > s\} = \exp(-\lambda t - \mu s - \theta ts), \quad t \geq 0, s \geq 0$$

in Example 1.3.1, then $\lambda^{\#}(t) = \lambda + \theta t$ and

$$P\{Y(t) = 1 | Y(t-c) = 1\} = \exp\{-c(\lambda + \mu) - \theta c(2t - c)\}.$$

T and U are independent when $\theta = 0$, and in that case the two compensators A_0 and A_c are given by

$$dA_0(t) = Y(t)\lambda dt$$

and

$$dA_c(t) = Y(t-c)\exp\{-c(\lambda + \mu)\}\lambda dt.$$

While the cases $c < 0$ and $c > t$ are not as interesting, it is easy to show that the values of $dA_c(t)$ are, respectively, $N(t) - N(t - dt) = dN(t)$ and $P\{Y(t) = 1\}\lambda^{\#}(t)dt$.

(b) Suppose now $\mathcal{F}_t^N = \sigma\{N(s) : 0 \leq s \leq t\}$, so that the information at time t contains the information in only the failure process; that is, censoring is ignored. Then using arguments similar to those above,

$$\begin{aligned} dA_N(t) &= E\{N(t) - N(t - dt) | \mathcal{F}_{t-}^N\} \\ &= E\{dA_0(t) | \mathcal{F}_{t-}^N\} \\ &= I_{\{N(t-)=0\}} P\{Y(t) = 1 | N(t-) = 0\}\lambda^{\#}(t)dt. \end{aligned}$$

The increment $dA_N(t)$ is constant on $\{N(t-) = 0\}$, and as above, is a smoothed value (or conditional mean) of the binary variable $dA_0(t)$, which has value 0 on $\{N(t-) = 0, N^U(t-) = 1\}$ and $\lambda^{\#}(t)dt$ on $\{N(t-) = 0, N^U(t-) = 0\}$.

When T and U are jointly distributed as in (a),

$$\begin{aligned} P\{Y(t) = 1 | N(t-) = 0\} &= P\{N^U(t-) = 0 | N(t-) = 0\} \\ &= \frac{\exp(-\lambda t - \mu t - \theta t^2)}{1 - \int_0^t \exp(-\lambda s - \mu s - \theta s^2)(\lambda + \theta s)ds}. \end{aligned}$$

When $\theta = 0$, the compensator A_N is given by

$$dA_N(t) = I_{\{N(t-)=0\}} \frac{\lambda + \mu}{\lambda + \mu \exp(\lambda t + \mu t)} \lambda \, dt \qquad \square$$

In Section 1.2 we showed directly that for N and A in Example 1.2.1, $E\{N(t)\} = E\{A(t)\}$. Since $N(0) = A(0) = 0$ and since any martingale must have constant expected value (Exercise 1.1), this result is also a direct corollary of Theorem 1.3.1. For any submartingale X with Doob-Meyer decomposition $X = M + A$, $M(0) = 0$ will imply $E\{X(t)\} = E\{A(t)\}$. If M is a martingale with $EM^2(t) < \infty$ for $t \geq 0$, Jensen's inequality implies that M^2 is a submartingale, and the decomposition of M^2 leads to computationally appealing formulas for the second moment $E\{M^2(t)\}$ when the compensator for M^2 has a simple form. This technique is so useful that a special name and notation, given in the next corollary, is reserved for the compensator of the square of a martingale.

Corollary 1.4.2. Let M be a right-continuous martingale with respect to a right-continuous filtration $\{\mathcal{F}_t : t \geq 0\}$ and assume $EM^2(t) < \infty$ for any $t \geq 0$.

Then there exists a unique increasing right-continuous predictable process $\langle M, M \rangle$, called the *predictable quadratic variation* of M, such that $\langle M, M \rangle(0) = 0$ a.s., $E\langle M, M \rangle(t) < \infty$ for each t, and $\{M^2(t) - \langle M, M \rangle(t) : t \geq 0\}$ is a right-continuous martingale. □

Frequently, $\langle M, M \rangle$ will also be called the *predictable variation process* of M. Corollary 1.4.2 has an important application to the martingale $M = N - A$ as long as $E\{N(t) - A(t)\}^2 < \infty$ for every t. Both A and $\langle M, M \rangle$ must be known, however, for the decomposition of M^2 to be useful. Example 1.2.1 illustrates that A has a natural form when N counts the occurrence of uncensored observations, and when A is known, it is not difficult to guess the form of $\langle M, M \rangle$ by considering the change in the martingale $M^2 - \langle M, M \rangle$ in infinitesimal intervals $(s - ds, s]$. By the martingale property of $M^2 - \langle M, M \rangle$, the predictability of $\langle M, M \rangle$, the equality $dM^2(s) = \{dM(s)\}^2 + 2M(s - ds)dM(s)$, and the (imprecise) property $E\{dM(s)|\mathcal{F}_{s-}\} = 0$,

$$\begin{aligned} d\langle M, M \rangle(s) &= E\{dM^2(s)|\mathcal{F}_{s-}\} \\ &= E[\{dM(s)\}^2|\mathcal{F}_{s-}] \\ &= \text{var}\{dM(s)|\mathcal{F}_{s-}\}. \end{aligned}$$

Thus $\langle M, M \rangle(t)$ can be thought of as the "sum" (as s ranges from 0 to t) of the conditional variance of $dM(s)$, given information up to time s. The process N has zero or one jump in any infinitesimal interval, with $E\{dN(s)|\mathcal{F}_{s-}\} = dA(s)$ and $\text{var}\{dN(s) - dA(s)|\mathcal{F}_{s-}\} = dA(s)\{1 - dA(s)\}$. Thus, when A is continuous, $d\langle M, M \rangle(s) = \text{var}\{dM(s)|\mathcal{F}_{s-}\} \approx dA(s)$, so one would anticipate $\langle M, M \rangle(t) = A(t)$, and this is verified in Section 2.5. The form of $\langle M, M \rangle$ when A is not continuous is explored in Section 2.6.

A *predictable covariation process* $\langle M_1, M_2 \rangle$, satisfying

$$d\langle M_1, M_2 \rangle(t) = \text{cov}\{dM_1(t), dM_2(t)|\mathcal{F}_{t-}\}$$

for two martingales M_1 and M_2, can be defined using similar arguments. Such a process is useful in calculating the covariance between martingales. The next theorem proves its existence.

Theorem 1.4.2. Let M_1 and M_2 be two right-continuous martingales with respect to a stochastic basis $(\Omega, \mathcal{F}, \{\mathcal{F}_t : t \geq 0\}, P)$, and assume $E\{M_i(t)\}^2 < \infty$ for $t \geq 0$ and $i = 1, 2$.

Then there exists a right-continuous predictable process $\langle M_1, M_2 \rangle$, called a *predictable covariation process*, with $\langle M_1, M_2 \rangle(0) = 0$, $E\langle M_1, M_2 \rangle(t) < \infty$, such that

1. $\langle M_1, M_2 \rangle$ is the difference of two increasing right-continuous predictable processes, and
2. $M_1 M_2 - \langle M_1, M_2 \rangle$ is a martingale. □

Note that Conclusion (1) implies that $\langle M_1, M_2\rangle$ has paths of bounded variation.

Though we do not prove it here, the predictable covariation process$\langle M_i, M_j\rangle$ is unique since $\{\mathcal{F}_t : t \geq 0\}$ is a right-continuous filtration (cf. Section 3.1 of Jacobsen, 1982).

Proof of Theorem 1.4.2. For any constants a and b, and time points $s \leq t$,

$$E\{aM_1(t) + bM_2(t)|\mathcal{F}_s\} = aM_1(s) + bM_2(s)$$

so both $M_1 + M_2$ and $M_1 - M_2$ are right-continuous martingales. Furthermore,

$$\begin{aligned} &E\{aM_1(t) + bM_2(t)\}^2 \\ &\qquad \leq a^2EM_1^2(t) + b^2EM_2^2(t) + 2ab\{EM_1^2(t) + EM_2^2(t)\} \\ &\qquad < \infty. \end{aligned}$$

By the Doob-Meyer Decomposition theorem, there exist two unique decompositions of the form

$$\{M_1(t) + M_2(t)\}^2 = M^{M_1+M_2}(t) + \langle M_1 + M_2, M_1 + M_2\rangle(t) \tag{4.2}$$

and

$$\{M_1(t) - M_2(t)\}^2 = M^{M_1-M_2}(t) + \langle M_1 - M_2, M_1 - M_2\rangle(t). \tag{4.3}$$

Now define

$$\langle M_1, M_2\rangle = 1/4\{\langle M_1 + M_2, M_1 + M_2\rangle - \langle M_1 - M_2, M_1 - M_2\rangle\}$$

and let

$$M(t) = M_1(t)M_2(t) - \langle M_1, M_2\rangle(t).$$

By subtracting (4.3) from (4.2) and dividing by 4, we see that

$$M(t) = 1/4\{M^{M_1+M_2}(t) - M^{M_1-M_2}(t)\},$$

which is a martingale.

Both (1) and (2) are now established. □

Since $M_1M_2 - \langle M_1, M_2\rangle$ is a martingale, for $0 \leq s \leq t$,

$$E\{M_1(t)M_2(t) - M_1(s)M_2(s)|\mathcal{F}_s\} = E\{\langle M_1, M_2\rangle(t) - \langle M_1, M_2\rangle(s)|\mathcal{F}_s\}. \tag{4.4}$$

Setting $s = 0$ and taking expectations yields

$$\begin{aligned} E\langle M_1, M_2\rangle(t) &= E\{M_1(t)M_2(t) - M_1(0)M_2(0)\} \\ &= [E\{M_1(t)M_2(t)\} - EM_1(t)EM_2(t)] - \\ &\qquad [E\{M_1(0)M_2(0)\} - EM_1(0)EM_2(0)]. \end{aligned}$$

Thus, if $M_1(0)$ and $M_2(0)$ are uncorrelated (often they are zero), then $E\langle M_1, M_2\rangle(t)$ is the covariance of $M_1(t)$ and $M_2(t)$.

Corollary 1.4.3. If M_1 and M_2 are two right-continuous $\mathcal{F}_t$-martingales with $E\{M_i(t)\}^2 < \infty$ for any $t \geq 0$, then the right-continuous process $M_1 M_2$ is a martingale if and only if $\langle M_1, M_2\rangle \equiv 0$. In this case, M_1 and M_2 are said to be *orthogonal*. □

Orthogonal martingales are uncorrelated if $M_1(0)$ and $M_2(0)$ are uncorrelated.

We will need some additional relationships later. Since M_1 and M_2 are martingales,

$$E\left[\{M_1(t) - M_1(s)\}\{M_2(t) - M_2(s)\}|\mathcal{F}_s\right] = E\left\{M_1(t)M_2(t) - M_1(s)M_2(s)|\mathcal{F}_s\right\}. \tag{4.5}$$

Combining Eqs. (4.4) and (4.5), we have

$$E\left[\{M_1(t) - M_1(s)\}\{M_2(t) - M_2(s)\}|\mathcal{F}_s\right] = E\left\{\langle M_1, M_2\rangle(t) - \langle M_1, M_2\rangle(s)|\mathcal{F}_s\right\}. \tag{4.6}$$

For any partition $0 = t_0 < t_1 < \ldots < t_n = t$ of $[0, t]$, we then will have

$$E\langle M_1, M_2\rangle(t) = \sum_{i=1}^{n} E\left[\{M_1(t_i) - M_1(t_{i-1})\}\{M_2(t_i) - M_2(t_{i-1})\}\right].$$

The predictable covariation process will be useful and easy to compute for martingale transforms, or stochastic integrals with respect to martingales, introduced in the next section.

1.5 THE MARTINGALE TRANSFORM $\int H dM$

As illustrated in Chapter 0, many censored data statistics are of the form $\sum_i \int H_i dM_i$ where $M_i = N_i - A_i$ for some counting process N_i, and H_i is a predictable process relative to the filtration making the M_i martingales. In some instances, these integral representations are an important step in the derivation of statistics such as maximum partial likelihood estimates of regression coefficients. Often, the following additional conditions hold:

1. H_i is a bounded predictable process.
2. $E\{N_i(t)\} < \infty$ for any t. (5.1)

We show that under the conditions of (5.1) the processes $\sum_i \int H_i dM_i$ are themselves martingales, allowing the use of formulas for quadratic variation and martingale central limit theorems. The martingale property of $L_i \equiv \int H_i dM_i$ is not surprising since

$$\begin{aligned} E\{dL_i(t)|\mathcal{F}_{t-}\} &= E\{H_i(t)dM_i(t)|\mathcal{F}_{t-}\} \\ &= H_i(t)E\{dM_i(t)|\mathcal{F}_{t-}\} \\ &= 0, \end{aligned}$$

where the second equality follows because H_i is predictable, and the last because M_i is a martingale.

Most statistics discussed here satisfy (5.1), but not all do. The local martingales discussed in Chapter 2 allow us to relax (5.1), and extend large sample distribution theory to more general inference problems for counting process data.

Another look at the logrank statistic helps explain the value of these integral representations.

Example 1.5.1 The Logrank Statistic. The logrank statistic was introduced in Example 0.2.2 as a significance test for the equality of two failure time distributions. In the notation of that example, (T_{ij}, U_{ij}) are $n_1 + n_2$ independent pairs of failure and censoring variables, and the observed variables are, respectively, $X_{ij} = \min(T_{ij}, U_{ij})$ and $\delta_{ij} = I_{\{T_{ij} \le U_{ij}\}}$. Assume that T_{ij} has an absolutely continuous distribution, and Condition (3.1) holds. With

$$N_{ij}(t) = I_{\{X_{ij} \le t, \delta_{ij}=1\}}, \qquad j = 1, 2, \ldots n_i,$$

and

$$Y_i(t) = \sum_{j=1}^{n_i} I_{\{X_{ij} \ge t\}},$$

the logrank statistic U can be written

$$U = \sum_{j=1}^{n_1} \int_0^\infty \frac{Y_2}{Y_1 + Y_2} dM_{1j} - \sum_{j=1}^{n_2} \int_0^\infty \frac{Y_1}{Y_1 + Y_2} dM_{2j} + \int_0^\infty \frac{Y_1 Y_2}{Y_1 + Y_2}(\lambda_1 - \lambda_2) ds, \tag{5.2}$$

where M_{ij} is the martingale $N_{ij} - \int I_{\{X_{ij} \ge s\}} \lambda_i(s) ds$. Under the hypothesis $H_0 : \lambda_1 = \lambda_2$, U has representation $\sum \int H_i dM_i$ and satisfies Condition (5.1).

Equation (5.2) provides some insight into the operating characteristics of U. Using the martingale structure, we will show that the first two terms on the right side of (5.2) have expectation zero, so

$$E(U) = E \int_0^\infty \frac{Y_1(s)Y_2(s)}{Y_1(s) + Y_2(s)} \{\lambda_1(s) - \lambda_2(s)\} ds.$$

We show later that under H_0, $[(n_1 + n_2)/(n_1 n_2)]^{1/2} U$ is asymptotically normally distributed with zero mean, and that under many alternatives to H_0 (e.g., $\lambda_1 > \lambda_2$), $[(n_1 + n_2)/(n_1 n_2)]^{1/2} U \to \infty$ a.s. A study of the asymptotic power of U can be based on local alternatives to H_0 chosen so that $(\lambda_1 - \lambda_2) \to 0$ at a rate which insures

$$\left(\frac{n_1 + n_2}{n_1 n_2} \right)^{1/2} \int_0^\infty \frac{Y_1(s)Y_2(s)}{Y_1(s) + Y_2(s)} \{\lambda_1(s) - \lambda_2(s)\} ds$$

approaches a non-zero constant. Since $Y_i(s) = \sum_{j=1}^{n_i} I_{\{X_{ij} \geq s\}}$, the strong law of large numbers implies that

$$\left(\frac{Y_1 Y_2}{Y_1 + Y_2}\right)\left(\frac{n_1 + n_2}{n_1 n_2}\right)$$

is asymptotically a constant, and consequently our interest will be in local alternatives for which

$$\lim_{n_1, n_2 \to \infty} \left(\frac{n_1 n_2}{n_1 + n_2}\right)^{1/2} \{\lambda_1(s) - \lambda_2(s)\} = k(s)$$

for some function k. The detailed discussion of operating characteristics for the logrank and similar tests under local alternatives is in Chapter 7.

Another form of the logrank statistic resembles statistics studied later. Let (X_i, δ_i, Z_i), $i = 1, \ldots, n$, be a sample of independent triplets, where Z_i is a 0,1 valued binary covariate denoting group membership. It is easy to show that

$$U = \sum_{i=1}^{n} \int_0^\infty \left\{ Z_i - \frac{\sum_{j=1}^n Y_j(s) Z_j}{\sum_{j=1}^n Y_j(s)} \right\} dN_i(s), \tag{5.3}$$

where $N_i(t) = I_{\{X_i \leq t, \delta_i = 1\}}$ and $Y_i(t) = I_{\{X_i \geq t\}}$. Under $H_0 : \lambda_0(s) = \lambda_1(s)$ ($= \lambda(s)$ unspecified), $N_i(s)$ can be replaced in Eq. (5.3) by $M_i(s) = N_i(s) - \int_0^s I_{\{X_i \geq u\}} \lambda(u) du$, again yielding the form $\sum_i \int H_i dM_i$. □

Example 1.5.2. The Cox proportional hazards model gives rise to statistics which generalize (5.3). Suppose $\{X_i, \delta_i, \mathbf{Z}_i : i = 1, \ldots, n\}$ is a sample of n independent triplets, where $\mathbf{Z}_i = \{\mathbf{Z}_i(s) : s \geq 0\}$ is a p-dimensional vector process of possibly time dependent covariates. For now, assume each component process of $\mathbf{Z}_i$ is uniformly bounded and predictable. If λ_i denotes the hazard function for the failure time variable of the ith observation and if λ_0 denotes an underlying "nuisance" hazard function, then

$$\lambda_i(s) = \lambda_0(s) \exp\{\boldsymbol{\beta}' \mathbf{Z}_i(s)\}$$

is the Cox (1972) proportional hazards model for the effect of a vector of time dependent covariates on failure rate. To test $H : \boldsymbol{\beta} = \boldsymbol{\beta}_0$, the Cox partial likelihood score vector was shown in Chapter 0 to be

$$\begin{aligned} \mathbf{U}(\boldsymbol{\beta}_0) &= \sum_{i=1}^{n} \int_0^\infty \left[\mathbf{Z}_i(s) - \frac{\sum_{j=1}^n Y_j(s) \mathbf{Z}_j(s) \exp\{\boldsymbol{\beta}_0' \mathbf{Z}_j(s)\}}{\sum_{j=1}^n Y_j(s) \exp\{\boldsymbol{\beta}_0' \mathbf{Z}_j(s)\}} \right] dN_i(s) \quad (5.4) \\ &= \sum_{i=1}^{n} \int_0^\infty \left[\mathbf{Z}_i(s) - \frac{\sum_{j=1}^n Y_j(s) \mathbf{Z}_j(s) \exp\{\boldsymbol{\beta}_0' \mathbf{Z}_j(s)\}}{\sum_{j=1}^n Y_j(s) \exp\{\boldsymbol{\beta}_0' \mathbf{Z}_j(s)\}} \right] dM_i(s), \end{aligned}$$

where $M_i(s) = N_i(s) - \int_0^s Y_i(u) \lambda_0(u) \exp\{\boldsymbol{\beta}_0' \mathbf{Z}_i(u)\} du$. □

To prove that $\sum_i \int H_i dM_i$ is a martingale when the H_i satisfy (5.1) we show that each of the summands $\int H_i dM_i$ is a martingale. For simplicity, we suppress the subscript i.

The basic process $M = N - A$ is right-continuous with left-hand limits and, since the nondecreasing processes N and A are finite a.s. at any point t, M is of bounded variation. The integrand H in $\int H dM$ is a measurable process, and consequently $\int H dM$ exists as a Lebesgue-Stieltjes integral for all paths of H and M in a set of probability one.

The proof here of the martingale property of $\int H dM$ for bounded predictable H is the most direct, but is somewhat abstract. A more constructive proof for bounded left-continuous integrands is outlined in Exercise 1.14.

The predictable processes are those processes measurable with respect to the smallest σ-algebra on $R^+ \times \Omega$ containing the predictable rectangles, so it is natural to begin with processes which are indicators of these sets.

Lemma 1.5.1. Let $\{\mathcal{F}_t : t \geq 0\}$ be a right-continuous filtration, and suppose $B \subset R^+ \times \Omega$ is a predictable rectangle of the form

$$[0] \times A, \qquad A \in \mathcal{F}_0,$$

or

$$(a, b] \times A, \qquad A \in \mathcal{F}_a.$$

Let M be a martingale with respect to $\{\mathcal{F}_t : t \geq 0\}$, with $E \int_0^t |dM(s)| < \infty$ for all t, and with $\Delta M(0) = 0$. Then the process L given by

$$L(t) = \int_0^t I_B(s) dM(s)$$

is an $\mathcal{F}_t$-martingale.

Proof. Since $\Delta M(0) = 0$ the result is obvious for $B = [0] \times A, A \in \mathcal{F}_0$.

The assumptions in the Lemma directly imply $E|L(t)| < \infty$ for any t. When $B = [0] \times A$, $A \in \mathcal{F}_0$, $L(t)$ is clearly $\mathcal{F}_t$-measurable for all $t \geq 0$, and when $B = (a, b] \times A$, $A \in \mathcal{F}_a$, it is easy to establish that L is adapted by considering the cases $t \leq a$, $a < t \leq b$, and $t > b$ separately.

Now write

$$\begin{aligned} L(t + s) &= \int_0^{t+s} I_B(u) dM(u) \\ &= \int_0^t I_B(u) dM(u) + \int_t^{t+s} I_B(u) dM(u). \end{aligned}$$

To show that L is a martingale when $B = (a, b] \times A, A \in \mathcal{F}_a$, we need only show

$$E\left\{\int_t^{t+s} I_B(u) dM(u) | \mathcal{F}_t\right\} = 0.$$

It is easiest to consider the cases $a \le t$ and $t < a$ separately. When $a \le t$, $A \in \mathcal{F}_a$ implies that $A \in \mathcal{F}_t$ and

$$\begin{aligned} E\left\{\int_t^{t+s} I_B(u)dM(u)|\mathcal{F}_t\right\} &= E\left[I_A\{M((t+s)\wedge b) - M(t\wedge b)\}|\mathcal{F}_t\right] \\ &= I_A E\left\{M((t+s)\wedge b) - M(t\wedge b)|\mathcal{F}_t\right\} \\ &= 0. \end{aligned}$$

When $t < a$,

$$\begin{aligned} &E\left\{\int_t^{t+s} I_B(u)dM(u)|\mathcal{F}_t\right\} \\ &= E\left[I_A\{M((t+s)\wedge b) - M((t+s)\wedge a)\}|\mathcal{F}_t\right] \\ &= E\left[E\left\{I_A\{M((t+s)\wedge b) - M((t+s)\wedge a)\}|\mathcal{F}_a\right\}|\mathcal{F}_t\right] \\ &= E\left[I_A\left[E\left\{M((t+s)\wedge b) - M((t+s)\wedge a)\right\}|\mathcal{F}_a\right]|\mathcal{F}_t\right] \\ &= 0. \end{aligned}$$

□

One additional technical lemma is needed before the full result can be proved. Lemma 1.5.2 is a consequence of the Monotone Class Theorem, and is proved in Appendix A.

Lemma 1.5.2. Let D_0 be a set and $\mathcal{S}$ a class of subsets of D_0 closed under finite intersections. Let $\mathcal{H}$ be a vector space of real valued functions with domain D_0 such that

1. $\mathcal{H}$ contains the constant functions, and $I_A \in \mathcal{H}$ for any $A \in \mathcal{S}$, where I_A is the indicator of the set A; and

2. If $\{X_n : n \ge 0\}$ is an increasing sequence of nonnegative functions in $\mathcal{H}$ such that $\sup_n X_n$ is finite (respectively, bounded), then $\sup_n X_n \in \mathcal{H}$.

Then $\mathcal{H}$ contains all real-valued (respectively, bounded) mappings measurable with respect to $\sigma(\mathcal{S})$. □

Theorem 1.5.1. Let N be a counting process with $EN(t) < \infty$ for any t. Let $\{\mathcal{F}_t : t \ge 0\}$ be a right-continuous filtration such that

1. $M = N - A$ is an $\mathcal{F}_t$-martingale, where $A = \{A(t) : t \ge 0\}$ is an increasing $\mathcal{F}_t$-predictable process with $A(0) = 0$;
2. H is a bounded, $\mathcal{F}_t$-predictable process.

Then the process L given by

$$L(t) = \int_0^t H(u)dM(u)$$

is an $\mathcal{F}_t$-martingale.

Proof. Let $\mathcal{S}$ denote the class of measurable rectangles

$$[0] \times A, \qquad A \in \mathcal{F}_0,$$

$$(a, b] \times A, \qquad A \in \mathcal{F}_a.$$

along with the empty set $\emptyset$. It is easy to show that $\mathcal{S}$ is closed under finite intersection.

Let $\mathcal{H}$ denote the vector space of bounded, measurable, and adapted processes H such that $\int H\, dM$ is a martingale with respect to $\{\mathcal{F}_t : t \geq 0\}$. $\mathcal{H}$ obviously contains the constant functions, and by Lemma 1.5.1, $\mathcal{H}$ contains I_B, for $B \in \mathcal{S}$. Now let $H_n, n = 1, 2, \ldots,$ be an increasing sequence of mappings from $R^+ \times \Omega$ to R in $\mathcal{H}$ such that $\sup_n H_n \equiv H$ is bounded on a set of probability one. We must show that $\int H\, dM$ is a martingale.

For any t,

$$E\left|\int_0^t H(u)dM(u)\right| < \infty.$$

Each process $\int H_n dM$ is a martingale and consequently an adapted process. Since $\int_0^t H(u)dM(u)$ is the pointwise limit of the $\mathcal{F}_t$-measurable variables $\int_0^t H_n(u)dM(u)$, $\int_0^t H(u)dM(u)$ is itself $\mathcal{F}_t$-measurable. Finally,

$$\begin{aligned} E\left\{\int_0^{t+s} H(u)dM(u)|\mathcal{F}_t\right\} &= E\left\{\int_0^{t+s} \lim_{n\to\infty} H_n(u)dM(u)|\mathcal{F}_t\right\} \\ &= \lim_{n\to\infty} E\left\{\int_0^{t+s} H_n(u)dM(u)|\mathcal{F}_t\right\} \\ &= \lim_{n\to\infty} \int_0^t H_n(u)dM(u) \\ &= \int_0^t H(u)dM(u), \end{aligned}$$

where all interchanges of limits with integrals or conditional expectations follow from the Monotone Convergence Theorem. The theorem now follows from Lemma 1.5.2. □

The above proof actually establishes that $\int H\, dM$ is a martingale whenever H is a bounded predictable process and M is a martingale whose paths on a set of probability one have total variation bounded by a constant, since it is only those properties of H and M that are used. Functional analytic methods can be used to extend Theorem 1.5.1 to predictable processes H with

$$E\left|\int_0^t H(u)dM(u)\right| < \infty \text{ for all } t \geq 0$$

(c.f. Brémaud, 1981, Chapter 1, Theorem T6), but we will not need that generality.

Stochastic integrals with respect to martingales that are not necessarily of bounded variation can also be defined, and with proper regularity assumptions integrals of predictable processes with respect to martingales are again martingales. The Itô integral with respect to a Wiener process is perhaps the most common example, and a complete account of this integral, as well as integrals with respect to arbitrary square integrable martingales, may be found in Liptser and Shiryayev (1977).

Suppose now $\sum_i \int H_i dM_i$ is a statistic where the H_i and M_i satisfy the conditions of Theorem 1.5.1. If there is a common filtration $\{\mathcal{F}_t : t \geq 0\}$ with respect to which each H_i is predictable and each M_i is a martingale, then $\sum_{i=1}^n \int H_i dM_i$ will be a martingale process with respect to $\{\mathcal{F}_t : t \geq 0\}$. In all the applications in the sequel, such a filtration will usually arise naturally, and sometimes be so obvious as to not require explicit definition.

1.6 BIBLIOGRAPHIC NOTES

Section 1.2

In addition to Chung (1974), there are several excellent introductions to measure theoretic treatments of probability, including those in Billingsley (1986) and Ash (1972).

Stochastic processes are used widely in many scientific disciplines, and there is a rich literature exploring these applications. A first course in stochastic processes usually emphasizes the computation of finite dimensional distributions, rather than the study of sample paths. The two volumes by Karlin and Taylor (1975,1981) give detailed treatments to the distribution theory approach for many specific processes, while the book by Ross (1983) has a more modern flavor. Both texts introduce martingales, and show how they can be used in applications. The recent book by Karr (1986) summarizes many of the theoretical results known about inference for complex counting processes that arise in scientific and engineering problems.

Section 1.3

A more general version of the sufficiency part of Theorem 1.3.1 can be found in Gill (1980, Theorem 3.1.1). The proof given here for Theorem 1.3.1 is a modification of the one found in Fleming and Harrington (1978). The joint distribution for (T, U) in Example 1.3.1 is discussed by Tsiatis (1975).

Section 1.4

Lemma 1.4.2 is often used as the definition of a predictable process, as in Liptser and Shiryayev (1977, Section 5.4). Another definition, based on the concept of predictable stopping times, can be found in Elliott (1982, Chapters 5 and 6).

The Doob-Meyer decomposition has a long history in the literature on the general theory of stochastic processes. It was originally established by Doob (1953) for discrete time processes, and its proof in that setting is surprisingly simple. Meyer (1966) established a version for continuous time processes, which is presented in this chapter. The discrete time version is given in Liptser and Shiryayev (1977, Section 2.4), and they prove continuous time versions in Section 3.3.

Section 1.5

Aalen (1978) was the first to exploit the stochastic integral representation for censored data statistics. His elegant approach replaced the long, algebraically complex methods of establishing properties of these statistics that had been used prior to his work. While Aalen discussed estimation of cumulative hazard rates and some test statistics, Gill (1980) was the first to use Aalen's approach to study the logrank statistic and other common two–sample censored data test statistics.

Andersen and Gill (1982) were among the first to use martingale methods to study the proportional hazards model. Their approach is discussed in detail in Chapter 8.

Theorem 1.5.1 has a converse: If N is a counting process adapted to a stochastic basis $\{\mathcal{F}_t : t \geq 0\}$ with compensator A and L is an $\mathcal{F}_t$-martingale, then there exists a predictable process H such that $L = \int H dM$. This result was established by Boel, Varaiya and Wong (1975), and has been used in applications to signal detection and stochastic control of counting processes.

CHAPTER 2

Local Square Integrable Martingales

2.1 INTRODUCTION

Chapter 1 introduced the counting process and martingale framework used with censored survival data statistics. Martingales, stochastic integrals and predictable quadratic variation processes were studied under regularity conditions strong enough to allow straightforward proofs of important properties yet weak enough to cover some widely used applications. Indeed, Theorem 1.3.1 shows that the integrated conditional hazard rate is the compensator process for the simple counting process denoting the time of an observed failure subject to censoring. In Section 1.4 we saw that the resulting martingale for the counting process was an instance of the Doob-Meyer decomposition theorem for a positive submartingale, and that this decomposition theorem implies the existence of predictable quadratic variation processes for square integrable martingales. For a compensated counting process $M = N - A$, Theorem 1.5.1 establishes that $\int H dM$ is a martingale as long as $E\{N(t)\} < \infty$ for all t and H is a bounded predictable process. The examples in Section 1.5 illustrate statistics that have this integral representation.

Before this approach can be useful in a range of applications some additional details must be completed: we have not established the claim in Section 1.4 that the predictable quadratic variation for a censored data counting process is also the integrated hazard function (i.e., the compensators for N and $(N - A)^2$ are identical when A is continuous); and no methods for computing predictable quadratic variation processes in more general cases have been given.

These remaining details could be worked out under the same conditions used in Chapter 1, that is, for counting processes with finite expectation and for stochastic integrals with bounded, predictable integrands. The ensuing results could then be applied without difficulty to examples such as the logrank statistic and Cox model score tests with well-behaved covariates. The resulting proofs are surprisingly delicate though, and not much easier than those covering more general situations. A relatively simple idea, localization of stochastic processes, allows these boundedness conditions to be relaxed, producing results which apply to more general situations than the examples given in Chapter 1. The more general setting used here will also be a natural one in which to establish central limit theorems for stochastic integrals of martingales in Chapter 5. We work with that more general setting in this chapter, first exploring the idea of localization and its use in extending the Doob-Meyer Decomposition (Section 2.2), next re-establishing the results from Chapter 1 for local martingales of the form $N - A$ (Section 2.3), then establishing properties of stochastic integrals with respect to local martingales (Section 2.4) and, finally, giving specific formulas for compensators and predictable quadratic variation processes (Sections 2.5 and 2.6). Since this chapter contains some lengthy technical arguments, Section 2.7 summarizes the main results to allow the reader to skip the detailed development and proceed directly to the applications in later chapters.

When integral signs appear in this chapter without limits of integration, the lower limit is assumed to be zero, and the upper limit is an arbitrary point t on the positive real axis.

2.2 LOCALIZATION OF STOCHASTIC PROCESSES AND THE DOOB-MEYER DECOMPOSITION

Often, results for functions defined on the nonnegative real line require certain regularity properties, such as boundedness or integrability, on compact subintervals of $[0, \infty)$. A function $f : [0, \infty) \longmapsto R$ is sometimes said to have a property *locally* if the property holds when f is restricted to $[0, s]$, for all $s < \infty$ (e.g., f is *locally bounded* if it is bounded on each interval $[0, s], s < \infty$). Because each path of a stochastic process $X = \{X(t) : t \geq 0\}$ is a real-valued function on $[0, \infty)$, the interval over which a path property holds, such as being bounded by a constant or being of bounded variation, may vary from path to path. It follows that a certain path property for X may hold on intervals whose right end point is a random time determined by each path.

A property is said to hold *locally* for a stochastic process if the property is satisfied by the stopped process $X_n = \{X(t \wedge \tau_n) : t \geq 0\}$ for each n, where the τ_n form an increasing sequence of random times, called *stopping times*, which satisfy $\lim_{n \to \infty} \tau_n = \infty$ a.s. and have properties specified below.

Definition 2.2.1. Let $\{\mathcal{F}_t : t \geq 0\}$ be a filtration on a probability space. A nonnegative random variable τ is a *stopping time* with respect to $\{\mathcal{F}_t\}$ if $\{\tau \leq t\} \in \mathcal{F}_t$ for all $t \geq 0$. □

If τ is thought of as the time an event occurs, then τ will be a stopping time if the information in $\mathcal{F}_t$ specifies whether or not the event has happened by time t.

Stopping times play an important role in the theory and applications of stochastic processes, but our use will be limited primarily to relaxing regularity conditions.

Definition 2.2.2. An increasing sequence of random times $\tau_n, n = 1, 2, \ldots$, is called *a localizing sequence with respect to a filtration* if the following hold true:

1. Each τ_n is a stopping time relative to the filtration, and
2. $\lim_{n\to\infty} \tau_n = \infty$ a.s. □

Definition 2.2.3.

1. A stochastic process $M = \{M(t) : t \geq 0\}$ is a *local martingale (submartingale) with respect to a filtration* $\{\mathcal{F}_t : t \geq 0\}$ if there exists a localizing sequence $\{\tau_n\}$ such that, for each n, $M_n = \{M(t \wedge \tau_n) : 0 \leq t < \infty\}$ is an $\mathcal{F}_t$-martingale (submartingale).
2. If M_n above is a martingale and a square integrable process, M_n is called a *square integrable martingale* and M is called a *local square integrable martingale.*
3. An adapted process $X = \{X(t) : t \geq 0\}$ is called *locally bounded* if, for a suitable localizing sequence $\{\tau_n\}, X_n = \{X(t \wedge \tau_n) : t \geq 0\}$ is a bounded process for each n. □

Any right-continuous adapted process X with left-hand limits, bounded jump sizes, and $X(0) = 0$ a.s. can be made locally bounded by using the localizing sequence $\tau_n = \sup\{t : \sup_{0\leq s\leq t} \mid X(s) \mid< n\} \wedge n, n = 1, 2, \ldots$, (Exercise 2.6) and the stopped processes X_n is given by

$$X_n(t) = X(t \wedge \tau_n).$$

See Figure 2.2.1.

Without loss of generality it may be assumed that a locally bounded process X with $X(0) = 0$ a.s. satisfies $P\{\sup_t |X(t \wedge \tau_n)| \leq n\} = 1$, and we will consistently do so. Locally bounded processes will not be needed until Section 2.3.

Theorem 2.2.3 is a version of the Doob-Meyer Decomposition Theorem that may be used for local martingales, and its proof uses the optional sampling theorem for martingales along with some technical results about localized processes. We begin with the technical preliminaries.

Lemma 2.2.1. Any martingale is a local martingale.

Proof. Simply take $\tau_n = n$. □

Lemma 2.2.2. An $\mathcal{F}_t$-local martingale M is a martingale if, for any fixed t, $\{M(t \wedge \tau_n) : n = 1, 2, \ldots\}$ is a uniformly integrable sequence, where $\{\tau_n\}$ is a localizing sequence for M.

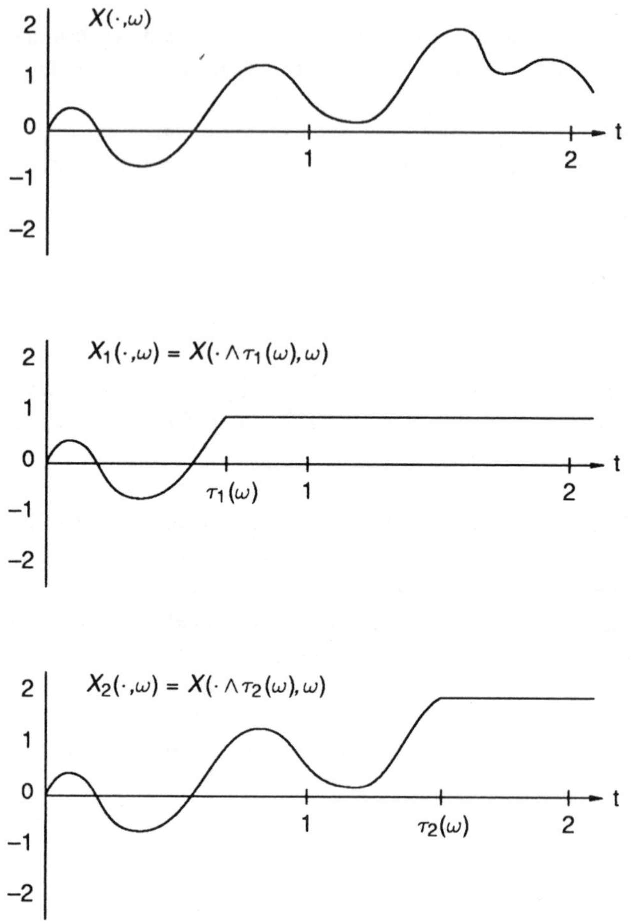

Figure 2.2.1 A sample path of a process X, and the corresponding paths of X "stopped" at the random times τ_1 and τ_2.

Proof. Since

$$
\begin{aligned}
M(t) &= \lim_{n\to\infty} M(t \wedge \tau_n) \\
&\equiv \lim_{n\to\infty} M_n(t),
\end{aligned}
$$

the $\mathcal{F}_t$-measurability of $M(t)$ follows from that of each $M_n(t)$. In addition,

$$
\begin{aligned}
E \mid M(t) \mid &= E \lim_{n\to\infty} \mid M_n(t) \mid \\
&= \lim_{n\to\infty} E \mid M_n(t) \mid \\
&< \infty,
\end{aligned}
$$

since the uniformly integrable sequence $M_n(t)$ converges in probability to $M(t)$ for each $t \geq 0$, (See Chung, 1974, Theorems 4.5.3 and 4.5.4).

It remains to show that $E\{M(t) \mid \mathcal{F}_s\} = M(s)$ a.s. for any $s \leq t$. We first establish that

$$\lim_{n\to\infty} E\{M_n(t) \mid \mathcal{F}_s\} = E\{M(t) \mid \mathcal{F}_s\} \qquad \text{in probability.} \tag{2.1}$$

Since $\{M_n(t)\}$ is uniformly integrable for each t and $E \mid M(t) \mid < \infty$, $\{M_n(t) - M(t)\}$ is uniformly integrable. Thus $\lim_{n\to\infty} M_n(t) = M(t)$ in probability implies

$$E \mid M_n(t) - M(t) \mid \longrightarrow 0,$$

and hence that

$$\begin{aligned} E \mid E\{M_n(t) \mid \mathcal{F}_s\} - E\{M(t) \mid \mathcal{F}_s\} \mid &= E \mid E\{M_n(t) - M(t) \mid \mathcal{F}_s\} \mid \\ &\leq EE\{\mid M_n(t) - M(t) \mid \mid \mathcal{F}_s\} \\ &= E \mid M_n(t) - M(t) \mid \\ &\rightarrow 0. \end{aligned}$$

Theorem 4.1.4 in Chung (1974) now establishes Eq. (2.1), which in turn implies the existence of a subsequence $\{n_k\}, k = 1, 2, \ldots$, of positive integers increasing to ∞ such that

$$E\{M_{n_k}(t) \mid \mathcal{F}_s\} \rightarrow E\{M(t) \mid \mathcal{F}_s\} \quad \text{a.s.}$$

(cf. Chung, Theorem 4.2.3).

Consequently,

$$\begin{aligned} E\{M(t) \mid \mathcal{F}_s\} &= \lim_{k\to\infty} E\{M_{n_k}(t) \mid \mathcal{F}_s\} \quad \text{a.s.} \\ &= \lim_{k\to\infty} M_{n_k}(s) \quad \text{a.s.} \\ &= M(s) \quad \text{a.s.} \end{aligned}$$

□

The main result of this section uses the fact that martingales and submartingales "sampled" at random times remain martingales and submartingales. A precise statement of this optional sampling requires a definition.

Definition 2.2.4. For any stopping time τ, the σ-algebra of events prior to τ, $\mathcal{F}_\tau$, consists of all sets $A \in \bigvee_{t\geq 0} \mathcal{F}_t$, satisfying

$$A \cap \{\tau \leq t\} \in \mathcal{F}_t \text{ for all } t \geq 0.$$

□

If τ is the time of some event, then heuristically $\mathcal{F}_\tau$ contains the information available by time τ. It is easy to show that $\mathcal{F}_\tau$ is in fact a σ-algebra (Exercise 2.3).

The conditions for the Optional Sampling Theorem take several forms. We will need only one of these, but for completeness state the most often cited versions. Proofs for both the discrete- and continuous-time versions of the theorem may be found in Liptser and Shiryayev (1977), Chapters 2 and 3.

Theorem 2.2.1 (Optional Sampling Theorem). Let $\{X(t) : t \geq 0\}$ be a right-continuous process adapted to a right-continuous filtration $\{\mathcal{F}_t : t \geq 0\}$, and let τ and τ^* be two $\mathcal{F}_t$-stopping times with $P(\tau \leq \tau^*) = 1$.

If $\{X(t) : t \geq 0\}$ is a martingale (respectively, submartingale), then

$$E\{X(\tau^*) \mid \mathcal{F}_\tau\} = X(\tau) \quad \text{a.s.}$$

$$(\text{respectively, } E\{X(\tau^*) \mid \mathcal{F}_\tau\} \geq X(\tau) \quad \text{a.s.})$$

when at least one of the following condition holds:

1. There exists a constant T such that $P\{\tau^* \leq T\} = 1$.
2. There exists an $\mathcal{F}_\infty = \bigvee_{t \geq 0} \mathcal{F}_t$ random variable $X(\infty)$ such that $\lim_{t\uparrow\infty} X(t) = X(\infty)$ a.s. and $\{X(t) : 0 \leq t \leq \infty\}$ is an $\mathcal{F}_t$-martingale (respectively, $\mathcal{F}_t$-submartingale).
3. The family of random variables $\{X(t) : 0 \leq t < \infty\}$ is uniformly integrable. □

The easiest condition for us to check will be (1). For martingales, (2) and (3) are equivalent.

The Optional Sampling Theorem gives rise to the Optional Stopping Theorem, which says that stopping a martingale or submartingale at a random time does not disturb its special structure.

Theorem 2.2.2 (Optional Stopping Theorem). Let $\{X(t) : 0 \leq t < \infty\}$ be a right-continuous $\mathcal{F}_t$-martingale (respectively, submartingale) and let τ be an $\mathcal{F}_t$-stopping time. Then $\{X(t \wedge \tau) : 0 \leq t < \infty\}$ is a martingale (respectively, submartingale).

Proof. We will establish the martingale case by applying the Optional Sampling Theorem to the process $\{X(u) : 0 \leq u \leq t\}$. Only obvious modifications are needed for the submartingale case.

Note first that $X(t \wedge \tau)$ is $\mathcal{F}_t$-measurable (Exercise 2.4). For $s \leq t$, it is easy to show that $t, t \wedge \tau$, and $s \wedge (\tau \wedge t)$ are all stopping times bounded by t. Applying the Optional Sampling Theorem to the martingale $\{X(u) : 0 \leq u \leq t\}$, we have

$$\begin{aligned} E \mid X(t \wedge \tau) \mid &= E \mid E\{X(t) \mid \mathcal{F}_{t\wedge\tau}\} \mid \\ &\leq E \mid X(t) \mid \\ &< \infty, \end{aligned}$$

so it remains only to show that, for any $0 \le s \le t$,

$$E\{X(t \wedge \tau) \mid \mathcal{F}_s\} = X(s \wedge \tau) \quad \text{a.s.}$$

Note first that

$$X(t \wedge \tau) = I_{\{\tau < s\}} X(t \wedge \tau) + I_{\{\tau \ge s\}} X(t \wedge \tau).$$

Since

$$I_{\{\tau < s\}} X(t \wedge \tau) = I_{\{\tau < s\}} X(s \wedge \tau)$$

is $\mathcal{F}_s$-measurable, we have

$$E\{I_{\{\tau < s\}} X(t \wedge \tau) \mid \mathcal{F}_s\} = I_{\{\tau < s\}} X(s \wedge \tau) \quad \text{a.s.}$$

The random variable $\tau^* = \max(\tau, s)$ is a stopping time, and

$$I_{\{\tau \ge s\}} X(t \wedge \tau) = I_{\{\tau \ge s\}} X(t \wedge \tau^*).$$

Since $I_{\{\tau \ge s\}}$ is $\mathcal{F}_s$-measurable (Exercise 2.1),

$$\begin{aligned} E\{I_{\{\tau \ge s\}} X(t \wedge \tau^*) \mid \mathcal{F}_s\} &= I_{\{\tau \ge s\}} E\{X(t \wedge \tau^*) \mid \mathcal{F}_s\} \\ &= I_{\{\tau \ge s\}} X(s) \qquad \text{(By Optional Sampling)} \\ &= I_{\{\tau \ge s\}} X(s \wedge \tau) \quad \text{a.s.} \end{aligned}$$

□

The next result shows that there is some flexibility in choosing localizing sequences.

Lemma 2.2.3. Suppose M is a right-continuous local square integrable martingale on $[0, \infty)$, with localizing sequence $\{\tau_n^*\}$. Let $\{\tau_n\}$ be another increasing sequence of stopping times with $\tau_n \uparrow \infty$ and $\tau_n \le \tau_n^*$ a.s.

Then $\{M(t) : t \ge 0\}$ also is a right-continuous local square integrable martingale on $[0, \infty)$, with localizing sequence $\{\tau_n\}$.

Proof. For each n, $M^2(\cdot \wedge \tau_n^*)$ is a submartingale on $[0, \infty)$, and $\tau_n \le \tau_n^*$. By the Optional Sampling Theorem

$$\begin{aligned} EM^2(t \wedge \tau_n) &\le E[E\{M^2(t \wedge \tau_n^*) \mid \mathcal{F}_{t \wedge \tau_n}\}] \\ &= EM^2(t \wedge \tau_n^*), \end{aligned}$$

and hence

$$\begin{aligned} \sup_t EM^2(t \wedge \tau_n) &\le \sup_t EM^2(t \wedge \tau_n^*) \\ &< \infty. \end{aligned}$$

To see that $M(\cdot \wedge \tau_n)$ is a martingale, define $M_n(\cdot) \equiv M(\cdot \wedge \tau_n^*)$ and observe that $M(t \wedge \tau_n) = M(t \wedge \tau_n \wedge \tau_n^*) = M_n(t \wedge \tau_n)$. An application of Theorem 2.2.2 to the martingale M_n completes the proof. □

We can now prove a version of the Doob-Meyer Decomposition (Theorem 1.4.1) for nonnegative local submartingales. This version implies the existence of a compensator A for any counting process N so that $N - A$ is a local martingale, and also implies that any local square integrable martingale has a unique predictable variation process $\langle M, M\rangle$ such that $M^2 - \langle M, M\rangle$ is a local martingale and $\langle M, M\rangle(0) = 0$. In the next two sections these two results will yield general formulas for first and second moments for statistics of the form $\sum_i \int H_i d(N_i - A_i)$ when H_i is locally bounded, and in Section 2.5 these general formulas will give way to more specific versions when the A_i are continuous processes.

The idea behind the extension of the Doob-Meyer Decomposition to nonnegative local submartingales is quite simple. Suppose $\{X(t) : t \geq 0\}$ is a nonnegative local submartingale with localizing sequence $\{\tau_n\}$. From Theorem 1.4.1, there exists a unique sequence of increasing predictable processes $\{A_n(t) : t \geq 0\}$ such that $A_n(0) = 0$ a.s. and $\{X(t \wedge \tau_n) - A_n(t) : t \geq 0\}$ are martingales. In the proof of Theorem 2.2.3, we construct a predictable increasing process A with $A(0) = 0$ such that $\{X(t \wedge \tau_n) - A(t \wedge \tau_n) : t \geq 0\}$ is a martingale for each n by taking $A = A_n$ on $[0, \tau_n]$, by making sure that $A_{n'} = A_n$ on $[0, \tau_n]$ when $n' \geq n$, and by letting $n \to \infty$.

Theorem 2.2.3 (Extended Doob-Meyer Decomposition). Let $X = \{X(t) : t \geq 0\}$ be a nonnegative right-continuous $\mathcal{F}_t$-local submartingale with localizing sequence $\{\tau_n\}$, where $\{\mathcal{F}_t : t \geq 0\}$ is a right-continuous filtration. Then there exists a unique increasing right-continuous predictable process A such that $A(0) = 0$ a.s., $P\{A(t) < \infty\} = 1$ for all $t > 0$, and $X - A$ is a right-continuous local martingale. At each t, $A(t)$ may be taken as the a.s. $\lim_{n\to\infty} A_n(t)$, where A_n is the compensator of the stopped submartingale $X(\cdot \wedge \tau_n)$.

Proof. Let A_n be the unique increasing right-continuous predictable process such that $A_n(0) = 0$ and $X(\cdot \wedge \tau_n) - A_n(\cdot)$ is a martingale. Since $\tau_n < \tau_{n'}$ a.s. when $n \leq n'$, the Optional Stopping Theorem applied to the martingale $X(\cdot \wedge \tau_{n'}) - A_{n'}(\cdot)$ and the stopping time τ_n implies that $X(\cdot \wedge \tau_n) - A_{n'}(\cdot \wedge \tau_n)$ is a martingale. ($A_{n'}(\cdot \wedge \tau_n)$ is predictable since $A_{n'}(\cdot)$ is predictable and τ_n is a stopping time, e.g., see Jacod and Shiryayev, 1987, p.16, Proposition 2.4.) By the uniqueness of the compensator in the Doob-Meyer Decomposition Theorem,

$$A_n(t) = A_{n'}(t \wedge \tau_n) \quad \text{a.s.,} \quad t \geq 0,\ n' \geq n. \tag{2.2}$$

Let $\mathcal{Q}$ denote the set of rational numbers, and take the sets C_n and D_n to be

$$C_n = \{\omega : A_n(s) = A_{n'}(s) \text{ for any } n' \geq n, s \in \mathcal{Q} \cap [0, \tau_n(\omega)]\},$$

$$D_n = \{\omega : A_n(s) = A_n(\tau_n(\omega)), s \in \mathcal{Q} \cap [\tau_n(\omega), \infty)\}.$$

On $\tilde{\Omega} = \bigcap_{n=1}^{\infty}(C_n \cap D_n)$, each A_n is constant on the rationals larger than τ_n, and no two processes $A_n, A_{n'}$ disagree at a rational time point s when $n' \geq n$ and $s \leq \tau_n$. Since

$$\begin{aligned} P\{\tilde{\Omega}\} &= P\left\{\bigcap_{n=1}^{\infty}(C_n \cap D_n)\right\} \\ &= 1 - P\left\{\bigcup_{n=1}^{\infty}(\bar{C}_n \cup \bar{D}_n)\right\} \\ &\geq 1 - \sum_{n=1}^{\infty}(P\{\bar{C}_n\} + P\{\bar{D}_n\}) \\ &= 1, \qquad \text{(by Eq. 2.2)}, \end{aligned}$$

$P\{\tilde{\Omega}\} = 1$. The right-continuity of the processes $A_n, n \geq 1$, thus implies that on a set $\tilde{\Omega}$ of probability 1, each A_n is constant after τ_n, and any two processes A_n and $A_{n'}$ coincide for $0 \leq s \leq \tau_n$ when $n' \geq n$.

Now $\tau_n \uparrow \infty$ a.s. implies that for any t, $\lim_{n\to\infty} A_n(t) = A(t)$ exists, and it is easily seen that $A(0) = 0$, A is predictable, and $EA_n(t) < \infty$. For any $t \geq 0$,

$$\begin{aligned} P\left[\bigcap_{n=1}^{\infty}\{A_n(t) < \infty\}\right] &= 1 - P\left[\bigcup_{n=1}^{\infty}\{A_n(t) = \infty\}\right] \\ &\geq 1 - \sum_{n=1}^{\infty} P\{A_n(t) = \infty\} \\ &= 1. \end{aligned}$$

Thus $A = A_n$ on $[0, \tau_n]$ for all paths in $\tilde{\Omega}$ and $\tau_n \uparrow \infty$ a.s. implies that $P\{A(t) < \infty\} = 1$ for all $t \geq 0$, and that A is right-continuous and increasing.

We must now show for A defined above that $X - A$ is a right-continuous local martingale, and that A is the unique such compensator for X. Since $X(\cdot \wedge \tau_n) - A_n(\cdot)$ is a martingale, A will be a compensator for X if $A(t \wedge \tau_n) = A_n(t)$ a.s. for any t. But we have already shown that $A = A_n$ on $[0, \tau_n]$ and that A_n is constant on $[\tau_n, \infty)$ for all paths in $\tilde{\Omega}$. Suppose now there exist two processes A^1 and A^2 that are increasing, right-continuous predictable processes with $A^i(0) = 0, P\{A^i(t) < \infty\} = 1$ for all $t > 0$, and that $X - A^i$ are local martingales. There must then exist two localizing sequences $\{\tau_n^i\}, i = 1, 2$, with $X(\cdot \wedge \tau_n^i) - A^i(\cdot \wedge \tau_n^i)$ being martingales for each n. For any n, $X(\cdot \wedge \tau_n)$ is a submartingale, and since the minimum of two stopping times is a stopping time (Exercise 2.5), the Optional Stopping Theorem implies both that $X(\cdot \wedge \tau_n \wedge \tau_n^1 \wedge \tau_n^2)$ is a submartingale and that $X(\cdot \wedge \tau_n \wedge \tau_n^1 \wedge \tau_n^2) - A^i(\cdot \wedge \tau_n \wedge \tau_n^1 \wedge \tau_n^2)$ is a martingale. The uniqueness portion of the standard Doob-Meyer Decomposition then implies $A^1(t \wedge \tau_n \wedge \tau_n^1 \wedge \tau_n^2) = A^2(t \wedge \tau_n \wedge \tau_n^1 \wedge \tau_n^2)$ a.s. for any t and n. As $n \to \infty, \tau_n \wedge \tau_n^1 \wedge \tau_n^2 \uparrow \infty$ a.s., and hence $A^1(t) = A^2(t)$ a.s. for every t. □

2.3 THE MARTINGALE $N - A$ REVISITED

The extended Doob-Meyer Decomposition (Theorem 2.2.3) can be used to represent an arbitrary counting process as the sum of a local martingale and a predictable increasing process. When the increasing process is locally bounded, the local martingale will be locally square integrable, and Theorem 2.2.3 can be used to establish the existence and some properties of predictable quadratic variation processes for local square integrable martingales.

Lemma 2.3.1 is a technical preliminary that provides convenient sufficient conditions for local boundedness.

Lemma 2.3.1. Let $\{\mathcal{F}_t : t \geq 0\}$ be a right-continuous filtration.

1. An $\mathcal{F}_t$-adapted process X with $X(0) = 0$ a.s. is locally bounded if and only if there is an adapted, left-continuous process H with finite right-hand limits such that $| X(t) | \leq H(t)$ for all $t \geq 0$, a.s.
2. Any $\mathcal{F}_t$-adapted, left-continuous process X with finite right-hand limits and with $X(0) = 0$ a.s. is locally bounded.

Proof.

1. ***Necessity.*** Suppose $\{\tau_n\}$ is a localizing sequence for X, such that

$$\sup_{0 \leq t} | X(t \wedge \tau_n) | \leq n \quad \text{a.s. for any } t,$$

which is equivalent to

$$\sup_{0 \leq t \leq \tau_n} | X(t) | \leq n \quad \text{a.s.}$$

Let $\tau_0 \equiv 0, H(0) = 1$ and for $t > 0$ let

$$H(t) = \sum_{j=1}^{\infty} j I_{\{\tau_{j-1} < t \leq \tau_j\}}.$$

The process H is left-continuous with finite right-hand limits, and is adapted since $H(t)$ is a sum of $\mathcal{F}_t$-measurable variables.

Sufficiency. We construct a localizing sequence for X by taking $\tau_n = n \wedge \sup\{t : \sup_{0 \leq s \leq t} H(s+) < n\}$, where $H(s+) = \lim_{t \downarrow s} H(t)$. The right-continuity of the filtration $\{\mathcal{F}_t : t \geq 0\}$ implies that each τ_n is a stopping time, and since H is left-continuous,

$$\begin{aligned} \sup_{0 \leq t} |X(t \wedge \tau_n)| &\leq \sup_{0 \leq t} H(t \wedge \tau_n) \\ &\leq n \quad \text{a.s.} \end{aligned}$$

The sequence $\{\tau_n\}$ is clearly increasing, and $\tau_n \uparrow \infty$ a.s. because the paths of H are left-continuous with finite right-hand limits.

2. Part (2) follows directly from (1). □

Theorem 2.3.1. Let N be an arbitrary counting process.

1. Then there exists a unique right-continuous predictable increasing process A such that $A(0) = 0$ a.s., $A(t) < \infty$ a.s. for any t, and the process $M = N - A$ is a local martingale.
2. If A in (1) is locally bounded, M is a local square integrable martingale.

Proof.

1. If $\tau_n = n \wedge \sup\{t : N(t) < n\}$, then N is a local submartingale with localizing sequence $\{\tau_n\}$ (Exercise 2.2). The result is now a direct application of Theorem 2.2.3.
2. For each n, let τ_n' be a stopping time such that

$$\sup_{0 \leq t} A(t \wedge \tau_n') < n \quad \text{a.s.}$$

Then $\tau_n^* \equiv \tau_n \wedge \tau_n' \uparrow \infty$ a.s. and

$$\begin{aligned} | M(t \wedge \tau_n^*) | &\leq N(t \wedge \tau_n^*) + A(t \wedge \tau_n^*) \qquad (3.1) \\ &\leq 2n. \end{aligned}$$

The proof is completed by applying the Optional Stopping Theorem to the martingale $N(\cdot \wedge \tau_n) - A(\cdot \wedge \tau_n)$ and the stopping time τ_n^* to establish that $N(\cdot \wedge \tau_n^*) - A(\cdot \wedge \tau_n^*)$ is a martingale, indeed a square integrable martingale by (3.1). □

Consistent with earlier terminology, the process A in the previous decomposition of an arbitrary counting process will be called a compensator. Somewhat surprisingly, Part (2) of Theorem 2.3.1 is always true, that is, if A is the compensator of an arbitrary counting process N, then A is locally bounded. This was established by Meyer (1976, Theorem IV.12). We summarize this result without proof in Theorem 2.3.2 along with another fact about the growth properties of counting process compensators. As mentioned earlier, all the statistical models we consider will directly or indirectly yield specific forms for A, and it will be easy to check that A is locally bounded in each case. In fact, A will most often be continuous, and its local boundedness will follow from Lemma 2.3.1.

Theorem 2.3.2. Let N be a counting process and let A be its unique compensator in the Extended Doob-Meyer Decomposition Theorem. Then

1. A is a locally bounded process, and
2. $\Delta A(t) \equiv A(t) - \lim_{s \uparrow t} A(s) \leq 1$ a.s. for all $t \geq 0$. □

A short proof of (2) due to Van Schuppen using the general theory of stochastic processes may be found in Gill (1980), Appendix 1.

The next result provides a method for determining $EN(t)$, and it establishes a condition under which $M = N - A$ is a martingale.

Lemma 2.3.2. Suppose N is a counting process. Then $EN(t) = EA(t)$ for any $t \geq 0$, where A is the compensator for N. If $EA(t) < \infty$ for all t, then $M = N - A$ is a martingale.

Proof. Let $\{\tau_n\}$ denote a localizing sequence for the local martingale $M = N - A$, so $M(\cdot \wedge \tau_n)$ is a martingale for any n. Since $M(0) = 0$ a.s.,

$$\begin{aligned} EM(t \wedge \tau_n) &= E\{EM(t \wedge \tau_n) \mid \mathcal{F}_0\} \\ &= E\{M(0)\} \\ &= 0 \end{aligned}$$

for any t, and thus,

$$EN(t \wedge \tau_n) = EA(t \wedge \tau_n)$$

for any t and n.

By the Monotone Convergence Theorem,

$$\begin{aligned} EN(t) &= \lim_{n \to \infty} E\{N(t \wedge \tau_n)\} \\ &= \lim_{n \to \infty} E\{A(t \wedge \tau_n)\} \\ &= EA(t) \end{aligned}$$

for any $t \geq 0$.

If $EA(t) < \infty$, then $EN(t) < \infty$. By Corollary 1.4.1 there exists a unique increasing right-continuous predictable process $\tilde{A}$ such that $\tilde{A}(0) = 0$ a.s., $E\tilde{A}(t) < \infty, t \geq 0$, and $\tilde{M} = N - \tilde{A}$ is a martingale. Since $\tilde{M}$ will also be a local martingale, it follows from Theorem 2.2.3 that $M = \tilde{M}$ a.s. and $A = \tilde{A}$ a.s., and consequently that M is a martingale. □

The predictable variation and covariation processes in Section 1.4 for martingales with finite second moments can be extended to local square integrable martingales. We begin with the predictable variation process $\langle M, M \rangle$.

Theorem 2.3.3. Suppose M is a right-continuous local square integrable martingale on $[0, \infty)$. Then there exists a unique predictable right-continuous increasing process $\langle M, M \rangle$ with $\langle M, M \rangle(0) = 0$ a.s. and $\langle M, M \rangle(t) < \infty$ a.s. for any t, such that $M^2 - \langle M, M \rangle$ is a right-continuous local martingale. If $\{\tau_n\}$ is a localizing sequence for M, then

$$\langle M, M \rangle(t) = \lim_{n \to \infty} \langle M(\cdot \wedge \tau_n), M(\cdot \wedge \tau_n) \rangle(t). \tag{3.2}$$

Proof. By assumption, there exists an increasing sequence of stopping times $\{\tau_n\}$ such that $\tau_n \uparrow \infty$ a.s. and $\{M(t \wedge \tau_n) : t \geq 0\}$ is a square integrable martingale for each n. Thus $M^2(\cdot \wedge \tau_n)$ is a submartingale. The proof is completed by applying Theorem 2.2.3 to $X = M^2$. □

The next two corollaries provide the second moment results for the locally square integrable martingale, M.

Corollary 2.3.1. Suppose M is a right-continuous local square integrable martingale on $[0, \infty)$, with $M(0) = 0$ a.s. Then

$$\begin{aligned} EM^2(t) &\leq \lim_{n \to \infty} EM^2(t \wedge \tau_n) \\ &= E\langle M, M\rangle(t). \end{aligned}$$

Proof. If $\{\tau_n\}$ is a localizing sequence for M then, for any t,

$$\lim_{n \to \infty} M^2(t \wedge \tau_n) = M^2(t) \quad \text{a.s.}$$

Fatou's Lemma implies $EM^2(t) \leq \lim_{n \to \infty} EM^2(t \wedge \tau_n)$. Since $\langle M, M\rangle(t \wedge \tau_n)$ increases to $\langle M, M\rangle(t)$ a.s. as $n \to \infty$,

$$\lim_{n \to \infty} E\langle M, M\rangle(t \wedge \tau_n) = E\langle M, M\rangle(t),$$

by the Monotone Convergence Theorem. The proof is completed by noting that the martingale property of

$$M^2(\cdot \wedge \tau_n) - \langle M, M\rangle(\cdot \wedge \tau_n)$$

and the assumption $M(0) = 0$ a.s. implies

$$EM^2(t \wedge \tau_n) = E\langle M, M\rangle(t \wedge \tau_n)$$

for any t and n. □

Corollary 2.3.2. If the local square integrable martingale M is a martingale with $M(0) = 0$ a.s., then for any $t \geq 0$,

$$EM^2(t) = E\langle M, M\rangle(t).$$

Proof. Suppose $EM^2(t) < \infty$. Then, by the Doob-Meyer Decomposition, $M^2 - \langle M, M\rangle$ is a martingale on $[0, t]$ and $EM^2(t) = E\langle M, M\rangle(t)$.
If $EM^2(t) = \infty$, $E\langle M, M\rangle(t) = \infty$ as well by Corollary 2.3.1. □

We complete this section by establishing existence of the predictable covariation process for two local square integrable martingales.

Theorem 2.3.4. Suppose M_1 and M_2 are right-continuous local square integrable martingales on $[0, \infty)$. Then there exists a predictable right-continuous process $\langle M_1, M_2\rangle$ with $\langle M_1, M_2\rangle(0) = 0$ a.s. and $\langle M_1, M_2\rangle(t) < \infty$ a.s. for any t, and such that $M_1 M_2 - \langle M_1, M_2\rangle$ is a right-continuous local martingale on $[0, \infty)$. In fact,

$$\langle M_1, M_2\rangle = \frac{1}{2}\{\langle M_1 + M_2, M_1 + M_2\rangle - \langle M_1, M_1\rangle - \langle M_2, M_2\rangle\}. \qquad \square$$

When the underlying filtration is complete and right-continuous, $\langle M_i, M_j\rangle$ is unique. We will not use the uniqueness, and so do not establish it here. A detailed proof of the uniqueness may be found in Jacobsen (1982, Section 3.1).

Proof of Theorem 2.3.4.. Suppose $\{\tau_n^1\}$ and $\{\tau_n^2\}$ are localizing sequences for M_1 and M_2 respectively, and set $\tau_n \equiv \tau_n^1 \wedge \tau_n^2$. By Lemma 2.2.3, M_1, M_2, and $M_1 + M_2$ are all local square integrable martingales with the localizing sequence $\{\tau_n\}$.

If we define $\langle M_1, M_2\rangle \equiv 1/2\{\langle M_1 + M_2, M_1 + M_2\rangle - \langle M_1, M_1\rangle - \langle M_2, M_2\rangle\}$, then

$$\begin{aligned}
&\{M_1 M_2 - \langle M_1, M_2\rangle\}(\cdot \wedge \tau_n) \\
&\quad = 1/2\{(M_1 + M_2)^2 - \langle M_1 + M_2, M_1 + M_2\rangle - (M_1^2 - \langle M_1, M_1\rangle) \\
&\qquad -(M_2^2 - \langle M_2, M_2\rangle)\}(\cdot \wedge \tau_n)
\end{aligned}$$

is a martingale, since each of

$$((M_1 + M_2)^2 - \langle M_1 + M_2, M_1 + M_2\rangle)(\cdot \wedge \tau_n),$$

$$(M_1^2 - \langle M_1, M_1\rangle)(\cdot \wedge \tau_n),$$

and

$$(M_2^2 - \langle M_2, M_2\rangle)(\cdot \wedge \tau_n)$$

are martingales.

By Theorem 2.3.3, $\langle M_1 + M_2, M_1 + M_2\rangle(t)$, $\langle M_1, M_1\rangle(t)$, and $\langle M_2, M_2\rangle(t)$ are finite a.s., so for each fixed t

$$\langle M_1 + M_2, M_1 + M_2\rangle(t) - \langle M_1, M_1\rangle(t) - \langle M_2, M_2\rangle(t)$$

is well defined. Finally, note that

$$\langle M_1, M_2\rangle(t) = \lim_{n\to\infty} \langle M_1(\cdot \wedge \tau_n), M_2(\cdot \wedge \tau_n)\rangle(t), \tag{3.3}$$

since

$$\begin{aligned}
&\langle M_1, M_2\rangle(t) \\
&\quad = 1/2\{\langle M_1 + M_2, M_1 + M_2\rangle(t) - \langle M_1, M_1\rangle(t) - \langle M_2, M_2\rangle(t)\} \\
&\quad = 1/2 \lim_{n\to\infty} \{\langle M_1(\cdot \wedge \tau_n) + M_2(\cdot \wedge \tau_n), M_1(\cdot \wedge \tau_n) + M_2(\cdot \wedge \tau_n)\rangle(t) \\
&\qquad -\langle M_1(\cdot \wedge \tau_n), M_1(\cdot \wedge \tau_n)\rangle(t) - \langle M_2(\cdot \wedge \tau_n), M_2(\cdot \wedge \tau_n)\rangle(t)\} \\
&\quad = \lim_{n\to\infty} \langle M_1(\cdot \wedge \tau_n), M_2(\cdot \wedge \tau_n)\rangle(t).
\end{aligned}$$

2.4 STOCHASTIC INTEGRALS WITH RESPECT TO LOCAL MARTINGALES

We have seen earlier that statistics for counting process data are often of the form

$$U = \sum_{i=1}^{n} \int H_i dM_i,$$

where the M_i are compensated counting processes, $N_i - A_i$. (Although U also depends on n, we suppress that dependence in the notation until we study asymptotic properties of sequences of statistics indexed by sample size.) Indeed, in Chapters 0 and 1, the two-sample logrank statistic and proportional hazards score statistics have this form. When the covariate vector Z is bounded, H_i is bounded and $EN_i(t) < \infty$ for any t, properties that were exploited in Chapter 1.

In this section, we will examine properties of U subject to the weaker condition:

$$H_i, i = 1, \ldots, n, \text{ is locally bounded and } \mathcal{F}_t\text{-predictable}, \tag{4.1}$$

and we remove the assumption $EN_i(t) < \infty$.

The process U turns out to be a local square integrable martingale with general formulas for its first and second moments. Section 2.5 derives specific versions of these formulas for continuous A_i. In Chapter 5, the local square integrability of U will be a key ingredient in establishing asymptotic normality.

Left-continuous, adapted processes with right-hand limits are the most common instances of $\mathcal{F}_t$-predictable processes, and are the easiest integrands to use to insure correct use of the stochastic integration theory for martingales. Each statistic studied in this book arises naturally with a left-continuous, adapted integrand or with a left-continuous modification of the integrand.

Several of the arguments in this section use the local boundedness of the compensator A which, though not proven here, is always true.

Theorem 2.4.1 begins with the basic components that are used to build U. Recall that for any counting process N, $P\{N(t) < \infty\} = 1$ for all $t \geq 0$.

Theorem 2.4.1. Let $(\Omega, \mathcal{F}, \{\mathcal{F}_t : t \geq 0\}, P)$ be a stochastic basis with right-continuous filtration $\{\mathcal{F}_t : t \geq 0\}$, H a locally bounded $\mathcal{F}_t$-predictable process, and N a counting process. Let $M = N - A$ be the local square integrable $\mathcal{F}_t$-martingale whose existence is established in Theorems 2.3.1 and 2.3.2. Then $\int H dM$ is a local square integrable martingale. □

The proof of Theorem 2.4.1 will use the following lemma.

Lemma 2.4.1. Let H be a locally bounded $\mathcal{F}_t$-predictable process and M a local $\mathcal{F}_t$-martingale of locally bounded variation with $\Delta M(0) = 0$. Then $\int H dM$ is a local square integrable martingale.

Proof of Lemma 2.4.1. The assumption on M implies the existence of a localizing sequence $\{\tau_n\}$ so that each stopped process $M(\cdot \wedge \tau_n)$ is a martingale with

$$\int_0^t |\, dM(\cdot \wedge \tau_n)\,| = \int_0^{t\wedge\tau_n} |\, dM(\cdot)\,|$$
$$< \infty \quad \text{a.s.}$$

Without loss of generality we may assume

$$\int_0^{t\wedge\tau_n} |\, dM \,| \leq n \quad \text{a.s.}$$

Let $\{\tau_n'\}$ be a localizing sequence of H so that $|\, H(\cdot \wedge \tau_n')\,| \leq \Gamma_n$ a.s. for some sequence of finite constants $\{\Gamma_n\}$. Then $\{\tau_n^*\} = \{\tau_n \wedge \tau_n'\}$ is a localizing sequence for both M and H and, in particular, M is a local square integrable martingale with localizing sequence $\{\tau_n^*\}$. Now

$$\left|\int_0^{t\wedge\tau_n^*} H dM\right| \leq \int_0^t |\, H(\cdot \wedge \tau_n^*)\,||\, dM(\cdot \wedge \tau_n^*)\,|$$
$$\leq \Gamma_n \int_0^t |\, dM(\cdot \wedge \tau_n^*)\,|$$
$$\leq n\Gamma_n \quad \text{a.s., for any } t.$$

The process $\int H(\cdot \wedge \tau_n^*)dM(\cdot \wedge \tau_n^*)$ is thus square integrable. Note also that $H(\cdot \wedge \tau_n^*)$ is predictable since H is predictable and τ_n^* is a stopping time (Jacod and Shiryayev, 1987, p. 16, Proposition 2.4). Since the proof of Theorem 1.5.1 used the form $M = N - A$ only to establish $E \mid \int HdM \mid < \infty$, the same proof establishes here that $\int_0^{t\wedge\tau_n^*} HdM$ is a martingale, and hence that $\int HdM$ is a local square integrable martingale with localizing sequence $\{\tau_n^*\}$. □

Proof of Theorem 2.4.1. We need only show that $M = N - A$ is of locally bounded variation. Since M is the difference of two increasing processes, we must establish the existence of a common localizing sequence making both N and A locally bounded. If τ_n^* is defined as

$$\tau_n^* = \sup_{t\geq 0}\left\{t : N(t) < \frac{n-1}{2}\right\},$$

and τ_n' satisfies

$$\sup_{t\geq 0} A(t \wedge \tau_n') \leq \frac{n}{2} \quad \text{a.s.,}$$

then $\tau_n = \tau_n^* \wedge \tau_n' \wedge n$ will suffice. □

Corollary 2.4.1. For any compensated counting process local martingale $M = N - A$, $\int M(s-)\, dM(s)$ is a local square integrable martingale.

Proof. As in the proof of Theorem 2.4.1, the paths of M are locally bounded, and $M(s-)$ is adapted and left-continuous (hence predictable and locally bounded). The result now follows directly from Theorem 2.4.1. □

We first establish the form of predictable variation and covariation processes for processes $\int H_i dM_i, i = 1, 2$, under enough boundedness conditions to allow a direct proof. Localization is then used in the usual way to relax these conditions. Theorem 2.4.2 is stated only in terms of covariation processes: the form of a predictable variation process for $\int H dM$ is given by setting $H_1 = H_2 = H$, and $M_1 = M_2 = M$. Since two pairs of processes are involved, some care must be taken about the underlying (common) filtrations.

Theorem 2.4.2. Let $(\Omega, \mathcal{F}, \{\mathcal{F}_t : t \geq 0\}, P)$ be a stochastic basis, and assume:

1. H_i is a bounded (by Γ_i) $\mathcal{F}_t$-predictable process.
2. N_i is a bounded (by K_i) counting process.
3. The $\mathcal{F}_t$-martingale $M_i = N_i - A_i$ satisfies $EM_i^2(t) < \infty$ for any t, where A_i is the unique compensator arising in Corollary 1.4.1.

Then

$$\left\langle \int H_1 dM_1, \int H_2 dM_2 \right\rangle = \int H_1 H_2 d\langle M_1, M_2\rangle . \tag{4.2}$$

Proof. By Theorem 1.5.1, $\int H_i dM_i$ is a martingale over $[0, \infty)$, and

$$\begin{aligned} E\left\{\int_0^t H_i dM_i\right\}^2 &\leq (\Gamma_i)^2 E\{N_i(t) + A_i(t)\}^2 \\ &= (\Gamma_i)^2 \left[E\{N_i(t) - A_i(t)\}^2 + 4EN_i(t)A_i(t)\right] \\ &\leq (\Gamma_i)^2\{EM_i^2(t) + 4K_i EA_i(t)\} \\ &< \infty. \end{aligned}$$

Thus, by Theorem 1.4.2 and the subsequent remark, $\langle \int H_1 dM_1, \int H_2 dM_2\rangle$ exists and is unique.

To prove Eq. (4.2) it suffices to show $\{L(t) : t \geq 0\}$ is a martingale, where

$$L(t) = \int_0^t H_1 dM_1 \int_0^t H_2 dM_2 - \int_0^t H_1 H_2 d\langle M_1, M_2\rangle.$$

We begin with the special case of simple, predictable H_i. For $i = 1, 2$, suppose B_i is a predictable rectangle of the form

$$[0] \times A_i, \qquad A_i \in \mathcal{F}_0,$$

or

$$(a_i, b_i] \times A_i, \qquad A_i \in \mathcal{F}_{a_i},$$

and suppose $H_i = I_{B_i}$. The process L is easily shown to be adapted. To show $E|L(t)| < \infty$, it is sufficient to show

$$E\left|\int_0^t H_1H_2d\langle M_1, M_2\rangle\right| < \infty.$$

From the definition of $\langle M_1, M_2\rangle$ in the proof of Theorem 1.4.2

$$E\left|\int_0^t H_1H_2d\langle M_1, M_2\rangle\right| \leq$$
$$\frac{1}{4}E\{\langle M_1+M_2, M_1+M_2\rangle(t) + \langle M_1-M_2, M_1-M_2\rangle(t)\}.$$

Since $M_1 + M_2$ and $M_1 - M_2$ are martingales with finite second moments at all t, Corollary 1.4.2 implies the right-hand side of the above inequality is finite.

To establish

$$E\{L(t)|\mathcal{F}_s\} = L(s), \qquad s \leq t,$$

we form the partition $0 = t_0 < t_1 < \cdots < t_{n-1} < t_n = \infty$ containing from 2 to 6 points: $0, \infty$, and $\{a_i, b_i\}$ for each i which has B_i of the form $(a_i, b_i] \times A_i$. For $i = 1$ and 2,

$$H_i(t) = H_{i0}I_{\{t=0\}} + \sum_{l=1}^{n} H_{il}I_{\{t_{l-1}<t\leq t_l\}},$$

where $H_{i0} \equiv I_{A_i}I_{\{B_i=[0]\times A_i\}}$ and, for $l \geq 1$, $H_{il} \equiv I_{A_i}I_{\{(t_{l-1},t_l]\subset(a_i,b_i]\}}$. Then H_{i0} is $\mathcal{F}_0$-measurable, and H_{il} is $\mathcal{F}_{t_{l-1}}$-measurable, $l = 1, \ldots n$. Let

$$X_i(t) = \int_0^t H_i(u)dM_i(u), \quad i = 1, 2,$$

and let j be such that $t_{j-1} \leq s < t_j$. Now

$$E\Big\{\int_0^t H_1dM_1 \int_0^t H_2dM_2 - \int_0^s H_1dM_1 \int_0^s H_2dM_2\Big|\mathcal{F}_s\Big\}$$
$$= E\{X_1(t)X_2(t) - X_1(s)X_2(s)|\mathcal{F}_s\}$$
$$= E\{X_1(t\wedge t_j)X_2(t\wedge t_j) - X_1(s)X_2(s)|\mathcal{F}_s\}$$
$$+E\left\{\sum_{l=j+1}^{n} E\left[\{X_1(t\wedge t_l)X_2(t\wedge t_l) - X_1(t\wedge t_{l-1})X_2(t\wedge t_{l-1})\}|\mathcal{F}_{t_{l-1}}\right]\Big|\mathcal{F}_s\right\}$$
$$= E[\{X_1(t\wedge t_j) - X_1(s)\}\{X_2(t\wedge t_j) - X_2(s)\}\,|\mathcal{F}_s]$$
$$+E\Big\{\sum_{l=j+1}^{n} E\Big[\{X_1(t\wedge t_l) - X_1(t\wedge t_{l-1})\}\{X_2(t\wedge t_l) - X_2(t\wedge t_{l-1})\}$$
$$\Big|\mathcal{F}_{t_{l-1}}\Big]\Big|\mathcal{F}_s\Big\} \qquad \text{(by Eq. (4.5) of Chapter 1)}$$

$$= H_{1j}H_{2j}E[\{M_1(t\wedge t_j)-M_1(s)\}\{M_2(t\wedge t_j)-M_2(s)\} \mid \mathcal{F}_s]$$

$$+E\left\{\sum_{l=j+1}^{n} H_{1l}H_{2l}E\Big[\{M_1(t\wedge t_l)-M_1(t\wedge t_{l-1})\}\{M_2(t\wedge t_l)-M_2(t\wedge t_{l-1})\} \Big|\mathcal{F}_{t_{l-1}}\Big]\Big|\mathcal{F}_s\right\}$$

$$= H_{1j}H_{2j}E\{\langle M_1,M_2\rangle(t\wedge t_j)-\langle M_1,M_2\rangle(s)|\mathcal{F}_s\}$$

$$+E\left\{\sum_{l=j+1}^{n} H_{1l}H_{2l}E\left[\{\langle M_1,M_2\rangle(t\wedge t_l)-\langle M_1,M_2\rangle(t\wedge t_{l-1})\}|\mathcal{F}_{t_{l-1}}\right]\Big|\mathcal{F}_s\right\}$$

(by Eq. (4.6) of Chapter 1)

$$= E\left\{\int_s^{t\wedge t_j} H_{1j}H_{2j}d\langle M_1,M_2\rangle\Big|\mathcal{F}_s\right\}$$

$$+E\left\{\sum_{l=j+1}^{n} E\left[\int_{t\wedge t_{l-1}}^{t\wedge t_l} H_{1l}H_{2l}d\langle M_1,M_2\rangle\Big|\mathcal{F}_{t_{l-1}}\right]\Big|\mathcal{F}_s\right\}$$

$$= E\left\{\int_s^t H_1H_2d\langle M_1,M_2\rangle\Big|\mathcal{F}_s\right\}.$$

The proof is complete for simple predictable processes I_{B_i}. Using the Monotone Class Theorem as in the proof of Theorem 1.5.1, we can now extend the result to bounded predictable H_i.

Let $\mathcal{H}_*$ denote the vector space of bounded measurable processes H, such that

$$L = \int HdM_1\int I_BdM_2 - \int HI_Bd\langle M_1,M_2\rangle$$

is a martingale, where I_B is any fixed simple predictable process. We have established that $\mathcal{H}_*$ contains the simple predictable processes. Now let $H_n, n = 1,2,\ldots$ be an increasing sequence of mappings from $R^+\times\Omega$ to R in H_* such that $\sup_n H_n(t,\omega)\equiv H(t,\omega)$ is bounded on a set of probability one. Since each H_n is $\mathcal{F}_t$-adapted, H is adapted and, by a simple extension of Theorem A.2.5 in Appendix A to integrators of locally bounded variation, L is adapted. Clearly $E|L(t)|<\infty$ for any t. If

$$E\{L(t)|\mathcal{F}_s\} = L(s), s\le t,$$

then $\mathcal{H}_*$ contains all bounded $\mathcal{F}_t$-predictable processes. This follows since

$$E\left\{\int_0^t HdM_1\int_0^t I_BdM_2 - \int_0^t HI_Bd\langle M_1,M_2\rangle\Big|\mathcal{F}_s\right\}$$

$$= E\left\{\int_0^t \lim_{n\to\infty} H_ndM_1\int_0^t I_BdM_2 - \int_0^t \lim_{n\to\infty} H_nI_Bd\langle M_1,M_2\rangle\Big|\mathcal{F}_s\right\}$$

$$= E \lim_{n\to\infty} \left\{ \int_0^t H_n dM_1 \int_0^t I_B dM_2 - \int_0^t H_n I_B d\langle M_1, M_2\rangle \middle| \mathcal{F}_s \right\}$$

$$= \lim_{n\to\infty} E \left\{ \int_0^t H_n dM_1 \int_0^t I_B dM_2 - \int_0^t H_n I_B d\langle M_1, M_2\rangle \middle| \mathcal{F}_s \right\}$$

$$= \lim_{n\to\infty} \left\{ \int_0^s H_n dM_1 \int_0^s I_B dM_2 - \int_0^s H_n I_B d\langle M_1, M_2\rangle \right\}$$

$$= \int_0^s H dM_1 \int_0^s I_B dM_2 - \int_0^s H I_B d\langle M_1, M_2\rangle.$$

Now let H_1 be any bounded $\mathcal{F}_t$-predictable process, and let $\mathcal{H}'$ be the vector space of bounded measurable processes H_2 such that

$$L' = \int H_1 dM_1 \int H_2 dM_2 - \int H_1 H_2 d\langle M_1, M_2\rangle$$

is a martingale. The space $\mathcal{H}'$ contains the simple predictable processes, $H_2 = I_B$. Applying the monotone class theorem again, we conclude $\mathcal{H}'$ contains all bounded $\mathcal{F}_t$-predictable processes, and the proof is complete. □

It is not difficult to generalize Theorem 2.4.2 to processes considered earlier in this chapter.

Theorem 2.4.3. Assume that on a stochastic basis $(\Omega, \mathcal{F}, \{\mathcal{F}_t : t \geq 0\}, P)$:

1. H_i is a locally bounded $\mathcal{F}_t$-predictable process;
2. N_i is a counting process.

Then for the local martingales $M_i = N_i - A_i$,

$$\left\langle \int H_1 dM_1, \int H_2 dM_2 \right\rangle = \int H_1 H_2 d\langle M_1, M_2\rangle; \tag{4.3}$$

that is, the process

$$\int H_1 dM_1 \int H_2 dM_2 - \int H_1 H_2 d\langle M_1, M_2\rangle$$

is a local martingale over $[0, \infty)$.

Proof. By Theorem 2.4.1, each $\int H_i dM_i$ is a local square integrable martingale on $[0, \infty)$. Consequently, Theorem 2.3.3 implies

$$\left\langle \int H_i dM_i, \int H_i dM_i \right\rangle$$

exists and is unique and is given almost surely by

$$\left\langle \int H_i dM_i, \int H_i dM_i \right\rangle(t) = \lim_{n\to\infty} \left\langle \int_0^{\cdot\wedge\tau_n^i} H_i dM_i, \int_0^{\cdot\wedge\tau_n^i} H_i dM_i \right\rangle(t), \quad (4.4)$$

where $\{\tau_n^i\}$ is a localizing sequence for $\int H_i dM_i$.

The localizing sequences $\{\tau_n^i\}$ can be taken to be

$$\tau_n^i = n \wedge \sup\{t : N_i(t) < n\} \wedge \tau_n^{0i} \wedge \tau_n^{1i},$$

where $\{\tau_n^{0i}\}$ is such that $\sup_{0 \le t} A_i(t \wedge \tau_n^{0i}) \le n$ a.s., and $\{\tau_n^{1i}\}$ is a sequence rendering H_i locally bounded. Since $H_i(\cdot \wedge \tau_n^i)$ and $N_i(\cdot \wedge \tau_n^i)$ are bounded and $M_i(\cdot \wedge \tau_n^i)$ is a square integrable martingale, it follows from Theorem 2.4.2 that

$$\begin{aligned}
&\left\langle \int_0^{\cdot\wedge\tau_n^i} H_i dM_i, \int_0^{\cdot\wedge\tau_n^i} H_i dM_i \right\rangle(t) \\
&= \int_0^t H_i^2(u \wedge \tau_n^i) d\langle M_i, M_i\rangle(u \wedge \tau_n^i) \\
&= \int_0^{t\wedge\tau_n^i} H_i^2(u) d\langle M_i, M_i\rangle(u) \quad \text{a.s.}
\end{aligned}$$

This last equation and Eq. (4.4) establishes Eq. (4.3) for the case when both $H_1 = H_2$ and $M_1 = M_2$.

When $i \ne j$, Theorem 2.3.4, Eq. (3.3), and the subsequent remark imply that $\langle \int H_i dM_i, \int H_j dM_j \rangle$ exists, is unique, and is given by

$$\lim_{n\to\infty} \left\langle \int_0^{\cdot\wedge\tau_n} H_i dM_i, \int_0^{\cdot\wedge\tau_n} H_j dM_j \right\rangle(t) \quad \text{a.s.,} \quad (4.5)$$

where $\{\tau_n^i\}$ is a localizing sequence for $\int H_i dM_i$, and $\tau_n \equiv \tau_n^i \wedge \tau_n^j$. Each of τ_n^i and τ_n^j can be taken as in the first part of this proof, and it follows from Theorem 2.4.2 that

$$\begin{aligned}
&\left\langle \int_0^{\cdot\wedge\tau_n} H_i dM_i, \int_0^{\cdot\wedge\tau_n} H_j dM_j \right\rangle(t) \\
&= \int_0^t H_i(u \wedge \tau_n) H_j(u \wedge \tau_n) d\langle M_i, M_j\rangle(u \wedge \tau_n) \\
&= \int_0^{t\wedge\tau_n} H_i(u) H_j(u) d\langle M_i, M_j\rangle(u) \quad \text{a.s.} \quad (4.6)
\end{aligned}$$

For $i \ne j$, the result now follows from Eqs. (4.5) and (4.6). □

Theorem 2.4.3 can be used to obtain an expression for $E\left(\int_0^t H_1 dM_1 \int_0^t H_2 dM_2\right)$ when the expectation exists.

Theorem 2.4.4. Let $H_i, M_i, i = 1,2$, and $\{\mathcal{F}_t : t \geq 0\}$ be as defined in Theorem 2.4.3. Suppose $E\int_0^t H_i^2 d\langle M_i, M_i\rangle < \infty$ for $i = 1,2$. Then

1. $\int H_i dM_i$ is a martingale over $[0,t]$;
2. $E\int_0^t H_i dM_i = 0$; and
3. For $i, j \in \{1,2\}$,

$$E\left(\int_0^t H_i dM_i \int_0^t H_j dM_j\right) = E\int_0^t H_i H_j d\langle M_i, M_j\rangle.$$

Proof.

1. For fixed i, Theorem 2.4.3 implies the existence of a localizing sequence $\{\tau_n^i\}$ making $\int H_i dM_i$ and $\left(\int H_i dM_i\right)^2 - \int H_i^2 d\langle M_i, M_i\rangle$ local martingales. Both processes equal zero a.s. at $t = 0$. Since $\int H_i^2 d\langle M_i, M_i\rangle$ is increasing, for any $s \leq t$,

$$\begin{aligned} E\left(\int_0^{s\wedge\tau_n^i} H_i dM_i\right)^2 &= E\left(\int_0^{s\wedge\tau_n^i} H_i^2 d\langle M_i, M_i\rangle\right) \\ &\leq E\int_0^s H_i^2 d\langle M_i, M_i\rangle \\ &\leq E\int_0^t H_i^2 d\langle M_i, M_i\rangle \\ &< \infty. \end{aligned}$$

The sequence $\left\{\int_0^{s\wedge\tau_n^i} H_i dM_i : n = 1,2,\ldots,\right\}$ is therefore uniformly integrable. Lemma 2.2.2 now implies that $\int H_i dM_i$ is a martingale over $[0,t]$.

2.
$$\begin{aligned} E\int_0^t H_i dM_i &= EE\left(\int_0^t H_i dM_i | \mathcal{F}_0\right) \\ &= E\int_0^0 H_i dM_i \\ &= 0. \end{aligned}$$

3. When $i = j$, the result follows from Corollary 2.3.2. When $i \neq j$

$$\begin{aligned} &2E\left(\int_0^t H_1 dM_1 \int_0^t H_2 dM_2\right) \\ &= E\left\{\left(\int_0^t H_1 dM_1 + \int_0^t H_2 dM_2\right)^2 - \left(\int_0^t H_1 dM_1\right)^2 \right. \\ &\qquad \left. - \left(\int_0^t H_2 dM_2\right)^2\right\}. \end{aligned} \tag{4.7}$$

Since $\int H_1 dM_1 + \int H_2 dM_2$ is a local square integrable martingale which is a martingale over $[0, t]$, Corollary 2.3.2 implies

$$E\left(\int_0^t H_1 dM_1 + \int_0^t H_2 dM_2\right)^2$$
$$= E\left(\left\langle \int H_1 dM_1 + \int H_2 dM_2, \int H_1 dM_1 + \int H_2 dM_2 \right\rangle (t)\right)$$
$$= E\left(\left\langle \int H_1 dM_1, \int H_1 dM_1 \right\rangle (t) + \left\langle \int H_2 dM_2, \int H_2 dM_2 \right\rangle (t)\right.$$
$$\left. + 2\left\langle \int H_1 dM_1, \int H_2 dM_2 \right\rangle (t)\right). \tag{4.8}$$

Combining Eqs. (4.7) and (4.8),

$$E\left(\int_0^t H_1 dM_1 \int_0^t H_2 dM_2\right) = E\left\langle \int H_1 dM_1, \int H_2 dM_2 \right\rangle (t),$$

and the result follows from Theorem 2.4.3. □

We complete this section with the result that processes $U^{(n)} = \sum_{i=1}^n \int H_i dM_i$ satisfying Eq. (4.1) are local square integrable martingales. Results on first and second moments (when they exist) are also summarized along with expressions for product moments. Theorem 2.4.5 follows directly from Theorems 2.4.1 and 2.4.4.

Theorem 2.4.5. Consider two processes $U_\ell \equiv \sum_{i=1}^n \int H_{i,\ell} dM_i, \ell = 1, 2$. Let $\{\mathcal{F}_t : t \geq 0\}$ be a right-continuous filtration such that for each i and ℓ, $H_{i,\ell}$ is a locally bounded $\mathcal{F}_t$-predictable process, and $M_i = N_i - A_i$ is the local square integrable martingale corresponding to the arbitrary counting process N_i.
Then

1. U_ℓ is a local square integrable martingale.

If $E \int_0^t H_{i,\ell}^2 d\langle M_i, M_i\rangle < \infty$ for any i, ℓ,

2. $EU_\ell(t) = 0$;
3. $E\{U_1(t)U_2(t)\} = E \sum_{i=1}^n \sum_{j=1}^n \int_0^t H_{i1} H_{j2} d\langle M_i, M_j\rangle$; and
4. U_ℓ is a martingale over $[0, t]$. □

Result (3) is useful when the processes $\langle M_i, M_j\rangle$ are known. These covariation processes are derived in Section 2.5 for the special case of continuous compensators A.

2.5 CONTINUOUS COMPENSATORS

Continuous compensators arise frequently in applications, especially in survival analysis. We have already seen (Theorem 1.3.1) that, for the counting process marking the occurrence of a censored failure time with an absolutely continuous distribution $F(t) = 1 - \exp\left\{-\int_0^t \lambda(s)ds\right\}$, the compensator A is continuous and has value at time t

$$A(t) = \int_0^t I_{\{X \geq u\}}\lambda(u)du.$$

Results about continuous compensators apply, then, to all statistics of the form $\sum_i \int H_i dM_i$ built from this model. Although the hazard function in the Cox regression model is more complicated than for homogeneous observations, the compensator A is still continuous when failure time distributions are absolutely continuous, and results about statistics for the Cox model rely on this section.

The most general statements of the results for continuous compensators are in Theorems 2.5.2 and 2.5.3. We begin with a more restrictive result used in the proof of Theorem 2.5.2, which obtains $\langle M, M\rangle$ constructively via Eq. (3.2).

Theorem 2.5.1. Let N be a counting process on $[0, \infty)$ with $EN(t) < \infty$ for all $t \geq 0$, and let A denote its compensator. Assume that almost all paths of A are continuous, and that $EM^2(t) < \infty$ for all t, where $M = N - A$. Then $\langle M, M\rangle = A$, that is, $M^2 - A$ is a right-continuous martingale.

Proof. By integration by parts for Lebesgue-Stieltjes integrals (Appendix A, Theorem A.1.2), for any right-continuous processes X_i, X_j of bounded variation on $[0, t]$

$$\begin{aligned}
X_i(t)X_j(t) &= X_i(0)X_j(0) + \int_0^t X_i(s-)dX_j(s) + \int_0^t X_j(s)dX_i(s) \\
&= X_i(0)X_j(0) + \int_0^t X_i(s-)dX_j(s) + \int_0^t X_j(s-)dX_i(s) + \sum_{s \leq t} \Delta X_i(s)\Delta X_j(s).
\end{aligned}$$

Set $X_i = X_j = M$. Since $M(0) = 0$ a.s.,

$$M^2(t) = 2\int_0^t M(s-)dM(s) + \sum_{s \leq t}\{\Delta M(s)\}^2,$$

and since $\Delta M(s) = \Delta N(s) - \Delta A(s) = \Delta N(s)$,

$$\begin{aligned}
\sum_{s \leq t}\{\Delta M(s)\}^2 &= \sum_{s \leq t}\{\Delta N(s)\}^2 \\
&= \sum_{s \leq t}\Delta N(s) \\
&= N(t).
\end{aligned}$$

Thus,

$$M^2(t) - A(t) = 2\int_0^t M(s-)dM(s) + M(t). \qquad (5.1)$$

The process $M(s-)$ is adapted, left-continuous, has right-hand limits, and is therefore locally bounded. By Corollary 2.4.1, the right-hand side of Eq. (5.1) is a local martingale. Now since A is locally bounded, M is a local square integrable martingale and any other predictable, right-continuous, increasing process Y such that $Y(0) = 0$ a.s., $Y(t) < \infty$ a.s., and $M^2 - Y$ is a local martingale must equal A up to indistinguishability (Theorem 2.3.3). Since the process $\langle M, M\rangle$ from Corollary 1.4.2 is such a process, $\langle M, M\rangle = A$ a.s. □

The key step in the above proof is the use of the local martingale property of $\int M(s-)dM(s)$. Thus even though M itself is a martingale with finite second moments, the proof still relies on the notion of local martingales.

Theorem 2.5.1 can be extended to situations where $EN(t) = \infty$ or $EM^2(t) = \infty$, and to results about predictable covariation processes for pairs of compensated counting processes. Such pairs often arise as components in multivariate counting processes.

Definition 2.5.1. A k-variate process $\{N_1, N_2, \ldots, N_k\}$ is called a *multivariate counting process* if

1. Each N_j, $j = 1, \ldots, k$, is a counting process, and
2. No two component processes jump at the same time. □

Theorem 2.5.2. Let $\{N_1, \ldots, N_k\}$ be a multivariate counting process and, for $j = 1, \ldots, k$, let A_j be the compensator of N_j. Assume that each A_j is a continuous process. Then

1. $\langle M_j, M_j\rangle = A_j$, that is, A_j is the unique predictable, right-continuous, increasing process with $A_j(0) = 0$ a.s. and $A_j(t) < \infty$ a.s. for any t, such that $M_j^2 - A_j$ is a local martingale, $j = 1, \ldots, k$.
2. If $i \neq j$, $\langle M_i, M_j\rangle(t) = 0$ a.s., that is, M_iM_j is a local martingale.

Proof.

1. There exists a localizing sequence $\{\tau_n^j\}$ for M_j such that

$$M_j(\cdot \wedge \tau_n^j) = N_j(\cdot \wedge \tau_n^j) - A_j(\cdot \wedge \tau_n^j)$$

is a martingale on $[0, \infty)$ with finite second moments. We can assume $N_j(\cdot \wedge \tau_n^j)$ is bounded by n. By Theorem 2.5.1,

$$\langle M_j(\cdot \wedge \tau_n^j), M_j(\cdot \wedge \tau_n^j)\rangle(t) = A_j(t \wedge \tau_n^j) \quad \text{a.s.}$$

for any t and, using Eq (3.2),

$$\begin{aligned}\langle M_j, M_j\rangle(t) &= \lim_{n\to\infty} A_j(t \wedge \tau_n^j) \\ &= A_j(t)\end{aligned}$$

for any t.

2. Let $\{\tau_n^i\}$ and $\{\tau_n^j\}$ be localizing sequences for M_i and M_j, respectively. Then $\tau_n \equiv \tau_n^i \wedge \tau_n^j$ is a localizing sequence for each of M_i and M_j by Lemma 2.2.3, and

$$\begin{aligned}M(\cdot \wedge \tau_n) &\equiv M_i(\cdot \wedge \tau_n) + M_j(\cdot \wedge \tau_n) \\ &= N_i(\cdot \wedge \tau_n) + N_j(\cdot \wedge \tau_n) - \{A_i(\cdot \wedge \tau_n) + A_j(\cdot \wedge \tau_n)\} \\ &\equiv N(\cdot \wedge \tau_n) - A(\cdot \wedge \tau_n)\end{aligned}$$

is a martingale with finite second moments. The process $N(\cdot \wedge \tau_n)$ is a counting process since N_i and N_j do not jump at the same time, and again we can assume $N(\cdot \wedge \tau_n)$ is bounded by n. By Theorem 2.5.1, when $N = N_i + N_j$ and $A = A_i + A_j$,

$$\langle M(\cdot \wedge \tau_n), M(\cdot \wedge \tau_n)\rangle(t) = A(t \wedge \tau_n) \quad \text{a.s.}$$

for any t. By applying Theorem 2.3.3 to the local square integrable martingale $M_i + M_j$, it follows from Eq. (3.2) that

$$\begin{aligned}\langle M_i + M_j, M_i + M_j\rangle(t) &= \lim_{n\to\infty} A(t \wedge \tau_n) \\ &= A(t) \\ &= A_i(t) + A_j(t) \quad \text{a.s.}\end{aligned}$$

for any t. By definition,

$$\begin{aligned}\langle M_i, M_j\rangle(t) &= 1/2\{\langle M_i + M_j, M_i + M_j\rangle(t) \\ &\quad -\langle M_i, M_i\rangle(t) - \langle M_j, M_j\rangle(t)\} \\ &= 1/2\{A_i(t) + A_j(t) - A_i(t) - A_j(t)\} \\ &= 0.\end{aligned}$$

□

Theorem 2.5.2 allows formulation of a sufficient condition for second moments for $M = N - A$.

Theorem 2.5.3. Let N be a counting process and A its compensator. If A is continuous then

$$EM^2(t) \le EA(t), \qquad t \ge 0.$$

If in addition, $EA(t) < \infty$ (or equivalently if $EN(t) < \infty$) for any t, then

$$EM^2(t) = EA(t), \qquad t \geq 0$$

and $M^2 - A$ is a martingale.

Proof. $M = N - A$ is a local square integrable martingale (Theorem 2.3.1), and the first part of the theorem follows from Theorem 2.5.2 and Corollary 2.3.1.

Suppose now $EA(t) < \infty, t \geq 0$. Then $M = N - A$ is a martingale (Lemma 2.3.2), and $EM^2(t) = EA(t)$ by Corollary 2.3.2. Corollary 1.4.2 now implies that $M^2 - A$ is a martingale. □

These results are easily applied to some of the examples discussed earlier. Suppose we have the censored survival data model of Example 1.5.1, with continuous $F(t) = 1 - \exp\{-\int_0^t \lambda(s)ds\}$. The logrank statistic can be written

$$V = \sum_{i=1}^{n} \int_0^\infty \left(Z_i - \frac{\sum_{j=1}^n Y_j(s)Z_j}{\sum_{j=1}^n Y_j(s)} \right) d\left\{ N_i(s) - \int_0^s I_{\{X_i \geq u\}} \lambda(u)du \right\}$$

(see Eq. (5.3) of Chapter 1, and the following sentence), where Z_j is a binary, sample membership indicator variable. The class of weighted logrank statistics are of the form $\sum_{i=1}^n \int_0^\infty H_i d(N_i - A_i)$, where $A_i = \int I_{\{X_i \geq u\}} \lambda(u)du$ is continuous and where

$$H_i = W \left(Z_i - \frac{\sum_{j=1}^n Y_j Z_j}{\sum_{j=1}^n Y_j} \right)$$

for some $\mathcal{F}_t$-predictable "weighting" stochastic process W. If W is bounded by one for these statistics and $Z_i \in \{0, 1\}$, H_i is bounded and, since $0 \leq N_i \leq 1$, $EN_i(t) < \infty$. The next theorem summarizes some properties of these statistics.

Theorem 2.5.4. Consider the censored survival data model of Examples 1.5.1 and 1.5.2, where $N_i(t) = I_{\{X_i \leq t, \delta_i = 1\}}$ and $N_i^U(t) = I_{\{X_i \leq t, \delta_i = 0\}}$. Set

$$\mathcal{F}_t = \sigma\{N_i(s), N_i^U(s), Z_i : 0 \leq s \leq t, i = 1, \ldots, n\}.$$

Let $U = \sum_{i=1}^n \int H_i d(N_i - A_i)$, with H_i being $\mathcal{F}_t$-predictable and bounded on $[0, \infty)$ and $A_i = \int I_{\{X_i \geq u\}} \lambda(u)du$.

Then

1. The process U is a martingale over $[0, \infty)$;
2. $EU(t) = 0, 0 \leq t \leq \infty$; and
3. $\operatorname{var} U(t) = \sum_{i=1}^n \int_0^t E\left\{ H_i^2(u) I_{\{X_i \geq u\}} \right\} \lambda(u)du, 0 \leq t \leq \infty$.

Proof. It is straightforward to verify that each $N_i - A_i$ is an $\mathcal{F}_t$-local square integrable martingale.

We also need to establish that $E\int_0^t H_i^2 d\langle M_i, M_i\rangle < \infty$. When $i \neq j$, N_i and N_j jump at the same time with probability zero, so $\{N_1, \ldots, N_n\}$ is a multivariate counting process. Since each A_i is continuous, Theorem 2.5.2 implies $\langle M_i, M_i\rangle = A_i$ and $\langle M_i, M_j\rangle = 0$. For s in $[0, \infty)$

$$\begin{aligned} E\int_0^s & H_i^2(u) I_{\{X_i \geq u\}} \lambda(u) du \\ &\leq \Gamma^2 \int_0^s P\{X_i \geq u\}\lambda(u)du \\ &\leq \Gamma^2 \int_0^s P\{T_i \geq u\}\lambda(u)du \\ &\leq \Gamma^2 \\ &< \infty, \end{aligned}$$

where for each $H_i, |H_i| < \Gamma$. By Theorem 2.4.5, (1) is true and (2) is true for $0 \leq t < \infty$.

If $\tau = \max_{1 \leq i \leq n} X_i$, then

$$\begin{aligned} \{\tau \leq s\} &= \{X_i \leq s : 1 \leq i \leq n\} \\ &= \{N_i(s) + N_i^U(s) = 1 : 1 \leq i \leq n\} \\ &\in \mathcal{F}_s, \end{aligned}$$

and τ is a stopping time. By the Optional Stopping Theorem, $\{U(s \wedge \tau) : s \geq 0\}$ is a martingale and, using the Monotone Convergence Theorem,

$$\begin{aligned} EA_i(\tau) &= \lim_{t\to\infty} EA_i(t \wedge \tau) \\ &= \lim_{t\to\infty} EN_i(t \wedge \tau) \\ &= EN_i(\tau) \\ &\leq 1. \end{aligned}$$

Since $|U(s \wedge \tau)| \leq \Gamma \sum_{i=1}^n \{N_i(\tau) + A_i(\tau)\}$ for all $s \geq 0$, the Dominated Convergence Theorem yields

$$EU(\tau) = \lim_{s\to\infty} EU(s \wedge \tau) = 0.$$

Since $U(\infty) = U(\tau)$, $EU(\infty) = 0$.

Theorem 2.4.5 and Corollary 1.4.2 imply that $U^2 - \langle U, U\rangle$ is a martingale on $[0, \infty)$, so using exactly the same arguments as above

$$\begin{aligned} \text{var}\, U(t) &= E\, U^2(t) \\ &= E \sum_{i=1}^n \sum_{j=1}^n \int_0^t H_i(u)\, H_j(u) d\langle M_i, M_j\rangle(u) \end{aligned}$$

for $0 \leq t \leq \infty$. Part (3) now follows immediately. □

We will see later that the asymptotic distribution theory for $U(\infty)$ is surprisingly delicate, and that it is sometimes only possible to establish the asymptotic normality for $U(t)$, where t is finite and satisfies certain regularity conditions. The moment formulas for $U(t)$, $t < \infty$, will be useful in those situations. It is clear from the proof of Theorem 2.5.4 that H need only be bounded on $[0, t]$ to establish first and second moments for $U(t)$.

2.6 COMPENSATORS WITH DISCONTINUITIES

Theorem 2.4.5 contains expressions for the variances and covariances of statistics of the form

$$U_l \equiv \sum_{i=1}^{n} \int H_{i,l} dM_i, \quad l = 1, 2,$$

where $H_{i,l}$ is a locally bounded $\mathcal{F}_t$-predictable process, and where $M_i = N_i - A_i$ for some almost surely finite counting process N_i. When the expectation exists,

$$\operatorname{cov}(U_1, U_2) = E \sum_{i=1}^{n} \sum_{j=1}^{n} \int H_{i,1} H_{j,2} d\langle M_i, M_j \rangle, \tag{6.1}$$

and when A_i is continuous $\langle M_i, M_j \rangle = I_{\{i=j\}} A_i$. In this section, we examine $\langle M_i, M_j \rangle$ and $\operatorname{cov}(U_1, U_2)$, when A_i has discontinuities.

Example 2.6.1. Suppose $\{(T_i, U_i) : i = 1, 2, \ldots, n\}$ is a set of n independent pairs of positive and a.s. finite failure and censoring time variables. Let $F_i(t) = P\{T_i \le t\}$ and $\Lambda(t) = \int_0^t \{1 - F(u-)\}^{-1} dF(u)$, and suppose Condition (3.7) of Chapter 1 holds. Set $X_i = T_i \wedge U_i$ and $\delta_i = I_{\{X_i = T_i\}}$. With $N_i(t) \equiv I_{\{X_i \le t, \delta_i = 1\}}$ and $N_i^U(t) = I_{\{X_i \le t, \delta_i = 0\}}$, let $\mathcal{F}_t \equiv \sigma\{N_i(u), N_i^U(u) : 0 \le u \le t, i = 1, \ldots, n\}$.

By Theorem 1.3.2, $A_i(t) = \int_0^t I_{\{X_i \ge u\}} d\Lambda_i(u)$ can have discontinuities whenever F_i has discontinuities over the support of X_i. By developing the form of $\langle M_i, M_j \rangle$ in this setting, we can use Eq. (6.1) to obtain expressions for covariances of censored survival data statistics in data having tied failure times. Such statistics include the logrank statistic in Example 0.2.2. □

In Section 1.4, we argue heuristically that

$$d\langle M, M \rangle(s) = \operatorname{var}\{dM(s) | \mathcal{F}_{s-}\} = dA(s)\{1 - dA(s)\}.$$

Part (2) in Theorem 2.6.1 establishes results motivated by that conjecture.

Theorem 2.6.1. Let $\{N_j : j = 1, \ldots, n\}$ be a collection of counting processes on a stochastic basis $(\Omega, \mathcal{F}, \{\mathcal{F}_t : t \ge 0\}, P)$, A_j the $\mathcal{F}_t$-compensator of N_j, and $M_j = N_j - A_j$. Then

1. $\int \Delta M_j dM_j - \int (1 - \Delta A_j) dA_j$ is an $\mathcal{F}_t$-local martingale.
2. $\langle M_j, M_j \rangle = \int (1 - \Delta A_j) dA_j$; that is, $\int (1 - \Delta A_j) dA_j$ is the unique predictable right-continuous increasing process which is a.s. zero at $t = 0$ and finite at any t, and such that $M_j^2 - \int (1 - \Delta A_j) dA_j$ is a local martingale.
3. If $\{N_1, \ldots, N_n\}$ is a multivariate counting process, $\langle M_i, M_j \rangle = -\int \Delta A_i dA_j$ when $i \neq j$.
4. Suppose $EN_i(t) < \infty$ for any t, i. Then the local martingales in (1), (2), and (3) are martingales.

Proof. Using the integration by parts formula for Lebesgue-Stieltjes integrals,

$$M_j^2(t) = 2 \int_0^t M_j(s-) dM_j(s) + \int_0^t \Delta M_j(s) dM_j(s). \tag{6.2}$$

In turn,

$$\begin{aligned} \int_0^t \Delta M_j(s) dM_j(s) &= \sum_{s \leq t} \Delta N_j(s) \{\Delta N_j(s) - \Delta A_j(s)\} - \int_0^t \Delta A_j(s) dM_j(s) \\ &= \sum_{s \leq t} \Delta N_j(s) - \int_0^t \Delta A_j(s) dA_j(s) - 2 \int_0^t \Delta A_j(s) dM_j(s), \end{aligned}$$

so,

$$\int \Delta M_j dM_j - \int (1 - \Delta A_j) dA_j = M_j - 2 \int \Delta A_j dM_j. \tag{6.3}$$

By Lemma 2.4.1 and Theorem 2.3.2 (1), $\int \Delta A_j dM_j$ is a local square integrable martingale, so (1) holds.

Corollary 2.4.1 implies

$$2 \int M_j(s-) dM_j(s) + M_j - 2 \int \Delta A_j(s) dM_j(s)$$

is a local square integrable martingale. The predictable process $\int (1 - \Delta A_j) dA_j$ is increasing by Theorem 2.3.2 (2), and consequently the Doob-Meyer Decomposition yields (2).

To prove (3), observe that for $i \neq j$, $N \equiv N_i + N_j$ is a counting process with compensator $A = A_i + A_j$. Thus by (2),

$$\langle M_i + M_j, M_i + M_j \rangle = A_i + A_j - \int (\Delta A_i + \Delta A_j) d(A_i + A_j),$$

so

$$\begin{aligned} \langle M_i, M_j \rangle &= \frac{1}{2} \{(A_i + A_j) - \int (\Delta A_i + \Delta A_j) d(A_i + A_j) \\ &\qquad - A_i + \int \Delta A_i dA_i - A_j + \int \Delta A_j dA_j\} \\ &= -\int \Delta A_i dA_j. \end{aligned}$$

Now assume $EN_j(t) < \infty$ for any j, t. By Lemma 2.3.2, M_j is a martingale. Since M_j also is a local square integrable martingale by Theorem 2.3.1 and the local boundedness of A_j, Corollary 2.3.2 implies

$$\begin{aligned} EM_j^2(t) &= E\langle M_j, M_j\rangle(t) \\ &= E\int_0^t \{1 - \Delta A_j\} dA_j \\ &\leq EA_j(t) \\ &< \infty. \end{aligned}$$

It now follows from Corollary 1.4.2 that $M_j^2 - \int(1 - \Delta A_j)dA_j$ is a martingale, implying that $M_i M_j + \int \Delta A_i dA_j$ also is a martingale.

Because M_j is a martingale and $|\Delta A_j(s)| < 1$, $M_j - 2\int \Delta A_j dM_j$ is a martingale by Theorem 1.5.1. Hence Eq. (6.3) implies $\int \Delta M_j dM_j - \int(1 - \Delta A_j)dA_j$ is a martingale. □

In many applications, such as Example 2.6.1, $\{N_1, \ldots, N_n\}$ is not a multivariate counting process since N_i and N_j can jump at the same time with positive probability. To obtain second moments of statistics in this situation, we must determine $\langle M_i, M_j\rangle$ for $i \neq j$, since Theorem 2.6.1 (3) does not apply. Fortunately, the following assumption often holds.

Assumption 2.6.1. For each $t \geq 0$, given $\mathcal{F}_{t-}$, $\{\Delta N_1(t), \ldots, \Delta N_n(t)\}$ are independent $0, 1$ random variables. □

As an exercise (Exercise 2.11), one can verify this assumption holds for Example 2.6.1.

By a simple computation, Assumption 2.6.1 implies for any $i \neq j$

$$E\{N_i(t)N_j(t)|\mathcal{F}_{t-}\} = E\{N_i(t)|\mathcal{F}_{t-}\}E\{N_j(t)|\mathcal{F}_{t-}\}.$$

In turn, this and the fact that $A(t)$ is $\mathcal{F}_{t-}$-measurable imply

$$\begin{aligned} E\{M_i(t)M_j(t)|\mathcal{F}_{t-}\} &= E[\{N_i(t) - A_i(t)\}\{N_j(t) - A_j(t)\}|\mathcal{F}_{t-}] \\ &= E\{N_i(t) - A_i(t)|\mathcal{F}_{t-}\}E\{N_j(t) - A_j(t)|\mathcal{F}_{t-}\} \\ &= M_i(t-)M_j(t-). \end{aligned} \tag{6.4}$$

This equality motivates the next lemma.

Lemma 2.6.1. Let $\{N_j : j = 1, \ldots, n\}$ be a collection of counting processes which satisfy Assumption 2.6.1. Set $M_j = N_j - A_j$, where A_j is the compensator for N_j. Then for any $i \neq j$ and $t \geq 0$,

$$\langle M_i, M_j\rangle(t) = 0 \quad \text{a.s.}$$

Proof. By integration by parts,

$$M_1(t)M_2(t) = M_1(0)M_2(0) + \int_0^t M_1(s-)dM_2(s) + \int_0^t M_2(s-)dM_1(s) + \int_0^t \Delta M_1(s)dM_2(s). \qquad (6.5)$$

The first term on the right-hand side of Eq. (6.5) is zero a.s. and, since $\{M_i(s-): s \geq 0\}$ is an adapted left-continuous process with right-hand limits, the next two are local square integrable martingales. By the almost sure finiteness of the counting process, we can write the last term as a finite sum,

$$\int_0^t \Delta M_1(s)dM_2(s) = \sum_{0<s\leq t} \Delta M_1(s)\Delta M_2(s).$$

It remains to show this last term is a local martingale. Let $\{\tau_n\}$ be a localizing sequence for M_1 and M_2, and $0 \leq u \leq t$. Then

$$\begin{aligned} E&\left\{\int_u^t \Delta M_1(s \wedge \tau_n)dM_2(s \wedge \tau_n)|\mathcal{F}_u\right\} \\ &= \sum_{u<s\leq t} E\left[E\{\Delta M_1(s \wedge \tau_n)\Delta M_2(s \wedge \tau_n)|\mathcal{F}_{s-}\}|\mathcal{F}_u\right] \\ &= 0, \end{aligned}$$

where this last equality follows directly from Eq. (6.4). □

The next theorem generalizes Theorem 2.5.4 to the setting in which ties in observed failure times can arise.

Theorem 2.6.2. Consider the censored survival data model of Example 2.6.1 and statistics of the form

$$U_l = \sum_{i=1}^n \int H_{i,l} dM_i, \; l = 1, 2,$$

such that $H_{i,l}$ is a bounded $\mathcal{F}_t$-predictable process on $[0, \infty)$, and such that $A_i = \int I_{\{X_i \geq u\}} d\Lambda(u)$. Then, for $l, l' \in \{1, 2\}$,

1. U_l is a martingale over $[0, \infty)$;
2. $EU_l(t) = 0$, $0 \leq t \leq \infty$; and
3. $\text{cov}\{U_l(t), U_{l'}(t)\} = \sum_{i=1}^n \int_0^t E\{H_{i,l}(u)H_{i,l'}(u)I_{\{X_i \geq u\}}\}\{1 - \Delta\Lambda(u)\}d\Lambda(u)$, $0 \leq t \leq \infty$.

Proof. Using Theorem 1.3.2, it is straightforward to verify M_i is a local square integrable martingale with respect to $\{\mathcal{F}_t : t \geq 0\}$, for $\mathcal{F}_t$ defined in Example 2.6.1. Theorem 2.6.1 (2) indicates that $\langle M_i, M_i \rangle = \int (1 - \Delta A_i) dA_i$, while Lemma 2.6.1 implies $\langle M_i, M_j \rangle = 0$ for $i \neq j$, since Assumption 2.6.1 holds. Since $H_{i,l}$ is bounded by a constant $\Gamma_{i,l}$, we have for any $s \in [0, \infty)$ (or $s \in [0, t]$),

$$
\begin{aligned}
E &\int_0^s H_{i,l}^2(u) d\langle M_i, M_i \rangle(u) \\
&= E \int_0^s H_{i,l}^2(u) I_{\{X_i \geq u\}} \{1 - \Delta\Lambda(u)\} d\Lambda(u) \\
&\leq \Gamma_{i,l}^2 \int_0^s P\{T_i \geq u\} d\Lambda(u) \\
&\leq \Gamma_{i,l}^2 \\
&< \infty.
\end{aligned}
$$

The rest of the proof proceeds exactly as in the proof of Theorem 2.5.4. □

2.7 SUMMARY

Chapters 1 and 2 explore the close relationship between martingales, predictable processes, and stochastic integrals, paying particular attention to these notions in the study of counting processes. It is easy to lose sight of the important ideas in these two chapters, and of some interpretations that help make the technically demanding results more transparent. This section collects the major results without some of the distracting details, and re-emphasizes their interpretations. We hope this section can also serve as a guide to the martingale methods used later for readers who cannot afford the time to study carefully the material in Chapter 2.

While some of the results in Chapters 1 and 2 are closely related (Chapter 2 uses localization for more general versions of many Chapter 1 results), Chapter 2 also introduces important material (covariation processes for stochastic integrals, formulas for compensators) not covered in Chapter 1. We begin with the material common to both chapters.

The concept of a martingale is central. Subject to measurability conditions (Definition 1.2.8), a process M with $E|M(t)| < \infty$ for any t is a martingale if the increments $M(t) - M(u), 0 \leq u \leq t$, have conditional mean zero given some history of information $\mathcal{F}_u$ up to time u. Thus

$$E\{M(t) - M(u) | \mathcal{F}_u\} = 0,$$

and

$$E\{\Delta M(t) | \mathcal{F}_{t-}\} = 0$$

(Proposition 1.2.1). It follows that, for any $s, t, u > 0$,

$$
\begin{aligned}
&E[\{M(t) - M(t-u)\}\{M(t+s) - M(t)\}|\mathcal{F}_t] \\
&= \{M(t) - M(t-u)\}E[\{M(t+s) - M(t)\}|\mathcal{F}_t] \\
&= 0.
\end{aligned}
$$

Taking one more expectation shows that a martingale has uncorrelated increments.

In infinitesimal notation,

$$E\{dM(t)|\mathcal{F}_{t-}\} = 0,$$

and this (imprecise) property motivates many results to be summarized.

Predictable processes also play an important role. While the formal definition (Definitions 1.4.1 and 1.4.2) seems indirect, predictable processes are essentially left-continuous processes adapted to a history $\mathcal{F}_t$ and are thus determined at time t by the past strictly prior to t, i.e., by $\mathcal{F}_{t-}$. Predictable processes arise as compensators in martingales, and as integrands in stochastic integrals.

Compensator processes are one part of the Doob-Meyer Decomposition (Theorem 1.4.1), which says that if X is a nonnegative submartingale (i.e. an adapted process with $EX(t) < \infty$ for any t for which $E\{dX(t)|\mathcal{F}_{t-}\} \geq 0$), then there is a unique right-continuous increasing predictable process A such that $A(0) = 0$, $EA(t) < \infty$ for any t, and $X - A$ is a martingale.

The compensator A of X is a process, determined at time t by the strict past, that carries the predictable component of X. The process $X - A$ is analogous to the residuals in a regression problem when model variation has been accounted for. The interpretation of the compensator when the submartingale is an integrable counting process, N, is helpful. In small intervals $dN(s)$ is either 1 or 0, and

$$
\begin{aligned}
0 &= E\{dN(t) - dA(t)|\mathcal{F}_{t-}\} \\
&= P\{dN(t) = 1|\mathcal{F}_{t-}\} - dA(t),
\end{aligned}
$$

since $A(t)$ is $\mathcal{F}_{t-}$-measurable. Thus $P\{dN(t) = 1|\mathcal{F}_{t-}\} = dA(t)$ and $dA(t)$ is the conditional rate at which N jumps.

If a predictable process H is the integrand in a stochastic integral $L = \int H dM$ with respect to a martingale M, then

$$
\begin{aligned}
E\{dL(t)|\mathcal{F}_{t-}\} &= E\{H(t)dM(t)|\mathcal{F}_{t-}\} \\
&= H(t)E\{dM(t)|\mathcal{F}_{t-}\} \\
&= 0,
\end{aligned}
$$

that is, we expect L to be a martingale. Theorem 1.5.1 establishes that L is a martingale whenever both M is a martingale and H is a bounded predictable process.

The predictable quadratic variation process $\langle M, M \rangle$ of a square integrable martingale M is the predictable compensator for M^2. It satisfies $d\langle M, M \rangle(t) =$

$E\{d\{M^2(t)\}|\mathcal{F}_{t-}\}$ (Corollary 1.4.2). The predictable covariation process $\langle M_1, M_2\rangle$ for two square integrable martingales M_1 and M_2 satisfies $d\langle M_1, M_2\rangle(t) = E\{d\{M_1(t)M_2(t)\}|\mathcal{F}_{t-}\}$. Since $M_1M_2 - \langle M_1, M_2\rangle$ is a martingale, when $M_i(0) = 0$ a.s. it follows that $EM_1(t)M_2(t) = E\langle M_1, M_2\rangle(t)$ and the predictable covariation process $\langle M_1, M_2\rangle$ can be used to calculate the covariance $E\{M_1M_2\}$.

The first part of Chapter 2 uses localization to extend the results in Chapter 1 to less restrictive settings, relaxing assumptions such as $EN(t) < \infty$, $EM^2(t) < \infty$, and boundedness of H. A localizing sequence for a process X is a nondecreasing sequence of stopping times (Definition 2.2.1) $\{\tau_n : n \geq 1\}$ with $\tau_n \uparrow \infty$ a.s. The process X has a property locally if each of the stopped processes $X(\cdot \wedge \tau_n)$ has that property. Thus, M is a local square integrable martingale if there is a localizing sequence $\{\tau_n\}$ such that $M(\cdot \wedge \tau_n)$ is a square integrable martingale for each n, H is a locally bounded process if $H(\cdot \wedge \tau_n)$ is bounded, etc.

The class of locally bounded processes includes all adapted processes with value zero at time zero which are either left-continuous with finite right-hand limits (Lemma 2.3.1), or are right-continuous with left-hand limits and have bounded jump sizes (Exercise 2.6). Thus an arbitrary adapted counting process N (i.e. one that only need satisfy $P\{N(t) < \infty\} = 1$ for all t) is locally bounded and, in turn, is a local submartingale. An extension of the Doob-Meyer Decomposition to arbitrary local submartingales in Theorem 2.2.3 shows that for any nonnegative local submartingale X there is always an increasing right-continuous predictable process A (with $A(0) = 0$ and $P\{A(t) < \infty\} = 1$ for any $t > 0$), such that $M = X - A$ is a local martingale. To obtain A constructively, let $\{\tau_n\}$ be any localizing sequence for X and A_n be the unique compensator for the submartingale $X(\cdot \wedge \tau_n)$. The process A is obtained by setting $A = A_n$ on $(\tau_{n-1}, \tau_n]$ and by letting $n \to \infty$.

The extended Doob-Meyer Decomposition implies that an arbitrary adapted counting process N always admits a decomposition $M = N - A$, with M a local square integrable martingale (Theorems 2.3.1 and 2.3.2). Since $EN(t) = EA(t)$, $EN(t) < \infty$ and $M = N - A$ is a martingale over $[0, t]$ whenever $EA(t) < \infty$ (Lemma 2.3.2). The Doob-Meyer extension also implies that unique predictable quadratic variation and covariation processes exist for arbitrary local square integrable martingales M so that $M^2 - \langle M, M\rangle$ and $M_1M_2 - \langle M_1, M_2\rangle$ are local martingales, (Theorems 2.3.3 and 2.3.4). These local martingales will be martingales as long as $EM_i^2(t) < \infty$ for any t. When $M(0) = 0$, $EM^2(t) \leq E\langle M, M\rangle(t)$, with equality holding when M is a martingale (Corollaries 2.3.1 and 2.3.2). Expressions for $EM^2(t)$ and its boundedness follow from the formulas for $\langle M, M\rangle$ discussed later in this section.

The results on integration with respect to counting process martingales have analogues for processes with local properties, so that if H is a locally bounded predictable process and M is a local martingale of locally bounded variation with $\Delta M(0) = 0$, (e.g., $M = N - A$), then $\int H dM$ is itself a local square integrable martingale (Theorem 2.4.1 and Lemma 2.4.1).

When $\int H dM$ is a martingale, $E\int H dM = 0$. This implies that $E\int H dN = E\int H dA$ when $M = N - A$, and first moments for counting process statistics become easier to calculate. Second moments require the predictable quadratic variation and covariation processes for stochastic integrals derived in Chapter 2. If $M_1 = N_1 - A_1$ and $M_2 = N_2 - A_2$ are counting process martingales with the N_i bounded and $EM_i^2 < \infty$, and H_1 and H_2 are bounded predictable processes, then $\langle \int H_i dM_i, \int H_i dM_i \rangle = \int H_i^2 d\langle M_i, M_i \rangle$ and $\langle \int H_1 dM_1, \int H_2 dM_2 \rangle = \int H_1 H_2 d\langle M_1, M_2 \rangle$, (Theorem 2.4.2). The result extends to arbitrary counting processes N_i and locally bounded predictable H_i, with the qualification that

$$\int H_1 dM_1 \int H_2 dM_2 - \int H_1 H_2 d\langle M_1, M_2 \rangle$$

is a local martingale instead of a martingale (Theorem 2.4.3).

The key result yielding the martingale structure, and first and second moments of $\int H_i d(N_i - A_i)$ for arbitrary counting processes N_i and locally bounded predictable H_i easily follows (Theorem 2.4.4). For $M_i \equiv N_i - A_i$, $\int H_i dM_i$ will be a martingale whenever

$$E \int_0^t H_i^2 d\langle M_i, M_i \rangle < \infty. \tag{7.1}$$

This in turn implies $E \int_0^t H_i dM_i = 0$, and together with Corollary 2.3.2 and Theorem 2.4.3, that

$$E\left(\int_0^t H_1 dM_1 \int_0^t H_2 dM_2 \right) = E \int_0^t H_1 H_2 d\langle M_1, M_2 \rangle.$$

Finiteness of the variance of $\int_0^t H_i dM_i$ has replaced the stronger conditions of $EN_i(t) < \infty, EM_i^2(t) < \infty$, or the boundedness of H_i.

It is clear that accessible formulas for $\langle M_i, M_i \rangle$ are needed. As noted earlier, $d\langle M, M \rangle(s) = E\{dM^2(s)|\mathcal{F}_{s-}\}$. Because

$$dM^2(s) = \{dM(s)\}^2 + 2M(s - ds)dM(s)$$

and $E(dM|\mathcal{F}_{s-}) = 0$, it follows that

$$d\langle M, M \rangle(s) = E[\{dM(s)\}^2 |\mathcal{F}_{s-}] = \text{var}\{dM(s)|\mathcal{F}_{s-}\}.$$

Thus $\langle M, M \rangle(t)$ is the sum over $s \in (0, t]$ of the conditional variance of $M(s)$, given information up to time s. Let $M = N - A$, and observe that $dN(s)$ is conditionally a Bernoulli random variable with $E\{dN(s)|\mathcal{F}_{s-}\} = dA(s)$ and $\text{var}\{dN(s) - dA(s)|\mathcal{F}_{s-}\} = \{1 - dA(s)\}dA(s)$. Thus, we would anticipate that $\langle M, M \rangle(t) = \int_0^t \{1 - \Delta A(s)\} dA(s)$, and that $\langle M, M \rangle(t) = A(t)$ when A is continuous.

These heuristic arguments about the form of the predictable quadratic variation process $\langle M, M \rangle$ are tightened in Sections 5 and 6 of Chapter 2. In Section 5, we explore a multivariate counting process $\mathbf{N}' = (N_1, \ldots, N_k)$, a k-dimensional process whose components are counting processes no two of which can jump simultaneously. Assume the paths of the associated compensators A_i are continuous with probability one. For the local martingales $M_i = N_i - A_i$, Theorem 2.5.2 establishes that $\langle M_i, M_i \rangle = A_i$ and $\langle M_i, M_j \rangle = 0$ when $i \neq j$. Thus the component processes are locally and conditionally Poisson-like with rate $dA_i(t)$, and are conditionally pairwise uncorrelated (i.e., orthogonal). When $A(t) = \int_0^t l(u)du$, N behaves in short time intervals, given its history, as a Poisson process with rate function l. Theorem 2.5.3 provides other consequences of the result $\langle M_i, M_i \rangle = A_i$: if $EA(t) < \infty$ for any t, then $M = N - A$ is a martingale (by Lemma 2.3.2), so $EM^2(t) = EA(t)$ (by Corollary 2.3.2), and $M^2 - A$ is a martingale (by Corollary 1.4.2).

Compensators with discontinuities are considered in Section 6 where, for the local martingale $M_i = N_i - A_i$, Theorem 2.6.1 establishes that

$$\langle M_i, M_i \rangle = \int (1 - \Delta A_i) dA_i.$$

The processes $\langle M_i, M_j \rangle$ for $i \neq j$ are more complex when the A_i have discontinuities, since $\mathbf{N}' = (N_1, \ldots, N_k)$ can have component processes which may jump simultaneously and will not be a multivariate counting process. However, it often is true that, for each $t \geq 0$, $\{\Delta N_1(t), \ldots, \Delta N_k(t)\}$ are conditionally independent $0, 1$ random variables given $\mathcal{F}_{t-}$ (Assumption 2.6.1), and then $\langle M_i, M_j \rangle = 0$ for $i \neq j$, (Lemma 2.6.1).

The results for counting process martingales are particularly useful in studying the censored failure time model of Example 2.6.1. When N is the indicator process for an observed failure, its compensator A is given by

$$A(t) = \int_0^t I_{\{X \geq u\}} d\Lambda(u),$$

where X is the minimum of failure and censoring, and Λ is the cumulative hazard for failure. In most applications the n processes $N_1, \ldots, N_n$ corresponding to observations on n cases or subjects, with compensators $A_1, \ldots, A_n$, either have continuous compensators or satisfy the conditional independence assumption of Lemma 2.6.1. Thus the corresponding martingales $M_1, \ldots, M_n$ are orthogonal. The statistical methods we emphasize in this book are based on statistics of the form

$$U_l = \sum_{i=1}^{n} \int H_{i,l} dM_i,$$

i.e., linear combinations of stochastic integrals with respect to orthogonal martingales. The integrands $H_{i,l}$ can easily be made predictable, and when these integrands are also bounded, the U_l will be zero mean martingales, (Theorem 2.6.2). For two such statistics, U_1, U_2,

$$\text{cov}\{U_1(t), U_2(t)\} = \sum_{i=1}^{n} \int_0^t E\{H_{i1}(u) H_{i2}(u) I_{\{X_i \geq u\}}\}\{1 - \Delta\Lambda(u)\} d\Lambda(u).$$

The remaining chapters make extensive use of this martingale structure and of the first and second moment results. We will also see in Chapter 5 that the use of localization provides an easy and natural means to obtain asymptotic distribution results for U_l under relaxed conditions on M_i and H_i.

2.8 BIBLIOGRAPHIC NOTES

Section 2.2

Stopping times are also called optional times or Markov times in the literature. Although they are used in this chapter for localizing sequences relaxing regularity conditions, they have been studied extensively in the context of Markov processes for a variety of theoretical and applied reasons. Interesting applications of stopping times and martingales to renewal theory, queuing problems, and the random walk can be found in Ross (1983).

The optional sampling theorem was first proved by Doob, and the standard reference is Doob (1953). Applications of the theorem to more traditional problems in applied probability can be found in Karlin and Taylor (1975). A thorough account of the modern theory of stopping times can be found in Dellacherie (1972).

Local martingales, and evidently the general idea of localization for stochastic processes, were introduced by Kunita and Watanabe (1967) to extend results about stochastic integrals with respect to Brownian motion, or Wiener processes.

Liptser and Shiryayev (1977) contain a thorough account of different versions of the Doob-Meyer decomposition theorems in Chapter 3.

Section 2.3

Section 2.3 uses constructive definitions whenever possible for the local martingales and predictable quadratic variation processes in this section, preserving the link with the material in Chapter 1. More abstract treatments are given in Elliott (1982), (cf., Chapter 10).

Section 2.4

Stochastic integration with respect to local martingales was introduced by Kunita and Watanabe (1967). Elliott (1982, Chapter 11) summarizes the extensions of the French school of probability, particularly the work of Dellacherie, Jacod and Meyer.

CHAPTER 3

Finite Sample Moments and Large Sample Consistency of Tests and Estimators

3.1 INTRODUCTION

In preceding chapters we saw that many censored survival data statistics can be written as $\Sigma_i \int H_i dM_i$. Such statistics are martingales, a structure yielding formulas for first and second moments. In this chapter, we will use these results to explore some finite-sample properties of estimators of survival and hazard functions, and of some censored data test statistics. An important inequality (Lenglart, 1977) will be used to prove a large-sample property (consistency).

In uncensored data, the Kaplan-Meier estimator reduces to the empirical cumulative distribution function, and the test statistics we will consider are generalizations of some often used nonparametric distribution-free rank statistics, including the Wilcoxon rank-sum statistic and the exponential scores test. While there are substantial results about exact finite-sample distributions of these uncensored data statistics, the presence of censoring makes it impractical to compute finite-sample distributions for the censored data versions. Without martingale methods, even the computation of first and second moments for these tests and estimators can be tedious. The methods used here, however, will allow a characterization of the bias of the Kaplan-Meier estimator, will show that the bias decays quickly, and will yield both expressions and natural estimators for the finite-sample variance. With a bit of extra work, these methods can be used to establish uniform convergence in probability of the Kaplan-Meier estimator to the underlying survivor function, and we present that as well in the last section of the chapter.

Since inference about a parameter requires both an estimator and its sampling distribution, inference using the Kaplan-Meier estimator and the rank test generalizations discussed here is delayed until Chapters 6 and 7, that is, until after the martingale central limit theorems in Chapter 5. Nevertheless, the techniques discussed later will use the mean and variance estimators derived here.

Throughout this and the later chapters, we will use the random censorship model. In many applications, especially in clinical research, this is the form of censoring commonly encountered. The results presented in this and later chapters do hold under some other models for censoring, most notably under the so called type II censoring found in reliability and life testing. The form of more general censoring models which incorporate type II censoring may be found in Gill (1980).

The notion of random censoring has been presented earlier, but we restate the definition here for convenience. This chapter deals with procedures for $r \geq 1$ homogeneous samples, and Definition 3.1.1 allows the failure and censoring time distributions to depend on a categorical grouping ($i = 1, \ldots, r$), and the censoring distribution to vary with each case ($j = 1, 2, \ldots, n_i$) within grouping. Varying the censoring distribution by case allows us to treat known, fixed censoring times with the random censorship model, as in Examples 3.2.1 and 3.2.2. The definition will be extended later to allow both distributions to depend on sample size when we derive asymptotic distributions and relative efficiencies in Chapters 6 and 7.

Definition 3.1.1. ***(Random Censorship Model)*** In the *random censorship model*, the ordered pairs (T_{ij}, U_{ij}), $j=1, \ldots, n_i$, $i=1, \ldots, r$, are n $(n \equiv \sum_{i=1}^{r} n_i)$ independent finite failure and censoring time random variables that satisfy Eq. (3.7) in Chapter 1, for each $n = 1, 2, \ldots$. The observable data are

$$X_{ij} = \min(T_{ij}, U_{ij}) \equiv T_{ij} \wedge U_{ij}$$

and

$$\delta_{ij} = I_{\{X_{ij}=T_{ij}\}}.$$

The notation for the underlying distributions in the random censorship model will be

$$S_i(t) = P\{T_{ij} > t\},$$
$$F_i(t) = 1 - S_i(t),$$
$$C_{ij}(t) = P\{U_{ij} > t\},$$
$$L_{ij}(t) = 1 - C_{ij}(t),$$

and

$$\pi_{ij}(t) = P\{X_{ij} \geq t\}.$$ □

The following stochastic processes have been introduced in previous chapters:

$$\overline{N}_i(t) \equiv \sum_{j=1}^{n_i} N_{ij}(t) \equiv \sum_{j=1}^{n_i} I_{\{X_{ij} \leq t, \delta_{ij}=1\}},$$

$$N_{ij}^U(t) = I_{\{X_{ij} \le t, \delta_{ij}=0\}},$$

$$\overline{Y}_i(t) \equiv \sum_{j=1}^{n_i} Y_{ij}(t)$$

$$\equiv \sum_{j=1}^{n_i} I_{\{X_{ij} \ge t\}},$$

$$M_{ij}(t) = N_{ij}(t) - \int_0^t Y_{ij}(s)\, d\Lambda_i(s),$$

and

$$M_i(t) = \sum_{j=1}^{n_i} M_{ij}(t)$$

$$= \overline{N}_i(t) - \int_0^t \overline{Y}_i(s) d\Lambda_i(s),$$

where, as in Definition 1.3.1,

$$\Lambda_i(t) = \int_0^t \{1 - F_i(s-)\}^{-1} dF_i(s).$$

All martingale properties depend on a specification of the way information accrues over time or, in other words, a filtration. Until specified otherwise, the filtration $\{\mathcal{F}_t : t \ge 0\}$ we will use will be given by

$$\mathcal{F}_t = \sigma\{N_{ij}(s), N_{ij}^U(s) : 0 \le s \le t, j = 1, \ldots, n_i, i = 1, \ldots, r\}.$$

This natural filtration specifies, at time t, which items have failed or have been censored up to and including that time and, in this setting, Assumption 2.6.1 holds for the counting processes $\{N_{ij} : j = 1, \ldots, n_i, i = 1, \ldots, r\}$.

3.2 NONPARAMETRIC ESTIMATION OF THE SURVIVAL DISTRIBUTION

In this section, we examine finite sample properties of estimators of the survival distribution in a single homogeneous sample ($r = 1$), and suppress the subscript i.

We first examine methods for estimating the cumulative hazard function, $\Lambda(t)$. From Theorem 1.3.2 and Exercise 1.11, $M_j(t) = N_j(t) - \int_0^t Y_j(s) d\Lambda(s)$ is a martingale for each j with respect to $\{\mathcal{F}_t : t \ge 0\}$. In turn, $M(t) = \overline{N}(t) - \int_0^t \overline{Y}(s) d\Lambda(s)$ is a martingale, where $\overline{N}(t) = \sum_{j=1}^n N_j(t)$ and $\overline{Y}(t) = \sum_{j=1}^n I_{\{X_j \ge t\}}$. Since the process given at time t by

$$\frac{I_{\{\overline{Y}(t)>0\}}}{\overline{Y}(t)} = \begin{cases} 1/\overline{Y}(t) & \text{if } \overline{Y}(t) > 0, \\ 0 & \text{if } \overline{Y}(t) = 0, \end{cases}$$

is a left-continuous adapted process with right-hand limits, $\{\mathcal{M}(t) : t \geq 0\}$ given by

$$\begin{aligned}\mathcal{M}(t) &= \int_0^t \frac{I_{\{\overline{Y}(s)>0\}}}{\overline{Y}(s)} dM(s)\\ &= \int_0^t \frac{d\overline{N}(s)}{\overline{Y}(s)} - \int_0^t I_{\{\overline{Y}(s)>0\}} d\Lambda(s)\end{aligned}$$

is a martingale. It follows, since $\mathcal{M}(0) = 0$, that

$$E\int_0^t \frac{d\overline{N}(s)}{\overline{Y}(s)} = E\int_0^t I_{\{\overline{Y}(s)>0\}} d\Lambda(s). \tag{2.1}$$

Let $\Lambda^*(t) = \int_0^t I_{\{\overline{Y}(s)>0\}} d\Lambda(s)$. Then, if $T = \inf\{t : \overline{Y}(t) = 0\}, \Lambda^*(t) = \int_0^{t\wedge T} d\Lambda(s) = \Lambda(t \wedge T)$. By Eq. (2.1), we might expect that $\hat{\Lambda}(t) \equiv \int_0^t d\overline{N}(s)/\overline{Y}(s)$ would be a good "estimator" for $\Lambda^*(t) = \Lambda(t \wedge T)$, but that it would not be possible to obtain an unbiased estimator of $\Lambda(t)$ without making parametric assumptions.

The following theorem summarizes some properties of $\hat{\Lambda}$, an estimator first proposed by Nelson (1969).

Theorem 3.2.1. Let $t \geq 0$ be such that $\Lambda(t) < \infty$. Then

1. $E\{\hat{\Lambda}(t) - \Lambda^*(t)\} = 0$,

2. $E\{\hat{\Lambda}(t) - \Lambda(t)\} = -\int_0^t \left[\Pi_{j=1}^n\{1 - \pi_j(s)\}\right] d\Lambda(s)$,

2$'$. if $\pi_j(s) = \pi(s)$ for all j, then

$$\begin{aligned}E\{\hat{\Lambda}(t) - \Lambda(t)\} &= -\int_0^t \{1 - \pi(s)\}^n d\Lambda(s)\\ &\geq -\{1 - \pi(t)\}^n \Lambda(t),\end{aligned}$$

and

3.

$$\begin{aligned}\sigma_*^2(t) &= E[\sqrt{n}\{\hat{\Lambda}(t) - \Lambda^*(t)\}]^2\\ &= E\left[n\int_0^t \frac{I_{\{\overline{Y}(s)>0\}}}{\overline{Y}(s)}\{1 - \Delta\Lambda(s)\} d\Lambda(s)\right].\end{aligned}$$

Proof. Equation (2.1) establishes (1), and hence implies

$$\begin{aligned}E\{\hat{\Lambda}(t) - \Lambda(t)\} &= E\{\Lambda^*(t) - \Lambda(t)\}\\ &= -E\int_0^t I_{\{\overline{Y}(s)=0\}} d\Lambda(s)\\ &= -\int_0^t P\{\overline{Y}(s) = 0\} d\Lambda(s),\end{aligned}$$

from which (2) and (2$'$) follow directly.

To prove (3), the equation

$$\langle M, M\rangle(t) = \int_0^t \overline{Y}(s)\{1 - \Delta\Lambda(s)\}d\Lambda(s)$$

follows from Theorem 2.6.1 and Lemma 2.6.1. Therefore,

$$\begin{aligned} E\{\hat{\Lambda}(t) - \Lambda^*(t)\}^2 &= E\left\{\int_0^t \frac{I_{\{\overline{Y}(s)>0\}}}{\overline{Y}(s)} dM(s)\right\}^2 \\ &= E\int_0^t \frac{I_{\{\overline{Y}(s)>0\}}}{\overline{Y}^2(s)}\, d\langle M, M\rangle(s) \\ &= E\int_0^t \frac{I_{\{\overline{Y}(s)>0\}}}{\overline{Y}(s)}\{1 - \Delta\Lambda(s)\}\, d\Lambda(s). \qquad \square \end{aligned}$$

Suppose $\pi_j(s) = \pi(s)$ for all j and s. If $\pi(t) > 0$, Theorem 3.2.1 indicates that $\hat{\Lambda}(t)$ is an asymptotically unbiased estimator of $\Lambda(t)$, with bias converging to zero at an exponential rate as $n \to \infty$. For the second moment,

$$\begin{aligned} \sigma_*^2(t) &= E\; n\{\hat{\Lambda}(t) - \Lambda^*(t)\}^2 \\ &= E\int_0^t \frac{n}{\overline{Y}(s)} I_{\{\overline{Y}(s)>0\}}\{1 - \Delta\Lambda(s)\}d\Lambda(s), \end{aligned}$$

which, for large n, should approach

$$\sigma^2(t) \equiv \int_0^t \{\pi(s)\}^{-1}\{1 - \Delta\Lambda(s)\}d\Lambda(s).$$

Since $E\; n\{\Lambda^*(t) - \Lambda(t)\}^2$ converges to zero when $\pi(t) > 0$, since

$$\sqrt{n}\{\hat{\Lambda}(t) - \Lambda^*(t)\} = \frac{1}{\sqrt{n}}\sum_{j=1}^{n}\int_0^t \frac{n}{\overline{Y}(s)} dM_j(s),$$

where $\{M_j\}$ is an independent and identically distributed collection, and since $n\{\overline{Y}(s)\}^{-1}$ converges to $\{\pi(s)\}^{-1}$, we might expect that $\sqrt{n}\{\hat{\Lambda}(t) - \Lambda(t)\}$ is approximately distributed as $N(0, \sigma^2(t))$ for large n. This is indeed the case, as can be shown using Theorem 6.2.1 in Chapter 6.

The precision of $\hat{\Lambda}$ at time t can be measured either by its variance, $E\{\hat{\Lambda}(t) - E\hat{\Lambda}(t)\}^2$ or, since it is biased, by its mean squared error, $E[\{\hat{\Lambda}(t) - \Lambda(t)\}^2]$. Since the squared bias satisfies

$$[E\{\hat{\Lambda}(t) - \Lambda(t)\}]^2 \le \{1 - \pi(t)\}^{2n}\{\Lambda(t)\}^2,$$

these will be nearly equal, and the variance of $\hat{\Lambda}(t)$ can be safely used except when $\Lambda(t)$ is large and n is small, or when $\pi(t)$ is zero. The variance is given by

$$\begin{aligned} \text{var}\hat{\Lambda}(t) &= n^{-1}\sigma_*^2(t) + 2\,E\,\left[\{\hat{\Lambda}(t) - \Lambda^*(t)\}\{\Lambda^*(t) - E\,\Lambda^*(t)\}\right] \\ &\quad + E\{\Lambda^*(t) - E\,\Lambda^*(t)\}^2 \qquad (2.2) \\ &\approx \frac{1}{n}\sigma_*^2(t), \end{aligned}$$

for even relatively small values of n. In estimating the variance of $\hat{\Lambda}(t)$, it is therefore sufficient to find a good estimator of $n^{-1}\sigma_*^2(t)$.

Theorem 3.2.2. Let $t \geq 0$ be such that $\Lambda(t) < \infty$. Define

$$\frac{1}{n}\hat{\sigma}^2(t) = \int_0^t \frac{I_{\{\overline{Y}(s)>0\}}}{\overline{Y}^2(s)} \left\{1 - \frac{\Delta\overline{N}(s) - 1}{\overline{Y}(s) - 1}\right\} d\overline{N}(s),$$

where $0/0 \equiv 0$ as usual. Then

$$E\left\{\frac{1}{n}\hat{\sigma}^2(t) - \frac{1}{n}\sigma_*^2(t)\right\} = \int_0^t P\{\overline{Y}(s) = 1\}\Delta\Lambda(s)d\Lambda(s). \qquad \square$$

When Λ is a continuous function of t, Theorem 3.2.2 indicates that $\hat{\sigma}^2(t)$ provides an unbiased estimator of $\sigma_*^2(t)$. In fact, the theorem suggests that bias arises from situations in which $\Delta\Lambda(s) > 0$ and $\overline{Y}(s) = 1$. The situation here is closely related to problems encountered in estimating a binomial parameter since, when $\Delta\Lambda(s) > 0$,

$$\Delta\Lambda(s) = P\{T = s|T \geq s\}.$$

Suppose $\Delta\Lambda(t_0) \equiv p > 0$ for some $t_0 < t$. The "binomial parameter" p has the usual estimator $\hat{p} \equiv \Delta\overline{N}(t_0)/\overline{Y}(t_0)$ with conditional variance, given $\overline{Y}(t_0) \equiv n$, $\tilde{\sigma}^2 \equiv p(1-p)/n$. When $n > 1$,

$$\begin{aligned} \hat{\tilde{\sigma}}^2 &\equiv \frac{n}{n-1}\,\frac{\hat{p}(1-\hat{p})}{n} \\ &= \frac{1}{\overline{Y}^2(t_0)}\left\{1 - \frac{\Delta\overline{N}(t_0) - 1}{\overline{Y}(t_0) - 1}\right\}\Delta\overline{N}(t_0) \end{aligned}$$

is an unbiased estimator of $\tilde{\sigma}^2$, while no unbiased estimator exists when $n = 1$. If we use $\hat{\tilde{\sigma}}^2 \equiv \hat{p} = \Delta\overline{N}(t_0)/\overline{Y}(t_0)$ when $n = 1$, a bias of $p^2 = \Delta\Lambda d\Lambda$ will arise in that special case. In summary, then, the classical setting of estimating a binomial parameter motivates $\hat{\Lambda}(t)$ as an "unbiased estimator" of $\Lambda^*(t)$, the form of $\sigma_*^2(t)$ and of its estimator $\hat{\sigma}^2$, and the bias of $\hat{\sigma}^2$ as an estimator of $\sigma_*^2(t)$.

Proof of Theorem 3.2.2.

$$E\{\frac{1}{n}\hat{\sigma}^2(t) - \frac{1}{n}\sigma_*^2(t)\} =$$

$$E\int_0^t \frac{I_{\{\overline{Y}(s)>0\}}}{\overline{Y}^2(s)}\{d\overline{N}(s) - \overline{Y}(s)d\Lambda(s)\}$$

$$- E\int_0^t \frac{I_{\{\overline{Y}(s)>1\}}}{\overline{Y}^2(s)\{\overline{Y}(s)-1\}}[\{\Delta\overline{N}(s) - 1\}d\overline{N}(s) - \overline{Y}(s)\{\overline{Y}(s)-1\}\Delta\Lambda(s)\, d\Lambda(s)]$$

$$+ E\int_0^t I_{\{\overline{Y}(s)=1\}}\Delta\Lambda(s)d\Lambda(s). \tag{2.3}$$

The first term on the right-hand side of Eq. (2.3) is zero since it is the expectation at time t of a zero mean martingale. The second term will be zero for the same reason if

$$\int [\{\Delta\overline{N}(s) - 1\}d\overline{N}(s) - \overline{Y}(s)\{\overline{Y}(s)-1\}\Delta\Lambda(s)\, d\Lambda(s)] \tag{2.4}$$

is also a zero mean martingale, where as before an integral without limits denotes a process whose value at time t is the integral over the interval $(0, t]$.

Now,

$$\int \{\Delta\overline{N}(s) - 1\}d\overline{N}(s) - \int\{\overline{Y}(s) - 1\}\overline{Y}(s)\Delta\Lambda(s)d\Lambda(s)$$

$$= \int\{\Delta\overline{N}(s) - 1\}d\overline{N}(s) - \int\{\Delta\overline{N}(s)\}\overline{Y}(s)\, d\Lambda(s) + \int \overline{Y}(s)\Delta\Lambda(s)\, d\Lambda(s)$$

$$+ \int \overline{Y}(s)\Delta\Lambda(s)\{d\overline{N}(s) - \overline{Y}(s)\, d\Lambda(s)\}$$

$$= \int\{\Delta\overline{N}(s) - \overline{Y}(s)\Delta\Lambda(s)\}\{d\overline{N}(s) - \overline{Y}(s)\, d\Lambda(s)\} - \int \overline{Y}(s)\{1 - \Delta\Lambda(s)\}\, d\Lambda(s)$$

$$+ \int\{2\overline{Y}(s)\Delta\Lambda(s) - 1\}\{d\overline{N}(s) - \overline{Y}(s)\, d\Lambda(s)\}$$

$$= \int \Delta M(s)\, dM(s) - \langle M, M\rangle(s) + \int\{2\overline{Y}(s)\Delta\Lambda(s) - 1\}dM(s),$$

where $M(s) = \overline{N}(s) - \int \overline{Y}(s)\, d\Lambda(s)$. Thus, to establish that Expression (2.4) is a zero mean martingale, it is sufficient to show $\int \Delta M(s)dM(s) - \langle M, M\rangle(s)$ is a zero mean martingale. This will follow from Theorem 2.6.1 if $\int \Delta M_i(s)dM_j(s)$ is a zero mean martingale whenever $i \neq j$.

We first show that $\int \Delta N_i(s)dM_j(s)$ is a zero mean martingale when $i \neq j$. For $s < t$,

$$E\left\{\int_s^t \Delta N_i(v)dM_j(v) \mid \mathcal{F}_s\right\}$$

$$= E\left[\sum_{s<v\leq t} \Delta N_i(v)(\Delta N_j - \Delta A_j)(v) \mid \mathcal{F}_s\right]$$

$$= E\left[\sum_{s<v\leq t} E\left\{\Delta N_i(v)\Delta N_j(v) \mid \mathcal{F}_{v-}\right\} \mid \mathcal{F}_s\right]$$

$$-E\left[\sum_{s<v\leq t} \Delta A_j(v)E\left\{\Delta N_i(v) \mid \mathcal{F}_{v-}\right\} \mid \mathcal{F}_s\right]$$

$$= E\left[\sum_{s<v\leq t} E\left\{\Delta N_i(v) \mid \mathcal{F}_{v-}\right\} E\left\{\Delta M_j(v) \mid \mathcal{F}_{v-}\right\} \mid \mathcal{F}_u\right]$$

$$= 0,$$

where the second equality follows from the predictability of A_j, and the third because Assumption 2.6.1 is true here (Exercise 2.11). Since ΔA_i is a bounded predictable process, $\int \Delta M_i dM_j = \int \Delta N_i dM_j - \int \Delta A_i dM_j$ is a zero mean martingale whenever $i \neq j$. □

The relationship between Λ and S provides a method for estimating survival functions. By definition,

$$\Lambda(t) = \int_0^t \{1 - F(s-)\}^{-1} dF(s),$$

so

$$d\Lambda(s) = \frac{dF(s)}{1 - F(s-)}.$$

Since $F(0) = 0$,

$$F(t) = \int_0^t dF(s)$$

$$= \int_0^t \{1 - F(s-)\} d\Lambda(s), \tag{2.5}$$

so F is uniquely determined by Λ.

To estimate S, one might insert Nelson's estimator $\hat{\Lambda} = \int \overline{Y}^{-1} d\overline{N}$ for Λ in Eq. (2.5), and define $\hat{S}$ recursively:

$$\hat{S}(t) = 1 - \int_0^t \hat{S}(s-)d\hat{\Lambda}(s). \tag{2.6}$$

Since

$$\begin{aligned}\hat{S}(t-) - \hat{S}(t) &= -\Delta\hat{S}(t)\\ &= \hat{S}(t-)\frac{\Delta\overline{N}(t)}{\overline{Y}(t)},\\ \hat{S}(t) &= \hat{S}(t-)\left\{1 - \frac{\Delta\overline{N}(t)}{\overline{Y}(t)}\right\},\end{aligned}$$

and

$$\hat{S}(t) = \prod_{s \leq t}\left\{1 - \frac{\Delta\overline{N}(s)}{\overline{Y}(s)}\right\},$$

which is the Kaplan-Meier estimator introduced in Chapter 0.

The next theorem provides a useful identity for investigating properties of $\hat{S}$.

Theorem 3.2.3. If $S(t) > 0$,

$$\frac{\hat{S}(t)}{S(t)} = 1 - \int_0^t \frac{\hat{S}(s-)}{S(s)}\left\{\frac{d\overline{N}(s)}{\overline{Y}(s)} - d\Lambda(s)\right\}. \tag{2.7}$$

Proof. We will use general versions of the formulas for integration by parts and the differential of a reciprocal. Suppose U, V, and W are right-continuous functions of locally bounded variation on any finite interval $[0, t]$. Then for any $t \in (0, \infty)$, by Theorem A.1.2 in Appendix A,

$$U(t)V(t) = U(0)V(0) + \int_0^t U(s-)dV(s) + \int_0^t V(s)dU(s) \tag{2.8}$$

and

$$d\{W(s)\}^{-1} = -\{W(s)W(s-)\}^{-1}dW(s). \tag{2.9}$$

Let $U(s) = \hat{S}(s)$, $W(s) = S(s)$, and $V(s) = \{S(s)\}^{-1} = \{W(s)\}^{-1}$, so U, V, and W are monotone, bounded, and right-continuous on $[0, t]$. By Eq. (2.8),

$$\frac{\hat{S}(t)}{S(t)} = \frac{\hat{S}(0)}{S(0)} + \int_0^t \hat{S}(s-)\,dV(s) + \int_0^t \frac{1}{S(s)}dU(s),$$

and by Eq. (2.9),

$$dV(s) = -\{S(s)S(s-)\}^{-1}dS(s).$$

Since $dU(s) = d\hat{S}(s)$, and $\hat{S}(0) = 1 = S(0)$, we have

$$\begin{aligned}\frac{\hat{S}(t)}{S(t)} &= 1 - \int_0^t \frac{\hat{S}(s-)}{S(s)S(s-)} dS(s) + \int_0^t \frac{1}{S(s)} d\hat{S}(s) \\ &= 1 + \int_0^t \frac{\hat{S}(s-)}{S(s)} d\Lambda(s) - \int_0^t \frac{\hat{S}(s-)}{S(s)} d\hat{\Lambda}(s), \qquad (2.10)\end{aligned}$$

where the last equality follows from Eqs. (2.5) and (2.6). Equation (2.7) now follows immediately from Eq. (2.10). □

Theorem 3.2.3 implies that, when $S(t) > 0$,

$$\begin{aligned}\hat{S}(t) - S(t) &= -S(t) \int_0^t \frac{\hat{S}(s-)}{S(s)} \left\{ \frac{d\overline{N}(s)}{\overline{Y}(s)} - I_{\{\overline{Y}(s)>0\}} d\Lambda(s) - I_{\{\overline{Y}(s)=0\}} d\Lambda(s) \right\} \\ &= -S(t) \int_0^t \frac{\hat{S}(s-)}{S(s)} \frac{I_{\{\overline{Y}(s)>0\}}}{\overline{Y}(s)} dM(s) + B(t),\end{aligned}$$

where

$$B(t) = S(t) \int_0^t \frac{\hat{S}(s-)}{S(s)} I_{\{\overline{Y}(s)=0\}} d\Lambda(s).$$

If $T = \inf\{s : \overline{Y}(s) = 0\}$,

$$\begin{aligned}B(t) &= S(t) I_{\{T<t\}} \int_T^t \frac{\hat{S}(s-)}{S(s)} d\Lambda(s) \\ &= S(t)\hat{S}(T) I_{\{T<t\}} \int_T^t \frac{-dS(s)}{S(s)S(s-)},\end{aligned}$$

by Eq. (2.5) and because $\Delta\hat{S}(s) = 0$ if $s > T$. Thus, by Eq. (2.9),

$$\begin{aligned}B(t) &= S(t)\hat{S}(T) I_{\{T<t\}} \left\{ \frac{S(T) - S(t)}{S(t)S(T)} \right\} \\ &= I_{\{T<t\}} \frac{\hat{S}(T)\{S(T) - S(t)\}}{S(T)}.\end{aligned}$$

We have proved the following Corollary to Theorem 3.2.3.

Corollary 3.2.1. For any t such that $S(t) > 0$,

$$\hat{S}(t) - S(t) = -S(t) \int_0^t \frac{\hat{S}(s-)}{S(s)} \frac{I_{\{\overline{Y}(s)>0\}}}{\overline{Y}(s)} dM(s) + B(t), \qquad (2.11)$$

where

$$B(t) = I_{\{T<t\}} \frac{\hat{S}(T)\{S(T) - S(t)\}}{S(T)}$$

and $T = \inf\{s : \overline{Y}(s) = 0\}$. □

Eq. (2.11) provides direct information about the bias of the Kaplan-Meier estimator.

Lemma 3.2.1. If $S(t) > 0$,

1.

$$\begin{aligned} E\{\hat{S}(t) - S(t)\} &= E\ B(t) \\ &= E\left[I_{\{T<t\}} \frac{\hat{S}(T)\{S(T) - S(t)\}}{S(T)}\right] \\ &\geq 0, \end{aligned}$$

and

2. if $\pi_j(t) = \pi(t)$ for each j,

$$E\{\hat{S}(t) - S(t)\} \leq \{1 - S(t)\}\{1 - \pi(t)\}^n.$$

The Kaplan-Meier estimator thus has nonnegative bias which converges to zero at an exponential rate as $n \to \infty$.

Proof. Claim (1) follows directly from Eq. (2.11) and because

$$E\left\{-S(t) \int_0^t \frac{\hat{S}(s-)}{S(s)} \frac{I_{\{\overline{Y}(s)>0\}}}{\overline{Y}(s)} dM(s)\right\} = 0.$$

To establish (2), observe that

$$\begin{aligned} E\{\hat{S}(t) - S(t)\} &= E\left[I_{\{T<t\}} \frac{\hat{S}(T)\{S(T) - S(t)\}}{S(T)}\right] \\ &= E\left[I_{\{T<t\}} \hat{S}(T)\left\{1 - \frac{S(t)}{S(T)}\right\}\right] \\ &\leq E[I_{\{T<t\}}\{1 - S(t)\}] \\ &= \{1 - S(t)\}P\{\overline{Y}(t) = 0\}. \end{aligned}$$

Thus, if $\pi_j(t) = \pi(t)$ for every j,

$$E\{\hat{S}(t) - S(t)\} \leq \{1 - S(t)\}\{1 - \pi(t)\}^n. \qquad \square$$

By Lemma 3.2.1, $\hat{S}(t)$ will be biased (but not very much so) only if there is a positive probability that $\hat{S}(T) > 0$ and $S(t) < S(T)$. In the random censorship model in uncensored data (i.e., when $P\{U_j \geq T_j\} = 1$), $\hat{S}(t) \equiv 0$ for $t > T$, so $\hat{S}$ is unbiased for all t. Of course, in uncensored data, the Kaplan-Meier estimator reduces to the empirical cumulative distribution function, and unbiasedness is easy to establish. For a less trivial illustration, if censoring can occur only on an interval $[a, b]$ over which S is constant, then $\hat{S}(t)$ is unbiased for all $t \leq b$.

We use a simple example to illustrate the formulas.

Example 3.2.1. The entry pattern of a typical clinical trial with a failure time end point is shown in Figure 3.2.1. Accrual takes place from the opening of the study until a planned target sample size n is reached at, say, a units of time. After additional follow up of $b-a$ time units, the data from the trial are analyzed. (Trials with sequential designs reach final analysis in a more complicated fashion.)

Conditional on the date of accrual termination and the time of analysis, the entry time E_i of the ith subject may be assumed to have probability density $f_E(t)$ on $[0, a]$, and the censoring survivor function for each subject is

$$C(t) = \begin{cases} 1, & 0 \leq t \leq b-a, \\ \int_t^b f_E(b-s)ds, & b-a < t \leq b, \\ 0, & b < t. \end{cases}$$

In most controlled clinical trials, patients are rarely lost to follow up during the trial, and censoring is caused by the analysis of the trial preceding observed failure in a subset of cases.

If $T = \max\{X_j : j = 1, \ldots, n\}$, then the version of Kaplan-Meier defined here has bias

$$E\left[I_{\{T<t\}} \frac{\hat{S}(T)\{S(T) - S(t)\}}{S(T)}\right] = 0$$

for $t < b-a$, since $\hat{S}(T) \equiv 0$ whenever $T < b-a$. The Kaplan-Meier estimator is unbiased for values of t at which censoring is impossible.

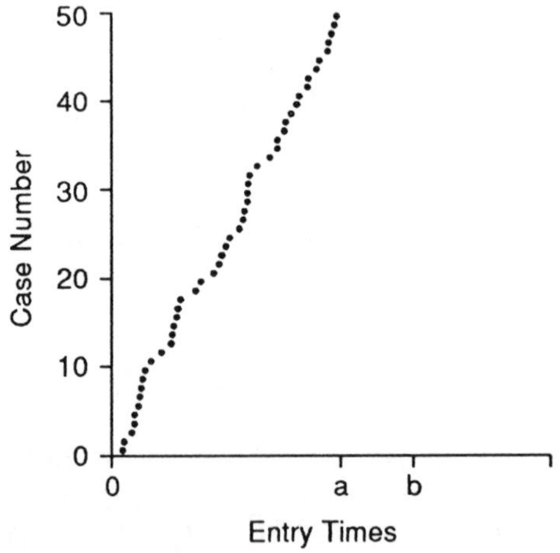

Figure 3.2.1 Typical entry pattern of a clinical trial.

For $t > b - a$, the expression for bias becomes complicated, even in small samples. A simple but useful bound for the bias can be found using (2) from Lemma 3.2.1 for specific situations. When the distribution of entry times is uniform on $[0, a]$ then

$$C(t) = \begin{cases} 1, & 0 < t \le b - a, \\ a^{-1}(b - t), & b - a < t \le b, \\ 0, & b < t < \infty. \end{cases}$$

$\hat{S}$ will be unbiased on $[0, b - a]$, and will have bias bounded by

$$\{1 - S(t)\} \left[1 - S(t)\left\{\frac{b - t}{a}\right\}\right]^n$$

on $b - a < t \le b$. Since there can be no data beyond $t = b$, the estimator would not be calculated for $t > b$, and the issue of bias is not relevant there. The "relative" bias on $[b - a, b]$ will be no larger than

$$\left\{\frac{1 - S(t)}{S(t)}\right\} \left[1 - S(t)\left\{\frac{b - t}{a}\right\}\right]^n.$$

Suppose a study is open for registration for 4 years, with 5 cases registered each year, and follow-up data are gathered for 2 additional years. Assume that an endpoint of interest is the time from registration to progression of a given disease. When the three-year progression-free rate is 35%, the relative bias in estimating $S(3)$ will be bounded by

$$\left(\frac{.65}{.35}\right) \left\{1 - (.35)\frac{3}{4}\right\}^{20}$$
$$= 4.21 \times 10^{-3}. \qquad \square$$

In Example 3.2.1, the censoring times are all determined by the entry times, E_i, and the time of analysis, b, and consequently are all known at the analysis time, even for the observed failures. The results of Corollary 3.2.1 and Lemma 3.2.1 also can be applied in this case when the analysis is viewed as conditional on the censoring times.

The Eastern Cooperative Oncology Group (EST 5477) study in unfavorable histology lymphoma fits the setting described above.

Example 3.2.2. The EST 5477 study involved randomization of 506 patients with specified types of non-Hodgkin's lymphoma in order to compare the ability of three induction regimens to induce disease remission and improve the chances of survival. Accrual of patients from 25 participating institutions began in March, 1978 and ended in November, 1983. In July, 1986, a subset of 332 cases with stage III or IV disease and specific histologic subtypes confirmed by pathology review were analyzed to investigate the effect of the three treatments on survival (O'Connell et al., 1987, 1988). At the update of the study file in 1986, only one patient was lost to follow up, after having been on study for 1192 days, or approximately 3.3 years. The minimum follow up time among the 134 surviving patients was 699 days (1.9 years). Although each surviving case was eligible for at least 31 months (2.6 years) of follow up, record submission in multicenter trials does not always coincide with a time of analysis, and routine follow-up visits are sometimes delayed or records are submitted late. Nevertheless, an estimate of survival curves for subsets of the population participating in the study will be unbiased for approximately the first two years of the registration period. In this study, treatment began soon after registration and, if the study participants are a random sample of the population-at-large with this disease, the treatment specific survival curves (Figure 3.2.2) give unbiased estimates of survival in the first two years with these therapies. □

The Mayo Clinic trial in primary biliary cirrhosis of the liver discussed in Chapter 0 had a slightly more complex censoring mechanism.

Example 3.2.3. Figure 3.2.3 displays survival curves for the patients receiving the drug D-penicillamine and those receiving placebo. Figure 3.2.4 is an estimated survival curve using all 312 cases, computed ignoring treatment.

The trial for histologic stages 3 and 4 showed the drug D-penicillamine to be ineffective in prolonging survival, so the pooled survival curve illustrates the

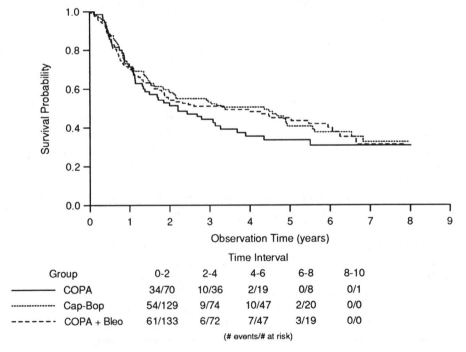

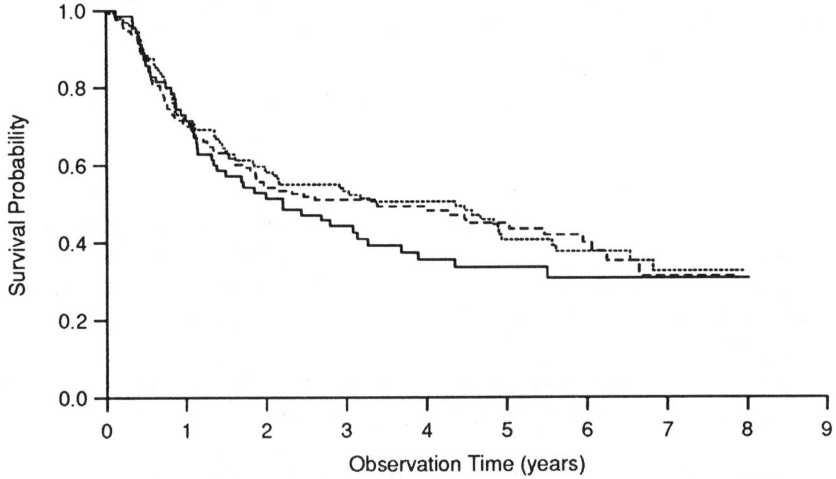

Group	Time Interval 0-2	2-4	4-6	6-8	8-10
——— COPA	34/70	10/36	2/19	0/8	0/1
········ Cap-Bop	54/129	9/74	10/47	2/20	0/0
- - - - - COPA + Bleo	61/133	6/72	7/47	3/19	0/0

(# events/# at risk)

Figure 3.2.2 Estimated survival curves for the three consolidation treatment groups in the Eastern Cooperative Oncology Group study EST 5477.

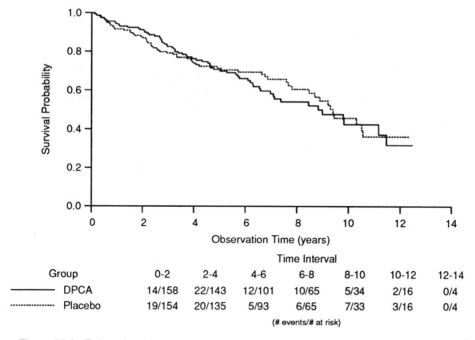

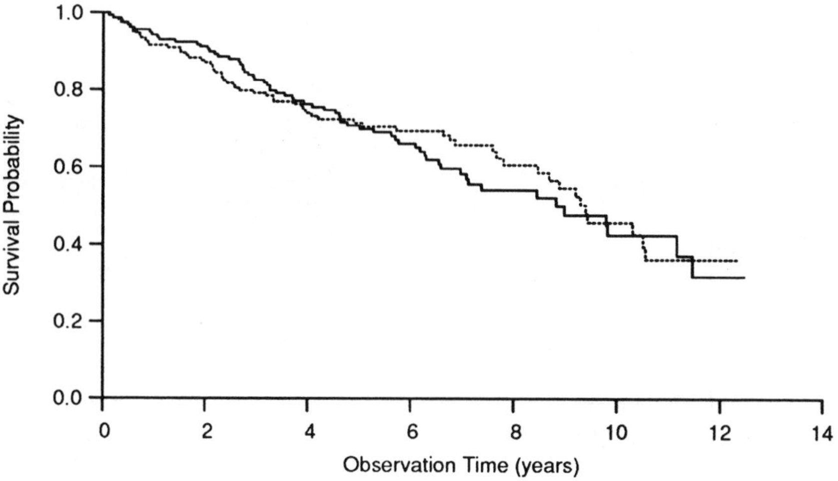

	Time Interval						
Group	0-2	2-4	4-6	6-8	8-10	10-12	12-14
———— DPCA	14/158	22/143	12/101	10/65	5/34	2/16	0/4
·········· Placebo	19/154	20/135	5/93	6/65	7/33	3/16	0/4

(# events/# at risk)

Figure 3.2.3 Estimated survival curves for the DPCA and placebo treatment groups in the Mayo Clinic study of primary bilary cirrhosis.

prognosis for untreated PBC. Of the 312 patients enrolled in the study, 187 were censored at the time of analysis. The large majority of the surviving cases (160) were in active follow up at the time of analysis, and this group had updated records as of January 1986. Although the last patient registration to the trial was in May 1984, the survival curve in Figure 3.2.4 is not necessarily unbiased over the first 20 months. Eight patients were lost to follow up, with a median and a minimum follow up, respectively, of 66 and 26 months. An additional 19 patients received a liver transplant and were considered censored. Transplantation may give rise to a censoring mechanism that does not satisfy Condition (3.1) of Chapter 1, and thus could cause bias in estimating underlying survival curves. Nevertheless, any bias here is likely to be very small.

A detailed analysis of the association of clinical characteristics with survival in untreated PBC is given in Chapter 4. □

An approximation to the variance of $\hat{S}(t)$ can be obtained using arguments similar to those used for the variance of the Nelson estimator. Again, suppose $\pi_j(t) = \pi(t) > 0$ for each j. Then

$$\text{var } \hat{S}(t) = E[\hat{S}(t) - \{S(t) + EB(t)\}]^2, \tag{2.12}$$

and

$$E[\hat{S}(t) - \{S(t) + EB(t)\}]^2 - E\{\hat{S}(t) - S(t) - B(t)\}^2 \to 0$$

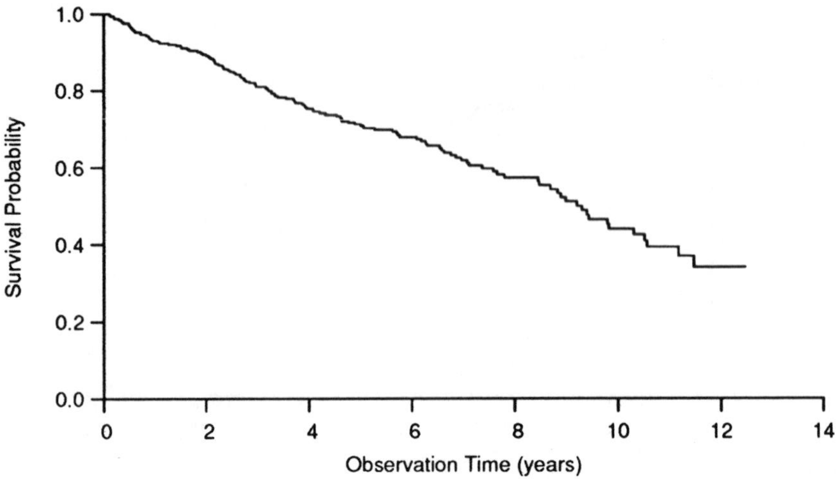

Time Interval

0-2	2-4	4-6	6-8	8-10	10-12	12-14
33/312	42/278	17/194	16/130	12/67	5/32	0/8

(# events/# at risk)

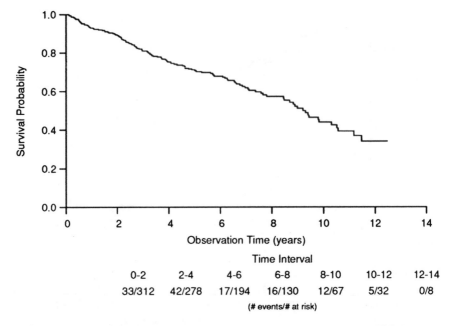

Figure 3.2.4 Estimated survival curve using all 312 randomized cases in the PBC data.

exponentially fast as $n \to \infty$, where the convergence follows easily from Lemma 3.2.1 and the bound $0 \le B(t) < 1$. In turn,

$$\begin{aligned} n^{-1}V(t) &\equiv E\{\hat{S}(t) - S(t) - B(t)\}^2 \\ &= E\left\{S(t)\int_0^t \frac{\hat{S}(s-)}{S(s)}\frac{I_{\{\overline{Y}(s)>0\}}}{\overline{Y}(s)}dM(s)\right\}^2 \\ &= S^2(t)\int_0^t E\left\{\frac{\hat{S}^2(s-)}{S^2(s)}\frac{I_{\{\overline{Y}(s)>0\}}}{\overline{Y}(s)}\right\}\{1-\Delta\Lambda(s)\}d\Lambda(s). \quad (2.13) \end{aligned}$$

By Eq. (2.12), we can estimate the finite-sample variance of the Kaplan-Meier estimator by obtaining an estimate $n^{-1}\hat{V}(t)$ of $n^{-1}V(t)$. After replacing S, $\Delta\Lambda$, and $d\Lambda$ in Eq. (2.13) by $\hat{S}$, $\Delta\overline{N}/\overline{Y}$, and $d\overline{N}/\overline{Y}$, respectively,

$$\begin{aligned} n^{-1}\hat{V}(t) &\equiv \hat{S}^2(t)\int_0^t \left[\frac{\prod_{v<s}\{1-\frac{\Delta\overline{N}(v)}{\overline{Y}(v)}\}}{\prod_{v\le s}\{1-\frac{\Delta\overline{N}(v)}{\overline{Y}(v)}\}}\right]^2 \frac{I_{\{\overline{Y}(s)>0\}}}{\overline{Y}(s)}\left\{1-\frac{\Delta\overline{N}(s)}{\overline{Y}(s)}\right\}\frac{d\overline{N}(s)}{\overline{Y}(s)} \\ &= \hat{S}^2(t)\int_0^t \left\{1-\frac{\Delta\overline{N}(s)}{\overline{Y}(s)}\right\}^{-2}\left\{1-\frac{\Delta\overline{N}(s)}{\overline{Y}(s)}\right\}\frac{d\overline{N}(s)}{\overline{Y}^2(s)} \\ &= \hat{S}^2(t)\int_0^t \frac{d\overline{N}(s)}{\{\overline{Y}(s)-\Delta\overline{N}(s)\}\overline{Y}(s)}. \end{aligned}$$

Note that $n^{-1}\hat{V}(t) = 0$ at t^* such that $\overline{Y}(t^*) = \Delta\overline{N}(t^*)$ because $\hat{S}(t^*) = 0$.

Using a different approach, Kaplan and Meier suggest in their 1958 paper that $\hat{V}(t)$ be used to estimate $\text{var}\{\sqrt{n}\hat{S}(t)\}$, and a similar formula appeared as early as 1926 (Greenwood, 1926) in estimating the variance of life table estimates from vital statistics.

Expression (2.13) for the approximate variance of $\hat{S}(t)$ yields some insight into the asymptotic distribution of $\hat{S}$. We have

$$V(t) = S^2(t)\int_0^t E\left\{\frac{\hat{S}^2(s-)}{S^2(s)}\frac{n}{\overline{Y}(s)}I_{\{\overline{Y}(s)>0\}}\right\}\{1-\Delta\Lambda(s)\}d\Lambda(s).$$

Since $V(t) - \text{var}\{\sqrt{n}\hat{S}(t)\} \longrightarrow 0$ exponentially fast, it seems reasonable (see Exercise 3.1) that for large n, $\sqrt{n}\{\hat{S}(t)-S(t)\}$ is approximately $N(0, \sigma^2(t))$, where

$$\sigma^2(t) = S^2(t)\int_0^t \frac{S^2(s-)}{S^2(s)}\{\pi(s)\}^{-1}\{1-\Delta\Lambda(s)\}d\Lambda(s).$$

The estimator $\hat{S}$ is indeed asymptotically normal, and the expression for $\sigma^2(t)$ simplifies considerably. Because of the relationship between Λ and F,

$$\begin{aligned} 1-\Delta\Lambda(s) &= 1-\frac{F(s)-F(s-)}{1-F(s-)} \\ &= \frac{1-F(s)}{1-F(s-)}, \end{aligned}$$

and

$$d\Lambda(s) = \frac{dF(s)}{1-F(s-)},$$

so

$$\begin{aligned} \sigma^2(t) &= -S^2(t)\int_0^t \frac{S^2(s-)}{S^2(s)}\{\pi(s)\}^{-1}\frac{S(s)}{S^2(s-)}dS(s) \\ &= -S^2(t)\int_0^t \frac{dS(s)}{\pi(s)S(s)}, \end{aligned}$$

a result first derived by Breslow and Crowley (1974).

There is no uniformly good way to extend $\hat{S}$ beyond the last observation time. The results derived here use the (mathematically) convenient convention of $\hat{S}(t) = \hat{S}(T)$ for $t \geq T$. In moderately or heavily censored data, this convention is difficult to defend in practice, and a data analyst usually considers $\hat{S}$ undefined on the (random) interval $(T, \infty]$, or at most unobservable with a value $0 \leq \hat{S}(t) \leq \hat{S}(T)$ for $t > T$. This latter approach is used in Fleming and Harrington (1984a) to study the mean squared error of $\hat{S}(t)$.

The counting process and martingale approach, then, provides a framework for proposing an estimator, $\hat{S}$, of S, and for deriving formulas for the exact variance of $n^{1/2}\hat{S}$, an approximation (V) for this variance, a corresponding estimator ($\hat{V}$), and the asymptotic variance (σ^2) of $n^{1/2}\hat{S}$.

Later we will need an additional finite-sample property of $\hat{S}$, a linear "in probability" upper bound. Similar results for the empirical distribution function can be found in Shorack and Wellner (1986, Theorem 9.1.2 and Inequality 10.4.1). The result does not require $\pi_j(t) = \pi(t)$, but the proof given here requires that S be continuous. The proof uses the following martingale inequality from Meyer (1966, p. 81), which we state without proof.

Lemma 3.2.2. If $\{X(s) : s \geq 0\}$ is a submartingale,

$$P\left\{\sup_{0\leq s\leq t} X(s) \geq \frac{1}{\beta}\right\} \leq \beta E \mid X(t) \mid . \qquad \square$$

Theorem 3.2.4. Define $T = \sup\{t : Y(t) > 0\}$, and assume F is continuous. For any $\beta > 0$,

$$P\{\hat{S} \leq \frac{1}{\beta} S \text{ on } [0, T]\} \geq 1 - \beta.$$

Proof. For any $t \leq 0$, define

$$\begin{aligned} Z(t) &\equiv \frac{\hat{S}(t \wedge T)}{S(t \wedge T)} \\ &= 1 - \int_0^{t\wedge T} \frac{\hat{S}(s-)}{S(s)} I_{\{\overline{Y}(s)>0\}} \left\{\frac{d\overline{N}(s)}{\overline{Y}(s)} - d\Lambda(s)\right\}. \end{aligned}$$

Then $\{Z(s) : 0 \leq s \leq t\}$ is a mean one martingale on $[0, t]$ as long as $S(t) > 0$. By Lemma 3.2.2,

$$\begin{aligned} P\left\{\sup_{0\leq s\leq t} Z(s) \geq \frac{1}{\beta}\right\} &\leq \beta E \mid Z(t) \mid \\ &= \beta E Z(t) \\ &= \beta E Z(0) \\ &= \beta. \end{aligned}$$

Thus,

$$P\{\hat{S} \leq \frac{1}{\beta} S \text{ on } [0, t \wedge T]\} \geq 1 - \beta.$$

Let $\tau = \sup\{t : S(t) > 0\}$. For any $t < \tau$,

$$P\{\hat{S} \leq \frac{1}{\beta} S \text{ on } [0, T]\} \geq P\{\hat{S} \leq \frac{1}{\beta} S \text{ on } [0, t \wedge T]\} - P\{t \leq T\}.$$

However,

$$\begin{aligned} P\{t \le T\} &= 1 - P\{T < t\} \\ &= 1 - \prod_{j=1}^{n} P\{X_j < t\} \\ &\le 1 - \prod_{j=1}^{n} P\{T_j < t\} \\ &= 1 - \{F(t)\}^n. \end{aligned}$$

Choose any $\epsilon > 0$. By taking t sufficiently close to τ, we have

$$P\{\hat{S} \le \frac{1}{\beta} S \text{ on } [0, T]\} \ge 1 - \beta - \epsilon. \qquad \square$$

3.3 SOME FINITE SAMPLE PROPERTIES OF LINEAR RANK STATISTICS

Martingale methods for censored data are useful when studying finite-sample properties of test statistics as well as estimators, and we will illustrate their use here for a class of two-sample tests. We continue to use the general random censorship model (Definition 3.1.1), but now with $r = 2$.

Suppose K is a bounded nonnegative predictable process adapted to $\{\mathcal{F}_t : t \ge 0\}$. Assume as well that

$$\overline{Y}_1(t) \wedge \overline{Y}_2(t) = 0 \text{ implies } K(t) = 0. \tag{3.1}$$

We shall study statistics W_K of the "class $\mathcal{K}$" (Gill, 1980),

$$W_K = \int_0^\infty K(s)\{d\hat{\Lambda}_1(s) - d\hat{\Lambda}_2(s)\},$$

used in testing the hypothesis $H : \Lambda_1 = \Lambda_2$. These statistics also can be written as

$$\begin{aligned} W_K &= \int_0^\infty K(s)\frac{d\overline{N}_1(s)}{\overline{Y}_1(s)} - \int_0^\infty K(s)\frac{d\overline{N}_2(s)}{\overline{Y}_2(s)} \qquad (3.2)\\ &= \int_0^\infty \frac{K(s)}{\overline{Y}_1(s)} dM_1(s) - \int_0^\infty \frac{K(s)}{\overline{Y}_2(s)} dM_2(s) + \int_0^\infty K(s)\{d\Lambda_1(s) - d\Lambda_2(s)\}. \end{aligned}$$

From Chapter 0, we know that the logrank statistic is of the class $\mathcal{K}$, with

$$K(s) = \left\{\frac{n_1 n_2}{n_1 + n_2}\right\}^{1/2} \frac{\overline{Y}_1(s)}{n_1} \frac{\overline{Y}_2(s)}{n_2} \frac{n_1 + n_2}{\overline{Y}_1(s) + \overline{Y}_2(s)}. \tag{3.3}$$

Since (3.1) implies the process $K(s)$ for any statistic of the class $\mathcal{K}$ can be written as

$$K(s) = W(s)\left\{\frac{n_1 n_2}{n_1 + n_2}\right\}^{1/2} \frac{\overline{Y}_1(s)}{n_1} \frac{\overline{Y}_2(s)}{n_2} \frac{n_1 + n_2}{\overline{Y}_1(s) + \overline{Y}_2(s)}, \tag{3.4}$$

these statistics will also be called weighted logrank statistics. The weighted logrank statistics have close ties to classical test statistics based on ranks, as the next example shows.

Example 3.3.1. The Wilcoxon rank-sum statistic is often used to test for equality of underlying distributions F_1 and F_2 in two samples of uncensored data. Let $T_1, T_2, \ldots, T_{n_1+n_2}$ denote a pooled sample $T_{11}, T_{12}, \ldots, T_{1n_1}$, $T_{21}, T_{22}, \ldots, T_{2n_2}$, and take $T_{(1)} < T_{(2)} < \ldots < T_{(n_1+n_2)}$ to be the order statistics of the combined sample, assuming there are no tied values. The rank R_j of item j in the first sample is defined by $R_j = i$ if $T_{1j} = T_{(i)}$. Under the hypothesis $H : F_1 = F_2$, the Wilcoxon rank-sum statistic, $\sum_{j=1}^{n_1} R_j$, has mean $n_1(n+1)/2$ and variance $n_1 n_2 (n+1)/12$, where $n = n_1 + n_2$. Tests of H can be based on the exact distribution of the Wilcoxon statistic in small samples, and on its approximately normal distribution in moderate to large samples.

A transformation of the Wilcoxon statistic can also be written as a weighted logrank test, with a simple weight function. After some algebra, the statistic

$$W_G = \int_0^\infty \frac{\overline{Y}(s)}{n+1} \frac{\overline{Y}_1(s)\overline{Y}_2(s)}{\overline{Y}(s)} \left\{\frac{d\overline{N}_1(s)}{\overline{Y}_1(s)} - \frac{d\overline{N}_2(s)}{\overline{Y}_2(s)}\right\}$$

reduces to

$$\begin{aligned}
&\int_0^\infty \frac{\overline{Y}(s)}{n+1} \left\{d\overline{N}_1(s) - \frac{\overline{Y}_1(s)}{\overline{Y}(s)} d\overline{N}(s)\right\} \\
&\qquad = (n+1)^{-1} \int_0^\infty \{\overline{Y}(s) d\overline{N}_1(s) - \overline{Y}_1(s) d\overline{N}(s)\} \\
&\qquad = (n+1)^{-1} \left[\int_0^\infty \overline{Y}(s) d\overline{N}_1(s) - \{n_1 - \overline{N}_1(s-)\} d\overline{N}(s)\right].
\end{aligned}$$

By Formula (2.8) for integration by parts,

$$\int_0^\infty \overline{N}_1(s-) d\overline{N}(s) = n_1 n - \int_0^\infty \overline{N}(s) d\overline{N}_1(s),$$

and W_G further reduces to

$$\begin{aligned}
&(n+1)^{-1} \left[\int_0^\infty \{\overline{Y}(s) - \overline{N}(s)\} d\overline{N}_1(s)\right] \\
&\qquad = (n+1)^{-1} \left[\int_0^\infty \{\overline{Y}(s) + \overline{Y}(s+) - n\} d\overline{N}_1(s)\right].
\end{aligned}$$

At a failure time s in uncensored data, $\Delta\overline{N}_1(s) = 1$, $\overline{Y}(s) = n - i + 1$, and $\overline{Y}(s+) = n - i$ if and only if there is a $T_{1j} = s$ with $R_j = i$. The last line above thus equals

$$(n+1)^{-1}\sum_{j=1}^{n_1}(n+1-2R_j) = 2(n+1)^{-1}\left\{\frac{n_1(n+1)}{2} - \sum_{j=1}^{n_1} R_j\right\},$$

which is a multiple of the centered rank-sum statistic.

The first generalization of the Wilcoxon statistic to censored data was proposed by Gehan (1965) and, although derived quite differently, (Exercise 3.2), his statistic reduces in uncensored data to W_G. Another approach may be used to generalize W_G to censored data, however. In uncensored data, the empirical survivor function $\hat{S}$ based on the combined data has left-continuous version

$$\begin{aligned}\hat{S}(t-) &= \frac{1}{n}\sum_{i=1}^{n} I_{\{T_i \geq t\}} \\ &= \frac{1}{n}\overline{Y}(t) \\ &\approx \frac{1}{n+1}\overline{Y}(t),\end{aligned}$$

and a reasonable censored data version of W_G replaces $(n+1)^{-1}\overline{Y}$ by the left-continuous version of the Kaplan-Meier estimator, $\hat{S}^-$, yielding

$$W_P = \int_0^\infty \hat{S}(s-)\frac{\overline{Y}_1(s)\overline{Y}_2(s)}{\overline{Y}(s)}\left\{\frac{d\overline{N}_1(s)}{\overline{Y}_1(s)} - \frac{d\overline{N}_2(s)}{\overline{Y}_2(s)}\right\}.$$

Statistics closely related to W_P were proposed by Peto and Peto (1972) and Prentice (1978), again using approaches quite different from stochastic integrals. □

There are formal derivations of W_G and W_P based on modifications of the theory of rank tests which suggest that these statistics yield efficient tests in some situations, but we do not give those derivations here. Instead, we will study their asymptotic efficiencies in Chapter 7 using martingale representations. W_P will be studied there as a member of a family of censored data rank tests $\{G^\rho : \rho \geq 0\}$ (Harrington and Fleming, 1982), where

$$G^\rho = \left[\frac{n_1+n_2}{n_1n_2}\right]^{1/2}\int_0^\infty \{\hat{S}(s-)\}^\rho\frac{\overline{Y}_1(s)\overline{Y}_2(s)}{\overline{Y}(s)}\left\{\frac{d\overline{N}_1(s)}{\overline{Y}_1(s)} - \frac{d\overline{N}_2(s)}{\overline{Y}_2(s)}\right\}.$$

In contrast to the G^ρ statistic, the Gehan-Wilcoxon statistic is given by

$$GW = \left[\frac{n_1+n_2}{n_1n_2}\right]^{1/2}\int_0^\infty \frac{\overline{Y}_1(s)\overline{Y}_2(s)}{n_1+n_2}\left\{\frac{d\overline{N}_1(s)}{\overline{Y}_1(s)} - \frac{d\overline{N}_2(s)}{\overline{Y}_2(s)}\right\}.$$

The next result gives an expression for finite-sample moments of statistics in the class $\mathcal{K}$ and, without any additional work, for the covariance between two members of $\mathcal{K}$ under the hypothesis $\Lambda_1 = \Lambda_2$. This result will prove useful later when we study the joint asymptotic distribution of statistics in $\mathcal{K}$.

Theorem 3.3.1. Consider the random censorship model and two statistics W_{K_ℓ} of the class $\mathcal{K}$, where

$$W_{K_\ell} \equiv \int_0^\infty K_\ell(s)\left\{\frac{d\overline{N}_1(s)}{\overline{Y}_1(s)} - \frac{d\overline{N}_2(s)}{\overline{Y}_2(s)}\right\}, \quad \ell = 1, 2.$$

Then

1.
$$EW_{K_\ell} = \int_0^\infty E\left[K_\ell(s)\{d\Lambda_1(s) - d\Lambda_2(s)\}\right].$$

Under $H_0 : \Lambda_1 = \Lambda_2$ (= Λ unspecified),

2.
$$E\left(W_{K_\ell}\right)^2 = E\sum_{i=1}^{2}\int_0^\infty \frac{K_\ell^2(s)}{\overline{Y}_i(s)}\{1 - \Delta\Lambda(s)\}d\Lambda(s),$$

and

3.
$$\text{cov}\ (W_{K_1} W_{K_2}) = E\sum_{i=1}^{2}\int_0^\infty \frac{K_1(s)K_2(s)}{\overline{Y}_i(s)}\{1 - \Delta\Lambda(s)\}d\Lambda(s).$$

Proof. Part (1) follows immediately from expression (3.2) for W_{K_ℓ}. Under H_0,

$$W_{K_\ell} = \int_0^\infty \sum_{j=1}^{n_1} \frac{K_\ell(s)}{\overline{Y}_1(s)} dM_{1j}(s) - \int_0^\infty \sum_{j=1}^{n_2} \frac{K_\ell(s)}{\overline{Y}_2(s)} dM_{2j}(s),$$

and (2) and (3) are consequences of Theorem 2.6.2. □

Theorem 3.3.1 can be applied to the Gehan-Wilcoxon (GW) and G^ρ statistics. For economy of notation, we drop the variable of integration in the expressions below. First, regardless of Λ_1 and Λ_2,

1.
$$EG^\rho = \int_0^\infty \left(\frac{n_1 n_2}{n_1 + n_2}\right)^{1/2} E\left\{(\hat{S}-)^\rho \frac{\overline{Y}_1}{n_1}\frac{\overline{Y}_2}{n_2}\frac{n_1 + n_2}{\overline{Y}_1 + \overline{Y}_2}\right\}(d\Lambda_1 - d\Lambda_2).$$

Under $H_0 : \Lambda_1 = \Lambda_2$,

2.
$$EG^\rho = 0 = E\ GW,$$

3.
$$E\{G^\rho\}^2 = E\int_0^\infty \frac{n_1 n_2}{n_1+n_2}(\hat{S}-)^{2\rho}\frac{\overline{Y}_1^2}{n_1^2}\frac{\overline{Y}_2^2}{n_2^2} \tag{3.5}$$
$$\frac{(n_1+n_2)^2}{(\overline{Y}_1+\overline{Y}_2)^2}\left\{\frac{1}{\overline{Y}_1}+\frac{1}{\overline{Y}_2}\right\}(1-\Delta\Lambda)d\Lambda$$
$$= E\int_0^\infty (\hat{S}-)^{2\rho}\frac{\overline{Y}_1}{n_1}\frac{\overline{Y}_2}{n_2}\frac{n_1+n_2}{\overline{Y}_1+\overline{Y}_2}(1-\Delta\Lambda)d\Lambda,$$

and

4.
$$\text{cov}(GW, G^\rho) = E\left[\int_0^\infty \frac{n_1 n_2}{n_1+n_2}(\hat{S}-)^{\rho}\frac{\overline{Y}_1^2}{n_1^2}\frac{\overline{Y}_2^2}{n_2^2}\right.$$
$$\left.\frac{n_1+n_2}{\overline{Y}_1+\overline{Y}_2}\left\{\frac{1}{\overline{Y}_1}+\frac{1}{\overline{Y}_2}\right\}(1-\Delta\Lambda)d\Lambda\right] \tag{3.6}$$
$$= E\int_0^\infty \{\hat{S}-\}^{\rho}\frac{\overline{Y}_1}{n_1}\frac{\overline{Y}_2}{n_2}(1-\Delta\Lambda)d\Lambda.$$

These expressions for moments suggest the form of the asymptotic moments of weighted logrank statistics. Assume

$$n = n_1 + n_2 \to \infty \text{ such that } n_i/n \to a_i \in (0,1); i = 1,2.$$

From Eq. (3.5) we would expect G^ρ to converge in distribution to a $N(0,\sigma^2)$ variate with

$$\sigma^2 = \int_0^\infty \{S(s-)\}^{2\rho}\frac{\pi_1(s)\ \pi_2(s)}{a_1\pi_1(s)+a_2\pi_2(s)}\{1-\Delta\Lambda(s)\}d\Lambda(s).$$

From Eq. (3.6), we might also anticipate that the joint distribution of $(GW,\ G^\rho)$ would approach a mean zero bivariate normal distribution with

$$\text{cov}(GW,\ G^\rho) = \int_0^\infty \{S(s-)\}^\rho \pi_1(s)\pi_2(s)\{1-\Delta\Lambda(s)\}d\Lambda(s).$$

Recall that under H_0 the variance, $\sigma^2_{(n)}$, of W_K is given by

$$\sigma^2_{(n)} = E\ W_K^2 = E\int_0^\infty \sum_{i=1}^2 \frac{K^2(s)}{\overline{Y}_i(s)}\{1-\Delta\Lambda(s)\}d\Lambda(s).$$

The next result proves that a natural estimator of $\sigma^2_{(n)}$ is unbiased.

Theorem 3.3.2. Let

$$\hat{\sigma}^2 = \int_0^\infty \sum_{i=1}^2 \frac{K^2(s)}{\overline{Y}_i(s)}\left\{1-\frac{\Delta\overline{N}_1(s)+\Delta\overline{N}_2(s)-1}{\overline{Y}_1(s)+\overline{Y}_2(s)-1}\right\}\frac{d\{\overline{N}_1(s)+\overline{N}_2(s)\}}{\overline{Y}_1(s)+\overline{Y}_2(s)}. \tag{3.7}$$

Then

$$E\,\hat{\sigma}^2 = \sigma^2_{(n)}. \qquad \square$$

When K is the logrank weight function in Eq. (3.3), $\hat{\sigma}^2$ is the variance estimator (Exercise 3.3) proposed by Mantel (1966) based on the hypergeometric conditional distribution of Fisher's exact test.

Proof of Theorem 3.3.2. Suppressing the variable of integration, we have

$$E\{\,\hat{\sigma}^2 - \sigma^2_{(n)}\} =$$

$$E\int_0^\infty \frac{K^2}{\overline{Y}_1\overline{Y}_2}\left\{\left(1 - \frac{\Delta\overline{N}_1 + \Delta\overline{N}_2 - 1}{\overline{Y}_1 + \overline{Y}_2 - 1}\right)(d\overline{N}_1 + d\overline{N}_2) - (\overline{Y}_1 + \overline{Y}_2)(1 - \Delta\Lambda)d\Lambda\right\}.$$

The right side above reduces to

$$E\int_0^\infty \frac{K^2}{\overline{Y}_1\overline{Y}_2}\left(d\overline{N} - \overline{Y}d\Lambda\right) -$$

$$E\int_0^\infty \frac{K^2}{\overline{Y}_1\overline{Y}_2(\overline{Y}_1 + \overline{Y}_2 - 1)}\{(\Delta\overline{N} - 1)d\overline{N} - \overline{Y}(\overline{Y} - 1)\Delta\Lambda d\Lambda\},$$

where $\overline{N} \equiv \overline{N}_1 + \overline{N}_2$, and $\overline{Y} \equiv \overline{Y}_1 + \overline{Y}_2$. Since expression (2.4) is a zero mean martingale, it follows that both terms on the right-hand side of the above equality are expectations of zero mean martingales, and that $E\,\hat{\sigma}^2 = \sigma^2_{(n)}$. $\square$

3.4 CONSISTENCY OF THE KAPLAN-MEIER ESTIMATOR

This chapter has emphasized expressions for the first and second moments of censored data statistics that do not rely on asymptotic approximations. On a few occasions, finite-sample means and variances have suggested the form of asymptotic moments, which, when the statistics are asymptotically normal, characterize their asymptotic distributions. While a thorough examination of the large-sample behavior of the tests and estimators considered here must be delayed until after the martingale central limit theorems in Chapter 5, the results in this chapter are nearly sufficient to study one large-sample property of the Nelson and Kaplan-Meier estimators, namely consistency. We will need one additional result, based on an inequality due to Lenglart.

An estimator of a finite-dimensional parameter is consistent if the Euclidean distance between the parameter and its estimator converges to zero in probability (or with probability one, for strong consistency) as the sample size becomes infinite. A nonparametric estimator of a function, such as a survival or cumulative hazard function, will be consistent if the distance in some metric between the function and its estimator converges to zero in probability. A natural metric to use is based on the supremum norm, and an estimator consistent in the supremum norm is called uniformly consistent.

Definition 3.4.1. Let f be a function defined on an interval $[a,b]$, where $-\infty \le a < b \le \infty$. The *supremum norm* of f is

$$\sup_{a\le s\le b} \mid f(s) \mid,$$

and the *distance* induced by this norm between two functions f and g is

$$\sup_{a\le s\le b} \mid f(s) - g(s) \mid . \qquad \square$$

The supremum norm has both practical and technical appeal in this setting. Practically speaking, a data analyst is more often interested in the behavior of, say, a survivor function estimator over an interval $[0,t]$, rather than at one or two time points. An estimator consistent in this norm will be uniformly "close" to the true survivor function on the whole interval for a sufficiently large sample. In Chapter 7, when we examine the large-sample distribution theory for some censored data rank tests using martingale central limit theorems, convergence of the product limit estimator in the supremum norm will be an important result.

We begin with a useful inequality due to Lenglart (Lenglart, 1977), stated without proof, and a corollary. A proof of the inequality may be found in Jacod and Shiryayev (1987, Lemma 3.30), and in Shorack and Wellner (1986, Inequality B.4.1).

Theorem 3.4.1 (Lenglart's Inequality). Let X be a right-continuous adapted process, and Y a nondecreasing predictable process with $Y(0) = 0$. Suppose, for all bounded stopping times T,

$$E\{|X(T)|\} \le E\{Y(T)\}.$$

Then for any stopping time T, and any $\epsilon, \eta > 0$,

$$P\left\{\sup_{t\le T} |X(t)| \ge \epsilon\right\} \le \frac{\eta}{\epsilon} + P\{Y(T) \ge \eta\}. \qquad \square$$

We will also use the following corollary:

Corollary 3.4.1. Let N be a counting process, and $M = N - A$ the corresponding local square integrable martingale. Suppose H is an adapted left-continuous process with right-hand limits or, more generally, a predictable and locally bounded precess. Then for any stopping time T such that $P\{T < \infty\} = 1$, and any $\epsilon, \eta > 0$,

$$P\left\{\sup_{t\le T}\left\{\int_0^t H(s)dM(s)\right\}^2 \ge \epsilon\right\} \le \frac{\eta}{\epsilon} + P\left\{\int_0^T H^2(s)d\langle M, M\rangle(s) \ge \eta\right\}.$$

Proof. Let $\{\tau_k : k = 1, 2, \ldots\}$ be a localizing sequence such that, for any k, $N(\cdot \wedge \tau_k)$, $A(\cdot \wedge \tau_k)$, and $H(\cdot \wedge \tau_k)$ are processes bounded by k (recall that A is always locally bounded), and $M(\cdot \wedge \tau_k)$ is a square integrable martingale.

By the Optional Stopping Theorem (Theorem 2.2.2) and Theorem 2.4.2,

$$E\{X_k(t \wedge T) - Y_k(t \wedge T)\} = 0 \text{ for any } t \geq 0, \tag{4.1}$$

where

$$X_k(t) \equiv \left\{ \int_0^{t \wedge \tau_k} H(s) dM(s) \right\}^2$$

and

$$Y_k(t) \equiv \int_0^{t \wedge \tau_k} H^2(s) d\langle M, M \rangle(s).$$

As $t \to \infty$, $X_k(t \wedge T) \to X_k(T)$ a.s., so, by the Dominated Convergence Theorem, $EX_k(t \wedge T) \to EX_k(T)$ and $EX_k(T) < \infty$. As $t \to \infty$, $Y_k(t \wedge T) \uparrow Y_k(T)$ a.s., so, by the Monotone Convergence Theorem, $EY_k(t \wedge T) \uparrow EY_k(T)$ and $EY_k(T) < \infty$. Thus, Eq. (4.1) implies that, for any k,

$$E\ X_k(T) = E\ Y_k(T).$$

By Lenglart's Inequality,

$$\begin{aligned} P_{1k} &\equiv P\left\{ \sup_{t \leq T} \left\{ \int_0^{t \wedge \tau_k} H(s) dM(s) \right\}^2 \geq \epsilon \right\} \\ &\leq \frac{\eta}{\epsilon} + P\left\{ \int_0^{T \wedge \tau_k} H^2(s) d\langle M, M \rangle(s) \geq \eta \right\} \\ &\equiv \frac{\eta}{\epsilon} + P_{2k}. \end{aligned}$$

By the Monotone Convergence Theorem, as $k \to \infty$,

$$P_{2k} \uparrow P\left\{ \int_0^T H^2(s) d\langle M, M \rangle(s) \geq \eta \right\} \equiv P_2,$$

so $P_{1k} \leq \eta/\epsilon + P_2$ for any k. By the Dominated Convergence Theorem, as $k \to \infty$,

$$P_{1k} \to P\left\{ \sup_{t \leq T} \left\{ \int_0^t H(s) dM(s) \right\}^2 \geq \epsilon \right\},$$

establishing the corollary. □

The Glivenko-Cantelli Theorem (Chung, 1974, Theorem 5.5.1) states that the empirical distribution function is a consistent estimator, uniformly on the whole real line, of an underlying arbitrary cumulative distribution function. While the

Kaplan-Meier estimator reduces to the empirical cumulative distribution function in uncensored data, it would be unreasonable to expect uniform consistency on the whole real line, regardless of censoring or failure time distributions. If $C(t) = 0$ but $S(t) > 0$ for some time t, and if $\delta > 0$, there will never be items at risk for at least $t + \delta$ time units, and only a correct parametric model could extrapolate from information at earlier times to estimate $S(t + \delta)$. The same will be true, of course, for the Nelson estimator. Indeed, Theorem 3.2.1 and Lemma 3.2.1 can be used to show that the estimators $\hat{S}$ and $\hat{\Lambda}$ satisfying $\hat{S}(t) = \hat{S}(t \wedge T)$ and $\hat{\Lambda}(t) = \hat{\Lambda}(t \wedge T)$ are asymptotically biased at any point $t+\delta$ for which $C(t) = 0$ but $S(t) > S(t+\delta)$.

It is natural to expect uniform consistency of $\hat{S}$ and $\hat{\Lambda}$ over $[0, t]$ for any $t < u$, where $u = \sup\{t : P\{X > t\} > 0\}$. Theorem 3.4.2 states that, subject to the continuity of S, uniform consistency of $\hat{S}$ actually holds over $[0, u]$.

Theorem 3.4.2 contains a stronger result than needed right now, namely, that consistency is uniform even when the underlying survivor or distribution functions are indexed by n. This result will be useful later in the treatment of asymptotic efficiencies of tests under sequences of alternatives converging to a null distribution. For clarity, we separate the simple and more general results in the statement of the theorem but, for economy, we give the proof for only the general result.

Theorem 3.4.2.

1. Let T be a failure time random variable with continuous distribution function $F(s) = P\{T \leq s\}$ and cumulative hazard function $\Lambda(s) = \int_0^s dF(v)/\{1 - F(v)\}$.
   a) If $t \in (0, \infty]$ is such that
   $$\overline{Y}(t) \xrightarrow{P} \infty \quad \text{as } n \to \infty, \tag{4.2}$$
   then
   $$\sup_{0 \leq s \leq t} \left| \int_0^s \frac{d\overline{N}(v)}{\overline{Y}(v)} - \Lambda(s) \right| \xrightarrow{P} 0 \quad \text{as } n \to \infty, \tag{4.3}$$
   and
   $$\sup_{0 \leq s \leq t} |\hat{F}(s) - F(s)| \xrightarrow{P} 0 \quad \text{as } n \to \infty, \tag{4.4}$$
   where $1 - \hat{F}$ is the Kaplan-Meier estimator.
   b) If $u \in (0, \infty]$ is such that (4.2) holds for all $t < u$, then (4.4) holds with t replaced by u.
2. Suppose the underlying continuous failure time distributions are indexed by increasing sample size n, i.e., $P\{T_j^n \leq s\} = F^n(s)$ for $j = 1, \ldots, n$.
   a) When Condition (4.2) holds, and when
   $$\limsup_{n \to \infty} F^n(t) < 1, \tag{4.5}$$

then, as $n \to \infty$,

$$\sup_{0 \le s \le t} \left| \int_0^s \frac{d\overline{N}(v)}{\overline{Y}(v)} - \Lambda^n(s) \right| \xrightarrow{P} 0 \tag{4.6}$$

and

$$\sup_{0 \le s \le t} |\hat{F}(s) - F^n(s)| \xrightarrow{P} 0. \tag{4.7}$$

b) If $u \in (0, \infty]$ is such that (4.2) and (4.5) hold for all $t < u$, and F is a continuous distribution function such that, for all $t \le u$,

$$\lim_{n \to \infty} F^n(t) = F(t), \tag{4.8}$$

then (4.7) holds with t replaced by u.

□

Figure 3.4.1 shows examples in which (4.2) fails to hold at u (in both Figure 3.4.1(a) and (b)) and (4.5) fails to hold (in Figure 3.4.1(b)).

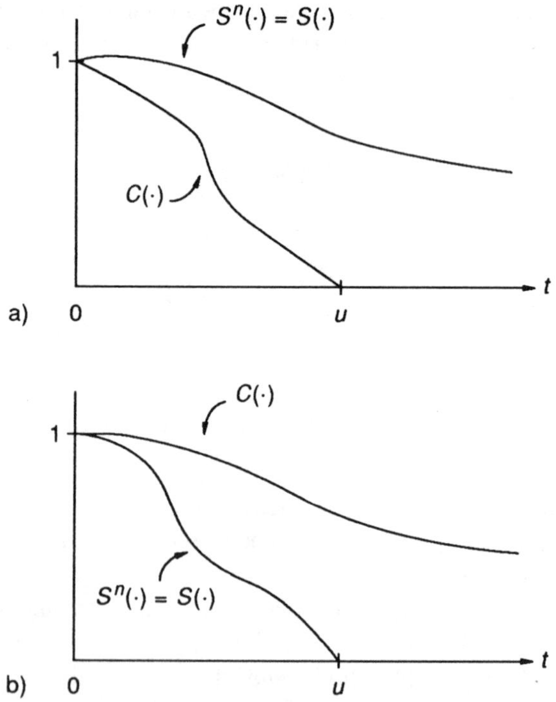

Figure 3.4.1 Equation (4.2) fails to hold at $t = u$ in both (a) and (b); inequality (4.5) is not true in (b).

When F^n does not depend on n, (4.2) implies that $F(t) < 1$, and Condition (4.5) is not needed. When F^n depends on n however, it is possible (Exercise 3.4) for (4.2) to be true and that

$$\limsup_{n\to\infty} F^n(t) = 1.$$

Proof. It is sufficient to prove (2), and we begin with (2)(a). Let $s \in [0, t]$. Then

$$\begin{aligned}
\left|\int_0^s \frac{d\overline{N}(v)}{\overline{Y}(v)} - \Lambda^n(s)\right| &\le \left|\int_0^s \frac{d\overline{N}(v)}{\overline{Y}(v)} - \int_0^s I_{\{\overline{Y}(v)>0\}}\, d\Lambda^n(v)\right| \\
&\quad + \left|\int_0^s I_{\{\overline{Y}(v)=0\}}\, d\Lambda^n(v)\right| \\
&\le \left|\int_0^s \frac{I_{\{\overline{Y}(v)>0\}}}{\overline{Y}(v)}\, dM(v)\right| + I_{\{\overline{Y}(t)=0\}}\Lambda^n(t).
\end{aligned}$$

By (4.2) and (4.5), $I_{\{\overline{Y}(t)=0\}}\Lambda^n(t) \xrightarrow{P} 0$, so to show (4.6) it suffices to show

$$\sup_{0\le s\le t}\left\{\int_0^s \frac{I_{\{\overline{Y}(v)>0\}}}{\overline{Y}(v)}\, dM(v)\right\}^2 \xrightarrow{P} 0.$$

By Corollary 3.4.1 to Lenglart's Inequality,

$$\begin{aligned}
P&\left\{\sup_{0\le s\le t}\left\{\int_0^s \frac{I_{\{\overline{Y}(v)>0\}}}{\overline{Y}(v)}\, dM(v)\right\}^2 \ge \epsilon\right\} \\
&\le \frac{\eta}{\epsilon} + P\left\{\int_0^t \frac{I_{\{\overline{Y}(v)>0\}}}{\overline{Y}(v)}\, d\Lambda^n(v) \ge \eta\right\} \\
&\le \frac{\eta}{\epsilon} + P\left\{\Lambda^n(t)/\overline{Y}(t) > \eta\right\}.
\end{aligned}$$

Conditions (4.2) and (4.5) imply that the second term on the right-hand side converges to zero as $n \to \infty$ for any $\eta > 0$, and convergence of the Nelson estimator is established.

For any fixed t such that (4.2) and (4.5) hold, Theorem 3.2.3 implies

$$P\left\{\frac{\hat{F} - F^n}{1 - F^n} = Z \text{ on } [0, t]\right\} \to 1 \text{ as } n \to \infty,$$

where

$$Z = \int \frac{\{1 - \hat{F}(v-)\}}{\{1 - F^n(v)\}} \frac{I_{\{\overline{Y}(v)>0\}}}{\overline{Y}(v)}\, dM(v).$$

Condition (4.5) implies

$$\liminf_{n\to\infty} \inf_{s\in[0,t]} \{1 - F^n(s)\} = \liminf_{n\to\infty} [1 - F^n(t)]$$
$$> 0.$$

Hence, to show $\sup_{0\le s\le t} |\hat{F}(s) - F^n(s)| \xrightarrow{P} 0$, it suffices to show

$$\sup_{0\le s\le t} \{Z(s)\}^2 \xrightarrow{P} 0.$$

By Corollary 3.4.1,

$$P\left\{\sup_{0\le s\le t} \{Z(s)\}^2 \ge \epsilon\right\} \le \frac{\eta}{\epsilon} + P\left\{\int_0^t \frac{\{1-\hat{F}(v-)\}^2}{\{1-F^n(v)\}^2} \frac{I_{\{\overline{Y}(v)>0\}}}{\overline{Y}(v)} d\Lambda^n(v) \ge \eta\right\}$$
$$\le \frac{\eta}{\epsilon} + P\left\{\frac{\Lambda^n(t)}{\{1-F^n(t)\}^2\overline{Y}(t)} \ge \eta\right\}.$$

Again, (4.2) and (4.5) imply the second term on the right-hand side above converges to zero as $n \to \infty$ for any $\eta > 0$. Since ϵ and η are arbitrary, (4.7) holds.

Suppose now u is such that (4.2) and (4.5) hold for all $t < u$, and that, in addition, Eq. (4.8) holds. To establish (2)(b) we must show

$$\sup_{0\le s\le u} |\hat{F}(s) - F^n(s)| \xrightarrow{P} 0 \text{ as } n \to \infty, \tag{4.9}$$

or, equivalently, for any $\epsilon > 0$, that

$$P\left\{\sup_{0\le s\le u} |\hat{F}(s) - F^n(s)| > \epsilon\right\} < \epsilon \tag{4.10}$$

for sufficiently large n.

For any point $t_0 < u$, part (2)(a) establishes uniform consistency on $[0, t_0]$, so the proof of (2)(b) depends on finding a particular $t_0 < u$ for which we can show uniform convergence on $[t_0, u]$. Since

$$|F^n(u) - F^n(t_0)| \le |F^n(u) - F(u)| + |F(u) - F(t_0)|$$
$$+ |F(t_0) - F^n(t_0)|,$$

the continuity of F and Eq. (4.8) imply the existence of a time point $t_0, 0 < t_0 < u$, and an integer $n' \ge 1$ such that

$$|F^n(u) - F^n(t_0)| < \epsilon/4$$

for all $n \ge n'$.

By Eq. (4.7), there is a second integer n'' such that

$$P\{|\hat{F}(t_0) - F^n(t_0)| > \epsilon/4\} < \epsilon/2 \tag{4.11}$$

for all $n \ge n''$.

For $t_0 \le s \le u$, the inequalities

$$\hat{F}(t_0) \le \hat{F}(s) \le \hat{F}(u)$$

and

$$F^n(t_0) \le F^n(s) \le F^n(u)$$

imply that

$$|\hat{F}(s) - F^n(s)| \le |\hat{F}(u) - F^n(t_0)| + |\hat{F}(t_0) - F^n(u)|.$$

Consequently,

$$\begin{aligned}\sup_{t_0 \le s \le u} |\hat{F}(s) - F^n(s)| &\le |\hat{F}(u) - F^n(t_0)| + |\hat{F}(t_0) - F^n(u)| \\ &\le |\hat{F}(u) - F^n(u)| + 2|F^n(u) - F^n(t_0)| \\ &\quad + |\hat{F}(t_0) - F^n(t_0)|.\end{aligned}$$

If $|\hat{F}(u) - F^n(u)| \to 0$ in probability, there will exist a third integer n''' so that for all $n \ge n'''$,

$$P\{|\hat{F}(u) - F^n(u)| > \epsilon/4\} < \epsilon/2.$$

Then, for $n \ge \max(n', n'', n''')$,

$$\begin{aligned}P\left\{\sup_{t_0 \le s \le u} |\hat{F}(s) - F^n(s)| > \epsilon\right\} &\le \epsilon/2 + \epsilon/2 \\ &= \epsilon.\end{aligned}$$

To complete the proof, we must show then that $|\hat{F}(u) - F^n(u)| \xrightarrow{P} 0$, and we consider separately the two cases $F(u) = 1$ and $F(u) < 1$. When $F(u) = 1$, there exist t_0 and n' such that $|F^n(t_0) - 1| < \epsilon/3$ for any $n \ge n'$. Now,

$$\begin{aligned}|\hat{F}(u) - F^n(u)| &\le |1 - \hat{F}(u)| + |1 - F^n(u)| \\ &\le |1 - \hat{F}(t_0)| + |1 - F^n(t_0)| \\ &\le 2|1 - F^n(t_0)| + |F^n(t_0) - \hat{F}(t_0)|.\end{aligned}$$

Since $|F^n(t_0) - \hat{F}(t_0)| \xrightarrow{P} 0$, there exists $n^* > n'$ such that, for any $n \ge n^*$, $P\{|F^n(t_0) - \hat{F}(t_0)| > \epsilon/3\} < \epsilon$, and

$$|\hat{F}(u) - F^n(u)| \xrightarrow{P} 0.$$

Suppose now $F(u) = 1 - \delta^* < 1$. Since

$$\begin{aligned}|\hat{F}(u) - F^n(u)| &\le |\hat{F}(u) - \hat{F}(t_0)| + |\hat{F}(t_0) - F^n(t_0)| \\ &\quad + |F^n(t_0) - F^n(u)|,\end{aligned} \tag{4.12}$$

Eqs. (4.7) and (4.8), and the continuity of F imply that the last two terms on the right side of the above inequality can be made negligible. We need only show that the first term can be made small, in probability, as $n \to \infty$. There exists n' such that, for any $n \geq n'$,

$$1 - F^n(t) > \delta^*/2 \text{ for any } t \in [0, u].$$

Then, for any $n \geq n'$, and for any $t_0 \leq u$,

$$\begin{aligned}
\int_{t_0}^{u} d\Lambda^n(t) &\leq \int_{t_0}^{u} \frac{dF^n(t)}{1 - F^n(t)} \\
&< \frac{2}{\delta^*} \int_{t_0}^{u} dF^n(t) \\
&= \frac{2}{\delta^*} \{F^n(u) - F^n(t_0)\} \\
&= \frac{2}{\delta^*} \{F^n(u) - F(u) + F(u) - F(t_0) + F(t_0) - F^n(t_0)\}.
\end{aligned}$$

Hence, there exist t_0 and $n'' > n'$ such that, for any $n \geq n''$,

$$\int_{t_0}^{u} d\Lambda^n(t) < \epsilon^3/108. \tag{4.13}$$

Now, for any $n \geq n''$,

$$\begin{aligned}
P\{|\hat{F}(u) - \hat{F}(t_0)| > \epsilon/3\} &= P\left\{\int_{t_0}^{u} I_{\{\overline{Y}(s)>0\}} \{1 - \hat{F}(s-)\} \frac{d\overline{N}(s)}{\overline{Y}(s)} > \epsilon/3\right\} \\
&\leq P\left\{\left|\int_{t_0}^{u} I_{\{\overline{Y}(s)>0\}} \left[\frac{d\overline{N}(s)}{\overline{Y}(s)} - d\Lambda^n(s)\right]\right| > \epsilon/6\right\} \\
&\quad + P\left\{\int_{t_0}^{u} d\Lambda^n(s) > \epsilon/6\right\} \\
&= P\left\{\left[\int_{t_0}^{u} \frac{I_{\{\overline{Y}(s)>0\}}}{\overline{Y}(s)} dM(s)\right]^2 > \epsilon^2/36\right\} \\
&\leq \eta \frac{36}{\epsilon^2} + P\left\{\int_{t_0}^{u} \frac{I_{\{\overline{Y}(s)>0\}}}{\overline{Y}(s)} d\Lambda^n(s) \geq \eta\right\}, \\
&= \epsilon/3,
\end{aligned}$$

where the last inequality follows from Corollary 3.4.1, and the last equality from setting $\eta = \epsilon^3/108$.

We can assume the choice of t_0 in (4.12) satisfies $|F^n(t_0) - F^n(u)| < \epsilon/3$ for any $n \geq n''$. Thus, since $|\hat{F}(t_0) - F^n(t_0)| \xrightarrow{P} 0$, $|\hat{F}(u) - F^n(u)| \xrightarrow{P} 0$, when $F(u) < 1$. □

When $F^n \equiv F$, Theorem 3.4.2 indicates that, if $\overline{Y}(t) \xrightarrow{P} \infty$ as $n \to \infty$ for any $t \in [0, u)$, the Kaplan-Meier estimator $\hat{S}(t)$ is a uniformly consistent estimator of $S(t) = \exp\left\{-\int_0^t \lambda(s)ds\right\}$ over $[0, u]$ as long as Condition (3.1) of Chapter 1 holds. Note that whether or not this condition holds, it follows from Theorem 3.4.2 that $\hat{S}(t)$ is a uniformly consistent estimator of $S^{\#}(t) = \exp\left\{-\int_0^t \lambda^{\#}(s)ds\right\}$ over $[0, u]$, where $\lambda^{\#}$ is given by the right-hand side of Condition (3.1) of Chapter 1.

3.5 BIBLIOGRAPHIC NOTES

Section 3.1

Variations of the random censorship model of Gill (1980) have been used in the literature for some time. Some of the early proofs of properties of censored data statistics assumed that censoring times were fixed and known in advance for all cases, a situation covered by Gill's general treatment. Kaplan and Meier (1958) used this approach. As in Example 3.2.1, censoring times are all hypothetically known in clinical trials with random entry and no loss to follow up.

The reliability and life-testing literature contains extensive studies of Type I and Type II censoring. Type I censoring arises when all items are put on test at the same time, and any failure times larger than a prespecified fixed time are censored. Type I censoring is an example of random censorship. Type II censoring arises when n items in a life test are put under observation at the same time, and any not failing by the rth ordered failure time, $r \leq n$, are censored at the rth failure. Censoring times among items are not independent in Type II censoring but many of the results in this chapter can be established in this case.

Mann, Schafer, and Singpurwalla (1974) discuss in detail models of censorship in engineering and reliability problems, and Gross and Clark (1975) illustrate the uses of these models in biomedical settings. Gill's general treatment (Theorem 3.1.1) of the possible dependence allowed between censoring and failure time observations incorporates Type II and other forms of censoring.

Section 3.2

The Nelson Estimator was first proposed in 1969, and was examined again in a later paper (Nelson, 1972). Since $\Lambda(t)$ is linear when the underlying failure time distribution is exponential, $\hat{\Lambda}(t)$ is often used as a graphical method for checking exponentiality, and that is the original reason it was proposed. The elegant martingale approach used here is due to Aalen (1978), who proposed the estimator for cumulative intensities in the multiplicative intensity model. Because Aalen's results can be used in a variety of special circumstances, the estimator is now sometimes called the Nelson-Aalen estimator. Johansen (1983) has derived the Nelson-Aalen estimator as a generalized maximum likelihood estimator, and Jacobsen (1982, 1984) has shown that the Nelson-Aalen estimator is asymptotically equivalent to a slightly different maximum likelihood estimator from an extended model. Karr (1987) has shown the Nelson-Aalen estimator can be obtained as a maximum likelihood estimator using the method of sieves.

Kaplan and Meier (1958) give an appealing probabilistic derivation of their estimator, and show the estimator can be obtained as a "nonparametric" maximum likelihood estimator.

The optimality of the Kaplan-Meier estimator was first established by Wellner (1982) and, recently, Gill (1989) has shown that the generalized likelihood approach can be used to obtain its efficiencies. Since the properties of generalized maximum likelihood estimators are difficult to establish, we have not given those derivations here. The integral representation in Theorem 3.2.3 for the Kaplan-Meier estimator appeared first in Gill (1980), who points out that it can be derived from an earlier, more general relationship in Aalen and Johansen (1978). Several authors have studied the asymptotic distribution theory for the Kaplan-Meier estimator, and those references will be given in Chapter 6.

Formulas for exact finite-sample moments of the Kaplan-Meier estimator were derived by Chen, et al. (1982) and by Wellner (1985) under the assumption that $S(t) = \{C(t)\}^{\alpha}$.

The estimator $e^{-\hat{\Lambda}}$ for the survivor function S has been studied by Altschuler (1970) and by Fleming and Harrington (1984a), who studied the finite-sample mean squared error of this estimator. Whittemore and Keller (1986) have examined the maximum likelihood estimate of S using splines, and have showed that their estimators are asymptotically equivalent to the Kaplan-Meier, but have higher finite-sample efficiencies.

Life table estimation provided an early approach to estimating S. Various related methods were proposed by Berkson and Gage (1950), Littel (1952), Elveback (1958), Kimball (1960), Chiang (1968) and Crowley (1970). Crowley also provides a comprehensive comparison of the properties of these estimators.

Section 3.3

The rank tests discussed here have been derived using a number of approaches. The methods used have generally adapted derivations of nonparametric statistics based on ranks, and these classical arguments are discussed in detail in Lehmann (1975) and, at a more mathematically advanced level, in Hájek and Šidák (1967) and Hettmansperger (1984). The logrank test was first proposed by Mantel (1966), using the arguments in Chapter 0 about combining statistics from independent contingency tables. Gehan (1965) proposed his version of the modified Wilcoxon statistic by adapting to censored data the Mann-Whitney scoring approach for the two-sample problem. Peto and Peto (1972) adapted the Wilcoxon test to censored data by generalizing methods of Hájek and Šidák for producing efficient rank tests invariant under monotonic transformations.

Prentice (1978) showed how the efficient score function, used to generate locally most powerful and fully efficient rank tests, could be adapted to formulate censored data versions of these tests, and Harrington and Fleming (1982) showed that the G^{ρ} statistics are asymptotically equivalent to statistics that can be derived using Prentice's approach.

Asymptotic theory for censored data rank tests under null and alternative hypotheses is discussed in Chapter 7, and appropriate references are discussed in that chapter.

Section 3.4

Peterson (1977) obtained strong consistency of the Kaplan-Meier estimator at a single point in time. Corollary 3.4.1 was established by Gill (1980) and then used to obtain the consistency results for $\hat{\Lambda}$ and $\hat{S}$ given in Eqs. (4.6) and (4.7) of Theorem 3.4.2(2). An alternative to the proof given here for the uniform consistency of $\hat{S}$ over $[0, u]$ is in Shorack and

Wellner (1986, Section 7.3). Wang (1987, Corollary 1, page 1316) established a modified version of Theorem 3.4.2 (1):

$$\sup_{0 \le s \le u} |\hat{F}(s) - F(s)| \xrightarrow{P} 0$$

as $n \longrightarrow \infty$, if and only if $C(u-) > 0$ and $\Delta F(u) = 0$. The crux of Wang's argument is the continuity of the product limit mapping established in Lemma 1 of Shorack and Wellner (1989).

CHAPTER 4

Censored Data Regression Models and Their Application

4.1 INTRODUCTION

The proportional hazards regression model provides a method for exploring the association of covariates with failure rates and survival distributions and for studying the effect of a primary covariate, such as treatment, while adjusting for other variables. The model is neither fully parametric nor fully nonparametric (it has been called semiparametric), and inference for the model is based on a likelihood type function that, in censored data, only approximates the probability or density function of any observed set of values.

The model has been a markedly successful way of analyzing failure time data for a variety of reasons. Empirically, it seems to provide a reasonable approximation to the association between covariates and failure time outcome variables often seen in medical and engineering studies. With proper restrictions, the covariates may themselves be deterministic or random functions of time, giving the model more flexibility than traditional linear models for analyzing follow-up data. The parameters in the model have simple and direct interpretations and, while the estimating equations for the parameters must be solved iteratively, standard numerical methods work quite well. In some important situations, inference procedures based on the model are nearly as efficient as those for a correct parametric model.

In this chapter we explore methods for obtaining parameter estimates and test statistics using the proportional hazards model and the closely related multiplicative intensity model which allows for recurrent outcome events. We outline the finite-sample and asymptotic distribution theory for these statistics, then illustrate their use in an extended data analysis. The martingale approach is heavily exploited to motivate the form of the statistics studied and to suggest their properties. Indeed, martingales are central to the definition of the multiplicative intensity model. The proofs of the results used here are given in detail in Chapter 8.

This chapter has a somewhat different flavor from the other applied chapters. In Chapters 3, 6, and 7, compensators of counting processes are derived from traditional statistical models for underlying failure time distributions in one or more groups. The proportional hazards model in Section 4.2 uses this approach. In the multiplicative intensity model introduced in Section 4.2, however, compensated counting processes are the starting point for a model, that is, the compensator of a counting process is assumed to have a particular form. This martingale approach to censored data regression models allows general censoring patterns and events that may occur more than once per case.

4.2 THE PROPORTIONAL HAZARDS AND MULTIPLICATIVE INTENSITY MODELS

The data available in regression problems with right-censored failure time data are independent observations on the triple $(X, \delta, \mathbf{Z})$, where X is the minimum of a failure and censoring time pair (T, U), $\delta = I_{\{T \le U\}}$ is the indicator of the event that failure has been observed, and $\mathbf{Z} = (Z_1, \ldots, Z_p)'$ is a p-dimensional column vector of covariates. The important inference questions in this setting are most often about the conditional distribution of failure, given the covariates.

Many parametric methods are available for analyzing failure time regression data with right-censoring. These models specify both the effect of the covariates and the form of baseline hazard rates up to unknown finite-dimensional parameters. Parametric methods for censored data are covered in detail by Kalbfleisch and Prentice (1980, Chapters 2 and 3) and by Lawless (1982, Chapter 6), and we do not add anything here to those treatments. We concentrate instead on semiparametric models in which the form of baseline hazard function is not specified. The martingale methods used to study the semiparametric models can also be used, however, to obtain results for the fully parametric models (cf. Borgan, 1984).

The information contained in the pair (X, δ) is equivalent to that contained in the processes N and Y defined in earlier chapters by

$$N(t) = I_{\{X \le t, \delta = 1\}}$$

and

$$Y(t) = I_{\{X \ge t\}}.$$

This leads to two approaches to censored data regression models, one based on a more traditional use of conditional distributions and hazard functions, and one based on the compensated martingale for N. In most applied settings, the two approaches are equivalent, as we will later see.

The traditional approach was used by Cox (1972). Assume for simplicity that no covariates vary with time, and let $S(t \mid \mathbf{Z})$ be the conditional survival function $P\{T > t \mid \mathbf{Z}\}$. The conditional hazard function is defined by

$$\lambda(t \mid \mathbf{Z}) = \lim_{h \downarrow 0} h^{-1} P\{t \le T < t + h \mid T \ge t, \mathbf{Z}\}.$$

Suppose that for any two covariate values $\mathbf{Z}_1$ and $\mathbf{Z}_2$ the associated conditional failure rates have a fixed ratio over time. Then $\lambda(t \mid \mathbf{Z}_1) = k(\mathbf{Z}_1, \mathbf{Z}_2)\lambda(t \mid \mathbf{Z}_2)$, where k is a nonnegative real valued function depending only on the covariates and not on time. If $\lambda_0(t)$ denotes the conditional hazard rate given $\mathbf{Z} = \mathbf{0}$ (usually called the baseline hazard rate) and $g(\mathbf{Z}) = k(\mathbf{Z}, \mathbf{0})$, then

$$\lambda(t \mid \mathbf{Z}) = \lambda_0(t)g(\mathbf{Z}).$$

Parsimonious models require a simple dependence of g on $\mathbf{Z}$, and so g is almost always taken to depend on $\mathbf{Z}$ through a linear combination $\beta_1 Z_1 + \cdots + \beta_p Z_p$, or $\boldsymbol{\beta}'\mathbf{Z}$, where $\boldsymbol{\beta} = (\beta_1, \ldots, \beta_p)'$ is a p-dimensional column vector of regression coefficients. Taking $g(x) = e^x$ insures that $g \geq 0$ and leads to the most common form of the proportional hazards regression model,

$$\lambda(t \mid \mathbf{Z}) = \lambda_0(t)\exp(\boldsymbol{\beta}'\mathbf{Z}).$$

For small values of Δt, the conditional hazard rate $\lambda(t \mid \mathbf{Z})$ satisfies

$$\lambda(t \mid \mathbf{Z})\Delta t \approx P\{t \leq T < t + \Delta t \mid T \geq t, \mathbf{Z}\},$$

and hence can be interpreted as an approximate conditional probability of observing a failure in $[t, t + \Delta t)$, given $\mathbf{Z}$ and no failure before t. It is not surprising, then, that the model is closely related to the logistic regression model for binary data. If T is a continuous failure time variable, we can "discretize" T by defining an infinite partition $0 = t_0 < t_1 < \ldots < t_m < \ldots$ of $[0, \infty)$ and a variable T^* by

$$T^* = t_i \quad \text{if} \quad t_i \leq T < t_{i+1}, \quad i = 0, 1, \ldots.$$

The conditional hazard function of the discrete distribution of T^* is given by

$$\lambda^*(t_i \mid \mathbf{Z}) = P\{T^* = t_i \mid T^* \geq t_i, \mathbf{Z}\} \qquad i = 0, 1, \ldots$$

and, in terms of the original distribution of T,

$$\lambda^*(t_i \mid \mathbf{Z}) = 1 - \exp\left\{-\int_{t_i}^{t_{i+1}} \lambda(u \mid \mathbf{Z})du\right\}.$$

If we assume that a case with arbitrary covariate $\mathbf{Z}$ has, for each t_i, conditional odds of failing in an interval $[t_i, t_{i+1})$ (i.e., $P\{t_i \leq T < t_{i+1} \mid T \geq t_i, \mathbf{Z}\}$ $\left[P\{T \geq t_{i+1} \mid T \geq t_i,\ \mathbf{Z}\}\right]^{-1}$) proportional to that with covariate $\mathbf{Z} = 0$, then for each t_i

$$\frac{\lambda^*(t_i \mid \mathbf{Z})}{1 - \lambda^*(t_i \mid \mathbf{Z})} = \frac{\lambda_0^*(t_i)}{1 - \lambda_0^*(t_i)} g(\mathbf{Z}),$$

for some function g, where

$$\lambda_0^*(t_i) = \lambda^*(t_i \mid \mathbf{Z} = \mathbf{0}).$$

A linear logistic model arises when $g(\mathbf{Z}) = e^{\beta'\mathbf{Z}}$. If this proportionality does not depend on the partition chosen, then the linear logistic model for T^* implies

$$\frac{1-\exp\left\{-\int_t^{t+\Delta t}\lambda(u\mid\mathbf{Z})du\right\}}{\exp\left\{-\int_t^{t+\Delta t}\lambda(u\mid\mathbf{Z})du\right\}} = \frac{1-\exp\left\{-\int_t^{t+\Delta t}\lambda_0(u)du\right\}}{\exp\left\{-\int_t^{t+\Delta t}\lambda_0(u)du\right\}}e^{\beta'\mathbf{Z}},$$

or

$$\frac{1-\exp\left\{-\int_t^{t+\Delta t}\lambda(u\mid\mathbf{Z})du\right\}}{1-\exp\left\{-\int_t^{t+\Delta t}\lambda_0(u)du\right\}} = \frac{\exp\left\{-\int_t^{t+\Delta t}\lambda(u\mid\mathbf{Z})du\right\}}{\exp\left\{-\int_t^{t+\Delta t}\lambda_0(u)du\right\}}e^{\beta'\mathbf{Z}}.$$

By letting $\Delta t \downarrow 0$ and using L'Hôpital's rule on the left-hand side, we obtain the proportional hazards model

$$\lambda(t\mid\mathbf{Z}) = \lambda_0(t)e^{\beta'\mathbf{Z}}.$$

Since $S(t\mid\mathbf{Z}) = \exp\left\{-\int_0^t\lambda(u\mid\mathbf{Z})du\right\}$, an alternative form of the proportional hazards model is

$$S(t\mid\mathbf{Z}) = \{S_0(t)\}^{\exp(\beta'\mathbf{Z})}.$$

Censoring plays much the same role in the proportional hazards model as it does in the case of homogeneous samples. The martingale approach in Chapter 1 required the weak dependence between T and U of independent censoring, i.e.,

$$P\{t \le T < t+dt \mid T \ge t, U \ge t\} = P\{t \le T < t+dt \mid T \ge t\},$$

for unbiased estimation of hazard rates, and the proportional hazards model requires the same condition in the presence of covariates, or

$$P\{t \le T < t+dt \mid T \ge t, U \ge t, \mathbf{Z}\} = P\{t \le T < t+dt \mid T \ge t, \mathbf{Z}\},$$

where $\mathbf{Z}$ is non-time-varying. In the partial likelihood approach to inference for the proportional hazards model (described in Section 4.3), the additional notion of uninformative censoring (defined as well in Section 4.3) will be introduced.

A more general model can be based on the processes N and Y and on a formal specification of the information accruing over time for a given case. Consider a stochastic basis with the right continuous filtration $\{\mathcal{F}_t : t \ge 0\}$ defined by

$$\mathcal{F}_t = \sigma\{\mathbf{Z}, N(u), Y(u+) : 0 \le u \le t\}.$$

The increasing process N satisfies the conditions of the Doob-Meyer Decomposition (Corollary 1.4.1), and hence with respect to this stochastic basis there is a unique predictable process A such that $N - A$ is a martingale. Heuristically, then,

$$\begin{aligned} E\{dN(s)\mid\mathcal{F}_{s-}\} &= E\{dA(s)\mid\mathcal{F}_{s-}\} \\ &= dA(s). \end{aligned}$$

The process $dA(s)$ specifies the way in which the conditional rate of growth for N depends on information available up to, but not including, each time point, and this dependence is most easily modeled if we assume $dA(s) = l(s)ds$ for some random function l, i.e.,

$$A(s) = \int_0^s l(u)du.$$

As mentioned earlier, the process A is also called the cumulative intensity process for N, and $A'(s) = l(s)$, when it exists, is called the intensity process for N.

In the counting process approach to censored failure time data, the intensity process for N is modeled, rather than conditional hazard rates for T. In the multiplicative intensity model (Aalen, 1978b), $l(s)$ is of the form

$$l(s) = \lambda_0(s)g(\mathbf{Z})Y(s),$$

where λ_0 is an arbitrary function and g depends only on the covariate vector $\mathbf{Z}$. As in the Cox model, taking $g(\mathbf{Z}) = e^{\beta'\mathbf{Z}}$ is a simple way to parameterize the effect of $\mathbf{Z}$ on l while insuring $l \geq 0$.

In the case of right-censored failure time data with non-time-varying covariates, the only additional information accruing over time is contained in N and Y, and the proportional hazards, and multiplicative intensity models coincide under independent censoring. Indeed, Theorems 1.3.1 and 1.3.2 established that the models were equivalent for smooth and discontinuous cumulative hazards, respectively, when there are no covariates. Those proofs apply here with the minor change of additional conditioning on the covariate vector $\mathbf{Z}$. The multiplicative intensity model is easily extended, however, to time-varying covariates and to more general censoring mechanisms which, for example, allow for recurrent events.

Suppose $\{N(t) : t \geq 0\}$ is a counting process, associated with a covariate $\mathbf{Z}$, denoting the number of events of a specified type that have occurred by time t. Whatever information is accruing over time, i.e., however the stochastic basis $\{\mathcal{F}_t : t \geq 0\}$ is specified, the increasing process N will always have a compensator, or cumulative intensity process A, and the only fundamental restriction on A is its predictability with respect to $\{\mathcal{F}_t : t \geq 0\}$. Suppose, for instance, the "at risk" process Y is simply a left-continuous process adapted to the information accruing, and that $Y(s) = 1$ at times s when failure can be observed and $Y(s) = 0$ otherwise. Such a process can be used to model situations in which cases enter and exit risk sets several times, as in random sampling of risk sets in large follow-up studies. Then the compensator process A can be modeled as

$$A(t) = \int_0^t \lambda_0(s)e^{\beta'\mathbf{Z}}Y(s)ds$$

in a multiplicative intensity model. When the covariate vector $\mathbf{Z}$ is a possibly random function of time, no conceptual complexity is added to the multiplicative intensity model by specifying

$$A(t) = \int_0^t e^{\beta'\mathbf{Z}(s)}Y(s)\lambda_0(s)ds.$$

The process A will be predictable as long as both $\mathbf{Z}$ and Y are predictable.

Time-dependent covariates can be incorporated into the proportional hazards model through a function

$$\lambda\{t \mid \mathbf{Z}(t)\} = \lim_{h \downarrow 0} h^{-1} P\{t \leq T < t + h \mid T \geq t, \mathbf{Z}(t)\},$$

as originally proposed by Cox (1972). Of course, λ as defined here inherits none of the usual properties of a traditional hazard function, and cannot be used to compute a conditional survival function $S\{t|\mathbf{Z}(t)\}$. There also is no direct way to incorporate general censoring patterns in the proportional hazards model.

In the multiplicative intensity model there is no restriction that N jump only once, i.e., counting processes more general than $N(t) = I_{\{T \leq t, \delta = 1\}}$ can be analyzed. In fact, the possibility of allowing repeated observed failures is obtained by taking $Y(t) = 1$, even after an observed failure. If $N(t)$ denotes the number of times a possibly recurrent event has been observed over time for a given case, then under the simple restrictions that $EN(t) < \infty$ (respectively, $P\{N(t) < \infty\} = 1$), N will still have a compensator A for which $N - A$ is a martingale (respectively, a local martingale), and A will still have the heuristic interpretation

$$E\{dN(s) \mid \mathcal{F}_{s-}\} = dA(s).$$

The effect of covariates on the rate at which N jumps can again be modeled through A.

A general multiplicative intensity model for an arbitrary counting process N, then, is a specification of the information accruing over time (i.e., the filtration $\{\mathcal{F}_t : t \geq 0\}$) and of the compensator A of N.

Definition 4.2.1. The *multiplicative intensity model* for observations from n independent items consists of n triples, $\{N_i, Y_i, \mathbf{Z}_i\}$, $i = 1, \cdots n$, of counting, censoring and covariate processes, a right-continuous filtration $\{\mathcal{F}_t : t \geq 0\}$ representing the statistical information accruing over time, and n intensity processes $l_i = \lambda_0 Y_i e^{\beta' \mathbf{Z}_i}$, $i = 1, \ldots, n$, along with the additional assumptions:

1. $\mathbf{N} = (N_1, \cdots, N_n)'$ is a multivariate counting process, from which it follows that $P\{\Delta N_i(t) = \Delta N_j(t) = 1\} = 0$ for any $t \geq 0$ and $i \neq j$.
2. For each i, $N_i - A_i$ is a local martingale with respect to $\{\mathcal{F}_t : t \geq 0\}$, where A_i is the continuous compensator

$$A_i = \int l_i(s) ds \quad .$$

3. Each of the censoring processes Y_i and covariate processes $\mathbf{Z}_i$ is predictable with respect to $\{\mathcal{F}_t : t \geq 0\}$, and the $\mathbf{Z}_i$ are locally bounded processes. □

The generality of Definition 4.2.1 comes with some cost. As in the case of right-censored data with non-time-varying covariates, the compensator must have a clear interpretation in the context of the data set being analyzed. Theorem 4.2.1 shows one link between multiplicative intensity, martingale-based models and more traditional approaches using instantaneous conditional rates.

Theorem 4.2.1. Let $\{N(t) : t \geq 0\}$ be a counting process and $\{A(t) : t \geq 0\}$ its compensator with respect to a right-continuous filtration $\{\mathcal{F}_t : t \geq 0\}$. Assume $A(t) = \int_0^t l(s)ds$ for some process l which is left-continuous, has right-hand limits, and is bounded by an integrable random variable Q, i.e., $l(t) \leq Q$ a.s. for all $t \geq 0$ and $EQ < \infty$.

Then

1. $$\lim_{h\downarrow 0} h^{-1}E\{N(t+h) - N(t) \mid \mathcal{F}_t\} = l(t+); \quad \text{and}$$

2. $$\begin{aligned}&\lim_{h\downarrow 0} h^{-1}P\{N(t+h) - N(t) = 1 \mid \mathcal{F}_t\}\\ &= \lim_{h\downarrow 0} h^{-1}\left[1 - P\{N(t+h) - N(t) = 0 \mid \mathcal{F}_t\}\right]\\ &= l(t+).\end{aligned}$$

Proof.

1. Since A is adapted to $\{\mathcal{F}_t : t \geq 0\}$ and

$$\begin{aligned} l(t) &= \lim_{h\downarrow 0} h^{-1} \int_{t-h}^{t} l(u)du \\ &= \lim_{h\downarrow 0} h^{-1}\{A(t) - A(t-h)\}, \end{aligned}$$

l must also be adapted. In fact, $l(t+)$ is adapted to $\{\mathcal{F}_t : t > 0\}$. To see this, let any $h > 0$ be given. Then

$$\begin{aligned} l(t+) &= \lim_{s\downarrow t} l(s) \\ &= \lim_{\substack{s\downarrow t\\ s<t+h}} l(s), \end{aligned}$$

which is $\mathcal{F}_{t+h}$-measurable. It follows that $l(t+)$ is $\mathcal{F}_{t+h}$-measurable for all positive h, and thus $\mathcal{F}_{t+} = \bigcap_{h>0} \mathcal{F}_{t+h}$-measurable and, by the right-continuity of $\{\mathcal{F}_t : t \geq 0\}$, $\mathcal{F}_t$-measurable.

By the Dominated Convergence Theorem,

$$\begin{aligned} \lim_{h\downarrow 0} h^{-1}E\{N(t+h) - N(t) \mid \mathcal{F}_t\} &= \lim_{h\downarrow 0} E\left\{h^{-1}\int_t^{t+h} l(s)ds \mid \mathcal{F}_t\right\} \\ &= E\left\{\lim_{h\downarrow 0} h^{-1}\int_t^{t+h} l(s)ds \mid \mathcal{F}_t\right\} \\ &= E\{l(t+) \mid \mathcal{F}_t\} \\ &= l(t+). \end{aligned}$$

2. For fixed t, let τ be the time of the first jump of N after time t. The random variable τ is a stopping time (Exercise 4.1), and the process $I_{(t,\tau]} = \{I_{\{t<s\leq\tau\}} : s \geq 0\}$ is adapted to $\{\mathcal{F}_s : s \geq 0\}$. Since $I_{(t,\tau]}$ is left-continuous, it is predictable and the process $\{J(s) : s \geq 0\}$ given by

$$J(s) = \int_t^{t+s} I_{(t,\tau]} d(N - A)$$

will be a martingale. Thus

$$\begin{aligned}
\lim_{h\downarrow 0} h^{-1} P\{N(t+h) - N(t) \geq 1 \mid \mathcal{F}_t\} &= \lim_{h\downarrow 0} h^{-1} E\left\{\int_t^{t+h} I_{(t,\tau]} dN \mid \mathcal{F}_t\right\} \\
&= E\left\{\lim_{h\downarrow 0} h^{-1} \int_t^{t+h} I_{(t,\tau]} l(s) ds \mid \mathcal{F}_t\right\} \\
&= E\{l(t+) \mid \mathcal{F}_t\} \\
&= l(t+).
\end{aligned}$$

To prove the first equality in (2), let τ' be the time of the second jump in N after time t. Proceeding as above,

$$\begin{aligned}
\lim_{h\downarrow 0} h^{-1} P\{N(t+h) - N(t) \geq 2 \mid \mathcal{F}_t\} &= \lim_{h\downarrow 0} h^{-1} E\left\{\int_t^{t+h} I_{(\tau,\tau']}(s) dN(s) \mid \mathcal{F}_t\right\} \\
&= E\left\{\lim_{h\downarrow 0} h^{-1} \int_t^{t+h} I_{(\tau,\tau']}(s) l(s) ds \mid \mathcal{F}_t\right\} \\
&= 0.
\end{aligned}$$

□

The simple proof above is due to Aalen (1975), who was one of the first to use martingale methods to study statistical methods for counting processes and censored data.

Loosely speaking, the assumptions of Theorem 4.2.1 hold for arbitrary counting processes as long as the conditional distribution of each jump time, given the times of previous jumps of the process, is absolutely continuous. That result was established by Boel, Varaiya, and Wong (1975a, b). The proof is somewhat technical and is not given here. The theorem implicitly uses the fact that the successive jump times $\tau_1, \tau_2, \ldots$ for a counting process are stopping times with respect to any stochastic basis to which a counting process is adapted.

Theorem 4.2.2. Suppose N is a counting process, and let $\tau_1, \tau_2, \ldots$ be the successive jump times for N, i.e., the times t such that $\Delta N(t) = N(t) - N(t-) = 1$.

Assume that N satisfies

1. $EN(t) < \infty$, for all $t \geq 0$, and
2. $P\{\tau_{n+1} - \tau_n \leq t \mid \mathcal{F}_{\tau_n}\}$ is absolutely continuous in t.

Then the compensator A of N may be written as $A(t) = \int_0^t l(u)du$, i.e., A is absolutely continuous. □

There are, then, two approaches to hazard based regression models which coincide in simple cases: the right-censored data proportional hazards model for conditional hazard rates $\lambda(t \mid \mathbf{Z})$ given covariates, and the multiplicative intensity model for the compensator A of a counting process N. When $\mathcal{F}_t = \sigma\{N(u), Y(u+), \mathbf{Z}(u+) : 0 \leq u \leq t\}$, both have the interpretation

$$P\{N(t + \Delta t) - N(t) = 1 \mid \mathcal{F}_t\} \approx \Delta t Y(t+) e^{\beta' \mathbf{Z}(t+)} \lambda_0(t+),$$

at least under the conditions of Theorems 4.2.1 and 4.2.2.

As mentioned earlier, Theorem 1.3.1 showed that for the simple counting process N used with right-censored data, the associated compensator A was always absolutely continuous when the failure distribution had a density. This is in part a special case of Theorems 4.2.2 and 4.2.1, but with an important additional insight. Indeed, Theorem 1.3.1 implies that although A will always be absolutely continuous, $A'(s)$ will only be statistically meaningful when the crude and net hazards agree. Theorems 4.2.1 and 4.2.2 give conditions under which $A = \int l$ in the multiplicative intensity model, but cannot insure that the intensity process l, which depends on the given stochastic basis, will be the statistical parameter of interest.

Examples 1.3.1 and 1.4.2 illustrated the effect of dependent censoring and changing filtrations on the hazard rate in the simple case of censored failure time data, and showed that the "observable" hazard could be quite different from the true hazard for failure. The next example explores a similar situation in a regression setting, and again makes use of the (imprecise) property $dA(t) = E\{dN(t)|\mathcal{F}_{t-}\}$.

Example 4.2.1. Let T and U be failure and censoring times, respectively, and assume the effect of a p-dimensional covariate vector $\mathbf{Z}$ on the hazard for T is given by $\lambda(t \mid \mathbf{Z}) = \lambda_0(t) e^{\beta' \mathbf{Z}}$.

1. Assume first that $p = 1$ and that Z is a binary covariate with values 0 and 1. When T and U are independent,

$$\begin{aligned} dA(t) &= E\{N(t) - N(t - dt) \mid N(s), N^U(s), Z : 0 \leq s < t\} \\ &= Y(t)\lambda_0(t) e^{\beta Z} dt \end{aligned}$$

but, when Z is omitted from the model,

$$\begin{aligned} dA(t) &= E\{N(t) - N(t - dt) \mid N(s), N^U(s) : 0 \leq s < t\} \\ &= Y(t)\lambda_0(t) E\{e^{\beta Z} | Y(t) = 1\} dt. \end{aligned}$$

In the second case, the observed hazard $\lambda(t) = \lambda_0(t)E\{e^{\beta Z}|Y(t) = 1\}$ in the (incorrectly) presumed homogeneous sample will depend both on the covariate effect (i.e., on β) and on the conditional distribution of Z, given $Y(t) = 1$.

2. Continue to consider the $p = 1$ setting in Part (1), but relax the assumption that T and U are independent when $Z = 1$. Assume survival is uncensored when $Z = 0$. Then

$$\begin{aligned} dA_0(t) &\equiv E\{N(t) - N(t - dt) \mid N(s), N^U(s), Z = 0 : 0 \leq s < t\} \\ &= Y(t)\lambda_0(t)dt, \end{aligned}$$

while

$$\begin{aligned} dA_1(t) &\equiv E\{N(t) - N(t - dt) \mid N(s), N^U(s), Z = 1 : 0 \leq s < t\} \\ &= Y(t)\lambda^{\#}(t \mid Z = 1)dt, \end{aligned}$$

where $\lambda^{\#}$ is the crude hazard function discussed in the remarks following Theorem 1.3.1. When $\lambda_0(t) = \lambda_0$, and the bivariate relationship arising in Example 1.3.1 holds,

$$dA_1(t) = Y(t)(\lambda_0 e^{\beta} + \theta t)dt,$$

so

$$\frac{dA_1(t \mid Y(t) = 1)}{dA_0(t \mid Y(t) = 1)} = e^{\beta} + \frac{\theta t}{\lambda_0}.$$

3. Assume T and U are independent. Suppose $p = 2$, and that both Z_1 and Z_2 are 0,1 valued covariates. Let

$$p_{ij} = P\{Z_1 = i, Z_2 = j\}, \qquad i, j \in \{0, 1\},$$

denote the joint distribution of (Z_1, Z_2). Such a situation might arise in a clinical trial in which Z_1 denotes treatment and Z_2 a dichotomous nuisance covariate.
When both covariates are observed,

$$\begin{aligned} dA(t) &\equiv E\{N(t) - N(t - dt) \mid N(s), N^U(s), Z_1, Z_2 : 0 \leq s < t\} \\ &= Y(t)\lambda_0(t)\exp(\beta_1 Z_1 + \beta_2 Z_2)dt. \end{aligned}$$

When Z_2 is omitted from the model,

$$\begin{aligned} dA(t) &\equiv E\{N(t) - N(t - dt) \mid N(s), N^U(s), Z_1 : 0 \leq s < t\} \\ &= Y(t)\lambda_0(t)e^{\beta_1 Z_1} E\left\{e^{\beta_2 Z_2} \mid Z_1, Y(t) = 1\right\} dt \\ &= Y(t)\lambda_0(t)e^{\beta_1 Z_1}\left[p(Z_1, t)e^{\beta_2} + \{1 - p(Z_1, t)\}\right] dt, \end{aligned}$$

where $p(Z_1, t) \equiv P\{Z_2 = 1 | Z_1, Y(t) = 1\}$. In this setting, the treatment hazard ratio

$$\frac{dA(t|Z_1 = 1, Y(t) = 1)}{dA(t|Z_1 = 0, Y(t) = 1)} = e^{\beta_1} \left\{ \frac{1 + (e^{\beta_2} - 1)p(1,t)}{1 + (e^{\beta_2} - 1)p(0,t)} \right\}$$

reduces to e^{β_1} if either

- $\beta_2 = 0$ (i.e., the nuisance covariate is not associated with outcome); or
- $p(Z_1, t)$ is independent of Z_1 (e.g., when $\beta_1 = 0$ *and* the treatment and nuisance covariate are balanced [i.e., p_{i1}/p_{i0} is independent of i], *and* U is independent of (Z_1, Z_2)).

For illustration, suppose U is independent of Z_1 and Z_2, and let $S_0(t) \equiv \exp\{-\int_0^t \lambda_0(s)ds\}$. Then

$$\frac{dA(t \mid Z_1 = 1, Y(t) = 1)}{dA(t \mid Z_1 = 0, Y(t) = 1)} = e^{\beta_1} \left\{ \frac{1 + (e^{\beta_2} - 1)[1 + \frac{p_{10}}{p_{11}}\{S_0(t)\}^{e^{\beta_1}(1-e^{\beta_2})}]^{-1}}{1 + (e^{\beta_2} - 1)[1 + \frac{p_{00}}{p_{01}}\{S_0(t)\}^{(1-e^{\beta_2})}]^{-1}} \right\},$$

instead of the true treatment hazard ratio

$$\frac{dA(t \mid Z_1 = 1, Z_2 = i, Y(t) = 1)}{dA(t \mid Z_1 = 0, Z_2 = i, Y(t) = 1)} = e^{\beta_1}. \qquad \square$$

Loosely speaking, a stochastic basis in the counting process setting is a model for the statistical information accruing over time. While insuring that the compensator of a counting process with respect to a stochastic basis is scientifically meaningful, a data analyst must also check the defining conditions of a stochastic basis (i.e., right-continuity, completeness and monotonicity) for any specific construction. Any filtration can be completed without affecting the martingale properties of adapted processes, and since families of σ-algebras modeling the information accruing over time are naturally increasing, only the right-continuity of the filtration needs to be verified. Theorem 4.2.3 will provide a starting point, and be used in the next section when building richer filtrations with multivariate counting processes.

Theorem 4.2.3 follows from T26, A2 in Bremaud (1981) and provides a method for verifying right-continuity of one often used filtration.

Theorem 4.2.3. Let $\{N_i, Y_i, \mathbf{Z}_i\}$, $i = 1, \cdots, n$, be n independent triples of counting, censoring, and covariate processes, and assume that the set of paths on which each $\mathbf{Z}_i$ is piecewise constant has probability one. Let

$$\mathcal{F}_t^i = \sigma\{N_i(u), Y_i(u+), \mathbf{Z}_i(u+) : 0 \le u \le t\},$$

and let

$$\mathcal{F}_t \equiv \bigvee_{i=1}^{n} \mathcal{F}_t^i$$

be the smallest σ-algebra containing $\mathcal{F}_t^i$ for all i. Then the family $\{\mathcal{F}_t : t \ge 0\}$ is right-continuous. $\square$

4.3 PARTIAL LIKELIHOOD INFERENCE

Inference procedures for the proportional hazards model and the multiplicative intensity model require both computational methods for finding estimates of the regression coefficients and the baseline hazard function, and at least approximate sampling distributions of those estimators. Both models rely heavily on the same likelihood-based approach: a heuristic interpretation of the conditional hazard function or intensity process is used to construct a partial likelihood; classical likelihood techniques are then applied to the partial likelihood to obtain estimators; and martingale representations of expressions closely related to the estimators are used to derive finite-sample first and second moments and asymptotic distributions.

Methods for establishing the asymptotic normality of censored data regression parameter estimates are quite direct when based on the martingale central limit theorems in Chapter 5. Nevertheless, the detailed proofs become somewhat long and we defer their presentation until Chapter 8. In this section we give the form of the partial likelihood function, show the way it is commonly adjusted in data with tied failure times, and derive the estimators and test statistics based on this likelihood. Later in the chapter we illustrate the use of these statistics in a complex data set.

While some progress has been made (Wong, 1986), no satisfactory general theory of partial likelihood has yet emerged. In this section, we provide a heuristic argument for its use similar to that in the introductory Chapter 0.

Partial likelihood based inference was proposed generally by D.R. Cox in 1975 for models containing high-dimensional nuisance parameters, after he had used the idea in the original presentation of the proportional hazards regression model. In proportional hazards regression, the baseline hazard function λ_0 is essentially an infinite-dimensional nuisance parameter when estimating the regression coefficients β. Models for censored data continue to yield the most widespread and successful examples of partial likelihood procedures. We sketch Cox's original idea here, and recommend that readers wishing more detail read the 1975 paper.

Suppose $\mathbf{X}$ is a random vector with a density $f_{\mathbf{X}}(\mathbf{x}, \theta)$, where θ is a vector parameter (ϕ, β), and that the primary inference question is about β, with ϕ simply a nuisance parameter. In some problems, one might jointly maximize the likelihood with respect to both ϕ and β, and use the appropriate section of the full covariance matrix for the maximum likelihood estimator $\hat{\theta}$ for inference about β. In other situations, it might be possible to condition on a sufficient statistic for ϕ and, as in the construction of similar tests, use the resulting conditional distribution for $\mathbf{X}$ in inference for β. When ϕ is infinite-dimensional, or when the likelihood for θ is complex, joint maximization with respect to the components of θ or the computation of conditional distributions given a sufficient statistic may not be feasible.

In some situations, $\mathbf{X}$ can be transformed into two components $\mathbf{V}$ and $\mathbf{W}$, and the density for $\mathbf{X}$, i.e., the likelihood for θ, can be written as a product of a marginal and a conditional likelihood,

$$f_{\mathbf{X}}(\mathbf{x}, \theta) = f_{\mathbf{W}|\mathbf{V}}(\mathbf{w}|\mathbf{v}, \theta) f_{\mathbf{V}}(\mathbf{v}, \theta), \tag{3.1}$$

where $\mathbf{x}' = (\mathbf{w}', \mathbf{v}')$. Even in complicated models, one of the factors on the right side of Eq. (3.1) may not involve ϕ, and can be used directly for inference on β. Since the other factor will usually depend on both ϕ and β, some information will be lost by using only part of the likelihood in Eq. (3.1), but the gain in simplicity, or in avoiding possible errors associated with applying standard likelihood methods to infinite-dimensional parameters, may compensate for any loss in efficiency. Of course, it is important in each case to explicitly compute, when possible, the exact loss in efficiency arising from the use of a marginal or conditional likelihood, and many results are available in the literature.

A common example of the decomposition in Eq. (3.1) is that of decomposing a vector of observations into vectors of order statistics and ranks of the original observations. We will later look at the marginal likelihood of the rank vector for observed failure times in the proportional hazards model.

Partial likelihood inference, as proposed by Cox, is based on an idea quite similar to that of conditional or marginal likelihoods. Suppose that a data vector $\mathbf{X}$ can be transformed into a sequence of pairs $(V_1, W_1, V_2, W_2, \ldots, V_N, W_N)$, instead of a single pair, where even the number of pairs N may depend randomly on $\mathbf{X}$.

The likelihood for θ can then be written as

$$\begin{aligned} f_{\mathbf{X}}(\mathbf{x}; \theta) &= f_{V_1, W_1, \ldots, V_N, W_N}(v_1, w_1, \ldots, v_N, w_N; \theta) \\ &= \prod_{n=1}^{N} f_{W_n | V_1, W_1, \ldots, V_n}(w_n | v_1, w_1, \ldots, v_{n-1}, w_{n-1}, v_n; \theta) \\ &\qquad f_{V_n | V_1, W_1, \ldots, V_{n-1}, W_{n-1}}(v_n | v_1, w_1, \ldots, v_{n-1}, w_{n-1}; \theta) \\ &= \left[\prod_{n=1}^{N} f_{W_n | \mathbf{Q}_n}(w_n | \mathbf{q}_n; \theta)\right] \left[\prod_{n=1}^{N} f_{V_n | \mathbf{P}_n}(v_n | \mathbf{p}_n; \theta)\right], \end{aligned} \tag{3.2}$$

where $P_1 = \{\emptyset\}, Q_1 = V_1$, and for $n = 2, \ldots, N$,

$$\mathbf{P}_n = (V_1, W_1, \ldots, V_{n-1}, W_{n-1})$$

and

$$\mathbf{Q}_n = (V_1, W_1, \ldots, W_{n-1}, V_n).$$

When the first product on the right–hand side of Eq. (3.2) depends only on β, Cox called this term the partial likelihood for β based on $\mathbf{W}$ in the sequence $(V_1, W_1, \ldots, V_N, W_N)$.

A number of questions arise about the use of a partial likelihood beyond the issue of a possible loss in efficiency. Since the partial likelihood is not a standard likelihood, one cannot expect *prima facie* that estimators based on the partial likelihood would be necessarily asymptotically normal or even consistent. Wong (1986) has shown that under certain regularity conditions, this is generally true, and martingale methods will be used in Chapter 8 to establish this for proportional hazards and multiplicative intensity models. In some situations, there may not be a unique decomposition for $f_{\mathbf{X}}(\mathbf{x}, \theta)$ of the form Eq. (3.2), or there may not be a natural way to construct any such decomposition. Partial likelihood methods seem most successful in problems in which a natural time ordering of observations gives rise to the sequence $(V_1, W_1, \ldots, V_N, W_N)$, as in stochastic processes or in the observations of ordered failure times during the course of a study or experiment.

Partial Likelihood for Failure Time Variables with Covariates

The construction of a general partial likelihood for failure time variables with covariates is not difficult. Initially, we restrict attention to the proportional hazards model with non-time-varying covariates, then later extend the development to the multiplicative intensity model. For simplicity, assume first that we have an absolutely continuous failure distribution and that there are no ties among the observed failure times. The full data of independent triplets $(X_i, \delta_i, \mathbf{Z}_i), i = 1, \ldots, n$, (where $X_i = \min(T_i, U_i)$, and (T_i, U_i) satisfy independent censoring), may be transformed into a sequence $(V_1, W_1, \ldots, V_N, W_N)$ as follows. Suppose there are L observed failures at the times $T_1^o < \ldots < T_L^o$ (setting $T_0^o = 0$ and $T_{L+1}^o = \infty$). Let (i) provide the label for the item failing at T_i^o (note $T_{(i)} = T_i^o$), so the covariates associated with the L failures are $\mathbf{Z}_{(1)}, \ldots, \mathbf{Z}_{(L)}$, and set $(L+1) \equiv n+1$. Suppose there are m_i items censored at or after T_i^o but before T_{i+1}^o, at times $T_{i1}^o, \ldots, T_{im_i}^o$. Let (i, j) provide the label for the item censored at T_{ij}^o, so the covariates associated with these m_i censored items are $\mathbf{Z}_{(i,1)}, \ldots, \mathbf{Z}_{(i,m_i)}$.

Now, in the following construction, condition on $\{\mathbf{Z}_i : i = 1, \ldots, n\}$. For $i = 0, \ldots, L$, take

$$V_{i+1} = \{T_{i+1}^o, T_{ij}^o, (i, j) : j = 1, \ldots, m_i\}$$

and

$$W_{i+1} = \{(i+1)\}.$$

In the nested sequence $(\mathbf{P}_i, \mathbf{Q}_i)$ defined earlier, $\mathbf{P}_i$ now contains the times of all censorings up to time T_{i-1}^o and the times of all failures through time T_{i-1}^o, along with corresponding labels for these items. $\mathbf{Q}_i$ additionally specifies T_i^o and specifies the times of all censorings at or after T_{i-1}^o and up to T_i^o, along with the corresponding labels for these censored items.

If the second product on the right in the partial likelihood Eq. (3.2) is ignored in inference for β, then, given the risk set at T_i^o and given that item (i) failed at T_i^o, the conditional probability of no failures in (T_i^o, T_{i+1}^o), of a failure at T_{i+1}^o, and of the censoring times and labels in $[T_i^o, T_{i+1}^o)$ will be ignored for all i. It is reasonable, though, that the information lost on β may not be substantial. First, no failures in (T_i^o, T_{i+1}^o) and a failure at T_{i+1}^o can be explained by a baseline hazard λ_0 which is arbitrarily small in (T_i^o, T_{i+1}^o), but rises sharply at T_{i+1}^o. Second, in many applications, censoring times and covariates of censored items should provide little or no information about covariate effects on failure rates. This latter assumption is informally presented, in Definition 4.3.1, using the counting process notation.

Definition 4.3.1. Consider the proportional hazards model given in Example 1.5.2. The full data consist of independent triplets $(X_i, \delta_i, \mathbf{Z}_i)$, $i = 1, \ldots, n$, where $X_i = \min(T_i, U_i)$ and $\delta_i = I_{(T_i \leq U_i)}$. Let the process Y_i^* be defined by

$$Y_i^*(s) = I_{\{N_i(s)=0, N_i^U(s-)=0\}},$$

where $N_i(s) = I_{\{X_i \leq s, \delta_i = 1\}}$ and $N_i^U(s) = I_{\{X_i \leq s, \delta_i = 0\}}$.

Censoring is called *uninformative* for the regression parameter β if the likelihood term

$$\prod_{i=1}^{n} \prod_{s \geq 0} \Big[P\{dN_i^U(s) = 1 | Y_i^*(s) = 1, \mathbf{Z}_i(s)\}^{Y_i^*(s)\Delta N_i^U(s)}$$
$$P\{dN_i^U(s) = 0 | Y_i^*(s) = 1, \mathbf{Z}_i(s)\}^{Y_i^*(s)\{1-\Delta N_i^U(s)\}} \Big]$$

does not depend on β. □

In words, censoring is uninformative if, at each t, the likelihood for items censored in $[t, t+dt)$, given the risk set at t and given the items failing at t, does not depend on β. Kalbfleisch and Prentice (1980, Section 5.2) provide further discussion of uninformative censoring, while Arjas and Haara (1984) provide rigorous details. A simple illustration of informative yet independent censoring arises when T and U are independent identically distributed variables, each having distributions depending on β.

The argument in the paragraph preceding Definition 4.3.1 suggests that inference for β could be based on the information in $\mathbf{P}_{i+1}$ given $\mathbf{Q}_i$; that is, on the partial likelihood

$$\prod_{i=1}^{L} P\{W_i = (i) | \mathbf{Q}_i\}. \tag{3.3}$$

To obtain an expression for $P\{W_i = (i) | \mathbf{Q}_i\}$, observe that the assumption of independent censoring implies that the probability that the item with label (i) fails at $T_i^o = t_i$, given the risk set $R_i \equiv \{j : X_j \geq T_i^o\}$ and given that one failure occurs at t_i, is

$$\begin{aligned}
P\{W_i = (i) | \mathbf{Q}_i\} \\
&= \frac{P\{T_{(i)} \in [t_i, t_i + dt) | \mathcal{F}_{t_i}\} \prod_{j \in R_i - (i)} P\{T_j \notin [t_i, t_i + dt) | \mathcal{F}_{t_i}\}}{\sum_{l \in R_i} \left[P\{T_l \in [t_i, t_i + dt) | \mathcal{F}_{t_i}\} \prod_{j \in R_i - l} P\{T_j \notin [t_i, t_i + dt) | \mathcal{F}_{t_i}\} \right]} \\
&= \frac{d\Lambda(t_i | \mathbf{Z}_{(i)}) \prod_{j \in R_i - (i)} \{1 - d\Lambda(t_i | \mathbf{Z}_j\}}{\sum_{l \in R_i} \left[d\Lambda(t_i | \mathbf{Z}_l) \prod_{j \in R_i - l} \{1 - d\Lambda(t_i | \mathbf{Z}_j)\} \right]} \qquad (3.4) \\
&= \frac{\lambda(t_i | \mathbf{Z}_{(i)})}{\sum_{l \in R_i} \lambda(t_i | \mathbf{Z}_l)} \\
&= \frac{\exp(\beta' \mathbf{Z}_{(i)})}{\sum_{l \in R_i} \exp(\beta' \mathbf{Z}_l)},
\end{aligned}$$

where the second to the last equality follows from the absolute continuity of the failure distribution, which allows higher order terms to be ignored.

Thus, using expression (3.3), the partial likelihood is given by

$$\prod_{i=1}^{L} \frac{\exp(\beta' \mathbf{Z}_{(i)})}{\sum_{l \in R_i} \exp(\beta' \mathbf{Z}_l)}. \tag{3.5}$$

This development can be generalized to allow for ties occurring in the observed times of failure. We first show, however, that (3.5) can also be derived as a marginal likelihood.

Marginal Likelihood for Failure Time Variables with Covariates

We continue to assume there are no ties and assume first the data $\{(X_i, \delta_i, \mathbf{Z}_i) : i = l, \ldots, n\}$ are uncensored; i.e., that $X_i \equiv \min(T_i, U_i) = T_i$, and $\delta_i \equiv 1$. Let $T_1^o < T_2^o < \ldots < T_n^o$ denote the ordered failure times, and let (i) denote the label of the item failing at T_i^o. The components of the vector $\mathbf{O} \equiv (T_1^o, T_2^o, \ldots, T_n^o)'$ are the "order statistics" while $\mathbf{r} \equiv ((1), (2), \ldots, (n))'$ will be called the "rank statistics" or, more correctly, the vector of anti-ranks. Of course, $T_i^o = T_{(i)}$. For illustration, if $(T_1, T_2, T_3, T_4) = (28, 15, 17, 6)$, then $\mathbf{O} = (6, 15, 17, 28)'$ and $\mathbf{r} = (4, 2, 3, 1)'$. There is a one-to-one correspondence between $\{\mathbf{O}, \mathbf{r}\}$ and the original data $\{T_i; i = 1, \ldots, n\}$.

In the proportional hazards regression model, $\lambda(t|\mathbf{Z}) = \lambda_0(t)\exp(\beta'\mathbf{Z})$ with λ_0 unspecified, $\mathbf{r}$ carries most of the information about β. Let G be the group of strictly increasing differentiable transformations of $(0, \infty)$ onto $(0, \infty)$ and $g \in G$. If $V \equiv g^{-1}(T)$ then, by a change of variables, V has hazard function

$$\begin{aligned}\lambda(v|\mathbf{z}) &= \lambda_0\{g(v)\}\frac{dg(v)}{dv}\exp(\beta'\mathbf{z}) \\ &\equiv \lambda_1(v)\exp(\beta'\mathbf{z}),\end{aligned}$$

where λ_1 is also an unspecified function. It is evident then that $\{(T_i, \mathbf{Z}_i) : i = 1, \ldots, n\}$ and $\{(V_i, \mathbf{Z}_i) : i = 1, \ldots, n\}$ should yield the same inference about β. In other words, inference on β should be invariant under G. One can find a $g \in G$ to map one realization of $\mathbf{O}$ into any other $\mathbf{O}$, yet $\mathbf{r}$ remains unchanged.

Inference about β can thus be based on the marginal likelihod of $\mathbf{r}$, or

$$\begin{aligned}P\{\mathbf{r} = ((1), \ldots (n)); \beta\} &= P\{T_{(1)} < \cdots < T_{(n)}; \beta\} \\ &= \int_0^\infty \int_{t_1}^\infty \cdots \int_{t_{n-1}}^\infty \prod_{i=1}^{n} f(t_i|\mathbf{Z}_{(i)})dt_n \ldots dt_1.\end{aligned} \tag{3.6}$$

Note

$$\begin{aligned}\int_{t_{n-1}}^\infty f(t_n|\mathbf{Z}_{(n)})dt_n &= S(t_{n-1}|\mathbf{Z}_{(n)}) \\ &= \frac{\exp(\beta'\mathbf{Z}_{(n)})}{\exp(\beta'\mathbf{Z}_{(n)})}\{S_0(t_{n-1})\}^{\exp(\beta'\mathbf{Z}_{(n)})},\end{aligned}$$

where

$$S(t|\mathbf{Z}) \equiv \left[\exp\left\{-\int_0^t \lambda_0(s)ds\right\}\right]^{\exp(\beta'\mathbf{Z})}$$
$$= \{S_0(t)\}^{\exp(\beta'\mathbf{Z})}.$$

Next,

$$\int_{t_{n-2}}^{\infty} f(t_{n-1}|\mathbf{Z}_{(n-1)})S(t_{n-1}|\mathbf{Z}_{(n)})dt_{n-1}$$
$$= \int_{t_{n-2}}^{\infty} \lambda(t_{n-1}|\mathbf{Z}_{(n-1)})$$
$$\{S(t_{n-1}|\mathbf{Z}_{(n-1)})\}^{[\exp(\beta'\mathbf{Z}_{(n)})+\exp(\beta'\mathbf{Z}_{(n-1)})]/\exp(\beta'\mathbf{Z}_{(n-1)})}dt_{n-1}$$
$$= \frac{\exp(\beta'\mathbf{Z}_{(n-1)})}{\exp(\beta'\mathbf{Z}_{(n-1)}) + \exp(\beta'\mathbf{Z}_{(n)})}\{S_0(t_{n-2})\}^{(\exp(\beta'\mathbf{Z}_{(n)})+\exp(\beta'\mathbf{Z}_{(n-1)}))}.$$

Proceeding recursively, Eq. (3.6) reduces to

$$\prod_{i=1}^{n} \frac{\exp(\beta'\mathbf{Z}_{(i)})}{\sum_{l\in R_i}\exp(\beta'\mathbf{Z}_l)},$$

where $R_i = \{(i), (i+1), \ldots, (n)\}$, which indeed does equal the partial likelihood (3.5) in uncensored data.

A generalization of the derivation of the marginal likelihood of the ranks to the censored data setting $\{(X_i, \delta_i, \mathbf{Z}_i) : i = 1, \ldots n\}$ (i.e., where $\delta_i \not\equiv 1$) is more difficult. Let $X_1^o < \ldots < X_n^o$ denote the ordered observation times, with $\mathbf{O}^* \equiv (X_1^o, \ldots, X_n^o)'$. Let $\mathbf{r}^* = ((1), \ldots (n))'$ denote the corresponding vector of anti-ranks so that $X_{(i)} = X_i^o$ for $i = 1, \ldots, n$. If we define $\boldsymbol{\delta}^* = (\delta_{(1)}, \ldots, \delta_{(n)})'$, then there is a one-to-one correspondence between the original data $\{(X_i, \delta_i) : i = 1, \ldots n\}$ and $\{\mathbf{O}^*, \mathbf{r}^*, \boldsymbol{\delta}^*\}$. For illustration, again suppose $(X_1, \ldots, X_4) = (28, 15, 17, 6)$, but now suppose $(\delta_1, \ldots, \delta_4) = (0, 0, 1, 1)$. Then $\mathbf{O}^* = (6, 15, 17, 28)'$, $\mathbf{r}^* = (4, 2, 3, 1)'$, and $\boldsymbol{\delta}^* = (1, 0, 1, 0)'$.

If we again consider the group G of transformations, it is clear that both $\mathbf{r}^*$ and $\boldsymbol{\delta}^*$ remain unchanged. Unfortunately, basing inference for β on both $\mathbf{r}^*$ and $\boldsymbol{\delta}^*$ is difficult due to their complex dependence on the relationship between failure and censoring distributions. (An interesting exception to this is type II censoring at the kth failure, where $\boldsymbol{\delta}^* = (1, \ldots, 1, 0, \ldots, 0)'$ is deterministic with k ones and $n - k$ zeros). Rather than constructing the marginal likelihood of $(\mathbf{r}^*, \boldsymbol{\delta}^*)$, we take the simpler approach of constructing the likelihood of all rankings $\mathbf{r}$ of $(T_1, \ldots, T_n)$ which are consistent with $(\mathbf{r}^*, \boldsymbol{\delta}^*)$. Returning to our illustration, these rankings would be $\mathbf{r} = (4, 2, 3, 1)'$, $\mathbf{r} = (4, 3, 2, 1)'$, and $\mathbf{r} = (4, 3, 1, 2)'$.

As in the derivation of the partial likelihood, let $T_1^o < \cdots < T_L^o$ represent the L observed failure times with corresponding labels $\{(i) : i = 1, \ldots L\}$, a different use of the notation (i) from that in the previous two paragraphs. Let $\{T_{ij}^o : j = 1, \ldots, m_i\}$ represent times of the m_i censorings occurring at or after T_i^o but before T_{i+1}^o, with corresponding labels $\{(i, j) : j = 1, \ldots, m_i\}$. Note that $T_{(i)} = T_i^o$ and $T_{(i,j)} > T_i^o$. The likelihood of all ranks $\mathbf{r}$ of $(T_1, \ldots, T_n)$ which are consistent with $(\mathbf{r}^*, \delta^*)$ is equivalent to

$$P\{T_{(1)} < \cdots < T_{(L)}, T_{(i)} < T_{(i,j)} : i = 1, \ldots, L; j = 1, \ldots, m_i\}. \tag{3.7}$$

It is apparent from (3.7) that the information about β which is discarded by this approach, relative to a likelihood based on $(\mathbf{r}^*, \delta^*)$, is the relative ordering of censorings which occur in each interval $[T_i^o, T_{i+1}^o)$.

Conditional upon $T_{(i)} = t_i$, the probability that the m_i individuals $\{(i, j) : j = 1, \ldots, m_i\}$ survive longer than t_i is

$$\begin{aligned} h(t_i) &= \prod_{j=1}^{m_i} \exp\left\{ -\int_0^{t_i} \lambda_0(u) \exp(\beta' \mathbf{Z}_{(i,j)}) du \right\} \\ &= \{S_0(t_i)\}^{\sum_{j=1}^{m_i} \exp(\beta' \mathbf{Z}_{(i,j)})}. \end{aligned}$$

Thus (3.7) reduces to

$$\int_0^\infty \int_{t_1}^\infty \cdots \int_{t_{L-1}}^\infty \prod_{i=1}^L f(t_i | \mathbf{Z}_{(i)}) h(t_i) dt_L \ldots dt_1 = \prod_{i=1}^L \frac{\exp(\beta' \mathbf{Z}_{(i)})}{\sum_{l \in R_i} \exp(\beta' \mathbf{Z}_l)},$$

where $R_i = \{((j), (j, k)) : j = i, \ldots, L, k = 1, \ldots, m_j\}$ is the risk set at T_i^o. Since inferences based on rank statistics in uncensored data are often efficient, the equivalence of the marginal and partial likelihoods suggests that, in uncensored untied data, partial likelihood inference should lead to efficient tests and estimators. That indeed is the case, as will be seen later. Efficiency considerations in censored data will be explored in Chapter 8.

Partial Likelihood: Ties in Observed Failure Times

To extend the development of partial likelihood to a setting allowing for ties in the observed times of failure, we relax the assumption that the underlying failure distribution is absolutely continuous. As before, suppose there are L observed failure times, $T_1^o < \cdots < T_L^o$, with m_i items censored at or after T_i^o but before T_{i+1}^o, at times $T_{i1}^o, \ldots, T_{im_i}^o$. Again let (i, j) provide the label for the item censored at T_{ij}^o. Now assume there are D_i observed failures at each time $T_i^o, i = 1, \ldots, L$, and let $\{(i)^j : j = 1, \ldots, D_i\}$ denote the labels for the D_i items failing at T_i^o. For $i = 0, \ldots, L$, let

$$V_{i+1} = \{T_{i+1}^o, D_{i+1}, T_{ij}^o, (i, j) : j = 1, \ldots, m_i\}$$

and

$$W_{i+1} = \{(i + 1)^1, \ldots, (i + 1)^{D_{i+1}}\}.$$

The partial likelihood based on information in $\mathbf{P}_{i+1}$ given $\mathbf{Q}_i$, for $i = 1, \ldots, L$, is

$$\prod_{i=1}^{L} P\{W_i = \{(i)^1, \ldots, (i)^{D_i}\} \mid \mathbf{Q}_i\}. \tag{3.8}$$

The ith term of this product is simply the conditional probability that the D_i failures should be those observed, given the risk set at T_i^o and given that D_i failures occur. To obtain an expression for this in the setting in which the underlying failure distribution need not be absolutely continuous, consider the discrete logistic model

$$\frac{d\Lambda(t|\mathbf{Z})}{1 - d\Lambda(t|\mathbf{Z})} = \frac{d\Lambda_0(t)}{1 - d\Lambda_0(t)} \exp(\beta'\mathbf{Z}); \tag{3.9}$$

We have observed earlier that when the discrete hazard contributions are small, model (3.9) converges to the Cox model $\lambda(t|\mathbf{Z}) = \lambda_0(t)\exp(\beta'\mathbf{Z})$. The right-hand side of (3.4) can be rewritten as

$$\frac{d\Lambda(t_i|\mathbf{Z}_{(i)})}{1 - d\Lambda(t_i|\mathbf{Z}_{(i)})} \Bigg/ \sum_{l \in R_i} \frac{d\Lambda(t_i|\mathbf{Z}_l)}{1 - d\Lambda(t_i|\mathbf{Z}_l)}. \tag{3.10}$$

Consequently,

$$P\{W_i = \{(i)^1, \ldots, (i)^{D_i}\} \mid \mathbf{Q}_i\} = \frac{\exp(\beta'\mathbf{S}_i)}{\sum_{\mathbf{l} \in R_{D_i}(T_i^o)} \exp(\beta'\mathbf{S}_{\mathbf{l}})},$$

where $\mathbf{S}_i = \sum_{j=1}^{D_i} \mathbf{Z}_{(i)^j}$ is the sum of the covariate vectors of the D_i individuals who fail at time T_i^o, $\mathbf{l} \equiv (l_1, \ldots, l_{D_i})$ is a set of D_i labels chosen without replacement from the risk set $R_i \equiv \{(j)^d, (j,k) : j = i, \ldots, L, d = 1, \ldots, D_j, k = 1, \ldots, m_j\}$, $\mathbf{S}_{\mathbf{l}} = \sum_{j=1}^{D_i} Z_{l_j}$, and $R_{D_i}(T_i^o)$ is the collection of all sets of D_i labels chosen from R_i without replacement. The resulting partial likelihood for censored tied data satisfies

$$\begin{aligned}\prod_{i=1}^{L} P\{W_i = \{(i)^1, \ldots, (i)^{D_i}\} \mid \mathbf{Q}_i\} &= \prod_{i=1}^{L} \frac{\exp(\beta'\mathbf{S}_i)}{\sum_{\mathbf{l} \in R_{D_i}(T_i^o)} \exp(\beta'\mathbf{S}_{\mathbf{l}})} \\ &\approx c \prod_{i=1}^{L} \frac{\exp(\beta'\mathbf{S}_i)}{\left\{\sum_{l \in R_i} \exp(\beta'\mathbf{Z}_l)\right\}^{D_i}},\end{aligned} \tag{3.11}$$

where c is a constant and where the approximation is accurate if $(D_i - 1)$ is small relative to the size of the risk set R_i, for all i.

When data from this proportional hazards model are expressed in counting process notation, the partial likelihood in (3.11) has the form

$$\prod_{i=1}^{n} \prod_{s \geq 0} \left\{ \frac{\exp(\beta'\mathbf{Z}_i)}{\sum_{j=1}^{n} Y_j(s)\exp(\beta'\mathbf{Z}_j)} \right\}^{\Delta N_i(s)}, \tag{3.12}$$

where $\Delta N_i(s) = 1$ if $N_i(s) - N_i(s-) = 1$ and $\Delta N_i(s) = 0$ otherwise. Although the partial likelihood expression in (3.12) was derived for the setting of the proportional hazards model with non-time-varying covariates, a conceptually analogous argument yields (3.12) as the partial likelihood in the multiplicative intensity model. In fact, in the setting of time dependent covariates in the multiplicative intensity model, the partial likelihood for n independent triplets $(N_i, Y_i, \mathbf{Z}_i), i = 1, \ldots, n,$ has the form

$$\prod_{i=1}^{n} \prod_{s \geq 0} \left\{ \frac{Y_i(s) \exp\{\beta' \mathbf{Z}_i(s)\}}{\sum_{j=1}^{n} Y_j(s) \exp\{\beta' \mathbf{Z}_j(s)\}} \right\}^{\Delta N_i(s)}. \tag{3.13}$$

To summarize, both the proportional hazards model and the multiplicative intensity model lead to the likelihood (3.13). In the proportional hazards model with random censoring, the likelihood can be constructed as a partial likelihood, or as a marginal likelihood of the ranks of the latent failure times. For the multiplicative intensity model, the partial likelihood can be obtained using the interpretation of the intensity process suggested by Theorem 4.2.1. We will take (3.13) as our basis for inference in both models.

An important generalization of the multiplicative intensity model in Definition 4.2.1 is the stratified regression model arising when an individual in stratum q has intensity process

$$l = \lambda_{oq} Y \exp(\beta' \mathbf{Z}), \tag{3.14}$$

where $\{\lambda_{oq}(t) : t \geq 0, q = 1, \ldots, Q\}$ are Q unspecified hazard functions. This generalization, for example, allows one to model predictive categorical covariates whose effect on the hazard is not proportional. Using similar arguments to those in the development of (3.11), we can obtain the partial likelihood for the stratified model

$$\begin{aligned} L(\beta) &= \prod_{q=1}^{Q} L_q(\beta) \\ &= \prod_{q=1}^{Q} \prod_{i=1}^{L_q} \frac{\exp\{\beta' \mathbf{S}_{qi}(T_{qi}^o)\}}{\sum_{l \in R_{D_{qi}}(T_{qi}^o)} \exp\{\beta' \mathbf{S}_\ell(T_{qi}^o)\}}, \end{aligned} \tag{3.15}$$

where $\{T_{qi}^o : i = 1, \ldots, L_q, q = 1, \ldots, Q\}$ is the set of observed failure times, $\mathbf{S}_{qi}(T_{qi}^o)$ is the sum of the covariate vectors of those D_{qi} individuals who fail at T_{qi}^o, and $R_{D_{qi}}(T_{qi}^o)$ is the collection of all sets of D_{qi} labels chosen without replacement from those at risk in stratum q at time T_{qi}^o.

The marginal and partial likelihood derivations all extend standard likelihoods with heuristic yet plausible arguments, but do not guarantee the usual optimality properties of likelihood based methods. Nevertheless, it is appealing to use these approximate likelihoods in standard ways, e.g., as the basis for maximum likelihood estimators, score statistics, and likelihood ratio tests.

In the remainder of this section, we will describe the mechanics of these methods for censored data, emphasizing their similarity to standard likelihood approaches.

While detailed proofs of the validity of these methods are given in Chapter 8, martingale representations for partial likelihood statistics will still provide insights into their properties.

Methods of Likelihood Based Inference

We first review some common methods of likelihood based inference. Suppose $\boldsymbol{\theta} = (\theta_1, \ldots, \theta_q)'$ is a q-dimensional parameter in a statistical model, and that $L(\boldsymbol{\theta}, Y_1, \ldots, Y_n)$ is a likelihood for $\boldsymbol{\theta}$ based on data $Y_1, \ldots, Y_n$. The maximum likelihood estimator (mle) $\hat{\boldsymbol{\theta}}$ of $\boldsymbol{\theta}$ is that member $\hat{\boldsymbol{\theta}}$ of the parameter space Θ for $\boldsymbol{\theta}$ satisfying

$$L(\hat{\boldsymbol{\theta}}, Y_1, \ldots, Y_n) = \max_{\boldsymbol{\theta} \in \Theta} L(\boldsymbol{\theta}, Y_1, \ldots, Y_n).$$

$\hat{\boldsymbol{\theta}}$ will also maximize the log-likelihood $\mathcal{L}(\boldsymbol{\theta}, Y_1, \ldots, Y_n) = \log L(\boldsymbol{\theta}, Y_1, \ldots, Y_n)$ and, subject to smoothness and concavity conditions on $\mathcal{L}$, will be the unique solution to the system of q equations

$$\frac{\partial}{\partial \boldsymbol{\theta}} \mathcal{L}(\boldsymbol{\theta}, Y_1, \ldots, Y_n) = \mathbf{0}. \tag{3.16}$$

The term

$$\mathbf{U}(\boldsymbol{\theta}) = \frac{\partial}{\partial \boldsymbol{\theta}} \mathcal{L}(\boldsymbol{\theta}, Y_1, \ldots, Y_n) = \left\{ \frac{\partial}{\partial \theta_i} \mathcal{L}(\boldsymbol{\theta}, Y_1, \ldots, Y_n), i = 1, \ldots, q \right\}'$$

is called the score statistic for $\boldsymbol{\theta}$ based on L, and Eq. (3.16) is the vector of score equations.

Likelihood based inference for $\boldsymbol{\theta}$ is usually based on the score statistic $\mathbf{U}$, on the mle $\hat{\boldsymbol{\theta}}$, or on a likelihood ratio statistic. All three methods can be used for estimation and for testing both simple and composite hypotheses. Although the three approaches are asymptotically equivalent, their properties in finite samples may differ slightly.

Consider first a test of the simple hypothesis $H : \boldsymbol{\theta} = \boldsymbol{\theta}_0$, against the composite alternative $A : \boldsymbol{\theta} \neq \boldsymbol{\theta}_0$. For independent observations $Y_1, \ldots, Y_n$

$$\mathbf{U}(\boldsymbol{\theta}_0) = \sum_{i=1}^{n} \frac{\partial}{\partial \boldsymbol{\theta}} \mathcal{L}_i(\boldsymbol{\theta}, Y_i) \bigg|_{\boldsymbol{\theta} = \boldsymbol{\theta}_0},$$

where $\mathcal{L}_i(\boldsymbol{\theta}, Y_i) = \log \mathrm{L}_i(\boldsymbol{\theta}, Y_i)$. A short calculation shows that

$$E_{\boldsymbol{\theta}_0} \left\{ \frac{\partial}{\partial \boldsymbol{\theta}} \mathcal{L}_i(\boldsymbol{\theta}, Y_i) \bigg|_{\boldsymbol{\theta} = \boldsymbol{\theta}_0} \right\} = \mathbf{0},$$

and thus, when the Y_i are identically distributed, $n^{-1/2}\mathbf{U}(\boldsymbol{\theta}_0)$ will be approximately multivariate normally distributed with mean zero, under the hypothesis $\boldsymbol{\theta} = \boldsymbol{\theta}_0$. The variance-covariance matrix for $n^{-1/2}\mathbf{U}(\boldsymbol{\theta}_0)$ is $n^{-1}E\{\mathbf{U}(\boldsymbol{\theta}_0)\mathbf{U}'(\boldsymbol{\theta}_0)\}$, and another short calculation shows that

$$E\{\mathbf{U}(\boldsymbol{\theta}_0)\mathbf{U}'(\boldsymbol{\theta}_0)\} = -E \frac{\partial^2}{\partial \boldsymbol{\theta}^2} \mathcal{L}(\boldsymbol{\theta}, Y_1, \ldots Y_n) \bigg|_{\boldsymbol{\theta} = \boldsymbol{\theta}_0},$$

where $\partial^2/\partial\theta^2$ denotes the matrix of second partial derivatives $\partial^2/\partial\theta_i\partial\theta_j$. The covariance matrix

$$\mathbf{I}(\theta) = -E\left\{\frac{\partial^2}{\partial\theta^2}\mathcal{L}(\theta, Y_1, \ldots, Y_n)\right\}.$$

is called the information matrix, and it can be estimated by the observed information matrix

$$\mathcal{I}(\theta_0) = -\frac{\partial^2}{\partial\theta^2}\mathcal{L}(\theta, Y_1, \ldots, Y_n)\bigg|_{\theta=\theta_0}$$

when $\theta = \theta_0$ and when $\mathbf{I}$ is difficult to compute.

Under H,

$$\mathbf{U}'(\theta_0)\{\mathcal{I}(\theta_0)\}^{-1}\mathbf{U}(\theta_0) \tag{3.17}$$

has approximately a χ^2 distribution with q degrees of freedom (denoted by χ^2_q), and tests of H can be conducted by comparing (3.17) with quantile points on the χ^2_q distribution.

Tests of $H : \theta = \theta_0$ can be based directly on the mle $\hat{\theta}$. The vector $\sqrt{n}(\hat{\theta} - \theta_0)$ has asymptotically a mean zero multivariate normal distribution with covariance matrix $n\mathbf{I}^{-1}$ (θ_0), which may be estimated by $n\mathcal{I}^{-1}(\theta_0)$ or $n\mathcal{I}^{-1}(\hat{\theta})$. The Wald statistic for testing H, $(\hat{\theta} - \theta_0)'\mathcal{I}(\hat{\theta})(\hat{\theta} - \theta_0)$, has approximately a χ^2_q distribution.

The likelihood ratio statistic for H is given by

$$\begin{aligned}\Lambda &= \frac{L(\theta_0)}{\max_{\theta\in\Theta} L(\theta)} \\ &= L(\theta_0)/L(\hat{\theta}).\end{aligned}$$

Under H,

$$-2\log\Lambda = 2\{\mathcal{L}(\hat{\theta}, Y_1, \ldots, Y_n) - \mathcal{L}(\theta_0, Y_1, \ldots, Y_n)\}$$

has, again, a χ^2_q distribution in large samples.

Example 4.3.1 Exponential Model. Let $X_i \equiv \min(T_i, U_i)$ and $\delta_i \equiv I_{\{X_i = T_i\}}$, $i = 1, \ldots, n$, be n observations in the general Random Censorship Model. Assume there are no covariates, and suppose T_i has an exponential distribution with failure rate λ so that $P\{T_i > t\} = \exp(-\lambda t)$ for $t \geq 0$. Then the ith individual's contribution to the likelihood, the score statistic, and the observed information are, respectively,

$$\begin{aligned}L_i(\lambda) &= \lambda^{\delta_i}\exp(-\lambda X_i), \\ U_i(\lambda) &= \frac{\partial}{\partial\lambda}\ln L_i(\lambda) \\ &= \frac{1}{\lambda}(\delta_i - \lambda X_i),\end{aligned}$$

and

$$\begin{aligned}\mathcal{I}_i(\lambda) &= -\frac{\partial^2}{\partial\lambda^2}\ln L_i(\lambda) \\ &= \frac{\delta_i}{\lambda^2}.\end{aligned}$$

The term $(\delta_i - \lambda X_i)$ in $U_i(\lambda)$ is the difference between the observed number of events for that individual and, given X_i, the conditionally expected number of events. When λ is replaced by the maximum likelihood estimate $\hat{\lambda} = \sum \delta_i / \sum X_i$, the difference $(\delta_i - \hat{\lambda} X_i)$ will be referred to as the ith individual's martingale residual.

In this example, the Rao or score statistic in (3.17), approximately distributed as χ_1^2, is

$$\frac{\{\sum_{i=1}^n (\delta_i - \lambda X_i)\}^2}{\sum_{i=1}^n \delta_i},$$

having the form $(\mathcal{O} - \mathcal{E})^2/\mathcal{O}$, where $\mathcal{O}$ and $\mathcal{E}$ denote the observed and conditionally expected number of events, respectively. The Wald statistic

$$\left(\frac{\sum_{i=1}^n \delta_i}{\sum_{i=1}^n X_i} - \lambda\right)^2 \frac{\sum_{i=1}^n \delta_i}{\lambda^2},$$

has approximately a χ_1^2 distribution. □

Tests for composite null hypotheses can be based on either the score vector, the Wald statistic, or the likelihood ratio statistic. While each of the methods can be used for general hypotheses expressed by restrictions on $\boldsymbol{\theta}$ to a subset of Θ, the most common composite hypotheses are those given by simple hypotheses about certain components of $\boldsymbol{\theta}$, and we will outline only those methods here.

Suppose, then, that $\boldsymbol{\theta}' = (\boldsymbol{\theta}_1', \boldsymbol{\theta}_2')$, where $\boldsymbol{\theta}_1$ is any $r \times 1$ vector, $1 \leq r < q$, and we wish to test $H : \boldsymbol{\theta}_1 = \boldsymbol{\theta}_{10}$. If $\mathbf{U}'(\boldsymbol{\theta}) = \{\mathbf{U}_1'(\boldsymbol{\theta}), \mathbf{U}_2'(\boldsymbol{\theta})\}$, with the components of $\mathbf{U}_1$ given by $\mathbf{U}_{1j}(\boldsymbol{\theta}) = \frac{\partial}{\partial \theta_j} \mathcal{L}(\boldsymbol{\theta}, Y_1, \ldots, Y_n)$, $1 \leq j \leq r$, a test can be based on the vector $\mathbf{U}_1((\boldsymbol{\theta}_{10}', \tilde{\boldsymbol{\theta}}_2')')$, where $\tilde{\boldsymbol{\theta}}_2$ is the restricted maximum likelihood estimate of $\boldsymbol{\theta}_2$ computed under the constraint $\boldsymbol{\theta}_1 = \boldsymbol{\theta}_{10}$.

The conditional distribution of $\mathbf{U}_1((\boldsymbol{\theta}_{10}', \tilde{\boldsymbol{\theta}}_2')')$, given $\boldsymbol{\theta}_2 = \tilde{\boldsymbol{\theta}}_2$, can be used for critical values for the test statistic and, when $\tilde{\boldsymbol{\theta}}_2$ is unique, this will be the same as the conditional distribution of $\mathbf{U}_1((\boldsymbol{\theta}_{10}', \tilde{\boldsymbol{\theta}}_2')')$, given $\mathbf{U}_2 = \mathbf{0}$. Since $\mathbf{U}(\boldsymbol{\theta})$ is asymptotically multivariate normal, this conditional distribution will be r-variate normal, with zero mean and covariance matrix $\sum_{10}\{(\boldsymbol{\theta}_{10}', \tilde{\boldsymbol{\theta}}_2')'\} = \mathbf{I}_{11} - \mathbf{I}_{12}\mathbf{I}_{22}^{-1}\mathbf{I}_{21}$, where

$$\mathbf{I}(\boldsymbol{\theta}) = \begin{pmatrix} \mathbf{I}_{11}(\boldsymbol{\theta}) & \mathbf{I}_{12}(\boldsymbol{\theta}) \\ \mathbf{I}_{21}(\boldsymbol{\theta}) & \mathbf{I}_{22}(\boldsymbol{\theta}) \end{pmatrix}.$$

(When $\mathbf{I}_{22}(\boldsymbol{\theta})$ is singular, $\{\mathbf{I}_{22}(\boldsymbol{\theta})\}^{-1}$ must be replaced by a generalized inverse.)

Tests of $\boldsymbol{\theta}_1 = \boldsymbol{\theta}_{10}$ can then be based on the quadratic form

$$\mathbf{U}_1'\{(\boldsymbol{\theta}_{10}', \tilde{\boldsymbol{\theta}}_2')'\} \sum_{10}^{-1} \mathbf{U}_1\{(\boldsymbol{\theta}_{10}', \tilde{\boldsymbol{\theta}}_2')'\},$$

which will have an approximate χ^2 distribution with r degrees of freedom.

In this situation, the Wald statistic is given by $(\hat{\boldsymbol{\theta}}_1 - \boldsymbol{\theta}_{10})' \sum_{10} (\hat{\boldsymbol{\theta}}_1 - \boldsymbol{\theta}_{10})$, which also has a χ_r^2 distribution. The likelihood ratio statistic Λ is given by

$$\Lambda = \frac{\max_{\boldsymbol{\theta} \in \Theta_0} L(\boldsymbol{\theta})}{\max_{\boldsymbol{\theta} \in \Theta} L(\boldsymbol{\theta})},$$

where $\Theta_0 = \{(\theta_1', \theta_2')' : \theta_1' = (\theta_{10}, \ldots, \theta_{r0})\}$. In the notation above

$$\Lambda = L(\theta_{10}', \tilde{\theta}_2')'/L(\hat{\theta}).$$

As with the other statistics, $-2\log\Lambda$ has an asymptotic χ_r^2 distribution.

Nearly all likelihood-based methods rely on asymptotic normality of the score vector which, as mentioned earlier, follows from the Central Limit Theorem for independent observations. The partial likelihood for regression parameters in the proportional hazards and multiplicative intensity models is not a full likelihood and, even if it were, we will see that the score vector for these parameters is not a sum of independent terms. We show in Chapter 8, however, that the partial likelihood score vector in the proportional hazards model is asymptotically multivariate normal, with covariance matrix

$$-E\frac{\partial}{\partial\theta}\mathbf{U}(\theta).$$

In the rest of this section, we illustrate the specific form of likelihood-based methods for censored data regression.

We emphasize here that the validity of asymptotic likelihood-based methods with these or any other models relies not only on the asymptotic normality of the score vector, but more fundamentally on the appropriateness of the chosen model. Methods for checking these conditions are discussed in later sections.

Likelihood Based Methods for Censored Data Regression

Some additional notation will make it easier to express partial likelihood based statistics. Let, as before, $(N_i, Y_i, \mathbf{Z}_i)$, $i = 1, \ldots, n$, denote n independent triplet processes from the multiplicative intensity model, with possibly time-dependent p-dimensional covariates $\mathbf{Z} = \{\mathbf{Z}(u) : 0 \le u < \infty\}$. Let $Y_i(t) = 1$ if item i is at risk and under observation at time t, and zero otherwise, and let

$$S^{(0)}(\beta, t) = n^{-1}\sum_{i=1}^{n} Y_i(t)\exp\{\beta'\mathbf{Z}_i(t)\}, \tag{3.18}$$

$$\mathbf{S}^{(1)}(\beta, t) = n^{-1}\sum_{i=1}^{n} \mathbf{Z}_i(t)Y_i(t)\exp\{\beta'\mathbf{Z}_i(t)\}, \tag{3.19}$$

and

$$\mathbf{S}^{(2)}(\beta, t) = n^{-1}\sum_{i=1}^{n} \{\mathbf{Z}_i(t)\}^{\otimes 2}Y_i(t)\ \exp\{\beta'\mathbf{Z}_i(t)\}. \tag{3.20}$$

(For any vector $\mathbf{X}$, $\mathbf{X}^{\otimes 2}$ denotes the outer product $\mathbf{XX}'$.)

In addition, let

$$\mathbf{V}(\beta,t) = \frac{\mathbf{S}^{(2)}(\beta,t)}{S^{(0)}(\beta,t)} - \left\{\frac{\mathbf{S}^{(1)}(\beta,t)}{S^{(0)}(\beta,t)}\right\}^{\otimes 2}. \tag{3.21}$$

The jth component of the $p \times 1$ vector

$$\mathbf{E}(\beta,t) = \frac{\mathbf{S}^{(1)}(\beta,t)}{S^{(0)}(\beta,t)} \tag{3.22}$$

can be thought of as the weighted empirical average of the jth covariate $Z_j(t)$ over all items observed to be at risk at time t, with weights proportional to each item's relative risk of failure at t, that is, weights proportional to $\exp\{\beta'\mathbf{Z}_i(t)\}$. One can show (Exercise 4.2) that

$$\mathbf{V}(\beta,t) = \frac{n^{-1}\sum_{i=1}^{n}\{\mathbf{Z}_i(t) - \mathbf{E}(\beta,t)\}^{\otimes 2} Y_i(t)\exp\{\beta'\mathbf{Z}_i(t)\}}{S^{(0)}(\beta,t)}, \tag{3.23}$$

and hence that $\mathbf{V}(\beta,t)$ is an empirical covariance matrix computed with the same weights.

The partial likelihood for the regression coefficients β is given by expression (3.13):

$$\prod_{i=1}^{n}\prod_{u\geq 0}\left[\frac{Y_i(u)\exp\{\beta'\mathbf{Z}_i(u)\}}{\sum_{j=1}^{n} Y_j(u)\exp\{\beta'\mathbf{Z}_j(u)\}}\right]^{\Delta N_i(u)}$$

and, since this likelihood is based on observations of (N_i, Y_i, Z_i) on $[0,\infty)$, we denote this by $L(\beta,\infty)$. The likelihood may also be thought of as the value at $t = \infty$ of the process

$$L(\beta,t) = \prod_{i=1}^{n}\prod_{0\leq u\leq t}\left[\frac{Y_i(u)\exp\{\beta'\mathbf{Z}_i(u)\}}{\sum_{j=1}^{n} Y_j(u)\exp\{\beta'\mathbf{Z}_j(u)\}}\right]^{\Delta N_i(u)}.$$

A short calculation shows that the score vector, $\mathbf{U}(\beta) = \frac{\partial}{\partial\beta}\log L(\beta,\infty)$, is the value at $t = \infty$ of the process $\mathbf{U}(\beta,\cdot)$ given at time t by

$$\mathbf{U}(\beta,t) = \sum_{i=1}^{n}\int_0^t \{\mathbf{Z}_i(u) - \mathbf{E}(\beta,u)\}dN_i(u). \tag{3.24}$$

This form of the score vector shows that it has an interpretation similar to that of the logrank test statistic. If we write

$$\mathbf{U}(\beta,t) = \sum_{i=1}^{n}\mathbf{U}_i(\beta,t),$$

then the contribution of the ith item to the score is a sum, over the ith item's event times, of the difference between the observed covariate vector for that item (i.e., group membership in the logrank statistic) and the conditionally expected value of that covariate vector, given the items at risk and their relative risk of failure at that time.

It is important to note that the $\mathbf{U}_i(\beta, t)$ are not independent, due to the term $\mathbf{E}(\beta, u)$. $\mathbf{U}(\beta, t)$ can, however, be written as a sum of uncorrelated terms, as can be seen from the martingale representation for $\mathbf{U}$.

Each term $\mathbf{Z}_i(u) - \mathbf{E}(\beta, u)$ may be thought of as a covariate centered by its empirical average calculated with the probability mass function which assigns a weight proportional to $Y_j(u)\exp\{\beta'\mathbf{Z}_j(u)\}$ to item j. The average of these centered terms with respect to this same discrete probability function is then zero. That is,

$$\frac{\sum_j \{\mathbf{Z}_j(u) - \mathbf{E}(\beta, u)\} Y_j(u)\exp\{\beta'\mathbf{Z}_j(u)\}}{\sum_j Y_j(u)\exp\{\beta'\mathbf{Z}_j(u)\}} = 0.$$

or, equivalently,

$$\sum_j \{\mathbf{Z}_j(u) - \mathbf{E}(\beta, u)\} Y_j(u)\exp\{\beta'\mathbf{Z}_j(u)\}\lambda_0(u) = 0.$$

This last equation implies that

$$\mathbf{U}(\beta, t) = \sum_{i=1}^{n} \int_0^t \{\mathbf{Z}_i(u) - \mathbf{E}(\beta, u)\} dM_i(u), \tag{3.25}$$

where

$$M_i(t) = N_i(t) - \int_0^t Y_i(u)\exp\{\beta'\mathbf{Z}_i(u)\}\lambda_0(u)du.$$

$\mathbf{U}(\beta, \cdot)$ thus has the same representation $\sum_i \int H_i dM_i$ as that which has been exploited earlier for survival distribution estimators and k-sample statistics.

While the martingale representation for $\mathbf{U}$ will be used in Chapter 8 to derive asymptotic properties of the score statistic, it can also be used for insight into finite-sample first and second moments. Indeed, with an appropriate filtration and simple regularity assumptions, $\mathbf{U}(\beta, \cdot)$ will be a stochastic integral of a predictable process with respect to a martingale and, for any t,

$$E\{\mathbf{U}(\beta, t)\} = \mathbf{0},$$

for $k = 1, \ldots, p$, and $i \neq j$,

$$E\left\{\int_0^t \{\mathbf{Z}_i(u) - \mathbf{E}(\beta, u)\}_k dM_i(u) \int_0^t \{\mathbf{Z}_j(u) - \mathbf{E}(\beta, u)\}_k dM_j(u)\right\} = 0,$$

and

$$E\left[\{n^{-1/2}\mathbf{U}(\beta, t)\}^{\otimes 2}\right]$$
$$= \frac{1}{n} E \sum_i \int_0^t \{\mathbf{Z}_i(u) - \mathbf{E}(\beta, u)\}^{\otimes 2} Y_i(u)\exp\{\beta'\mathbf{Z}_i(u)\}\lambda_0(u)du. \tag{3.26}$$

The two equations above Eq. (3.26) imply, respectively, that while the partial likelihood is not a real likelihood, the partial likelihood score statistic has mean zero and can be written as a sum of uncorrelated terms. It is not difficult to show (Exercise 4.3) that

$$-\frac{\partial}{\partial \beta}\mathbf{U}(\beta, t) = \int_0^t \sum_{i=1}^n \mathbf{V}(\beta, u)dN_i(u),$$

and that Eq. (3.26) is equal to

$$\int_0^t E\{\mathbf{V}(\beta, u)S^{(0)}(\beta, u)\}\lambda_0(u)du.$$

Since each component of $\int \sum_{i=1}^n \mathbf{V}(\beta, \cdot)dM_i$ is a martingale, Eq. (3.26) then implies that

$$n^{-1}E\left[-\frac{\partial}{\partial \beta}\mathbf{U}(\beta, t)\right] = E\left[\{n^{-1/2}\mathbf{U}(\beta, t)\}^{\otimes 2}\right]. \tag{3.27}$$

Just as in standard likelihood theory, then, the covariance matrix of the score statistic for β is the negative of the expected value of the second derivative of the log likelihood, i.e., the information matrix. From now on, we will let

$$\begin{aligned}\mathcal{I}(\beta, t) &= -\frac{\partial}{\partial \beta}\mathbf{U}(\beta, t)\\ &= -\left\{\frac{\partial^2}{\partial \beta_i \partial \beta_j}\mathcal{L}(\beta, t)\right\}_{p\times p}.\end{aligned}$$

In the exercises, we explore the form of the observed information matrix $\mathcal{I}(\beta, \infty)$ in the $(s+1)$-sample problem where $(\mathbf{Z}^{s\times 1})_j$ is an indicator for membership in sample $j, j = 1, \ldots, s$. In that setting, the score vector $\mathbf{U}(\mathbf{0}, \infty)$ is the $(s+1)$ sample logrank vector. Its estimated covariance matrix $\mathcal{I}(\mathbf{0}, \infty)$ has hypergeometric variance structure when the exact version of the partial likelihood in (3.11) is employed, and binomial variance structure (overestimating the actual information in the presence of ties) when the approximate version in (3.11) is used.

Estimation of the Hazard Function, λ_0.

The baseline hazard λ_0 does not appear in the partial likelihood, and hence is not estimated by the solutions to the likelihood equations. Several methods have been proposed for estimating parameters related to λ_0, and one appealing estimate of $\Lambda_0 = \int \lambda_0(s)ds$ is due to Breslow (1972, 1974). The estimator is a natural generalization of the Nelson estimator $\hat{\Lambda} = \int \overline{Y}^{-1} I\{\overline{Y} > 0\}d\overline{N}$ used in homogeneous samples and discussed in Chapter 3. When there are no covariates being modeled, $\hat{\Lambda}_0$ can be written

$$\hat{\Lambda}_0(t) = \int_0^t \left\{\sum_{i=1}^n Y_i(s)\right\}^{-1}\left\{\sum_{i=1}^n dN_i(s)\right\}. \tag{3.28}$$

(Note that $\sum_{i=1}^n Y_i(s) = 0$ implies $\sum_{i=1}^n dN_i(s) = 0$ and, as always, we set $0/0 = 0$). When regression coefficients have been estimated by, say, $\hat{\beta}$, the ith item in the sample has estimated failure rate $Y_i(s)\exp\{\hat{\beta}'\mathbf{Z}_i(s)\}\lambda_0(s)$. One item in the risk set failing at rate $\exp\{\hat{\beta}'\mathbf{Z}_i(s)\}\lambda_0(s)$ will produce an observed failure with the same probability as $\exp\{\hat{\beta}'\mathbf{Z}_i(s)\}$ cases, each failing with instantaneous rate $\lambda_0(s)$. The heterogeneous sample can then be thought of as one with $\sum_{i=1}^n Y_i(s)\exp\{\hat{\beta}'\mathbf{Z}_i(s)\}$ cases at risk at time s, all failing with rate $\lambda_0(s)$ and, consequently, $\Lambda_0 = \int \lambda_0(s)ds$ can be estimated by

$$\hat{\Lambda}_0(t) = \int_0^t \left[\sum_i Y_i(s)\exp\{\hat{\beta}'\mathbf{Z}_i(s)\}\right]^{-1} \left\{\sum_{i=1}^n dN_i(s)\right\}. \tag{3.29}$$

The martingale calculus can again be used to find expressions for the first and second moments of $\hat{\Lambda}_0$ (see Exercise 4.4).

In the stratified model in Eq. (3.14), one must estimate Q functions

$$\left\{\Lambda_{0_q}(t) = \int_0^t \lambda_{0_q}(s)ds : t \geq 0\right\},$$

for $q = 1, \ldots, Q$. By Eq. (3.15), the maximum partial likelihood estimate β is the solution of

$$\begin{aligned}
\mathbf{U}(\beta) &= \frac{\partial}{\partial\beta} \ln L(\beta) \\
&= \sum_{q=1}^{Q} \frac{\partial}{\partial\beta} \ln L_q(\beta) \\
&\equiv \sum_{q=1}^{Q} \mathbf{U}_q(\beta) \\
&= 0,
\end{aligned}$$

while the estimated covariance matrix for $\mathbf{U}$ satisfies

$$\begin{aligned}
\mathcal{I}(\beta) &\equiv -\frac{\partial}{\partial\beta}\mathbf{U}(\beta) \\
&= \sum_{q=1}^{Q} -\frac{\partial}{\partial\beta}\mathbf{U}_q(\beta) \\
&\equiv \sum_{q=1}^{Q} \mathcal{I}_q(\beta).
\end{aligned}$$

Given the data $\{(N_{qi}, Y_{qi}, \mathbf{Z}_{qi}) : i = 1, \ldots, n_q, q = 1, \ldots Q\}$, the Breslow estimate of Λ_{0_q} is

$$\hat{\Lambda}_{0q}(t) = \int_0^t \left[\sum_{i=1}^{n_q} Y_{qi}(s) \exp\{\hat{\beta}'\mathbf{Z}_{qi}(s)\}\right]^{-1} \left\{\sum_{i=1}^{n_q} d\right.$$

Section 4 illustrates the use of partial likelihood based data regression models with a complex data set.

4.4 APPLICATIONS OF PARTIAL LIKELIHOOD METHODS

The next three sections illustrate the usefulness of proportional hazards and multiplicative intensity regression model tools in the analysis of censored survival data. This section provides background for the liver disease natural history data set and illustrates the use of partial likelihood based inference procedures. An investigation of gamma interferon in chronic granulomatous disease also is presented to show an application of the multiplicative intensity model to a data set having repeated outcome events. In Section 4.5, graphical methods and methods for analysis of residuals are discussed, using the structure of case specific martingales and martingale transforms. These methods allow a graphical approach to model building, variable transformation, checking the proportional hazards assumption, and finding leverage points and outliers. Section 4.6 then displays the application of these graphical methods to data sets, including the liver disease natural history data. Special routines were added to the UNIX statistical language S for these analyses.

Background: Liver Data

The Mayo Clinic has established a database of 424 patients having primary biliary cirrhosis (PBC). These 424 form the complete collection of all PBC patients, referred to Mayo between January 1974 and May 1984, who met standard eligibility criteria for a randomized, double-blinded, placebo-controlled, clinical trial of the drug D-penicillamine (DPCA). The patient and treating physician agreed to randomization in 312 of the 424 cases. For each of the 312 clinical trial patients, clinical, biochemical, serologic, and histologic parameters were collected. For this analysis, complete follow up to July, 1986, was attempted on all patients. By this date, 125 of the 312 had died, with only 11 deaths not attributable to PBC. Only eight were lost to follow up, and 19 had undergone liver transplantation. Appendix D contains the survival data and the entry values of the important covariates.

Because PBC is a rare disease (the prevalence of the disease has been estimated to be 50 cases-per-million population), this database is a valuable resource to liver specialists. PBC is a fatal chronic liver disease of unknown cause. The primary pathologic event appears to be destruction of interlobular bile ducts, which may be mediated by immunological mechanisms. Results of the clinical trial of 312 patients established that DPCA is not effective in PBC, in spite of the drug's immunosuppressive properties. Until recently, effective treatments for PBC did not exist, and the approach to patients with the disease was limited to supportive care.

early to mid 1980s, the rate of liver transplantation in patients with advanced ge PBC increased substantially, largely due to the improvement in transplantation results through the use of immunosuppressive agents such as cyclosporine and OKT3.

We first show that DPCA has a negligible effect on prognosis, then use the data from the 312 randomized cases to develop a natural history model. Such a model will be useful not only in counseling patients and in understanding the course of PBC in untreated patients, but also in providing historical control information to evaluate the efficacy of new therapeutic interventions such as liver transplantation. These evaluations of liver transplantation will be important since randomized trials comparing transplantation with non-surgical management will not be performed and since PBC is one of the most common indicators for liver transplantation in adults. In this Chapter, we present the analyses to evaluate DPCA, to develop a natural history model, and to illustrate the model's use in evaluating liver transplantation. The data set of 112 nonrandomized patients is used in model validation of the natural history model and to illustrate its use in survival prediction. Appendix D contains the data for 106 of the 112 patients, since six were lost to followup soon after their initial visit to the Mayo Clinic. Of the 106, 36 had died by July 1986, with six others having undergone transplantation.

In the database of 418 patients, the 25 transplanted patients were considered censored at the date of transplantation. This induces a small bias in a natural history model. Transplantation occurred after a median followup of 66 months for the 19 transplanted clinical trial patients and 50 months for the 6 transplanted non-trial patients.

Effect of DPCA on Patient Survival

Figure 4.4.1 presents the Kaplan-Meier estimates of survival of PBC patients following randomization to either DPCA or placebo. The curves show little separation. The median survival time of the pooled group is just under 10 years.

Under the proportional hazards assumption, the Cox regression model can be used to measure treatment effect. If treatment is coded by $Z = 0$: DPCA, $Z = 1$: Placebo, then in the model

$$\lambda(t|Z) = \lambda_0(t)\exp(\beta Z), \tag{4.1}$$

$\lambda_0(\cdot)$ represents the hazard function for death while being treated with DPCA, and β is the log of the hazard ratio; i.e., if $\lambda_1(t) \equiv \lambda(t|Z = 1)$ then, for all t,

$$\lambda_1(t) / \lambda_0(t) = e^{\beta} .$$

If $L(\beta)$ is the Cox partial likelihood for β based on the censored survival data and $\mathcal{L} \equiv \ln L$, then the score statistic is given by

$$U(\beta) \equiv \frac{d}{d\beta}\mathcal{L}(\beta),$$

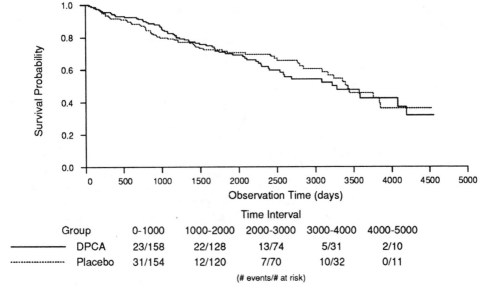

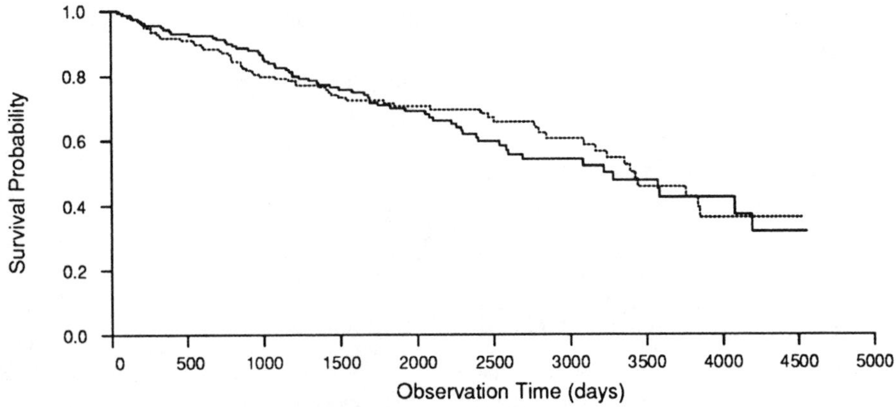

		Time Interval				
	Group	0-1000	1000-2000	2000-3000	3000-4000	4000-5000
————	DPCA	23/158	22/128	13/74	5/31	2/10
········	Placebo	31/154	12/120	7/70	10/32	0/11

(# events/# at risk)

Figure 4.4.1 Estimated survival curves in DPCA and placebo groups, PBC data.

and Fisher's observed information is

$$\mathcal{I}(\beta) = -\frac{d^2}{d\beta^2}\mathcal{L}(\beta).$$

For these data $U(0) = -1.781115$, $\mathcal{I}(0) = 31.19845$, and $\mathcal{L}(0) = -639.9799$. Hence the standardized score statistic or Rao test statistic is

$$\{U(0)\}^2 / \mathcal{I}(0) = 0.10168.$$

Later it will be established that this statistic is distributed asymptotically as a chi-square with one degree of freedom when $H : \beta = 0$. Since there are no nuisance covariates in this model, this score statistic is identical to the logrank statistic for no treatment effect.

The maximum partial likelihood estimate for β is $\hat{\beta} = -0.0571242$ and $\mathcal{L}(-0.0571242) = -639.9290$. Hence, the likelihood ratio statistic for the hypothesis $H : \beta = 0$ is

$$-2\{\mathcal{L}(0) - \mathcal{L}(\hat{\beta})\} = 0.10193.$$

Since the standard error for $\hat{\beta}$ is estimated by $\{\mathcal{I}(\hat{\beta})\}^{-1/2}$, where $\mathcal{I}(\hat{\beta}) = 31.1525$, the Wald statistic for $H : \beta = 0$ is

$$\hat{\beta}^2 \mathcal{I}(\hat{\beta}) = 0.10166.$$

As expected, the Rao, Wald, and likelihood ratio statistics yield nearly identical results.

Under the proportional hazards assumption, the hazard ratio

$$r \equiv \lambda_1(t) / \lambda_0(t) = e^{\beta}$$

is independent of t. In large samples,

$$\hat{\beta} \sim N(\beta, \{\mathcal{I}(\hat{\beta})\}^{-1});$$

thus $\hat{r} = e^{\hat{\beta}} = 0.94448$, and a 95% confidence interval for r is

$$\exp\{\hat{\beta} \pm 1.96\{\mathcal{I}(\hat{\beta})\}^{-1/2}\} = (0.66479, 1.34184).$$

We estimate the failure rate on placebo to be 94.4% that on DPCA, and there is evidence against it being more than 134% that on DPCA. If DPCA must improve patient survival by more than a factor of 1/2 to offset the drug's expense, toxicity and inconvenience of administration, then this trial supports not using the drug in this disease.

An analysis of subsets defined by clinical, biochemical and histological variables failed to yield evidence of important survival differences between the drug and the placebo in patient subgroups.

Natural History Model for PBC

The data in Appendix D on the 312 PBC randomized patients can be used to build a statistical model for the influence of covariates on disease outcome. Table 4.4.1 provides the distributions of 14 clinical, biochemical and histological variables. With the exception of 4 missing platelet counts and two missing urine copper values, the data are complete.

For the remainder of this section, we use the model

$$\lambda(t|\mathbf{Z}) = \lambda_0(t)\exp(\beta'\mathbf{Z}), \tag{4.2}$$

where $\mathbf{Z}' \equiv (Z_1, Z_2, \ldots, Z_K)$ is a vector of K predictors and $\beta' \equiv (\beta_1, \beta_2, \ldots, \beta_K)$ are the regression coefficients. Each predictor Z_i could be defined in a variety of ways, such as using the variables in Table 4.4.1, transformations or crossproducts of these variables, etc. In model (4.2), each individual patient is given a risk score $R \equiv \beta_1 Z_1 + \ldots + \beta_K Z_K$. Let $S(t|\mathbf{Z})$ denote the probability that patient with risk factors given by $\mathbf{Z}$ (and with risk score R) is still alive t years after time 0, and let $S_0(t)$ denote the survival function for individuals having risk score $R = 0$. Then

$$\begin{aligned} S(t|\mathbf{Z}) &= \{S_0(t)\}^{\exp(R)} \\ &= \{e^{-\Lambda_0(t)}\}^{\exp(R)}, \end{aligned}$$

where time $t = 0$ usually denotes the time the measurements in the covariate vector $\mathbf{Z}$ are obtained. In the PBC data set in Appendix D, time $t = 0$ is the date of treatment randomization. One can estimate $S(t|\mathbf{Z})$ by

$$\hat{S}(t|\mathbf{Z}) = \{e^{-\hat{\Lambda}_0(t)}\}^{\exp(\hat{R})}, \tag{4.3}$$

where the maximum partial likelihood estimate vector $\hat{\beta}$ is used to obtain $\hat{R} = \hat{\beta}_1 Z_1 + \cdots + \hat{\beta}_K Z_K$, and where $\hat{\Lambda}_0(\cdot)$ is the Breslow estimator in Eq. (3.29). In the proportional hazards model in Eq. (4.2), each coefficient β_k has the simple interpretation that every unit increase in the kth covariate, Z_k, changes the hazard function by the multiplicative factor $\exp(\beta_k)$.

Initially model (4.2) was fit to the data with $\mathbf{Z}' = (Z_1, \ldots, Z_{14})$ chosen to be the 14 variables in Table 4.4.1. If $\mathbf{U}(\beta)$ and $\mathcal{I}(\beta)$ denote the score vector and Fisher's observed information matrix, respectively, the collection of univariate Rao or logrank statistics, $\{[\{\mathbf{U}(0)\}_k]^2/\{\mathcal{I}(0)\}_{kk} : k = 1, \ldots, 14\}$, are listed in the right-hand column of Table 4.4.1. The term $\{\mathcal{I}(0)\}_{jk}$ is the estimated covariance of $U_j(0)$ and $U_k(0)$, and since $\mathcal{I}(0) = \int_0^\infty \sum_{i=1}^n \mathbf{V}(0, t) dN_i(t)$ for $\mathbf{V}$ defined by (3.23), $\{\mathcal{I}(0)\}_{jk}$ is the sum (over death times) of the covariances of Z_j and Z_k among those at risk at each death time. Thus inspection of

$$\{c_{jk} \equiv \{\mathcal{I}(0)\}_{jk}/[\{(\mathcal{I}(0)\}_{jj}\{\mathcal{I}(0)\}_{kk}]^{1/2} : j \neq k\}$$

Table 4.4.1 Prognostic Factors: Summary of Univariate Statistics (312 Patients in the PBC Clinical Trial of DPCA)

Demographic	min	1st Q	med	3rd Q	max	Missing	Rao χ^2(1 d.f.)
Age (years)	26.3	42.1	49.8	56.7	78.4	0	20.86
Sex	male:	36	female:	276		0	4.27
Clinical		**Absent**		**Present**		**Missing**	**Rao χ^2(1 d.f.)**
Ascites		288		24		0	104.02
Hepatomegaly		152		160		0	40.18
Spiders		222		90		0	30.31
Edema[1]	0: 263	1/2: 29		1: 20		0	97.89
Biochemical	**min**	**1st Q**	**med**	**3rd Q**	**max**	**Missing**	**Rao χ^2(1 d.f.)**
Bilirubin	0.3	0.8	1.35	3.45	28.0	0	190.62
Albumin	1.96	3.31	3.55	3.80	4.64	0	70.83
Urine Copper	4	41	73	123	588	2	84.35
Pro Time	9.0	10.0	10.6	11.1	17.1	0	51.76
Platelet Count	62	200	257	323	563	4	12.15
Alkaline Phos	289	867	1259	1985	13862	0	2.58
SGOT	26	81	115	152	457	0	29.59
Histologic	**1**	**2**	**3**	**4**		**Missing**	**Rao χ^2(1 d.f.)**
Stage	16	67	120	109		0	46.49

[1] Edema 0: No edema and no diuretic therapy for edema

$\frac{1}{2}$: Edema but no diuretics, or edema resolved by diuretics

1: Edema despite diuretic therapy

Rao statistics computed after six missing values were replaced by median values (i.e., 4 missing Platelet Counts, 257; 2 missing Urine Copper, 73)

is a method for finding co-linearities among the K components of $\mathbf{Z}$. The largest values of c_{jk} are 0.47 between hepatomegaly and stage, 0.37 between bilirubin and SGOT, and 0.37 between bilirubin and urine copper. Bilirubin is the strongest univariate predictor of survival. One would expect and can verify that the predictive strength of the variables SGOT and urine copper are reduced in models which adjust for bilirubin. In building a parsimonious natural history model based on easily accessible variables, there is hope that readily available measurements on hepatomegaly and bilirubin will contain much of the predictive information from the invasive variable histologic stage (which requires a liver biopsy), and in the frequently unmeasured variables, urine copper and SGOT.

The score statistics in Table 4.4.1 show that nearly all 14 variables are highly significantly associated with patient survival. The Kaplan-Meier plot in Figure 4.4.2 indicates that bilirubin levels distinguish patients with good and poor prognosis.

Parsimonious but accurate models based on inexpensive, non-invasive and readily available measurements are useful in clinical science, and so the variables stage, urine copper, and SGOT were eliminated temporarily from the variable selection process. The untransformed versions of the remaining 11 variables were inserted into Eq. (4.2), and a step-down procedure was employed to eliminate variables, using the Wald statistic as a criterion for deletion of the least predictive variable. Table 4.4.2 displays the first step of the procedure, which led to the elimination of

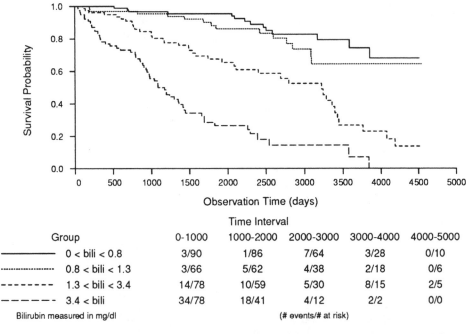

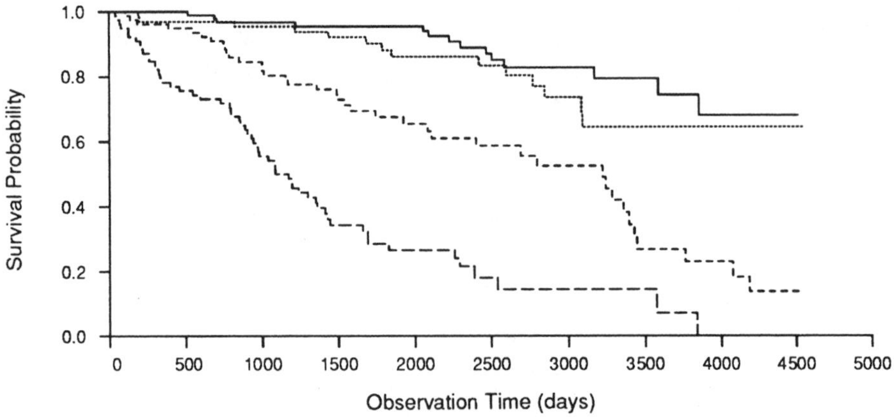

	Time Interval				
Group	0-1000	1000-2000	2000-3000	3000-4000	4000-5000
0 < bili < 0.8	3/90	1/86	7/64	3/28	0/10
0.8 < bili < 1.3	3/66	5/62	4/38	2/18	0/6
1.3 < bili < 3.4	14/78	10/59	5/30	8/15	2/5
3.4 < bili	34/78	18/41	4/12	2/2	0/0

Bilirubin measured in mg/dl (# events/# at risk)

Figure 4.4.2 Estimated survival curves for four groups determined by serum bilirubin levels, PBC data.

the variable alkaline phosphatase, and the sixth step, at which each of the remaining variables has a Wald statistic exceeding 6.0. The likelihood ratio test for the five eliminated variables has the value

$$-2(-554.237 + 550.603) = 7.268,$$

and has an approximate chi-square distribution with 5 degrees of freedom. There is little evidence to retain the variables alkaline phosphatase, ascites, platelet count, sex, or presence of spiders.

In the model in Table 4.4.2(b), all variables were entered untransformed. In such a model, an increase in the value of the ith covariate, Z_i, from x to $(x + d)$ will lead to a multiplicative increase in the hazard by a factor $\exp(d\beta_i)$, independent of the value x. However, the clinical literature suggests that changes from x to $(x + d)$ in the values of variables such as bilirubin should have a greater impact on prognosis when x is small. To evaluate the need for transformations of the four continuous variables in the six variable model in Table 4.4.2(b), the variables log(age), log(albumin), log(protime), and log(bilirubin) were added. The resulting

Table 4.4.2 Results of variable selection procedure in 312 randomized cases with PBC.

(a) First Step, log likelihood −550.603

	Coef.	Std. Err.	Z stat.
Age	2.819 e-2	9.538 e-3	2.96
Albumin	−9.713 e-1	2.681 e-1	−3.62
Alk. Phos	1.445 e-5	3.544 e-5	0.41
Ascites	2.813 e-1	3.093 e-1	0.91
Bilirubin	1.057 e-1	1.667 e-2	6.34
Edema	6.915 e-1	3.226 e-1	2.14
Hepatomegaly	4.853 e-1	2.913 e-1	2.21
Platelets	−6.063 e-4	1.025 e-3	−0.59
Prothrombin Time	2.428 e-1	8.420 e-2	2.88
Sex	−4.769 e-1	2.643 e-1	−1.80
Spiders	2.889 e-1	2.093 e-1	1.38

(b) Last Step, log likelihood −554.237

	Coef.	Std. Err.	Z stat.
Age	0.0338	0.00925	3.65
Albumin	−1.0752	0.24103	−4.46
Bilirubin	0.1070	0.01528	7.00
Edema	0.8072	0.30775	2.62
Hepatomegaly	0.5903	0.21179	2.79
Prothrombin Time	0.2603	0.07786	3.34

Results computed after the four patients with missing values for platelets were assigned the median count, 257, from Table 4.4.1.

ten variable model in Table 4.4.3(a) provides a significantly better fit than the model in Table 4.4.2(b); i.e., the likelihood ratio statistic having 4 d.f. is $-2(-554.237 + 538.274) = 31.926$. It is apparent that the logarithmic transformation of bilirubin provides a substantial improvement and, interestingly, that the dichotomous variable hepatomegaly is no longer independently predictive. Table 4.4.3(b) presents the log likelihood and regression coefficients for the five variable model containing age, albumin, log(bilirubin), edema, and protime. The score statistic for hepatomegaly in that model is only 1.38.

The square and logarithmic transformations of albumin, age, and protime were considered by proceeding "stepwise" in the order of the Z statistics in Table 4.4.3(b) for those untransformed variables. In the model containing age, log(bilirubin), edema and protime, the score statistics for albumin, log(albumin) and $(\text{albumin})^2$

Table 4.4.3 Regression models with log transformations of continuous variables, 312 randomized cases with PBC.

(a) Log likelihood −538.274

	Coef.	Std. Err.	Z stat.
Age	−0.0289	0.07141	−0.41
log(age)	3.2248	3.71828	0.87
Albumin	1.0068	1.73450	0.58
log(Albumin)	−5.8629	5.42315	−1.08
Bilirubin	−0.0461	0.03547	−1.30
log(Bilirubin)	1.0774	0.21127	5.10
Edema	0.8238	0.30386	2.71
Prothrombin Time	−0.6175	1.14523	−0.54
log(Pro Time)	10.1928	13.36131	0.76
Hepatomegaly	0.1964	0.22628	0.87

(b) Log likelihood −541.064

	Coef.	Std. Err.	Z stat.
Age	0.0337	0.00864	3.89
Albumin	−0.9473	0.23713	−3.99
log(Bilirubin)	0.8845	0.09854	8.98
Edema	0.8006	0.29914	2.68
Prothrombin Time	0.2463	0.08426	2.92

(c) Log likelihood −540.412

	Coef.	Std. Err.	Z stat.
Age	0.0333	0.00866	3.84
log(Albumin)	−3.0553	0.72408	−4.22
log(Bilirubin)	0.8792	0.09873	8.90
Edema	0.7847	0.29913	2.62
log(Prothrombin Time)	3.0157	1.02380	2.95

were 15.94, 17.78 and 14.10 respectively. This led to the choice of a logarithmic transformation of albumin. In the model containing log(albumin), log(bilirubin), edema and protime, the score statistics for age, log(age) and $(\text{age})^2$ were 15.00, 14.57 and 14.73. In the model with age, log(albumin), log(bilirubin) and edema, the score statistics for protime, log(protime) and $(\text{protime})^2$ were 8.34, 8.51 and 8.01 respectively.

The log likelihood, coefficients, and standard errors for the final model with transformed variables age, log(albumin), log(bilirubin), edema, and log(protime) are in Table 4.4.3(c). The additional model refinement steps which involved adding variables to this five variable model by considering either transformations of these variables or interaction terms failed to yield significantly improved prediction. The possible benefits of adding stage, SGOT or urine copper can be explored as an exercise.

The final model in Table 4.4.3(c) is biologically reasonable. The negative coefficient for albumin is consistent with the fact that, as the progression of PBC leads to increased hepatocellular damage, the liver's ability to produce albumin is diminished. The increasing damage to bile ducts reduces the liver's ability to excrete the normal amount of bilirubin from the blood, which leads to an increase in serum bilirubin, a bile pigment. Prothrombin, a protein in the plasma, is decreased, which leads to an increase in blood coagulation time. The accumulation of fluids in tissue, referred to as edema, often is associated with later stages of the disease.

The Breslow estimate of Λ_0 and Eq. (4.3) provide patient specific survival estimates. For an individual with risk score $\hat{R} = 5.24$, the median risk score in the 312 trial patients, the corresponding one- and five-year survival estimates are $\hat{S}(1) = 0.982$ and $\hat{S}(5) = 0.845$.

Consider a low-risk patient with serum total bilirubin, 0.5 mg/dl, serum albumin, 4.5 g/dl, age, 52 years, prothrombin time, 10.1 seconds, no edema, and no history of diuretic therapy (i.e., edema = 0). Her risk score is

$$\hat{R} = 0.879 * \log(0.5) - 3.053 * \log(4.5) + 0.033 * 52 + 3.016 * \log(10.1) + 0.785 * 0.0,$$

so $\hat{R} = 3.49$. Her estimated 5-year survival probability is

$$\hat{S}(5) = (0.845)^{\exp(3.49-5.24)} = 0.97,$$

indicating very low risk of death in the next five years, even without liver transplant.

In a high-risk patient with serum total bilirubin, 13.9 mg/dl, serum albumin, 2.8 g/dl, age, 52 years, prothrombin time, 13.8 sec., edema responding to diuretic therapy (i.e., edema 0.5), $\hat{R} = 9.19$ and her estimated one-year survival probability is

$$\hat{S}(1) = (0.982)^{\exp(9.19-5.24)} = 0.39.$$

Under most circumstances, such a high-risk patient would be considered a candidate for liver transplantation.

Table 4.4.4 Adjusted estimation of treatment effect, 312 randomized cases with PBC.

(a) Log likelihood −540.144

	Coef.	Std. Err.	Z stat.
Age	0.0347	0.00891	3.89
log(Albumin)	−3.0771	0.71899	−4.28
log(Bilirubin)	0.8840	0.09871	8.96
Edema	0.7859	0.29647	2.65
log(Prothrombin Time)	2.9707	1.01588	2.92
Treatment	0.1360	0.18543	0.73

A score test of the hypothesis that treatment has no effect on survival, when adjusting for the variables in Table 4.4.3(c), yields the chi-square 0.54. By Table 4.4.4, the adjusted 95% confidence interval for the ratio of placebo to DPCA hazard functions is $\exp\{0.136 \pm (0.185)(1.96)\} = (0.797, 1.646)$, which is shifted to the right of the unadjusted confidence interval obtained earlier.

Study of Gamma Interferon in Chronic Granulomatous Disease

Chronic Granulomatous Disease (CGD) is a group of inherited rare disorders of the immune function characterized by recurrent pyogenic infections which usually present early in life and may lead to death in childhood. Phagocytes from CGD patients ingest microorganisms normally but fail to kill them, primarily due to the inability to generate a respiratory burst dependent on the production of superoxide and other toxic oxygen metabolites. Thus, it is the failure to generate microbicidal oxygen metabolites within the phagocytes of CGD patients which confers the greatly increased susceptibility to these severe or even life threatening infections.

There is evidence establishing a role for gamma interferon as an important macrophage activating factor which could restore superoxide anion production and bacterial killing by phagocytes in CGD patients. In order to study the ability of gamma interferon to reduce the rate of serious infections, that is, the rate of infections requiring hospitalization for parenteral antibiotics, a double-blinded clinical trial was conducted in which patients were randomized to placebo vs. gamma interferon. Between October 1988 and March 1989, 128 eligible patients with CGD were accrued by the International CGD Cooperative Study Group. Since the study required delivering placebo injections three times weekly for a twelve month period to one-half of the patients, most being children, there was particular interest in achieving early termination of the trial if early results were extreme. A single interim analysis was to be performed as soon as patient followup was available through July 1989, six-months after the date on which one-half of the patients had been accrued.

At the time of interim analysis, twenty of 65 placebo patients and seven of 63 patients on gamma interferon each had experienced at least one serious infection. Initially an analysis of time to first infection was performed by fitting a Cox regression model

$$\lambda(t|Z) = \lambda_0(t)\exp(\beta Z),$$

with $Z = 0$: placebo, $Z = 1$: gamma interferon. The methods illustrated earlier in this section yielded a maximum partial likelihood estimate $\hat{\beta} = -1.2062$ with standard error $\{\mathcal{I}(\hat{\beta})\}^{-1/2} = 0.4398$. Thus, the 95% confidence interval for the hazard ratio, e^{β}, was (0.1264, 0.7088). The two-sided test of $H : \beta = 0$ based on the likelihood ratio statistic yielded $p = 0.0032$, meeting the O'Brien-Fleming (1979) guideline ($p \leq 0.005$) for early stopping.

Because all patients were followed through July 15, 1989 to record recurrent infections, additional data were available on the relative rate of serious infections on the two treatments. Of the twenty placebo patients who experienced at least one infection, four experienced two, two others experienced three, and one had five. Of the seven gamma interferon patients with at least one infection, two had a second event. Overall, a total of 32 serious infections were recorded among those on placebo compared to only 9 among patients on gamma interferon. The Andersen-Gill multiplicative intensity model with a single treatment indicator covariate was fit to these recurrent infection data by assuming the patient's intensity for a new serious infection at t satisfied $l_i(t) = Y_i(t)\lambda_0(t)\exp(\beta Z_i)$ and was not further altered by the pattern of previous infections, where $Y_i(t)$ indicated whether the ith patient was followed up to t months post randomization. This analysis yielded a partial likelihood estimate $\hat{\beta} = -1.2765$, with standard error $\{\mathcal{I}(\hat{\beta})\}^{-1/2} = 0.3774$. Thus the additional information on recurrent events reduced the variability in the estimate of treatment effect. This resulted in a narrower 95% confidence interval for the hazard ratio, e^{β}, given by (0.1332, 0.5846), and a likelihood ratio test of $H : \beta = 0$ being even more significant ($p = 0.0002$). By making full use of the recurrent infection data, the multiplicative intensity model based analysis strengthened the evidence justifying early termination of the CGD study. Since $\exp(\hat{\beta}) = 0.2790$, the results of this analysis suggest that gamma interferon reduces the rate of serious infection in CGD patients by 72.1%.

The model development in this section illustrates likelihood procedures based on the partial likelihood. The next section examines graphical methods based upon martingale residuals. These graphical methods are useful for identifying outliers and leverage points, and for providing insight in model building, variable transformation, and checking the proportional hazards assumption.

4.5 MARTINGALE RESIDUALS

In the multiplicative intensity model, the intensity function for the counting process N_i is given at time t by

$$l_i(t) = Y_i(t)\lambda\{t|\mathbf{Z}_i(t)\}, \tag{5.1}$$

where $\mathbf{Z}_i$ is a covariate vector process, Y_i is a predictable 0,1 valued process, and

$$\lambda\{t|\mathbf{Z}_i(t)\}dt = \exp\{\beta'\mathbf{Z}_i(t)\}d\Lambda_0(t) \tag{5.2}$$

for an unspecified cumulative hazard function $\Lambda_0(\cdot) = \int_0^{\cdot} \lambda_0(t)dt$ and vector of regression coefficients β. It follows that

$$M_i = N_i - \int_0^{\cdot} Y_i(s)\exp\{\beta'\mathbf{Z}_i(s)\}d\Lambda_0(s) \tag{5.3}$$

is a "case-specific" (local) square integrable martingale with respect to a stochastic basis discussed in Section 4.2. The Andersen-Gill proportional hazards model (AGPHM) arises if

$$\mathbf{Z}_i(t) = \mathbf{Z}_i \text{ for all } t \geq 0. \tag{5.4}$$

The Cox proportional hazards model without time-varying covariates arises as a special case of the AGPHM when $Y_i(t) = I_{\{X_i \geq t\}}$ and $N_i(t) = I_{\{X_i \leq t; \delta_i = 1\}}$, where X_i usually denotes the event or censorship time for the ith individual and δ_i is a 0, 1 variable indicating whether X_i is the event time ($\delta_i = 1$) or censorship time ($\delta_i = 0$). In this setting, the martingale in Eq. (5.3) evaluated at $t = \infty$ reduces to

$$M_i(\infty) = \delta_i - \int_0^{X_i} \exp(\beta'\mathbf{Z}_i)d\Lambda_0(s). \tag{5.5}$$

As illustrated in the previous two sections, the vector of regression parameters β and the unspecified baseline cumulative hazard $\Lambda_0(\cdot)$ commonly are estimated by the maximum partial likelihood estimator $\hat{\beta}$ and the Breslow estimate Eq. (3.29),

$$\hat{\Lambda}_0(\cdot) = \sum_{i=1}^{n} \int_0^{\cdot} \left\{ \sum_{k=1}^{n} Y_k(s)e^{\hat{\beta}'\mathbf{Z}_k(s)} \right\}^{-1} dN_i(s),$$

and standard methods are used for confidence intervals and tests of hypotheses. Diagnostic tools are important in such regression analyses for assessing model adequacy and the influence of individual cases on parameter estimation. In the remainder of this chapter, certain types of residuals which are useful diagnostic tools are discussed, focusing in particular on their graphical applications. We illustrate the use of residuals to assess (i) transformations of a covariate in a model already accounting for other covariates; (ii) the leverage exerted by each subject in parameter estimation; (iii) the accuracy of the model in predicting the failure rate of a given case; and (iv) the validity of the proportional hazards assumptions.

The martingales defined in Eq. (5.3) for the Andersen-Gill model can be interpreted, at each t, as the difference over $[0, t]$ between the observed number of events for the ith individual and a conditionally expected number of events. Replacing β by $\hat{\beta}$ and Λ_0 by $\hat{\Lambda}_0$ produces a type of residual given by

$$\widehat{M}_i \equiv N_i - \int_0^{\cdot} Y_i(s)e^{\hat{\beta}'\mathbf{Z}_i(s)}d\hat{\Lambda}_0(s). \tag{5.6}$$

We will refer to $\widehat{M}_i(\infty) \equiv \widetilde{M}_i$ and $\widehat{M}_i(t)$ as the "martingale residual," although the latter should be called the "martingale residual at t." In the Cox proportional hazards model, by Eq. (5.5),

$$\widetilde{M}_i = \delta_i - \int_0^{X_i} \exp(\hat{\beta}'\mathbf{Z}_i)d\hat{\Lambda}_0(s)$$
$$\equiv \delta_i - \hat{\Lambda}_i(X_i),$$

a residual proposed from a different perspective by Kay (1977).

Martingale residuals have some desirable properties. For any $t \in [0, \infty]$,

$$\sum_{i=1}^n \widehat{M}_i(t) = \sum_{i=1}^n \left\{N_i(t) - \int_0^t Y_i(s)e^{\hat{\beta}'\mathbf{Z}_i(s)}d\hat{\Lambda}_0(s)\right\}$$
$$= \sum_{i=1}^n \left[\int_0^t dN_i(s) - \int_0^t Y_i(s)e^{\hat{\beta}'\mathbf{Z}_i(s)}\left\{\frac{\sum_{j=1}^n dN_j(s)}{\sum_{k=1}^n Y_k(s)e^{\hat{\beta}'\mathbf{Z}_k(s)}}\right\}\right]$$
$$= 0,$$

so the martingale residuals sum to zero. Also, $\text{cov}(\widehat{M}_i, \widehat{M}_j) = 0 = E\widehat{M}_i$ asymptotically (Exercise 4.8). We will now begin to explore the use of these residuals in model building.

Covariate Transformations

Consider the model specified by Eqs. (5.1) and (5.2). A key component of such a model is the functional form $\exp\{\beta'\mathbf{Z}(t)\}$ specified for the covariates. Perhaps one of the variables X should be replaced by $X^{1/2}$, or by $I_{\{X>c\}}$, or by some other transformation. Consider the model,

$$\lambda\{t \mid X, \mathbf{Z}^{p\times 1}(t)\}dt = h(X)\exp\{\beta'\mathbf{Z}(t)\}d\Lambda_0(t)$$
$$\equiv \exp\{f(X)\}\exp\{\beta'\mathbf{Z}(t)\}d\Lambda_0(t), \qquad (5.7)$$

where the functional form of the covariate vector $\mathbf{Z}(t)$ is known, but where the positive function h for a non time-dependent covariate X is unknown. Plots of martingale residuals against X can be used to obtain estimates of h or f.

Suppose $\widehat{M}(t)$ denotes the martingale residual arising under Eq. (5.7) in a model omitting X,

$$\lambda\{t \mid \mathbf{Z}(t)\}dt = \exp\{\beta^{*\prime}\mathbf{Z}(t)\}d\Lambda_0^*(t).$$

Then, writing $\mathbf{Z}(t)$ as $\mathbf{Z}$,

$$E\{\widehat{M}(t)|X\} = E\{N(t)|X\} + E\left\{-\int_0^t Y(s)e^{\beta'\mathbf{Z}}\bar{h}(s, \mathbf{Z})d\Lambda_0(s)|X\right\}$$
$$+E\left\{\int_0^t Y(s)[d\Lambda(s|\mathbf{Z}) - e^{\hat{\beta}'\mathbf{Z}}d\hat{\Lambda}_0(s)]|X\right\} \qquad (5.8)$$
$$\equiv \text{term 1} + \text{term 2} + \text{term 3},$$

where

$$d\Lambda(s|\mathbf{Z}) = e^{\beta'\mathbf{Z}} \frac{E\{e^{f(X)}Y(s)|\mathbf{Z})\}}{E\{Y(s)|\mathbf{Z}\}} d\Lambda_0(s)$$
$$\equiv e^{\beta'\mathbf{Z}} \bar{h}(s, \mathbf{Z}) d\Lambda_0(s)$$

is now the failure rate of $N(s)$ when X is ignored. Note that, given $\mathbf{Z}$ and ignoring X, $N - \int Y(s) d\Lambda(s|\mathbf{Z})$ is a martingale.

Term 3 in the decomposition of $E\{\hat{M}(t)|X\}$can be rewritten

$$-E\left\{\sum_{i=1}^{n} \int_0^t \frac{Y(s)e^{\hat{\beta}'\mathbf{Z}}}{\sum_{j=1}^{n} Y_j(s)e^{\hat{\beta}'\mathbf{Z}_j}} \left[dN_i(s) - Y_i(s)\frac{e^{\beta'\mathbf{Z}}}{e^{\hat{\beta}'\mathbf{Z}}} e^{\hat{\beta}'\mathbf{Z}_i} \bar{h}(s, \mathbf{Z}) d\Lambda_0(s)\right] \middle| X\right\}.$$

As n increases, this term converges to zero in probability by the martingale structure, when

$$\frac{e^{\beta'\mathbf{Z}}}{e^{\hat{\beta}'\mathbf{Z}}} e^{\hat{\beta}'\mathbf{Z}_i} \bar{h}(s, \mathbf{Z}) \approx e^{\beta'\mathbf{Z}_i} \bar{h}(s, \mathbf{Z}_i).$$

Using the results in Bickel et al., (1990), $\hat{\beta}$ is a consistent estimate of β^*, where $\beta^* \approx \beta$ when β is small and X is ignorable (e.g., see (3) of Example 4.2.1). If we set $\bar{h}(s) = E\bar{h}(s, \mathbf{Z})$ and t_0 is a fixed time,

$$\begin{aligned} \text{term 2} &= -E\left\{\int_0^t Y(s)e^{\beta'\mathbf{Z}} \frac{\bar{h}(t_0)}{h(X)} h(X) d\Lambda_0(s)|X\right\} \\ &\quad + E\left\{\int_0^t Y(s)e^{\beta'\mathbf{Z}} [\bar{h}(t_0) - \bar{h}(s, \mathbf{Z})] d\Lambda_0(s)|X\right\} \\ &\equiv -\frac{\bar{h}(t_0)}{h(X)} E\{N(t)|X\} + R(t, X). \end{aligned}$$

Combining terms 1 and 2 yields

$$E\{\widehat{M}(t)|X\} \approx \left(1 - \frac{\bar{h}(t_0)}{h(X)}\right) E\{N(t)|X\} + R(t, X), \tag{5.9}$$

where $R(t, X)$ can be treated as a remainder term.

Equation (5.9) has the interesting interpretation

$$E(\text{\# excess failures}) \approx (1 - \text{hazard ratio}) E(\text{\# events}).$$

Inverting Eq. (5.9) yields

$$-\log\left\{1 - \frac{E(\widehat{M}(t)|X)}{E(N(t)|X)}\right\} \tag{5.10}$$

$$\approx [f(X) - \log\{\bar{h}(t_0)\}] - \log\left[1 - \frac{h(X)R(t, X)}{\bar{h}(t_0)E\{N(t)|X\}}\right]$$

$$= f(X) - \log \left[\frac{E\left\{\int_0^t \bar{h}(s,\mathbf{Z})Y(s)h(X)e^{\beta'\mathbf{Z}}d\Lambda_0(s)|X\right\}}{E\left\{\int_0^t Y(s)h(X)e^{\beta'\mathbf{Z}}d\Lambda_0(s)|X\right\}} \right]$$

$$\equiv f(X) - \log \bar{h}$$

$$\equiv f(X) - \bar{f}.$$

The left-hand side will be an accurate estimate of f whenever the function $\bar{f}$ has small variation compared to that of $f(X)$. This lack of dependence of $\bar{f}$ on X is particularly likely if X and $\mathbf{Z}$ are independent. When X and Y are jointly independent of $\mathbf{Z}$ (or $\mathbf{Z}$ is null), the ratio $\bar{h}$ can be simplified to

$$\frac{\int \bar{h}(s)P\{Y(s) = 1|X\}h(X)d\Lambda_0(s)}{\int P\{Y(s) = 1|X\}h(X)d\Lambda_0(s)},$$

which is the expected value of h for subjects in the risk set, at the time when a subject with covariate value X has an event. In the general case, $\bar{h}$ involves an average, over the expected event time of a subject with given value of X, of a function which is itself an average of h over the expected risk set at that time. The function $\bar{h}$ depends in a complex way upon both the censoring pattern and the true functional form, and will in general not be constant in X. When X is independent of $\mathbf{Z}$, however, $\bar{h}(s,\mathbf{Z})\exp(\beta'\mathbf{Z})$ will be independent of X, and smoothing provided by the integral will cause $\bar{f}$ to have small variation compared to f. In contrast, when X and $\mathbf{Z}$ are strongly correlated, $\bar{f}$ can vary substantially with X and could affect estimation of f, similar to the situation arising in classical linear models, (e.g., see Mansfield and Conerly (1987)).

Since $\widehat{M}_i$ and N_i are both available, Eq. (5.10) can be used to obtain

$$f(X) - \bar{f} \approx -\log\left\{1 - \frac{\text{smooth}(\widehat{M}, X)}{\text{smooth}(N, X)}\right\}, \tag{5.11}$$

where smooth$(\widehat{M}, X)$ is a smoothed estimate of $E\{\widehat{M}(t)|X\}$, perhaps obtained by smoothing a scatterplot of $\widehat{M}_i$ versus X_i, as is smooth(N, X). One such scatterplot smooth is LOWESS (Cleveland, 1979).

For most data sets (e.g., when $\beta EN < 2$), the approximation in Eq. (5.9) at $t = \infty$ may be replaced by one of two simpler approximations:

$$E\{\widetilde{M}|X\} \approx \{f(X) - \bar{f}\}c \tag{5.12}$$

or

$$E\{\widetilde{M}|X\} \approx \{h(X) - \bar{h}\}c, \tag{5.13}$$

where c =(total number of events)/(total number of subjects). Relationship (5.12) is a first order Taylor approximation, applicable when the dependence of f on X is not extreme and when the dependence of $E\{N|X\}$ on X is weak, such as in uncensored to moderately censored Cox model data. The approximation (5.13) can be used with Poisson-like data, i.e., when $Y_i(t) = 1$ for nearly all subjects over some interval of time.

In summary, when X is not strongly correlated with $\mathbf{Z}$, term 3 and the variation of $\bar{f}$ in X are usually small, implying that a smoothed plot of the $\widetilde{M}_i$ versus a covariate X will show approximately the correct functional form for X. This is quite useful in model building applications. For example, in an AGPHM containing p variables $\mathbf{Z}^{p\times 1}$, one can plot the martingale residuals against transformations of a $(p+1)$st variable, X. A nearly linear scatterplot smooth provides evidence for the appropriate transformation of X. A horizontal smooth is consistent with X providing little additional predictive value to a model containing $\mathbf{Z}$. The scatterplot smoother can also help identify whether a variable X has a threshold effect which would suggest entering a dichotomized version of X into the model. One advantage to plotting the "raw" martingale residuals in (5.12) and (5.13) rather than the transformed function in (5.11) is interpretability: the vertical axis represents excess failures.

Smoothed plots of the martingale residuals have the desirable property that the graphical representation of the residuals on the plot provide some indication as to the variability of and the influence of individuals on the estimate of a covariate transformation. Another type of residual, based on the deviance of an estimated model, is particularly well suited for finding outliers.

Deviance Residuals

Martingale residuals have skewed distributions, especially in the single event setting of the Cox model. When $N_i(\infty) \le 1$, the maximum value of $\widetilde{M}_i$ is one while the minimum is $-\infty$. Transformations to achieve a more normal shaped distribution are helpful, particularly when the accuracy of the predictions for individual subjects is to be assessed. One such transformation is motivated by the deviance residuals from the generalized linear models literature (McCullagh and Nelder (1989)). The deviance D is

$$D = 2\{\log \text{likelihood(saturated)} - \log \text{likelihood}(\hat{\beta})\},$$

where a saturated model is one in which β is allowed to vary with each observation. There may be other nuisance parameters θ which are held constant across the two models, such as σ^2 in a normal error linear model.

Consider the AGPHM specified by Eqs. (5.1), (5.2) and (5.4) and having $\Lambda_0 = \int \lambda_0(s)ds$, where the nuisance parameter is Λ_0. To obtain expressions for the deviance D in the following calculations, estimation of the nuisance parameter Λ_0 should not be performed separately for each model. In this setting,

$$D = 2\sup_h \sum_{i=1}^{n} \left\{ \int_0^\infty \left(\log e^{\mathbf{h}_i'\mathbf{Z}_i} - \log e^{\hat{\beta}'\mathbf{Z}_i} \right) dN_i(s) \right.$$
$$\left. - \int_0^\infty Y_i(s)(e^{\mathbf{h}_i'\mathbf{Z}_i} - e^{\hat{\beta}'\mathbf{Z}_i}) d\Lambda_0(s) \right\},$$

where $\mathbf{h}_i$ is the value of β for the ith case.

Because terms separate, the likelihood can be maximized with respect to each $\mathbf{h}_i$ separately. By a simple Lagrange multiplier argument, the maximum occurs when

$$\int_0^\infty Y_i(s) e^{\hat{\mathbf{h}}_i'\mathbf{Z}_i} d\Lambda_0(s) = \int_0^\infty dN_i(s).$$

If $\widetilde{\widetilde{M}}_i = N_i(\infty) - \int_0^\infty Y_i(s)e^{\beta'\mathbf{Z}_i}d\Lambda_0(s)$,

$$D = 2\sum_{i=1}^{n}\left\{-\log\left(e^{\beta'\mathbf{Z}_i}/e^{\hat{\mathbf{h}}_i'\mathbf{Z}_i}\right)\int_0^\infty dN_i(s) - \widetilde{\widetilde{M}}_i\right\}$$

$$= 2\sum_{i=1}^{n}\left[-N_i(\infty)\log\left\{\frac{N_i(\infty) - \widetilde{\widetilde{M}}_i}{N_i(\infty)}\right\} - \widetilde{\widetilde{M}}_i\right].$$

When the nuisance parameter Λ_0 is estimated by Breslow's estimate, $\widetilde{\widetilde{M}}_i$ is replaced by $\widetilde{M}_i$. The deviance residual, d_i, is defined as the signed square root of the ith summand in the expression for D, with $\widetilde{\widetilde{M}}_i$ replaced by $\widetilde{M}_i$; specifically,

$$d_i = \text{sign}(\widetilde{M}_i)\sqrt{2}\left[-\widetilde{M}_i - N_i(\infty)\log\left\{\frac{N_i(\infty) - \widetilde{M}_i}{N_i(\infty)}\right\}\right]^{1/2}. \qquad (5.14)$$

Equation (5.14) is equivalent to the deviance formula for a Poisson model found in McCullagh and Nelder (1989), with y_i replaced by $N_i(\infty)$ and $u = \hat{\lambda}t_i$ replaced by the observed cumulative hazard $\int Y_i(s)\exp(\hat{\beta}'\mathbf{Z}_i)d\hat{\Lambda}_0$. Note that $d_i = 0$ if and only if $\widetilde{M}_i = 0$, and that for the Cox model,

$$d_i = \text{sign}(\widetilde{M}_i)\sqrt{2}\left\{-\widetilde{M}_i - \delta_i\log(\delta_i - \widetilde{M}_i)\right\}^{1/2}.$$

The square root shrinks the large negative martingale residuals, while the logarithmic transformation expands those residuals that are close to unity. Plots by individual case of the deviance residuals for a given model provide evidence as to the accuracy of the model in predicting the failure rate of each case.

Martingale Transform Residuals and Score Residuals

The martingale residual, $\widetilde{M}_i$, arises naturally as a residual in that it is the difference between the observed and model-predicted expected number of events for the ith case. For $0 = t_0 < t_1 < t_2 < \cdots$, where $t_j \to \infty$ as $j \to \infty$,

$$\widetilde{M}_i = N_i(\infty) - \int_0^\infty Y_i(s)e^{\hat{\beta}'\mathbf{Z}_i(s)}d\hat{\Lambda}_0(s)$$

$$= \sum_{j=1}^{\infty}\left[\{N_i(t_j) - N_i(t_{j-1})\} - \int_{t_{j-1}}^{t_j} Y_i(s)e^{\hat{\beta}'\mathbf{Z}_i(s)}d\hat{\Lambda}_0(s)\right]$$

$$\equiv \sum_{j=1}^{\infty}\Delta\widehat{M}_{i,j}.$$

Specifically, $\Delta\widehat{M}_{i,j}$ represents the "observed minus conditionally expected" events for the ith case over $(t_{j-1}, t_j]$. One approach to obtaining a class of residuals would be to weight the $\Delta\widehat{M}_{i,j}$ terms in the sum and let the mesh of the partition approach zero. If the process of weights $W_i = \{W_i(s) : s \geq 0\}$ is bounded, predictable and adapted with respect to the stochastic basis, then $\int W_i(s)dM_i(s)$ is a martingale transform and hence also is a martingale. In turn, $\int_0^\infty W_i(s)d\widehat{M}_i(s)$ is a "martingale-transform" residual. The score residual below provides an important illustration of a residual from this class.

The partial likelihood for the model specified by Eqs. (5.1) and (5.2) is given by

$$L(\beta) = \prod_{i=1}^{n}\prod_{t\geq 0}\left[\frac{Y_i(t)\exp\{\beta'\mathbf{Z}_i(t)\}}{\sum_{l=1}^{n} Y_l(t)\exp\{\beta'\mathbf{Z}_l(t)\}}\right]^{\Delta N_i(t)},$$

so the score vector (using information over [0,t]) is

$$\mathbf{U}^{p\times 1}(\beta, t) = \sum_{i=1}^{n}\int_0^t \left[\mathbf{Z}_i(s) - \frac{\sum_{l=1}^{n} Y_l(s)\mathbf{Z}_l(s)\exp\{\beta'\mathbf{Z}_l(s)\}}{\sum_{l=1}^{n} Y_l(s)\exp\{\beta'\mathbf{Z}_l(s)\}}\right] dM_i(s).$$

If each component of the random covariate process $\mathbf{Z}_i$ is bounded, and $EN_i(t) < \infty$, we saw in Section 4.3 that each component of $\underset{\sim}{\mathrm{U}}(\beta, \cdot)$ is a martingale. If $\hat{\beta}$ is the maximum partial likelihood estimate of β, then

$$\begin{aligned} \mathbf{0} &= \mathbf{U}(\hat{\beta}, \infty) \\ &= \sum_{i=1}^{n}\int_0^\infty \left[\mathbf{Z}_i(s) - \frac{\sum_{l=1}^{n} Y_l(s)\mathbf{Z}_l(s)\exp\{\hat{\beta}'\mathbf{Z}_l(s)\}}{\sum_{l=1}^{n} Y_l(s)\exp\{\hat{\beta}'\mathbf{Z}_l(s)\}}\right] d\widehat{M}_i(s) \\ &\equiv \sum_{i=1}^{n} \mathbf{S}_i. \end{aligned}$$

For the ith subject, $\mathbf{S}_i$ is a p-dimensional "score vector residual," with the sum over subjects being identically zero for each component of the vector. The role of these residuals in diagnostics will now be demonstrated.

The Assessment of Influence

A direct approach to assessing the influence of the ith individual on the parameter estimates is to recompute $\hat{\beta}$ while leaving out that individual from the data set. This technique, referred to as the jackknife, can require considerable computation since it requires a recomputation of parameter estimates n times. Two computationally efficient methods provide approximations to the jackknife.

The influence of an observation on model fit depends on both the residual from the fit and on the extremity of its covariate value, roughly $(\mathbf{Z}_i - \bar{\mathbf{Z}}) \times$ residual. In the Andersen-Gill model specified by Eq. (5.2), both $\mathbf{Z}_i$ and $\bar{\mathbf{Z}}$ are functions of time, where the jth component of $\bar{\mathbf{Z}}$ is

$$\bar{Z}_j(t) = \frac{\sum_{i=1}^n Y_i(t) Z_{ij}(t) \exp\{\hat{\beta}'\mathbf{Z}_i(t)\}}{\sum_{i=1}^n Y_i(t)\exp\{\hat{\beta}'\mathbf{Z}_i(t)\}}.$$

This suggests using the score residual

$$S_{ij} = \int_0^\infty \{Z_{ij}(t) - \bar{Z}_j(t)\} d\widehat{M}_i(t)$$

as an influence measure.

To formalize this, the approach of Cain and Lange (1984) can be used by defining a weighted score vector

$$\underset{\sim}{\mathrm{U}}(\beta, \mathbf{w}) \equiv \sum_{i=1}^n w_i \int_0^\infty Y_i(t)\{\mathbf{Z}_i(t) - \tilde{\mathbf{Z}}(t)\} dN_i(t),$$

where $\tilde{\mathbf{Z}}$ is the reweighted mean

$$\tilde{\mathbf{Z}}(t) = \frac{\sum_{i=1}^n Y_i(t) w_i \mathbf{Z}_i(t) \exp\{\beta'\mathbf{Z}_i(t)\}}{\sum_{i=1}^n Y_i(t) w_i \exp\{\beta'\mathbf{Z}_i(t)\}}.$$

Then

$$\frac{\partial\beta}{\partial w_i} = \frac{\partial\beta}{\partial\mathbf{U}}\frac{\partial\mathbf{U}}{\partial w_i} = \{-\mathcal{I}(\beta)\}^{-1}\frac{\partial\mathbf{U}}{\partial w_i}, \tag{5.15}$$

where $\mathcal{I}(\beta)^{p\times p}$ is the observed information matrix. Evaluation of Eq. (5.15) at $w_i = 1, i = 1, \ldots, n$, and at $\beta = \hat{\beta}$ yields the infinitesimal jackknife estimate of influence. In this setting

$$\frac{\partial\mathbf{U}}{\partial w_i} = \int_0^\infty Y_i(t)\{\mathbf{Z}_i(t) - \tilde{\mathbf{Z}}(t)\} dN_i(t)$$
$$- \sum_{l=1}^n w_l \int_0^\infty Y_l(t) Y_i(t)\{\mathbf{Z}_i(t) - \tilde{\mathbf{Z}}(t)\}\left[\frac{\exp\{\beta'\mathbf{Z}_i(t)\}}{\sum_{k=1}^n Y_k(t) w_k \exp\{\beta'\mathbf{Z}_k(t)\}}\right] dN_l(t).$$

Thus, if $\hat{\Lambda}_0$ again denotes the Breslow estimate,

$$\left.\frac{\partial\mathbf{U}}{\partial w_i}\right|_{\beta=\hat{\beta}; w=1} = \int_0^\infty Y_i(t)\{\mathbf{Z}_i(t) - \bar{\mathbf{Z}}(t)\}\{dN_i(t) - e^{\hat{\beta}'\mathbf{Z}_i(t)} d\hat{\Lambda}_0(t)\}$$
$$= \mathbf{S}_i$$

In the special case of the Cox model, this reduces to Equation 4 of Cain and Lange. The influence of the ith case on the estimation of β then is approximately

$$\Delta^i\beta \equiv [\{-\mathcal{I}(\hat{\beta})\}^{-1}\mathbf{S}_i]^{p\times 1}. \tag{5.16}$$

This measure of influence may underestimate the true jackknife, especially for extreme values of $\mathbf{Z}$, because $\mathcal{I}^{-1}$ also changes when the observation is removed. An alternative approach is to compute the one-step update in $\hat{\beta}$ when a single covariate Z_{p+1} is added, with Z_{p+1} equal to unity for subject i and equal to zero for all others. This is explored for the Cox model by Storer and Crowley (1985). For the Andersen-Gill model, at $(\hat{\beta}'^{\,1\times p}, 0)$,

$$\bar{Z}_{p+1}(t) = \frac{Y_i(t)\exp\{\hat{\beta}'\mathbf{Z}_i(t)\}}{\sum_{k=1}^n Y_k(t)\exp\{\hat{\beta}'\mathbf{Z}_k(t)\}}.$$

If $U_j \equiv (\mathbf{U})_j$, then $U_j\left\{(\hat{\beta}'^{\,1\times p}, 0)'\right\} = 0$ for $j = 1, \ldots, p$, while

$$\begin{aligned} U_{p+1}\left\{(\hat{\beta}'^{\,1\times p}, 0)'\right\} &= \sum_{l=1}^{n} \int_0^\infty Y_l(t)\{Z_{l,p+1}(t) - \bar{Z}_{p+1}(t)\} dN_l(t) \\ &= \int_0^\infty dN_i(t) - \int_0^\infty Y_i(t) e^{\hat{\beta}'\mathbf{Z}_i(t)} d\hat{\Lambda}_0(t) \\ &= \widetilde{M}_i. \end{aligned}$$

The new observed information matrix evaluated at $(\hat{\beta}', 0)'$ is given by

$$\mathcal{I}_{\text{NEW}} = \begin{pmatrix} \mathcal{I}(\hat{\beta})^{p\times p} & \gamma_i^{p\times 1} \\ \gamma_i' & \eta_i \end{pmatrix},$$

where

$$\gamma_i = \int_0^\infty Y_i(t)\{\mathbf{Z}_i(t) - \bar{\mathbf{Z}}(t)\} \exp\{\hat{\beta}'\mathbf{Z}_i(t)\} d\hat{\Lambda}_0(t)$$

and

$$\eta_i = \int_0^\infty Y_i(t)\{1 - \bar{Z}_{p+1}(t)\} \exp\{\hat{\beta}'\mathbf{Z}_i(t)\} d\hat{\Lambda}_0(t).$$

The influence of the ith subject on the estimation of β is approximately

$$\{-\mathcal{I}_{\text{NEW}}\}^{-1} \mathbf{U}^{(p+1)\times 1}\left\{(\hat{\beta}', 0)'\right\}.$$

Using a standard formula for the inverse of a partitioned matrix, this reduces to

$$\widetilde{\Delta}^i\beta = \left[\frac{-\{\mathcal{I}(\hat{\beta})\}^{-1}\gamma_i}{\eta_i - \gamma_i'\{\mathcal{I}(\hat{\beta})\}^{-1}\gamma_i}\right] \widetilde{M}_i. \tag{5.17}$$

The two formulas in Eqs. (5.16) and (5.17) are not very different; high leverage points are highlighted by both. However the score residual based measure of influence given in Eq. (5.16) has several technical advantages, including: (i) simplicity of interpretation as components of the score statistic; (ii) availability for all values of β, not just $\hat{\beta}$, (for instance, at $\beta = \mathbf{0}$, they are components of the logrank statistic); and (iii) a structure that allows results from stochastic integrals to be used for deriving variance estimators.

Assessing Model Adequacy

We will briefly discuss assessing a model assumption specified by Eq. (5.1), then show the use of martingale residuals in checking the proportional hazards assumption implicit in Eqs. (5.2) and (5.4).

One subtle model assumption involves the interpretability of the function $\lambda\{t \mid \mathbf{Z}_i(t)\}$ in (5.1). By Eq. (5.1), heuristically,

$$E\{N_i(t) - N_i(t - dt)|\mathcal{F}_{t-}\} = Y_i(t)\lambda\{t|\mathbf{Z}_i(t)\}dt. \tag{5.18}$$

The interpretation of λ is intrinsically tied to the censoring process Y. Consider the special case in which Eq. (5.4) holds and U_i and T_i represent a censoring time and an absolutely continuously distributed survival time for the ith individual. Suppose $Y_i(t) \equiv I_{\{U_i \geq t; T_i \geq t\}}$, $N_i(t) = I_{\{T_i \leq t; T_i \leq U_i\}}$, and

$$\mathcal{F}_t = \sigma\{N_i(s), I_{\{U_i \leq s, U_i < T_i\}}, \mathbf{Z}_i : 0 \leq s \leq t\}.$$

In this setting, λ in (5.18) is the "crude hazard",

$$\lambda_c(t|\mathbf{Z}_i) = \frac{-\partial}{\partial x} P\{T \geq x, U \geq t | T \geq t, U \geq t, \mathbf{Z}_i\}\Big|_{x=t},$$

and the Breslow estimate is a consistent estimator for $\int_0^{\cdot} \lambda_c(s)ds$. The problem of interpretability arises when λ in Eq. (5.18) is assumed to be the "net hazard,"

$$\lambda_n(t|\mathbf{Z}_i) = \frac{-\partial}{\partial x} P\{T \geq x | T \geq t, \mathbf{Z}_i\}\Big|_{x=t}.$$

The assumption $\lambda_c(\cdot) = \lambda_n(\cdot)$ imposes additional structure on Eq. (5.1), and is untestable using martingale residuals or any other approach. As in Chapter 1, the assumption $\lambda_c(\cdot) = \lambda_n(\cdot)$ implies

$$M_i = N_i - \int_0^{\cdot} Y_i(s)\lambda_n(s|\mathbf{Z}_i)ds$$

is a martingale.

Two important model assumptions in the AGPHM are the specification of the form of the "link" function through Eq. (5.2) and the proportional hazards property which arises when Eq. (5.4) also holds. In the remainder of this section, we examine diagnostics of the proportional hazards assumption, which states that the hazard ratio $\lambda(t|\mathbf{Z}_{i'})/\lambda(t|\mathbf{Z}_i)$ is independent of t, for any individuals i, i'. Although we focus on methods based on the score process and martingale residuals, we initially consider some alternative approaches.

If Eqs. (5.2) and (5.4) hold, $\lambda(t|\mathbf{Z}_i) = \lambda_0(t)e^{\beta'\mathbf{Z}_i}$. In an analytical approach to assessing proportional hazards, one can consider the model

$$\lambda(t|\mathbf{Z}_i) = \lambda_0(t)e^{\beta'\mathbf{Z}_i}e^{\gamma'\mathbf{Z}_i(t)},$$

where, for $j = 1, \ldots, p$, $\{\mathbf{Z}_i(t)\}_j = Z_{ij}Q_j(t)$ for a specified function Q_j. A test of $H : \gamma = \mathbf{0}$ provides an assessment of proportional hazards which is sensitive to the type of departure specified by the functions $\{Q_j : j = 1, \ldots, p\}$.

In a second approach, the time axis can be divided into K intervals $[0, t_1)$, $[t_1, t_2), \ldots, [t_{K-1}, t_K)$, with $t_K = \infty$. To assess proportionality with respect to the jth variable, the Andersen-Gill model can then be fit with the term $\beta_j Z_j$ replaced by $\sum_{k=1}^{K} \beta_{jk} Z_j \; I_{\{t_{k-1} \leq t < t_k\}}$. Heterogeneity in a plot of $\{\hat{\beta}_{j1}, \hat{\beta}_{j2}, \ldots, \hat{\beta}_{jK}\}$ is evidence of non-proportionality.

A third approach can be used for the Cox model. Since

$$\lambda(t|\mathbf{Z}_i) = \lambda_0(t)\exp(\beta'\mathbf{Z}_i^{p\times 1}),$$

and $S(t|\mathbf{Z}_i = \mathbf{0}) \equiv S_0(t)$, then

$$S(t|\mathbf{Z}_i) = \{S_0(t)\}^{e^{\beta'\mathbf{Z}_i}}$$

and

$$\log\{-\log S(t|\mathbf{Z}_i)\} = \beta'\mathbf{Z}_i + \log\{-\log S_0(t)\}. \tag{5.19}$$

Suppose one wishes to assess the proportional hazards property for a discrete variable $Z_{i,p+1}$ in a proportional hazards model with variables $\mathbf{Z}_i^{p\times 1} \equiv (Z_{i1}, \ldots, Z_{ip})'$. Stratifying on the Q levels of $Z_{i,p+1}$ leads to the stratified model $\lambda_q(t|\mathbf{Z}_i) = \lambda_{0q}(t)$ $\exp(\beta'\mathbf{Z}_i)$ and

$$S_q(t|\mathbf{Z}_i) = \{S_{0q}(t)\}^{e^{\beta'\mathbf{Z}_i}}.$$

The estimates $\hat{S}_{0q}(t) = \exp\{-\hat{\Lambda}_{0q}(t)\}$, where $\hat{\Lambda}_{0q}$, is defined by Eq. (3.30), can be used to plot $\{\log[-\log \hat{S}_{0q}(t)] : t \geq 0\}, q = 1, \ldots, Q$. By Eq. (5.19), approximate parallelism in these Q plots is evidence that Z_{p+1} satisfies the proportional hazards assumption. If plots are parallel but are not equally spaced, a transformation of Z_{p+1} might be appropriate. Observe that a plot of $\log\{-\log \hat{S}_{0q}(\cdot)\}$ is equivalently a plot of $\log\{\hat{\Lambda}_{0q}(\cdot)\}$.

A fourth approach, developed in Gill and Schumacher (1987) for the two-sample problem, where $Z^{1\times 1} \in \{1, 2\}$, is related to the weighted logrank statistics. In the notation from Section 3 of Chapter 3, these statistics have the form

$$W_K = \int_0^\infty K(s)\{d\hat{\Lambda}_1(s) - d\hat{\Lambda}_2(s)\},$$

where $d\hat{\Lambda}_k(s) = \{\overline{Y}_k(s)\}^{-1} d\overline{N}_k(s)$ and K is a bounded nonnegative predictable process. When the proportional hazards assumption holds, the ratio

$$\hat{\theta}_K \equiv \frac{\int_0^\infty K(s)d\hat{\Lambda}_2(s)}{\int_0^\infty K(s)d\hat{\Lambda}_1(s)}$$

is a consistent estimate of $\theta = \lambda(t|Z = 2)/\lambda(t|Z = 1)$ for any weight function K. Specific departures from proportional hazards can be detected by a statistic $\hat{\theta}_{K_1} - \hat{\theta}_{K_2}$ through a judicious choice of weight functions K_1 and K_2. For example, if $\lambda(t|Z = 2)/\lambda(t|Z = 1)$ decreases monotonically in t, then $\hat{\theta}_{K_1} >> \hat{\theta}_{K_2}$ when K_1 and K_2 are the Prentice-Wilcoxon ($G^\rho : \rho = 1$) and logrank ($G^\rho : \rho = 0$) weight functions, respectively, defined in Chapter 3.

We illustrate approaches based on martingale and score residuals, first in the case of a single dichotomous covariate, i.e., $Z_i = \pm 1$. In this setting, the hazard ratio $\lambda(t|Z_i = 1)/\lambda(t|Z_i = -1)$, estimated by $\exp(2\hat{\beta})$, should be independent of t. Two non-proportional hazards situations are illustrated in Figure 4.5.1. Because the martingale residuals sum to zero, it follows for $j = 0, 1$, or 2 that

$$\sum_{i=1}^n I_{\{Z_i=1\}}\{\widehat{M}_i(j+1) - \widehat{M}_i(j)\} = -\sum_{i=1}^n I_{\{Z_i=-1\}}\{\widehat{M}_i(j+1) - \widehat{M}_i(j)\}$$
$$\equiv Y_j.$$

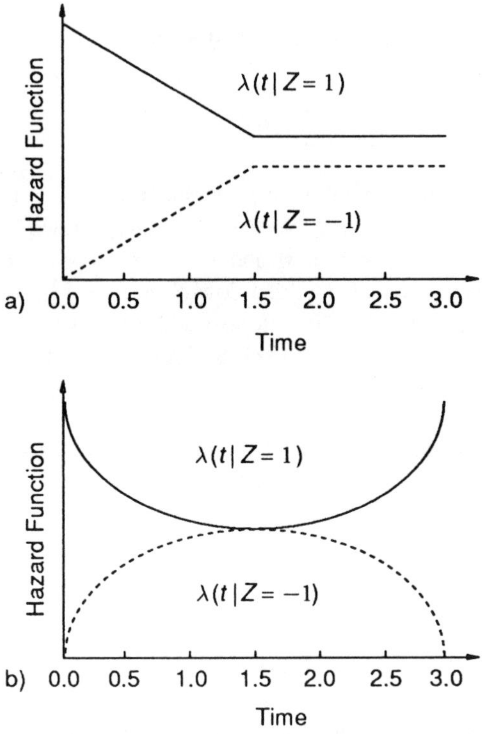

Figure 4.5.1 Two examples of pairs of hazard functions that do not have constant ratios.

In either Figure 4.5.1a or 4.5.1b, $|Y_j|$ will be stochastically larger than when the hazard ratio is constant. Setting

$$\bar{Z}(t) = \frac{\sum_{i=1}^{n} Z_i Y_i(t) \exp(\hat{\beta} Z_i)}{\sum_{i=1}^{n} Y_i(t) \exp(\hat{\beta} Z_i)},$$

the proportional hazards assumption is questionable if a large positive or negative value is obtained for $\sup_t \sum_{i=1}^{n} S_i(t)$, where

$$S_i(t) \equiv \int_0^t \{Z_i - \bar{Z}(s)\} d\widehat{M}_i(s). \tag{5.20}$$

If Z is any discrete or continuous covariate, this statistic should be sensitive to alternatives for which covariates have a monotone effect over time, specifically for alternatives satisfying the following assumption.

Assumption 4.5.1. The hazard ratio $\lambda(t|Z = y)/\lambda(t|Z = x)$ is strictly monotone decreasing in t, for all $x < y$, or is strictly monotone increasing for all $x < y$. □

An approximate sampling distribution for $\sup_t \sum_{i=1}^n S_i(t)$ can be derived by observing that

$$\sum_{i=1}^n S_i(t) = U(\hat{\beta}, t),$$

where $U(\beta, t)$ is the partial likelihood score statistic using information over $[0, t]$. Since $\hat{\beta}$ is the maximum partial likelihood estimate based on information over $[0, \infty]$, $\sum_{i=1}^n S_i(t) = 0$ for $t = 0$ and $t = \infty$. The next lemma establishes that a standardized version of the "tied down" process $\sum_{i=1}^n S_i(\cdot)$ converges to a Brownian bridge process. The lemma also addresses the $p > 1$ covariate vector situation and generalizes an earlier result by Wei (1984). The proof is given at the end of Section 8.4.

Lemma 4.5.1. Let $\mathbf{U}(\boldsymbol{\beta}, \cdot)$ denote the score vector process, and $\hat{\boldsymbol{\beta}}$ denote the maximum partial likelihood estimate of $\boldsymbol{\beta}$. Define the observed information matrix by

$$\mathcal{I}(\boldsymbol{\beta}, \cdot) \equiv \sum_{j=1}^n \int_0^{\cdot} \left\{ \frac{\sum_{j=1}^n Y_j(s)\mathbf{Z}_j\mathbf{Z}_j' e^{\boldsymbol{\beta}'\mathbf{Z}_j}}{\sum_{j=1}^n Y_j(s) e^{\boldsymbol{\beta}'\mathbf{Z}_j}} - \left(\frac{\sum_{j=1}^n Y_j(s)\mathbf{Z}_j e^{\boldsymbol{\beta}'\mathbf{Z}_j}}{\sum_{j=1}^n Y_j(s) e^{\boldsymbol{\beta}'\mathbf{Z}_j}} \right)^{\otimes 2} \right\} dN_i(s),$$

where $\mathbf{a}^{\otimes 2} = \mathbf{a}\mathbf{a}'$ for any vector $\mathbf{a}$. Assume the regularity conditions specified by Theorem 8.4.4, so that the in probability limit of $n^{-1}\mathcal{I}(\boldsymbol{\beta}, \cdot)$ is given by $\boldsymbol{\Sigma}(\cdot) = \int_0^{\cdot} \mathbf{v}(\boldsymbol{\beta}, x) s^{(0)}(\boldsymbol{\beta}, x)\lambda_0(x)dx$, where $s^{(0)}$ and $(\mathbf{v})_{jk}$ are the in probability limits of $S^{(0)}$ in Eq. (3.18) and of $(\mathbf{V})_{jk}$ in (3.23), for $j, k \in \{1, \ldots, p\}$.

1. Let $\mathbf{B}(\cdot)$ be a mean zero vector of Gaussian processes having independent increments and covariance matrix $\boldsymbol{\Sigma}(\cdot)$. Then,

$$\frac{1}{\sqrt{n}}\mathbf{U}(\hat{\boldsymbol{\beta}}, \cdot) \Longrightarrow \mathbf{B}(\cdot) - \boldsymbol{\Sigma}(\cdot)\{\boldsymbol{\Sigma}(\infty)\}^{-1}\mathbf{B}(\infty), \tag{5.21}$$

   where $\Longrightarrow$ denotes weak convergence over the relevant interval.

2. For some $j = 1, \ldots, p$, if $\{\mathbf{v}(\boldsymbol{\beta}, \cdot)\}_{jk}$ is proportional to $\{\mathbf{v}(\boldsymbol{\beta}, \cdot)\}_{jj}$ for every k, then

$$\{\mathcal{I}(\hat{\boldsymbol{\beta}}, \infty)_{jj}\}^{-1/2}\{\mathbf{U}(\hat{\boldsymbol{\beta}}, \cdot)\}_j \Longrightarrow W^o(\sigma_{jj}(\cdot)/\sigma_{jj}(\infty)), \tag{5.22}$$

   where $\sigma_{jj}(t) \equiv \{\boldsymbol{\Sigma}(t)\}_{jj}$ and where $\{W^o(t) : 0 \leq t \leq 1\}$ is distributed as a Brownian Bridge. □

When the jth component of the covariate vector satisfies the proportional hazards assumption, Eq. (5.22) indicates that the proportional hazards test statistic

$$\{\mathcal{I}(\hat{\boldsymbol{\beta}}, \infty)_{jj}\}^{-1/2} \sup_t \sum_{i=1}^n S_{ij}(t) \tag{5.23}$$

asymptotically has the distribution of $\sup_{0 \leq t \leq 1} W^o(t)$, as long as

$$\{\mathbf{v}(\boldsymbol{\beta}, t)\}_{jk}/\{\mathbf{v}(\boldsymbol{\beta}, t)\}_{jj} \text{ does not depend on } t, \tag{5.24}$$

for each k. Suppose $\{T_\ell^o : \ell = 1, \ldots, L\}$ denotes the set of distinct observed failure times. In applications, one can assess the validity of (5.24) by inspecting values of $\mathbf{V}(\hat{\beta}, T_\ell^o)$ defined in Eq. (3.23), the empirical covariance matrix for the covariate vector $\mathbf{Z}$ over all items at risk at T_ℓ^o, with weights proportional to each item's relative risk for failure at T_ℓ^o, $\exp(\hat{\beta}'\mathbf{Z}_i)$. Such terms are available since the observed information has form

$$\mathcal{I}(\hat{\beta}, \infty) = \sum_{\ell=1}^{L} D_\ell\{\mathbf{V}(\hat{\beta}, T_\ell^o)\},$$

where D_ℓ indicates the number of observed failures at T_ℓ^o.

An important special case of (5.24) arises when $\{\mathbf{v}(\beta, t)\}_{jk} = 0$ for each $k \neq j$. This restriction requires $(\mathbf{Z})_j$ to be orthogonal to other covariates over time. This would apply, for example, to intervention studies in which the jth covariate represents a randomly assigned treatment and strong treatment by factor interactions do not exist. Further research is necessary to address the robustness of the convergence in Eq. (5.22) to violations of condition (5.24).

If adequate data are available, graphical and analytical methods can be used to detect more general proportional hazards departures not characterized by Assumption 5.1, such as the alternative in Figure 4.5.1(b). By choosing band widths Δ and δ, graphical assessments are available by plotting

$$f_{\Delta,\delta}(x,t) \equiv \sum_{i=1}^{n} I_{\{x-\Delta \leq Z_{ij} \leq x+\Delta\}} \int_{t-\delta}^{t+\delta} \left\{dN_i(s) - Y_i(s)e^{\hat{\beta}'\mathbf{Z}_i} d\hat{\Lambda}_0(s)\right\}$$

as a function of t, for selected values of x. For discrete covariates, $\Delta = 0$ can be used. Trends in the plots of $f_{\Delta,\delta}(x, \cdot)$ show the nature of the departure from proportional hazards. For inference purposes, one can obtain an expression for the conditional distribution of any term

$$T_A(s,t) \equiv \sum_{i \in A} \int_s^t \{dN_i(u) - Y_i(u)e^{\hat{\beta}'\mathbf{Z}_i} d\hat{\Lambda}_0(u)\},$$

where A is any subset of $\{1, 2, \ldots, n\}$ and $s \leq t$. Specifically, $T_A(s,t)$ can be thought of as a sum over the L distinct failure times occurring over the interval $(s, t]$. At the lth of these L failure times, $t_{(l)}$, $\sum_{i \in A} \Delta N_i(t_{(l)})$ is the number of failures occurring in the set A. In turn, $\sum_{i \in A} \Delta N_i(t_{(l)})$ has the distribution arising from sampling $\sum_{i \in A} Y_i(t_{(l)})$ items without replacement from a set of $\sum_{i=1}^n Y_i(t_{(l)})$ items, which includes $\sum_{i=1}^n \Delta N_i(t_{(l)})$ total failures, and where each item has probability $Y_i(t_{(l)})e^{\hat{\beta}'\mathbf{Z}_i} \big/ \sum_{k=1}^n Y_k(t_{(l)})e^{\hat{\beta}'\mathbf{Z}_k}$ of being sampled. In particular then, $\sum_{i \in A} \Delta N_i(t_{(l)})$ has expectation

$$\sum_{i \in A} \left[\left\{\frac{Y_i(t_{(l)})e^{\hat{\beta}'\mathbf{Z}_i}}{\sum_{k=1}^n Y_k(t_{(l)})e^{\hat{\beta}'\mathbf{Z}_k}}\right\} \sum_{k=1}^n \Delta N_k(t_{(l)})\right]$$

$$= \sum_{i \in A} Y_i(t_{(l)})e^{\hat{\beta}'\mathbf{Z}_i} \Delta\hat{\Lambda}_0(t_{(l)}),$$

so indeed $T_A(s,t)$ has zero expectation in this sampling framework. Finally, the distribution of $T_A(s,t)$ is obtained by taking $\{\sum_{i \in A} \Delta N_i(t_{(l)}) : l = 1, \ldots, L\}$ to be a collection of independent random variables.

Many other methods for testing proportional hazards have been proposed, notably by Schoenfeld (1980), Andersen (1982), Aranda-Ordaz (1983) and Harrell (1986). One advantage of the statistic in (5.23) is that it does not require an arbitrary discretization of the time axis.

In this section, we have seen how inspection of martingale residuals plotted against a covariate can help estimate the functional form for the influence of a covariate on the hazard function, and how the validity of the proportional hazards assumption can be checked through plots against time of martingale and score residuals. An interesting contrast with fully parametric models or with the classical linear models setting is that little or no model-validation information is contained in the ordered martingale residuals or, *a fortiori*, in statistics such as the sum of squared residuals. Indeed, if one fits a no covariate model in uncensored data, the ordered martingale residuals are deterministically unity minus the expected value of exponential order statistics, regardless of the true model. This lack of model validation information in the ordered martingale residuals is due to the flexibility of the model in Eq. (5.2), which arises through the arbitrary nature of $\Lambda_0(\cdot)$.

4.6 APPLICATIONS OF RESIDUAL METHODS

This secton uses the natural history PBC data from Section 4.4 and an investigation of nuclear DNA ploidy patterns studied by flow cytometry in prostatic adenocarcinoma to illustrate the application of martingale residual based methods.

Study of DNA Ploidy Patterns in Prostatic Adenocarcinoma

Patients with prostatic carcinoma who have biopsy-proven metastatic deposits in the pelvic lymph nodes but no evidence of tumor dissemination on radioactive bone scans are commonly classified as having stage D1 prostatic carcinoma. Obtaining a reliable prognosis for such patients has been especially difficult. Thus the type and timing of surgical, radiation, and hormonal treatment are controversial issues in urologic oncology. Histologic grading performed with standard light microscopic techniques has been helpful in assessing prognosis, but the analysis of tumor nuclear DNA content by flow cytometry is considered a potentially useful source of new prognostic information. To obtain control information, human benign prostatic hyperplasia tissue from 60 patients was analyzed. The nuclear DNA ploidy pattern revealed a steep "2C peak" and a small "4C peak." The proportion of nuclei in the 4C peak or the G2 phase of growth, referred to as "G2%," was computed. A preliminary hypothesis was that patients with G2% exceeding three standard deviations above the mean from these 60 normals (i.e., exceeding 7.87 + 3(1.63) = 12.76 ≈ 13%) would have more rapidly progressing disease and early death. To assess this hypothesis, flow cytometry analyses were performed on 91 Mayo Clinic patients with D1 prostatic adenocarcinoma who had potential 5–14 years of follow up (Winkler, et al., 1988). Of these, 79 were classified as DNA diploid (2C and 4C peaks) or tetraploid (2C, 4C and 8C peaks). Twelve others with erratic peak structures were classified as DNA aneuploid and had a poor prognosis.

The relationship of G2% to progression of disease and with death was studied using the 79 DNA diploid and tetraploid patients. Plots of the martingale residuals against G2% appear in Figure 4.6.1 (a) and (b) for progression and survival, where the residuals were obtained by fitting a Cox model having no covariates. The plots suggest a threshold effect near 10–15 G2%. The scatterplot smoother was obtained

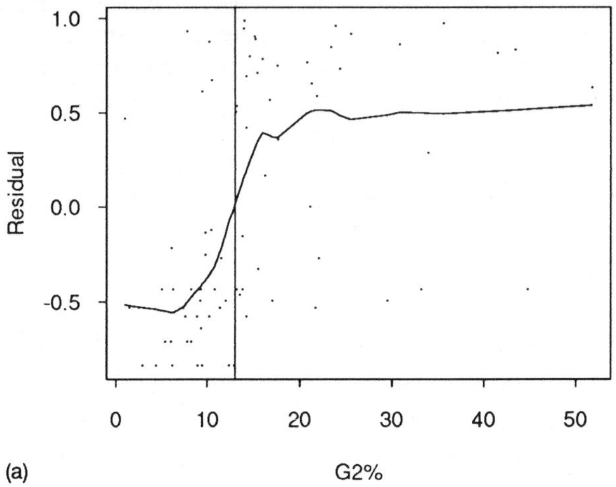

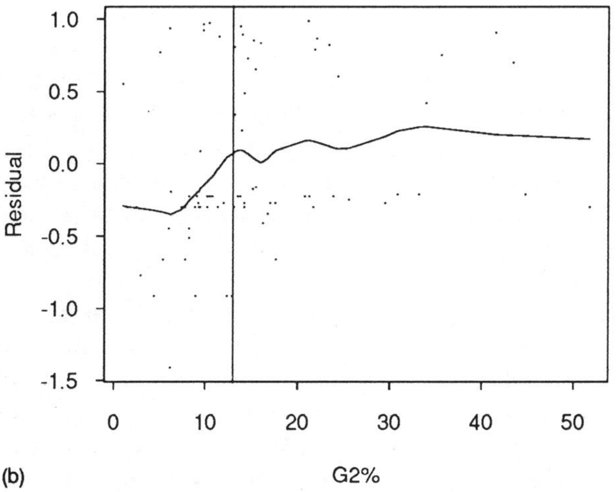

Figure 4.6.1 Martingale residuals in stage D1 prostate cancer data. Residuals are from models with no covariates. Failure time is time to disease progression in (a), time to death in (b). LOWESS smooths use 50% of the data for each local fit, i.e., a span of .5.

using LOWESS in the language S. Data suggest that patients with G2% exceeding approximately 13% have a much higher rate of progression and death. Entering a dummy indicator $Z_i = 0(1)$ for G2% $<$ 13% ($\geq$ 13%) into the Cox model for time to progression yielded strong evidence for the association (likelihood ratio statistic = 30.48), and a maximum partial likelihood estimate $\hat{\beta} = 2.27$ with standard error 0.5. Thus a 95% confidence interval for the ratio of progression rates in the $\geq$ 13% group to the $<$13% group is (5.87,15.96). A plot of martingale residuals against G2% from a Cox model containing the indicator variable Z_i reveals (in Figure 4.6.2 on page 181) that the dichotomy at 13% appears to explain much of the association.

To illustrate the assessment of influence, a Cox model with G2% as a linear covariate was fit to the progression data of the 79 patients. Figure 4.6.3, on page 181, shows the standardized score residuals, the Storer-Crowley one-step influence measure, and the actual jackknife. In these data, there are a few subjects for which the Storer-Crowley (SC) influence has the wrong sign. In all these cases the SC is accurately tracking the behavior of the first iteration step after removal of subject i; i.e., a one-step jackknife also has the incorrect sign. Both the score residual and the SC measure accurately approximate the true jackknife.

Checking Functional Form in the Liver PBC Model

In Section 4.4, Cox partial likelihood methods were used to develop a survival model based on the variables age, log(albumin), log(bilirubin), edema, and log(protime). A sixth variable, hepatomegaly, had been independently predictive until the logarithmic transformation of bilirubin was introduced, (see Tables 4.4.2(b) and 4.4.3(a)). The plots of martingale residuals in Figure 4.6.4, on page 182, provide some insight into what is happening in the data. The set of 312 martingale residuals used to produce the 6 plots were obtained from a single Cox model fitting the variables age, log(albumin), edema, and log(protime). It is obvious that a linear fit to the martingale residuals is much more appropriate when the residuals are plotted against log(bilirubin) in Figure 4.6.4(a) than when they are plotted against bilirubin in Figure 4.6.4(b). In each of these two plots, a running smooth using LOWESS is displayed, along with a least squares simple linear regression line. The rise at the left-hand edge of the LOWESS smooth to the residual plot against log(bilirubin) is due entirely to three patients (#8, #108, and #163) with low bilirubin levels who died early. There is no biological support for such a pattern.

To study the role of hepatomegaly, the subset of 152 martingale residuals for patients *without* enlarged livers are plotted against log(bilirubin) and bilirubin in Figures 4.6.4(a0) and 4.6.4(b0), respectively, while the subset of 160 martingale residuals for patients *with* enlarged livers are plotted against log(bilirubin) and bilirubin in Figures 4.6.4(a1) and 4.6.4(b1), respectively. In Figures 4.6.4(a0) and 4.6.4(a1), LOWESS scatterplot smoothers and least squares lines are compared to the least square line from Figure 4.6.4(a). When the residuals are plotted against log(bilirubin), it is evident that subsetting by hepatomegaly has negligible effect on the regression lines over the range of the data. In contrast, consider when the

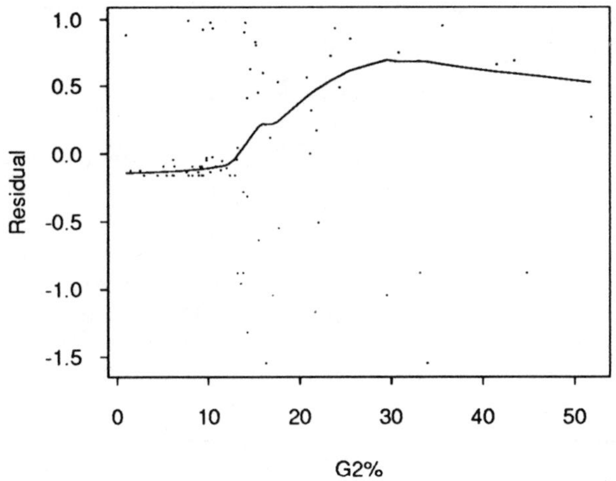

Figure 4.6.2 Martingale residuals in stage D1 prostate cancer. Residuals are from a model with a single binary covariate indicating whether G2% $\geq$ 13%. Failure time is time to disease progression. LOWESS smooths use a span of .5.

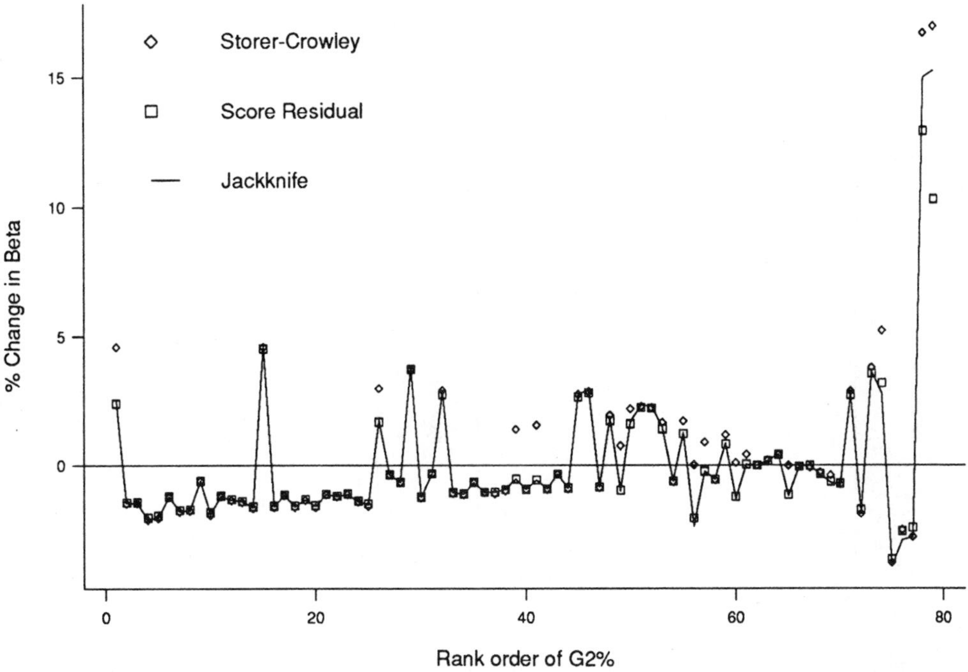

Figure 4.6.3 Influence measures for coefficient of G2% in stage D1 prostate cancer data. Failure time is time to disease progression

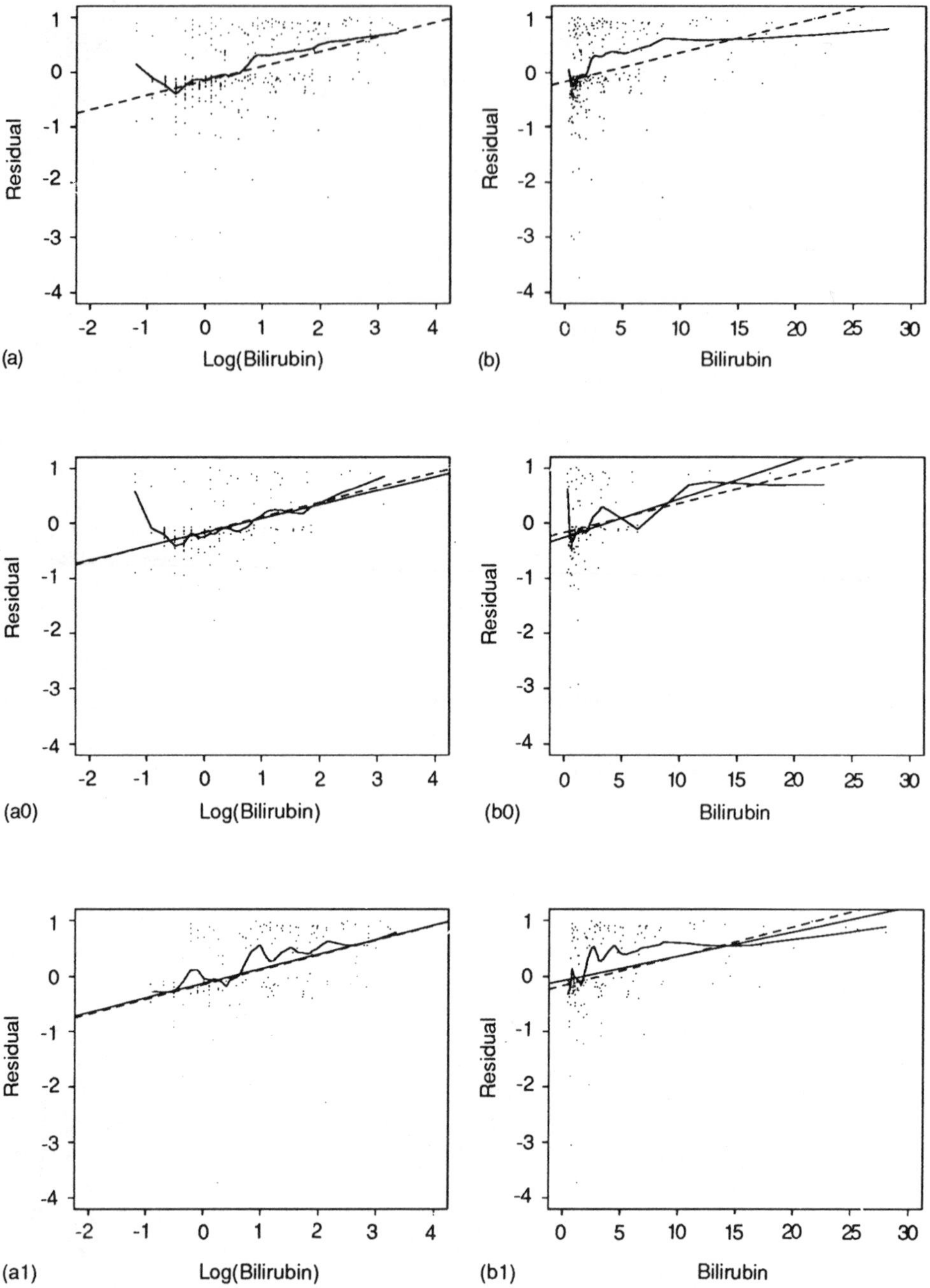

Figure 4.6.4 Martingale residuals in PBC data from a model with the covariates age, log(albumin), log(protime) and edema. Residuals are plotted against log(bilirubin) in these plots on left, and against bilirubin on the right. The top two plots use all 312 randomized cases, the middle two use the 152 cases without enlarged livers, and the bottom two use the 160 cases with enlarged livers. LOWESS smooths use a span of .2.

martingale residuals are plotted against bilirubin. For patients with hepatomegaly, Figure 4.6.4(b1) reveals the least squares line is pulled above the line from 4.6.4(b), especially over the region with bilirubin < 7, where the bulk of the data occur. For patients without hepatomegaly, the least squares line in Figure 4.6.4(b0) is pulled below the line from 4.6.4(b) over the region containing the bulk of the data. It is visually apparent that hepatomegaly is independently predictive when the variable bilirubin rather than log(bilirubin) is added to the model.

The appropriateness of the selected transformations for the four continuous variables in the five variable model [age, log(albumin), log(bilirubin), edema, log(protime)] is shown in Figure 4.6.5. For each continuous variable, martingale residuals

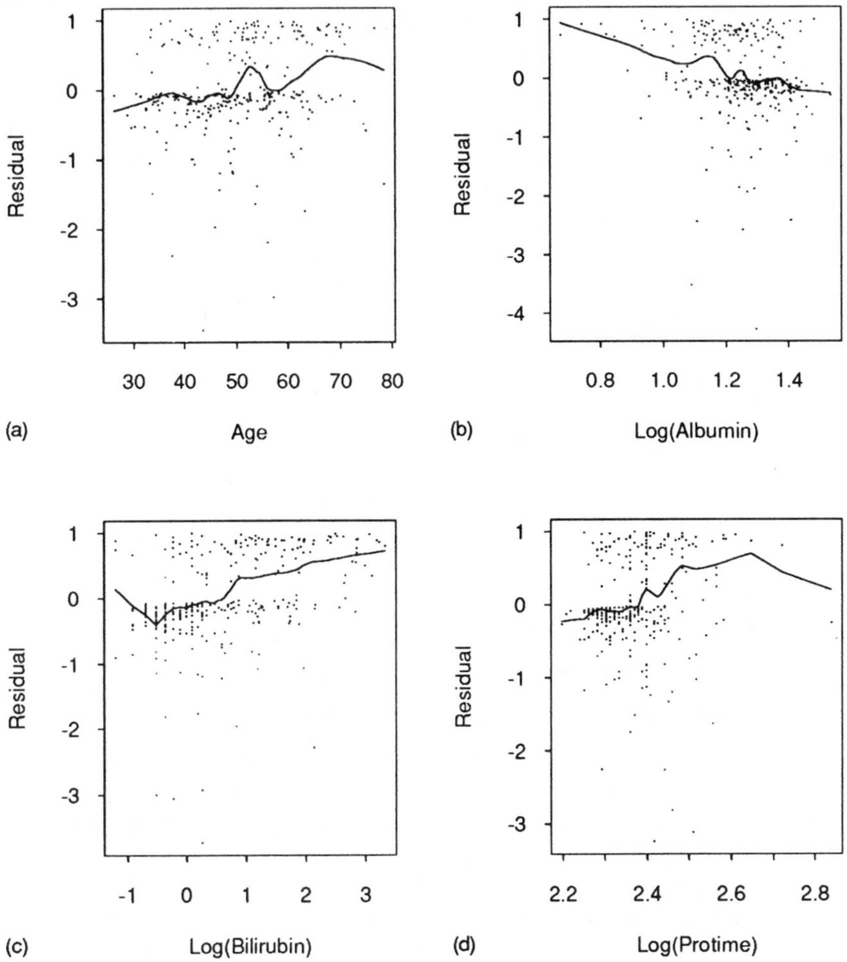

Figure 4.6.5 Martingale residuals in the PBC data. Residuals from a model with the covariate edema and three of the four continuous variables (age, log(albumin), log(bilirubin), and log(protime)) are plotted against the omitted variable. LOWESS smooths use a span of .2.

from the Cox model containing edema and the other three continuous variables are plotted against the transformed variable. Approximate linearity of each of the four LOWESS smooths provides additional support for the transformations selected in Section 4.4. The LOWESS smooth of the martingale residual plot against log(protime) departs from a linear fit in the right-hand tail, with a similar but less noticeable effect seen in the age plot as well. It appears that each of these departures is entirely due to the negative residual (i.e., unexpectedly long survival) of the two cases with extremely long protime, and high age, respectively. Later, we will inspect the influence of these two cases to see if they are largely responsible for the logarithmic transformation of the variable protime and the lack of a square transformation of the variable age.

Plots of martingale residuals in Figure 4.6.5 did not provide strong evidence for changing the transformations selected in Section 4.4 using partial likelihood based model fitting techniques. Certainly this is not always the case. For illustration, a less mature version of the PBC data was analyzed a number of years earlier in an attempt to develop a parsimonious model which would provide survival estimates $\{\hat{S}(t|\mathbf{Z}) : 0 \leq t \leq 18 \text{ months}\}$ for an 18 month period. Variable selection procedures led to the choice of variables log(albumin), edema, and log(protime). Table 4.6.1(a) provides the score statistics, adjusting for the three variables in the model, for a number of additional variables. Log(platelets) appeared to be independently predictive and its addition made the variable platelets strongly independently predictive (see Table 4.6.1(b)). The opposite signs of their coefficients in the model in Table 4.6.1(b) make it clear that each variable is correcting for the effect of the other. To assess visually the relationship of platelets with survival, martingale residuals from the three variable model are plotted in Figure 4.6.6(a). A threshold effect is apparent; patients with serum platelet counts below 130,000 (the lower limit of the Mayo Clinic published normal range) continue to be at higher risk for death even after adjustment for albumin, edema, and protime levels. When the variable platelets is entered into the model as a dichotomous variable, Table 4.6.2 reveals the continuous variables platelets and log(platelets) provide negligible additional predictive information. Furthermore, a plot of martingale residuals against platelets from the Cox model containing the dichotomous platelet variable (in Figure 4.6.6(b)) reveals the dichotomy appears to adequately explain the association.

Looking for Influence Points in the Liver PBC Model

Returning to the five variable PBC natural history model developed in Section 4.4 and displayed in Table 4.4.3(c), one can apply the standardized score residuals in Eq. (5.16) to search for cases which are influential on parameter estimates and choice of transformation. The score residual plots for each of the five variables appear in Figure 4.6.7. The bilirubin and edema plots each have an interesting case. For bilirubin, individual #81 was a 63 year old woman with no edema, good albumin (3.65), and rather high protime (11.7). In spite of very high bilirubin (14.4), she lived seven years before dying. For the variable edema, individual #293 was a 57-year-old woman with quite poor prognostic status. In spite of low albumin (2.98), high bilirubin (8.5) and protime (12.3), and edema resistent to

Table 4.6.1 Regression models for predicting 18 month survival in an earlier version of the PBC data.

(a) Three variable model

Variable	Coef.	Std. Err.	Chi-Square
log(Albumin)	−4.755	1.3385	12.62
log(Prothrombin Time)	9.430	1.7365	29.49
Edema	2.023	0.4546	19.80

Rao score statistics for additional variables

Variable	Chi-Square
log(Platelets)	6.47
Alk.Phos.	0.07
Platelets	2.73
Ascites	0.13
Age	0.74
log(Bilirubin)	1.83

(b) Model including linear and logarithmic platelet terms

Variable	Coef.	Std. Err.	Chi-Square
log(Albumin)	−4.764	1.4495	10.80
log(Prothrombin Time)	8.456	1.9950	17.97
Edema	2.335	0.4706	24.64
log(Platelets)	−5.423	1.5845	11.71
Platelets	0.021	0.0071	8.46

Table 4.6.2 A four-variable regression model for predicting 18 month survival in an earlier version of the PBC data.

(Dich. Plat.: 0 =Platelets$<$ 130.000; 1 =Platelets$\geq$ 130,000)

Variable	Coef.	Std. Err.	Chi-Square
log(Albumin)	−4.318	1.4029	9.47
log(Prothrombin Time)	8.721	1.9670	19.65
Edema	2.615	0.4993	27.43
Dich. Platelet	−1.791	0.4928	13.20

Rao score statistics for additional variables

Variable	Chi-Square
log(Platelets)	0.01
Alk.Phos.	0.44
Platelets	0.16
Ascites	0.12
Age	0.51
log(Bilirubin)	1.01

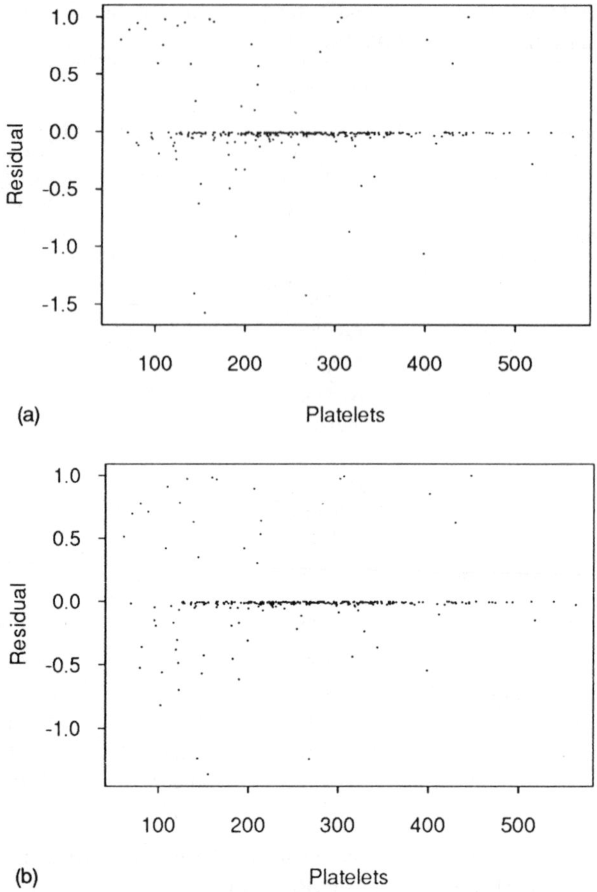

Figure 4.6.6 Martingale residuals plotted against platelets (divided by 10^3) in an earlier version of the PBC data. Residuals in (a) are from a model with the covariates log(albumin), edema, and log(protime). Residuals in (b) are from a model with these three covariates and a binary covariate indicating whether platelets $\geq 130 \times 10^3$.

diuretics, she remains alive after more than 3.5 years. After rechecking, no data errors were found for these two cases.

The most interesting score residual plots in Figure 4.6.7 are those corresponding to the variables log(protime) and age. The fanning out in the left- and right-hand tails of these two plots is a reflection of the earlier observation that influence on model fit depends on both the residual from the fit and on the extremity of its covariate value, roughly $(\mathbf{Z}_i - \bar{\mathbf{Z}})\times$ residual. The conjecture, supported by the martingale residual plots in Figure 4.6.5, that the two individuals with the highest protime and highest age might be influential points is confirmed in Figure 4.6.7. Individual #253, listed as the oldest patient in the study at 78.4 years of age, remained alive after nearly five years follow up despite high bilirubin (7.1), low

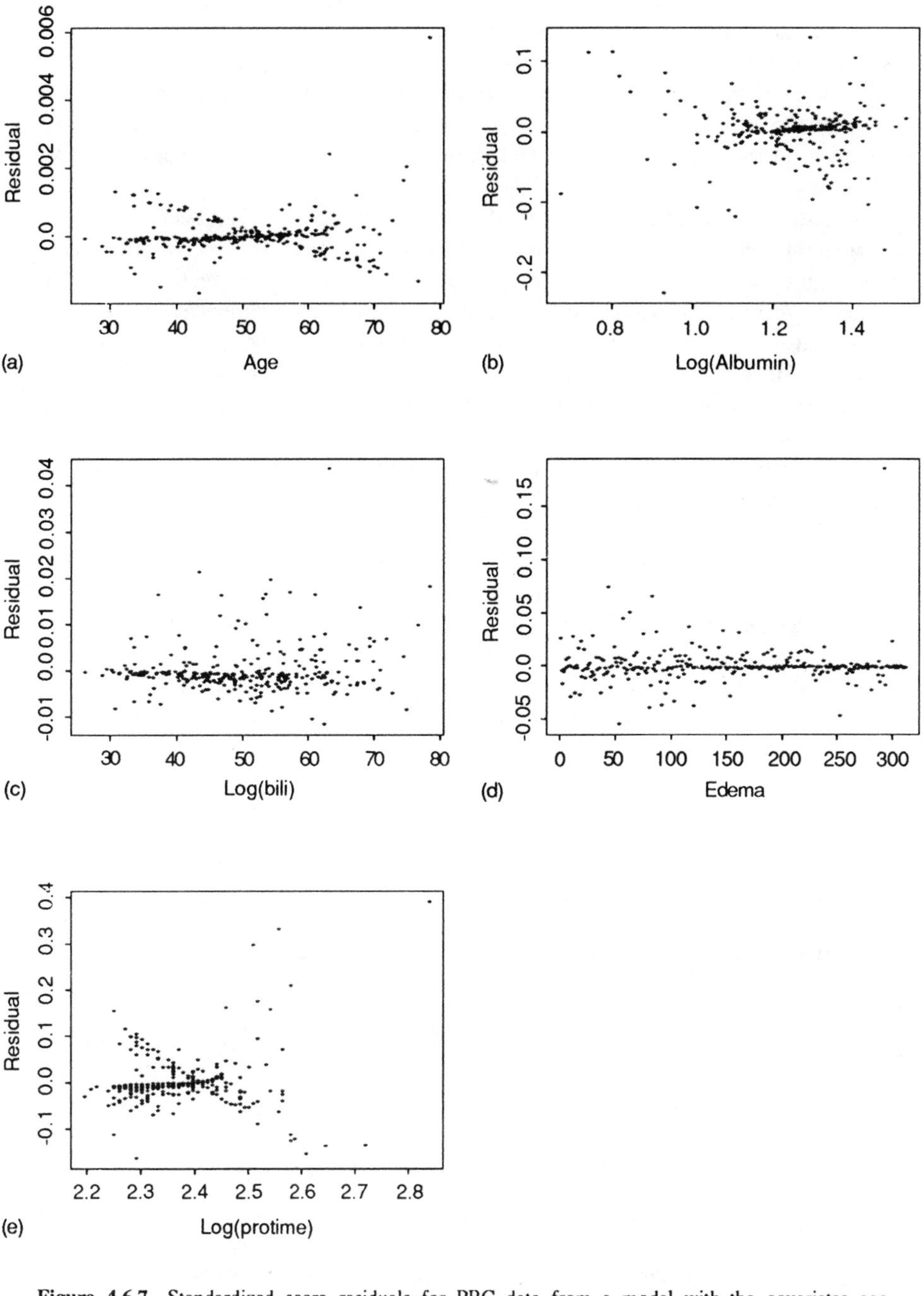

Figure 4.6.7 Standardized score residuals for PBC data from a model with the covariates age, log(albumin), log(bilirubin), edema, and log(protime). Residuals for edema are plotted against case number.

albumin (3.03), and advanced age. Individual #107, with the highest prothrombin time, has an unusual covariate profile. This 62-year-old female had excellent albumin (4.03) and bilirubin (0.6) levels and no edema. In contrast, her protime value of 17.1 was well above the range of other prothrombin time values. The fact that she was alive after more than 9 years follow-up gave her considerable influence on the variable protime and raised concerns about the accuracy of the value 17.1

To assure the highest data quality, the entire PBC data set had been subject to considerable scrutiny throughout the lengthy period of its collection. In this sense, it was quite surprising when a rechecking of original medical records revealed database errors in the age of individual #253 (78.4 years should have been 54.4 years) and in the protime of individual #107 (17.1 seconds should have been 10.7 seconds). Since considerable additional data rechecking led to no further identification of errors, the martingale and score residual plots were effective in identifying data inaccuracies.

After correcting the protime value for #107, martingale residuals from the model containing the variables age, log(albumin), log(bilirubin), and edema were plotted against log(protime) and protime in Figure 4.6.8(a) and (b), respectively. As anticipated, the drop in the right-hand side of the martingale residual plot for log(protime) in Figure 4.6.5 is eliminated in Figure 4.6.8(a). Comparison of the smooths in Figure 4.6.8(a) and (b) provides slightly more evidence of a linear fit to the residuals when they are plotted against protime rather than against its logarithmic transformation. This graphical observation was confirmed analytically by the score statistics 10.085 and 10.428 for the variables log(protime) and protime, respectively, obtained from the model containing the variables age, log(albumin), log(bilirubin), and edema.

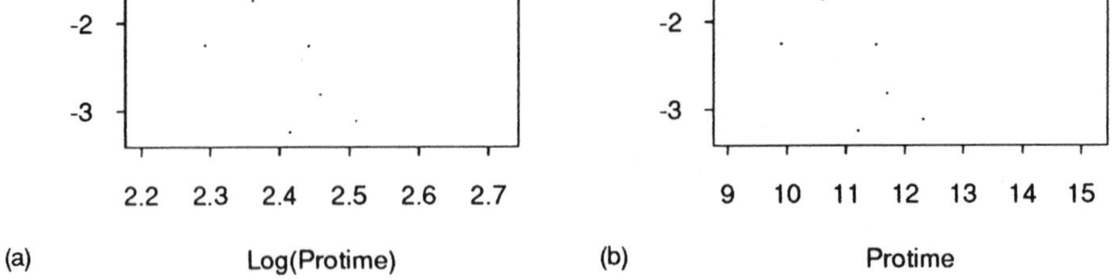

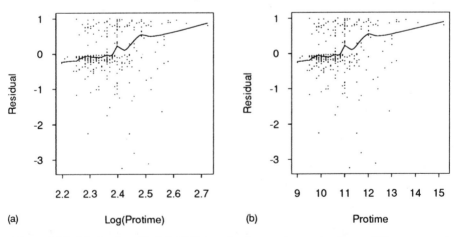

Figure 4.6.8 Martingale residuals in PBC data after the covariate protime in case 107 was corrected from 17.1 seconds to 10.7 seconds. Residuals are from a model with the covariates age, log(albumin), log(bilirubin), and edema.

When the age of patient #253 is changed from 78 to 54, the coefficient for age in the five variable model in Table 4.4.3(c) increases from 0.0333 $\pm$ 0.0087 to 0.0405 $\pm$ 0.0094. Thus each decade increase in age increases the hazard function by a multiplicative constant 1.500 rather than 1.395. Figures 4.6.9(a) and (b) provide martingale residuals from the model containing log(albumin, log(bilirubin), edema, and log(protime), plotted against age and $(\text{age})^2$, where the age of patient #253 has been corrected. The score statistic for the variables age and $(\text{age})^2$ in this four variable model are 19.09 and 20.44, respectively. This graphical and analytical information provides slightly more evidence in favor of the $(\text{age})^2$ transformation.

The data corrections identified through the inspection of martingale and score residual plots could alter the transformations selected for the variables age and protime. Since such changes would be motivated by corrections (albeit very significant corrections) in one data element in each of only two cases, the evidence in favor of the new transformations is not convincing. The choice of linear vs. logarithmic term for protime and square vs. linear term for age in the PBC natural history model would have meaningful impact on survival predictions only for individuals whose covariate values are outliers relative to the distribution of covariate values for the 312 Mayo Clinic PBC patients. Thus, one should be very cautions about "extrapolating," e.g., using the model to predict survival of PBC patients in the very advanced stage of the disease.

Assessing Model Accuracy for Individual Subjects

Plots of residuals provide a graphical assessment of the cases whose outcome is poorly predicted by the model. Those with large positive values for $\widetilde{M}_i$ have more

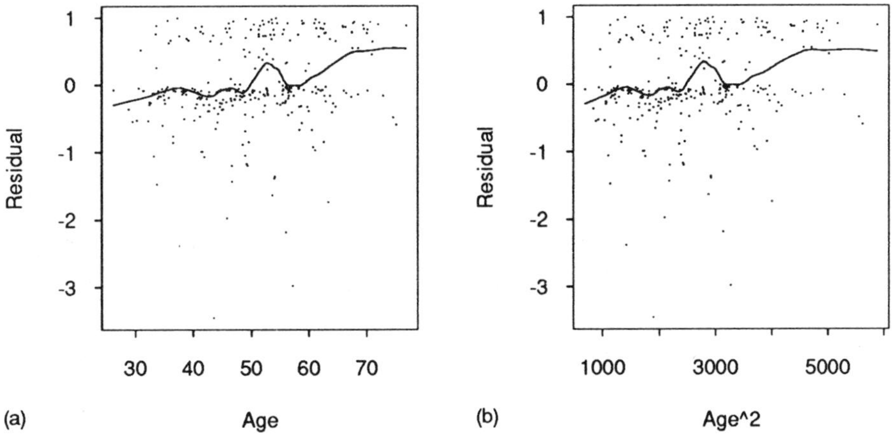

Figure 4.6.9 Martingale residuals in PBC data after the covariate age in Case 253 was corrected from 78.4 years to 54.4 years. Residuals are from a model with the covariates log(albumin), log(bilirubin), edema, and log(protime). LOWESS smooth based on a span of .2.

events than predicted, while those with large negative values have fewer. In one event models such as the Cox model, the martingale residuals are skewed, taking on values from $+1$ to $-\infty$. This skewness makes it difficult to detect outliers who "died too early," and may artificially create the appearance of some outliers who "died too late." The deviance transform symmetrizes the martingale residuals and helps alleviate this problem. In fact, the deviance residuals seem to be very nearly normally distributed when censoring is light.

Figure 4.6.10(a) provides the martingale residuals for the 312 patients in the five variable PBC natural history model displayed in Table 4.4.3(c). Each residual is plotted against the risk score $\hat{\beta}'\mathbf{Z}_i$. In contrast, Figure 4.6.10(b) provides the deviance residuals for the same model. The deviance transform suggests that three apparent outliers in the martingale plot are, in fact, not outliers. The heavy

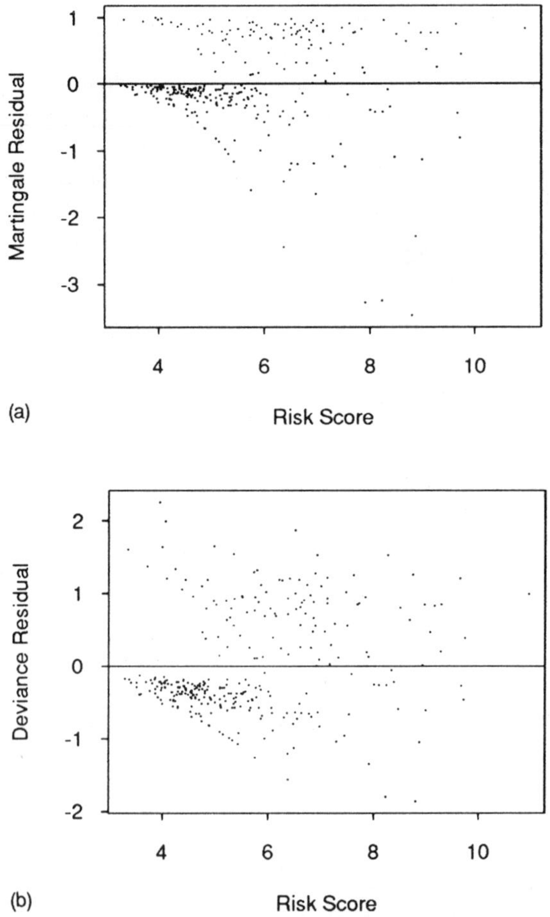

Figure 4.6.10 Residuals plotted against risk score in the PBC data. Residuals from a model with the covariates age, log(albumin), log(bilirubin), edema, and log(protime).

62% censoring leads to some non-normality in the deviance residuals, apparent by a large cluster of residuals near zero. The results in the deviance residual plot suggest examining the patients with the two largest and two smallest residuals.

Assessing Proportional Hazards in the Liver PBC Model

Throughout the model development in Section 4.4, all of the variables were assumed to satisfy the proportional hazards property. The graphical methods from Section 4.5 can be used to assess the validity of this assumption with respect to each of the five variables in Table 4.4.3(c). Figure 4.6.11 presents $\log(-\log S)$ plots (5.19) for each variable in a model containing the other four variables. For each plot, the variable being assessed (if continuous) was split into four levels so that each group had nearly 1/4 of the total number of deaths. The plots provide some evidence that log(protime) and in particular edema do not satisfy the proportional hazards assumption. For both variables, substantial early differences in the plots disappear later in time. This lack of parallelism contradicts the proportional hazards assumption.

Evidence that the variable edema violates the proportional hazards assumption is less convincing than Figure 4.6.11 might imply. Only a few patients in the two edema groups are followed beyond three years, due to the small numbers in each of these groups initially, and due to the very poor prognosis of these patients. For further insight, the standardized score residual plots, defined by the left-hand side of Eq. (5.22), appear in Figure 4.6.12. The steep rise in the plots for edema and protime suggests that the ratio of hazard functions of edema vs. non-edema patients and of high vs. low protime patients is decreasing over time, as was also implied by Figure 4.6.11. The edema plot nearly reached one of the lines $y = \pm 1.36$, while the protime plot exceeded one of the lines $y = \pm 1.63$. Thus the formal test for proportional hazards based on these plots is nearly significant at the $\alpha = 0.05$ level for edema, and is significant at less than the $\alpha = 0.01$ level for protime. In summary, re-modeling allowing for a non-proportional hazards effect for protime might be considered, while the data are inconclusive as to whether or not the variable edema violates the proportional hazards assumption. Larger numbers of patients with edema would need to be followed for a longer period of time for any graphical or analytic method to yield a definitive conclusion.

Validation of the PBC Model using Independent Data

The PBC natural history data set, described in Section 4.4 and provided in Appendix D, contains an independent set of 106 non-randomized PBC patients. These were used by Dickson et al. (1989) to assess the accuracy of the model developed using the 312 randomized patients. Two of the 106 patients had missing values for protime. Patients #359 and #368 were given imputed values 10.55 and 11.12, obtained through a regression on other covariates. The risk scores $R = Z_1\hat{\beta}_1 + \ldots + Z_K\hat{\beta}_K$ then were computed for each of the 106 cross-validation patients using $\hat{\beta}$ from Table 4.4.3(c). Using these scores, patients were divided into low, medium, and high risk subgroups. Within each subgroup, the average of the predicted

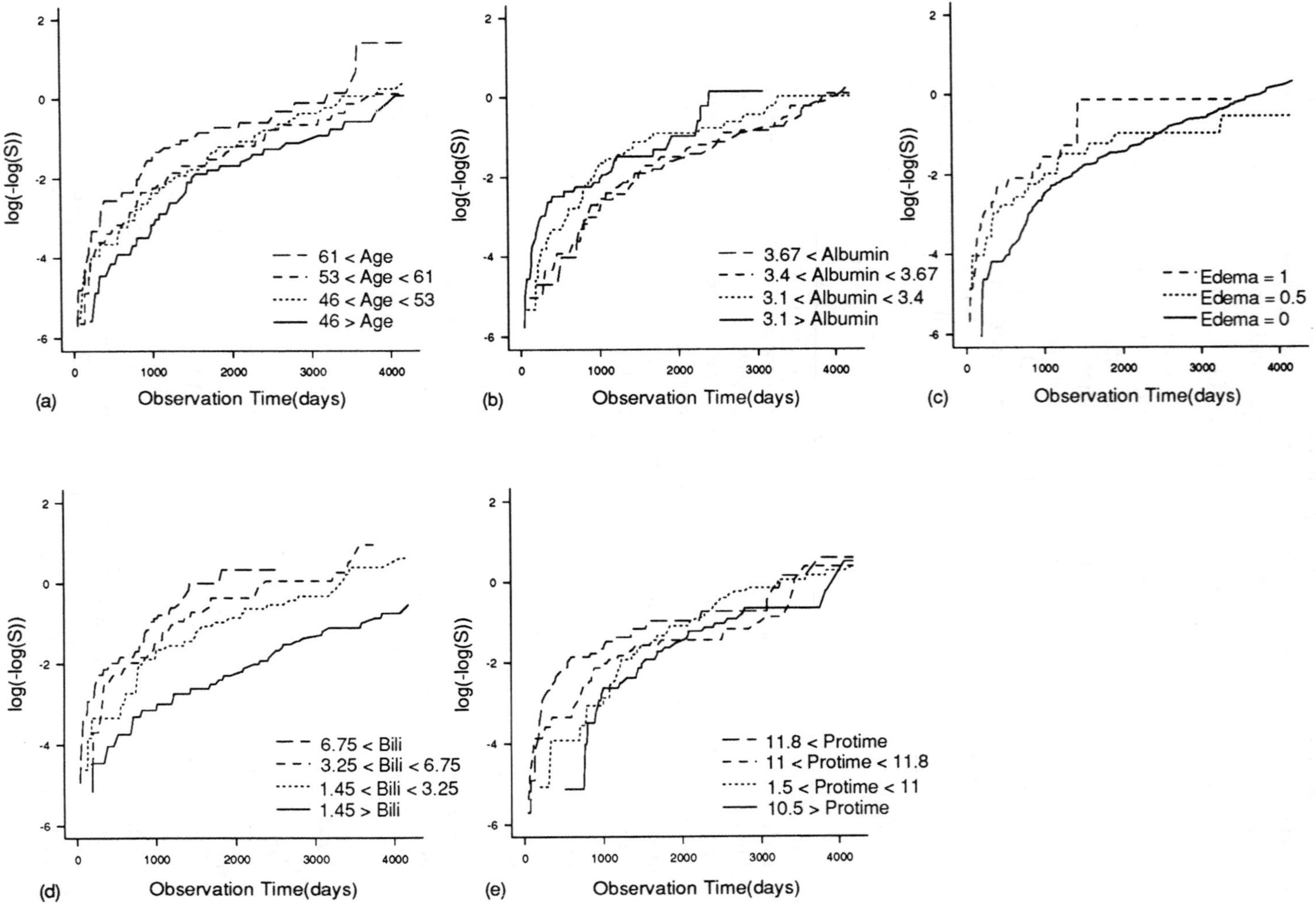

Figure 4.6.11 Logarithm of estimated baseline cumulative hazard functions in stratified models. Each model uses one of the five variables age, log(albumin),

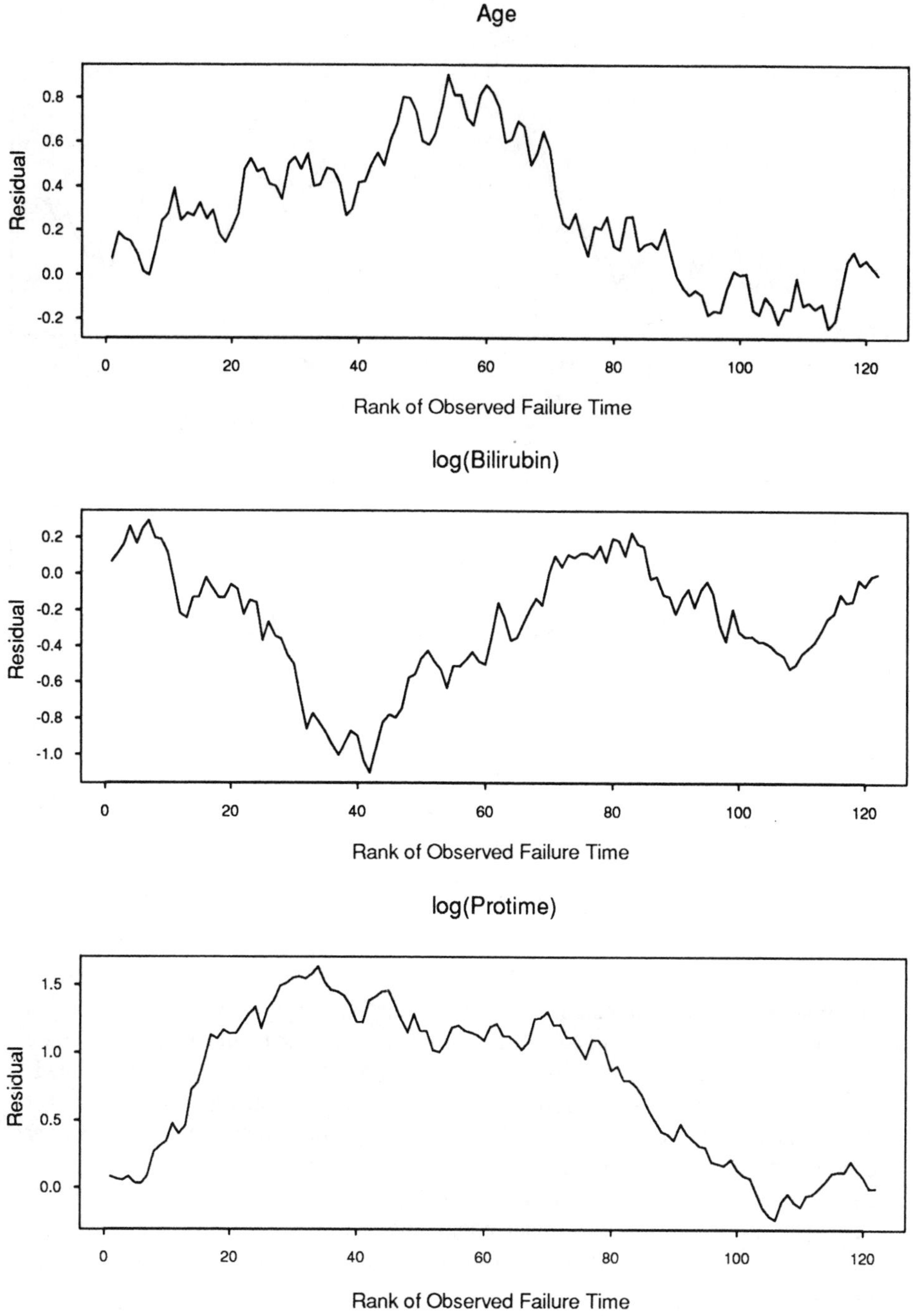

Figure 4.6.12a Standardized score residuals in PBC data. Residuals for each of the covariate compoents are plotted against rank of the distinct observed failure time.

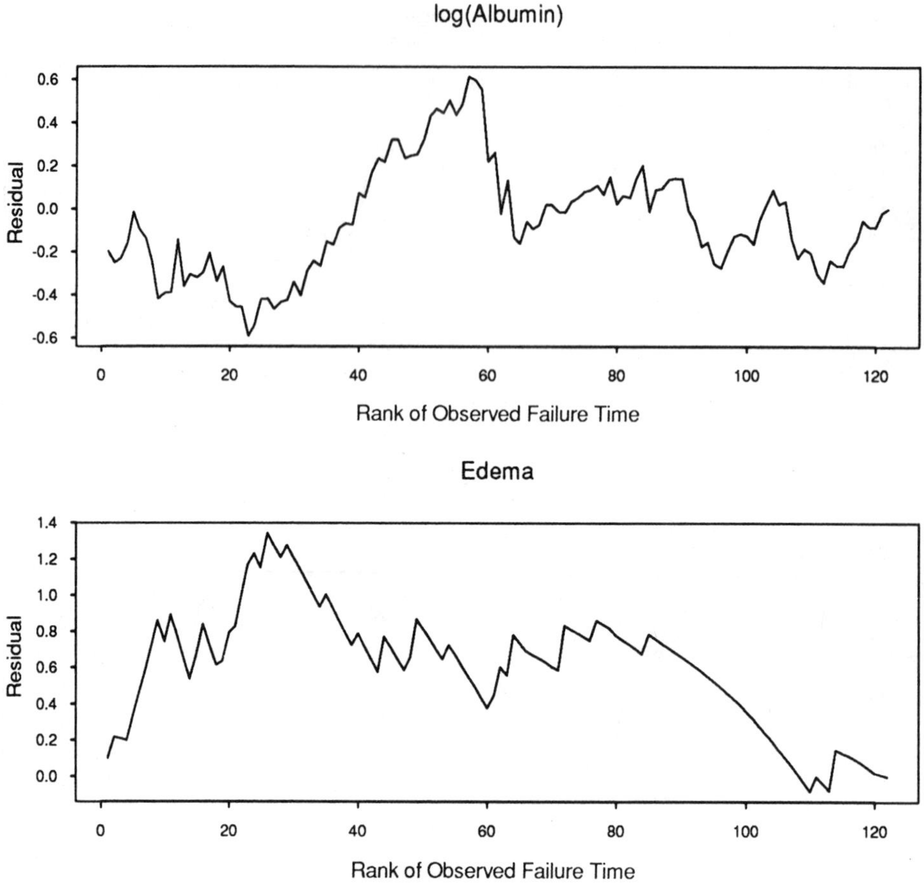

Figure 4.6.12b (*continued*)

survival curves was compared to the actual survival experience represented by a Kaplan-Meier curve. Figure 4.6.13 reveals good prediction by the model. Accuracy in the right-hand tail is difficult to assess due to the variability of the Kaplan-Meier plot. Within each subgroup, each individual's survival was compared to her own model predicted curve, with these individual comparisons then pooled over each subgroup using the one-sample logrank statistic. These statistics yielded chi-squares of approximately 0.5 in each of the three risk groups. Table 4.6.3 compares parameter estimates for the model containing age, log(albumin), log(bilirubin), edema, and log(protime) when the original 312 and the total 418 patients are used in estimation.

Additional model validation studies have revealed good prediction when the model in Table 4.6.3 using 418 patients is applied to untreated PBC patients who are in more advanced stages of disease or to patients who were managed by other

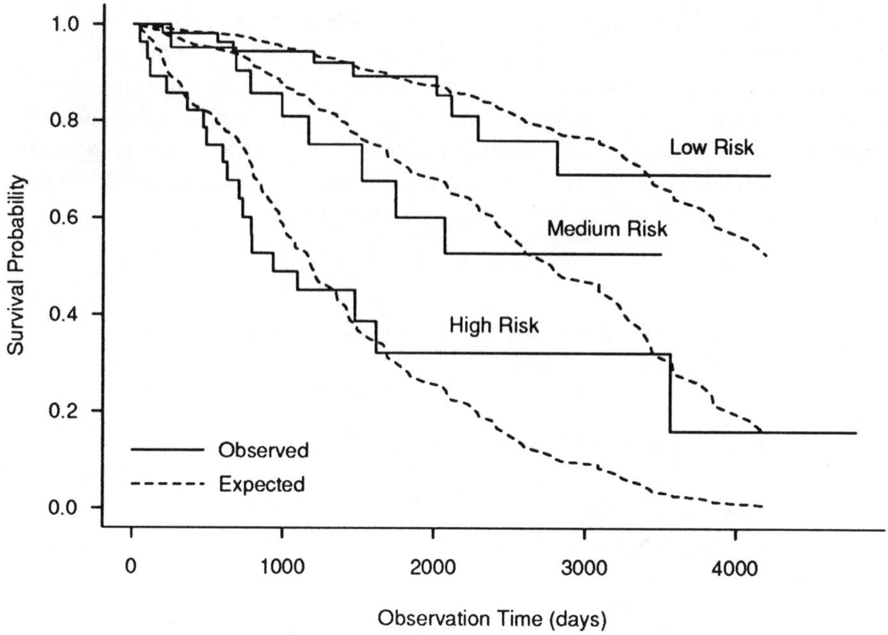

Figure 4.6.13 Predicted and observed survival curves by risk group in PBC validation data set.

clinical centers. Model validation using data from studies conducted by research centers other than the Mayo Clinic is described in Grambsch et al. (1989).

Evaluating Efficacy of Liver Transplantation

To illustrate the application of the natural history model, the covariate and outcome data were obtained for all 161 PBC patients who received liver transplantation at the University of Pittsburgh and its Baylor University satellite in Dallas between April 1980 and June 1987 (Markus et al., 1989). The 161 patients were divided into three categories by their risk scores, with category break points defined to yield equal numbers of deaths per category. The descriptive statistics for these cat-

Table 4.6.3 Regression Coefficients for Cox Regression Survival Models, PBC Data Set

	Initial Model (n = 312)		Final Model (n = 418)	
Variable	Coef.	Std. Err.	Coef.	Std.Err.
log(Bilirubin)	0.8792	0.0987	0.8707	0.0826
log(Albumin)	−3.053	0.724	−2.533	0.648
Age	0.0333	0.0087	0.0394	0.0077
log(Protime)	3.016	1.024	2.380	0.767
Edema	0.7847	0.2991	0.8592	0.2711

egories appear in Table 4.6.4. For each category, Figure 4.6.14 compares the post transplantation survival curve (using a Kaplan-Meier estimate) with the category's average predicted curve for survival had patients not been transplanted (estimated using the natural history model). The results reveal evidence that transplantation lengthens survival, although one should be cautious about issues such as generalizability of the model prediction and extrapolation required to use the model in the higher one or two risk groups in the transplantation data set.

Table 4.6.4 Descriptive Statistics of PBC Patients and Subgroups According to Risk Category

Group[1]	Variable[2]	Mean	Standard Deviation	Minimum	Maximum
1	Age	46.5	8.5	24.7	65.1
($N = 98$)	Bilirubin	12.1	8.2	1.1	41.6
	Albumin	3.1	0.5	1.9	4.5
	Protime	13.4	1.4	10.3	19.3
	Edema	0.36	0.42	0	1.0
	Risk score	7.36	1.04	4.37	8.65
2	Age	47.8	6.8	34.9	64.8
($N = 41$)	Bilirubin	24.1	12.2	6.3	52.7
	Albumin	2.7	0.6	1.8	4.5
	Protime	15.1	2.5	11.8	23.0
	Edema	0.77	0.37	0	1.0
	Risk score	9.17	0.26	8.67	9.11
3	Age	53.8	9.0	39.9	76.6
($N = 22$)	Bilirubin	27.8	11.7	4.5	58.5
	Albumin	2.5	0.4	1.6	3.0
	Protime	19.5	5.7	13.4	35.7
	Edema	0.93	0.23	0	1.0
	Risk score	10.44	0.47	9.95	11.58
All	Age	47.8	8.5	24.7	76.6
($N = 161$)	Bilirubin	17.3	11.8	1.1	58.5
	Albumin	2.9	0.6	1.6	4.5
	Protime	14.6	3.4	10.3	35.7
	Edema	0.54	0.45	0	1.0
	Risk score	8.24	1.44	4.37	11.58

[1]Stratification into risk groups was performed according to Mayo Clinic model risk scores, with cutoffs chosen to produce roughly equal numbers of deaths in the first three months. Group 1: 98 patients with 10 deaths in the first three months; group 2: 41 patients with 10 deaths; group 3: 22 patients with 9 deaths.

[2]Variables are: age (years); bilirubin (mg/dl); albumin (g/dl); prothrombin time (protime, seconds); severity of edema; and risk score according to the Mayo model.

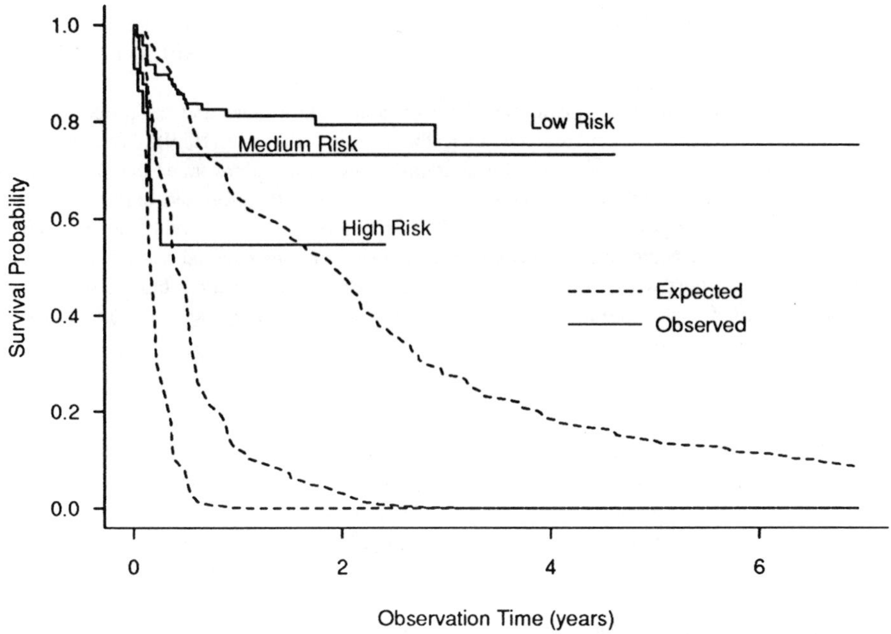

Figure 4.6.14 Predicted and observed survival curves by risk group in the University of Pittsburgh and Baylor University transplant series. Expected curves have the same ordering by risk group as the observed curves.

4.7 BIBLIOGRAPHIC NOTES

The literature on proportional hazards regression has grown quickly over the last twenty years, and a complete discussion of the contributions in this area would require more space than is appropriate. This brief summary mentions some of the influential papers on this topic.

Section 4.2

The roots of proportional hazards regression with an unspecified baseline hazard function can be found in parametric models, which often assume that the hazard function is constant (an exponential model) or is proportional to a power of the time variable (a Weibull model). Feigl and Zelen (1965) were among the first to discuss clinical applications of exponential regression; the leukemia data cited in their paper has frequently been used in the literature to illustrate new methods. Kalbfleisch and Prentice (1980), Lawless (1982), and Cox and Oakes (1984) all discuss a variety of parametric regression models, some of which allow hazard ratios to change over time. In accelerated failure time models a covariate produces a scale transformation of the time axis; we have not discussed those models here at all. Parametric accelerated failure time models are discussed in the three references above. Nonparametric inference for accelerated failure time models is based on a marginal likelihood function for the ranks of observed failure times. This approach presents some technically difficult problems for computing estimators of regression coefficients and calculating their asymptotic sampling distributions. These problems are discussed in Kalbfleisch and Prentice (1980); recent progress in this area may be found in Cuzick (1988), Tsiatis (1990), Lai and Ying (1990), and Wei, Ying and Lin (1990).

The connection between a logistic model for discrete failure times and the proportional hazards model is explored in more detail in Kalbfleisch and Prentice (1980). Prentice and Gloeckler (1978) have explored proportional hazards models in failure times that are made discrete by grouping observations. Pierce et al., (1979) have also proposed nonparametric regression methods for grouped survival data. Self and Prentice (1982) provide an extended discussion of the link between the multiplicative intensity and proportional hazards models.

Some of the earliest work examining the formal relationship between a hazard function and the intensity for a counting process, as in Theorem 4.2.1, may be found in Dolivo (1974).

Section 4.3

The seminal papers for partial likelihood are, of course, those by Cox (1972, 1975). Oakes (1981) has summarized some of the many implications of partial likelihood inference for survival data, and pointed out a number of unsolved problems at the time of that manuscript. Wong (1986) put the partial likelihood in a more general context, and established conditions under which regression or other parameters estimated from a partial likelihood are consistent and asymptotically normal.

The connection between the marginal and partial likelihoods in the proportional hazards model was first noticed by Kalbfleisch and Prentice (1973). Jacobsen (1984) and Bailey (1979, 1980) have shown that partial likelihood estimators are very similar but not identical to those obtained by generalized maximum likelihood.

The Bibliographic Notes to Chapter 8 contain additional references to work on asymptotic sampling distributions for regression parameter estimators in the proportional hazards model.

Section 4.4

The scientific literature contains many interesting examples of applications of proportional hazards regression. The model is widely used in cancer research and in studies of cardiovascular disease, so the literature on those topics is a rich source of examples. Crowley and Hu (1977) illustrated an interesting use of time-dependent covariates in an analysis of data from a Stanford heart transplant study, and the use of that data to illustrate novel survival analysis methods has become nearly obligatory. In a study of a large series of breast cancer patients, Gore et al., (1984) have illustrated techniques for model selection useful in large samples. Andersen (1986) explored application of multi-state time dependent covariate models. A more detailed discussion of the clinical background for the PBC and prostate data discussed in Section 4.6 can be found in Dickson et al., (1985), Dickson et al., (1989) and Winkler et al., (1988). The CGD data discussed in Section 4.4 were presented by the International Chronic Granulomatous Disease Cooperative Study Group (1991).

The social science literature also contains applications of proportional hazards regression. Tuma et al., (1979) applied two state Markov process models for social processes that specified a proportional effect of covariates on transition rates, while Lancaster and Nickell (1980) used parametric proportional hazards models for the analysis of employment data. Schmidt and Witte (1988) used parametric methods for the study of recidivism rates among a cohort of prisoners released during the same year. Andersen (1985) used the counting process methods discussed in this chapter in a longitudinal study of labor-market data.

Section 4.5

Regression diagnostics for the proportional hazards and multiplicative intensity models are not as well-developed as the diagnostics for linear models (cf. Belsley et al. (1980) and Cook and Weisberg (1982)), and a good deal of work remains to be done in this area. Broadly speaking, diagnostics for proportional hazards regression are used for variable selection, choosing variable transformations, checking the time-dependent behavior of hazard ratios, and detecting influence points and outliers.

Stepwise regression methods are commonly used for variable selection, and efficient algorithms for this are discussed in Peduzzi et al., (1980) and Harrell and Lee (1986). In general, these methods work in much the same way as in linear regression, with changes in the log partial likelihood used to add or drop variables instead of changes in residual sums of squares. Of course, the methods cannot escape the problems of interpretation that arise in linear models.

Research into the selection of variable transformations is currently proceeding in two directions. One approach uses residuals from a fitted model in much the same way those residuals are used in linear regression. Early contributions to the notion of generalized residuals for likelihood based regression models can be found in Cox and Snell (1968). A detailed exploration of current results is contained in McCullagh and Nelder (1989). The approach used in this chapter for adapting these methods to proportional hazards regression with censored data was explored in Therneau et al., (1990) and Barlow and Prentice (1988). A second approach uses nonparametric or flexible parametric methods to estimate directly the graph of a transformation. Hastie and Tibshirani (1986) have illustrated the use of local likelihoods and local scoring techniques for estimating transformations in generalized additive models that include the Cox model. Durrleman and Simon (1989) and Sleeper and Harrington (1990) have used regression splines to estimate transformations in generalized additive models. The latter paper showed how these methods yield much the same result when applied to the PBC data as those given in this section.

Many authors have discussed methods for checking the proportional hazards assumptions. Tests for the hypothesis of proportional hazards are mentioned in Section 4.5, along with some more standard graphical methods. Additional graphical methods can be based on the work of Lagakos (1980), Tanner and Wong (1983, 1984), Ramlau-Hansen (1983), Mau (1986), and Gray (1990). Kay (1984) contains a summary of the earlier work.

Relatively little has been done for detecting influence points or outliers in proportional hazards regression. Important work here includes Cain and Lange (1984), Storer and Crowley (1985), Barlow and Prentice (1988), and Therneau et al., (1990).

CHAPTER 5

A Martingale Central Limit Theorem

5.1 PRELIMINARIES

In this chapter, we use the martingale structure of $U^{(n)} \equiv \sum_{i=1}^{n} \int H_i^{(n)} dM_i^{(n)}$ to establish asymptotic distribution results for sequences of statistics which can be represented in this form. We first summarize the properties of the processes we study, then give a heuristic motivation for the main result. For clarity, the notation will indicate the dependence of processes on the sample size n.

We consider the large-sample joint distribution of r statistics $(U_1^{(n)}, \ldots, U_r^{(n)})$. For each n, assume

$$\{ N_{i,\ell}^{(n)} : i = 1, \ldots, n, \ell = 1, \ldots, r\} \text{ is a multivariate counting process with stochastic basis } (\Omega, \mathcal{F}, \{\mathcal{F}_t : t \geq 0\}, P); \tag{1.1}$$

$$\text{the compensator } A_{i,\ell}^{(n)} \text{ of } N_{i,\ell}^{(n)} \text{ is continuous.} \tag{1.2}$$

Although we assume Condition (1.2) holds throughout this chapter, the theorems here can be generalized to processes with discontinuous compensators (c.f. Helland, 1982). By Theorems 2.3.1 and 2.5.2, $M_{i,\ell}^{(n)} \equiv N_{i,\ell}^{(n)} - A_{i,\ell}^{(n)}$ is a locally square integrable martingale, $\{M_{i,\ell}^{(n)}\}^2 - A_{i,\ell}^{(n)}$ is a local martingale, and $M_{i,\ell}^{(n)} M_{i',\ell'}^{(n)}$ is a local martingale if $i \neq i'$ or $\ell \neq \ell'$. For each i, ℓ we also assume

$$H_{i,\ell}^{(n)} \text{ is a locally bounded } \mathcal{F}_t\text{-predictable process.} \tag{1.3}$$

Define

$$U_{i,\ell}^{(n)}(t) = \int_0^t H_{i,\ell}^{(n)}(s) dM_{i,\ell}^{(n)}(s),$$

$$U_\ell^{(n)}(t) = \sum_{i=1}^{n} \int_0^t H_{i,\ell}^{(n)}(s) dM_{i,\ell}^{(n)}(s),$$

and, for any $\epsilon > 0$,

$$U_{i,\ell,\epsilon}^{(n)}(t) = \int_0^t H_{i,\ell}^{(n)}(s) I_{\{|H_{i,\ell}^{(n)}(s)| \geq \epsilon\}} dM_{i,\ell}^{(n)}(s)$$

and

$$U_{\ell,\epsilon}^{(n)}(t) = \sum_{i=1}^{n} U_{i,\ell,\epsilon}^{(n)}(t).$$

By Theorem 2.4.1, $U_{i,\ell}^{(n)}, U_{\ell}^{(n)}, U_{i,\ell,\epsilon}^{(n)}$, and $U_{\ell,\epsilon}^{(n)}$ are local square integrable martingales. The process $U_{\ell,\epsilon}^{(n)}$ contains all jumps of the process $U_{\ell}^{(n)}$ which are at least ϵ in size. By the linearity of $\langle \cdot, \cdot \rangle$ and by Theorem 2.4.3,

$$\langle U_{\ell}^{(n)}, U_{\ell}^{(n)} \rangle(t) = \sum_{i=1}^{n} \int_0^t \{H_{i,\ell}^{(n)}(s)\}^2 dA_{i,\ell}^{(n)}(s)$$

and

$$\langle U_{\ell,\epsilon}^{(n)}, U_{\ell,\epsilon}^{(n)} \rangle(t) = \sum_{i=1}^{n} \int_0^t \{H_{i,\ell}^{(n)}(s)\}^2 I_{\{|H_{i,\ell}^{(n)}(s)| \geq \epsilon\}} dA_{i,\ell}^{(n)}(s).$$

The central result of this chapter gives the asymptotic distribution of $\{U_{\ell}^{(n)} : \ell = 1, \ldots, r\}$ as the sample size $n \to \infty$, and relies on the notion of weak convergence, or convergence in distribution, of stochastic processes. Appendix B introduces this topic, and summarizes some of the main results. Readers not familiar with the weak convergence of processes should begin there. When a sequence of processes $\{X_n\}$ converges weakly to a limit process X, we write $X_n \Longrightarrow X$. Convergence in distribution for a sequence of random variables, written $X_n \xrightarrow{\mathcal{D}} X$, is also reviewed in Appendix B. Some of the results here use convergence in the L^p norm or in probability for random variables. We write that $X_n \xrightarrow{L^p} X$ when the sequence of random variables $\{X_n\}$ satisfies

$$\lim_{n \to \infty} E(|X_n - X|^p) = 0,$$

and $X_n \xrightarrow{P} X$ when

$$\lim_{n \to \infty} P(|X_n - X| \geq \epsilon) = 0$$

for any $\epsilon > 0$.

The weak limit of $U_{\ell}^{(n)}$ is easy to anticipate, and, to make the arguments simpler, we first fix ℓ and then suppress it in the notation. Initially, we will assume that, for each n, there exists a constant $C^{(n)} < \infty$ (which may increase with n to ∞) such that, with probability one,

$$N_i^{(n)}, A_i^{(n)}, \text{ and } H_i^{(n)}, i = 1, \ldots, n, \text{ are bounded by } C^{(n)}. \tag{1.4}$$

It follows, then, by Theorem 2.4.4 that $U_i^{(n)}, U^{(n)}, U_{i,\epsilon}^{(n)}$ and $U_\epsilon^{(n)}$ are square integrable martingales. Later, we will remove Condition (1.4) by identifying a localizing sequence $\{\tau^{(n)}\}$ so that the stopped processes $N_i^{(n)}(\cdot \wedge \tau^{(n)})$, $A_i^{(n)}(\cdot \wedge \tau^{(n)})$, and $H_i^{(n)}(\cdot \wedge \tau^{(n)})$ satisfy Condition (1.4), and by noting that $U^{(n)}(\cdot \wedge \tau^{(n)})$ and $U^{(n)}$ converge weakly to the same limit as $\tau^{(n)} \to \infty$.

We will establish, under special conditions, the weak convergence of $U^{(n)}$ to the following process.

Definition 5.1.1. The standard *Wiener* or *Brownian motion* process, W, is a stochastic process satisfying:

1. $W(0) = 0$, and $EW(t) = 0$ for any t;
2. W has independent increments, therefore, $W(t) - W(u)$ is independent of $W(u)$ for any $0 \le u \le t$;
3. $W(t)$ has variance t; and
4. W is a Gaussian process with continuous sample paths.

If f is a measurable nonnegative function and $\alpha(t) = \int_0^t f^2(s)ds$, then $\int f dW$ denotes the process satisfying (1), (2), and (4), but with

$$\text{var}\left\{\int_0^t f(s)dW(s)\right\} = \alpha(t). \qquad \square$$

The statement of Property (4) could be simplified, since any process satisfying (1), (2) and (3) is a Gaussian process if and only if it has continuous sample paths.

The process $U^{(n)}$ satisfies (1) for all n. By the martingale property, $U^{(n)}$ has uncorrelated increments, so that (2) is plausible for large n. By Theorem 2.4.4,

$$\text{var}\{U^{(n)}(t)\} = E\langle U^{(n)}, U^{(n)}\rangle(t).$$

If

$$\langle U^{(n)}, U^{(n)}\rangle(t) \xrightarrow{P} \alpha(t)$$

for some integrand f^2, $U^{(n)}$ should satisfy (3) as $n \to \infty$ with t replaced by $\alpha(t)$. Finally, $U_\epsilon^{(n)}$ contains all jumps in $U^{(n)}$ of size ϵ or larger, $EU_\epsilon^{(n)} = 0$ and

$$\text{var}\{U_\epsilon^{(n)}(t)\} = E\langle U_\epsilon^{(n)}, U_\epsilon^{(n)}\rangle(t).$$

If $\langle U_\epsilon^{(n)}, U_\epsilon^{(n)}\rangle(t) \xrightarrow{P} 0$ as $n \to \infty$ for any $\epsilon > 0$, the sample paths of $U^{(n)}$ should become smooth for large n, and (4) also should hold.

Meyer (1971) has shown that, if $U_1, U_2, \ldots, U_r$ are local square integrable martingales, zero at time zero, with continuous sample paths such that $\langle U_i, U_j\rangle(t) = \delta_{ij}\alpha_i(t)$ for some increasing right continuous functions α_i satisfying $\alpha_i(0) = 0$, then the $\{U_1, U_2, \ldots, U_r\}$ are independent Gaussian processes, zero at time zero, with independent increments and $\text{var}\{U_i(t)\} = \alpha_i(t)$. Dropping the term "local" implies $\alpha_i(t) < \infty$ for all t.

The following is a formal statement of the result (in the special case of $r = 1$) we have sketched.

Theorem 5.1.1. Assume that Conditions (1.1) through (1.4) are true. Let f be a measurable nonnegative function and $\alpha(t) = \int_0^t f^2(s)ds < \infty$ for all $t > 0$. Suppose for all $t > 0$ that, as $n \to \infty$,

$$\langle U^{(n)}, U^{(n)}\rangle(t) \xrightarrow{P} \alpha(t) \tag{1.5}$$

and

$$\langle U_\epsilon^{(n)}, U_\epsilon^{(n)}\rangle(t) \xrightarrow{P} 0, \text{ for any } \epsilon > 0. \tag{1.6}$$

Then

$$U^{(n)} \Longrightarrow U^\infty \equiv \int f dW \text{ on } D[0,\infty) \text{ as } n \to \infty,$$

where W denotes Brownian motion, and where $D[0,\infty)$ denotes the space of functions on $[0,\infty)$ which are right-continuous with finite left-hand limits, endowed with the Skorohod topology. □

In this chapter, we prove Theorem 5.1.1 and a multivariate extension, when Condition (1.4) is removed. The multivariate version is contained in Theorem 5.3.4 for readers who wish to avoid the technical details of the proof and proceed directly to applications in Chapters 6 through 8.

The main result is Theorem 5.2.3, which establishes that, under certain conditions, normed sums S_n of triangular martingale difference arrays $\{X_{n,j}\}$ (defined in Section 5.2) converge in distribution to a normal variate. The result is due to McLeish (1974). Theorems 5.2.4 and 5.2.5 extend Theorem 5.2.3 to weak convergence of continuous time processes W_n built from martingale difference arrays. These processes have values that coincide with S_n at discrete time points, and sample paths that are constant between these points. Theorem 5.1.1, the convergence of the sequence $\{U^{(n)}\}$, is proved in Section 5.3 by showing that $U^{(n)}$ can be approximated arbitrarily closely, as $n \to \infty$, by a process $W^{(n)}$ that has the same structure and hence convergence properties of W_n.

The sufficient conditions given in Theorem 5.1.1 are descendants of those in Theorem 5.2.3. The development in Sections 5.2 and 5.3 shows that these conditions are closely tied to the requirement that useful limiting distributions have finite, positive variances, and to the Lindeberg condition appearing in central limit theorems for sums of independent random variables.

For continuity, some of the more technical arguments needed here have been relegated to Appendix C.

5.2 CONVERGENCE OF MARTINGALE DIFFERENCE ARRAYS

We begin with the asymptotic distribution for a "discretized" continuous-time version, $W^{(n)}$, of $U^{(n)} = \sum_{i=1}^{n} \int H_i^{(n)}(s)dM_i^{(n)}(s)$. A special sequence of partitions of $[0, \infty)$, as defined in Helland (1982), will be used to formulate $W^{(n)}$ in a manner which will facilitate establishing the asymptotic results.

Definition 5.2.1. A sequence $\{Q(k) : 1 \leq k < \infty\}$ of infinite partitions of $[0, \infty)$ is called an R-sequence if, for each k,

1. $Q(k) = \{t_{0,k}, t_{1,k}, \ldots\}$ with $0 = t_{0,k} < t_{1,k} < \ldots$;
2. $\lim_{j\to\infty} t_{j,k} = \infty$;
3. $Q(k) \subset Q(k+1)$; and
4. $\max_{j:t_{j,k}\leq t}(t_{j+1,k} - t_{j,k}) \to 0$ as $k \to \infty$ for all $t \geq 0$. □

We actually will employ a subsequence $\{Q(k_n)\}$ of an R-sequence of partitions in defining $W^{(n)}$. Later, we will specify the rate at which $k_n \to \infty$, as $n \to \infty$.

For any $t > 0$, let $k_n(t) = \max\{j : t_{j,k_n} \leq t\}$, so that $k_n(\cdot)$ is a right-continuous integer valued step function. For each $j \geq 0$, set

$$X_j^{(n)} = U^{(n)}(t_{j+1,k_n}) - U^{(n)}(t_{j,k_n}). \tag{2.1}$$

The variables $X_j^{(n)}$ are elements of a doubly infinite array, with rows indexed by n. The process $W^{(n)}$ is the sum of terms in the nth row, i.e.,

$$W^{(n)}(t) = \sum_{j=0}^{k_n(t)} X_j^{(n)}. \tag{2.2}$$

Figure 5.2.1 illustrates the relationship between the original process $U^{(n)}$ and $W^{(n)}$ whose paths, $W^{(n)}(\cdot, \omega)$, are right-continuous step functions which are constant on the subintervals of the partition $Q(k_n)$.

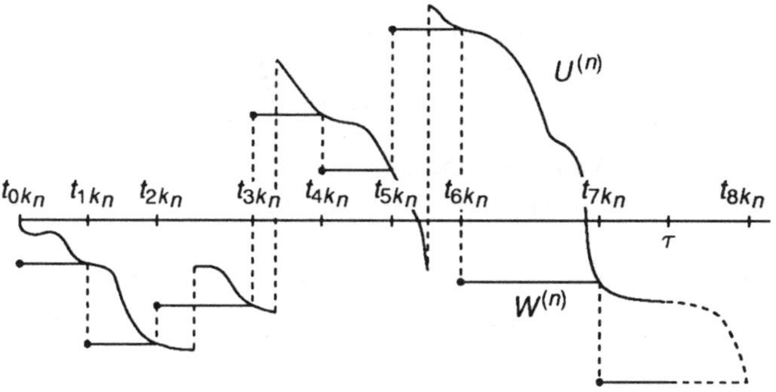

Figure 5.2.1 An example of the process $U^{(n)}$ and an approximating process $W^{(n)}$ over an interval $[0, \tau]$.

The process $W^{(n)}$ has two important characteristics. It is a step function approximation to $U^{(n)}$ which becomes increasingly accurate as $k_n \uparrow \infty$, and its value at t is the sum of the nth row in the "triangular" array $\{X_j^{(n)} : 0 \leq j \leq k_n(t), n \geq 1\}$. By proceeding as in McLeish (1974), we can obtain conditions sufficient for weak convergence of $W^{(n)}$. McLeish's limit theorems use the notion of a martingale difference array.

Definition 5.2.2. Let $\{X_{n,j} : j \geq 1, n \geq 1\}$ and $\{\mathcal{F}_{n,j} : j \geq 1, n \geq 1\}$ be doubly infinite arrays of random variables and sub-σ-algebras of a σ-algebra $\mathcal{F}$, and let $E_j(Z)$ denote $E\{Z|\mathcal{F}_{n,j}\}$ for any random variable Z. Then $\{X_{n,j}\}$ is called a *martingale difference array* if

1. $EX_{n,j}^2 \equiv \sigma_{n,j}^2 < \infty$, and, for any j and n, $X_{n,j}$ is $\mathcal{F}_{n,j}$ measurable,
2. $\mathcal{F}_{n,j-1} \subset \mathcal{F}_{n,j}$, and
3. $E_{j-1}(X_{n,j}) = 0$ a.s. for any n and j. □

Let $\{r_n\}$ be a sequence of integer-valued nondecreasing right-continuous functions defined on $[0, \infty)$. The random function $W_n(t) \equiv \sum_{j=1}^{r_n(t)} X_{n,j}$ has sample paths which are right-continuous step functions. The process $W^{(n)}$ has the same structure as W_n. This can be seen by setting $X_{n,j} = X_{j-1}^{(n)}$, and $\mathcal{F}_{n,j} = \mathcal{F}_{t_{j,k_n}}^{(n)}$ where, for each n, $\{\mathcal{F}_t^{(n)} : t \geq 0\}$ is a filtration with respect to which $U^{(n)}$ is adapted. When $U^{(n)}$ is a square integrable martingale, $EX_{j-1}^{(n)} = 0$ and $E\{X_{j-1}^{(n)}\}^2 < \infty$. Since

$$X_{j-1}^{(n)} = U^{(n)}(t_{j,k_n}) - U^{(n)}(t_{j-1,k_n})$$

is $\mathcal{F}_{t_{j,k_n}}^{(n)}$-measurable and

$$E_{j-1}X_{n,j} = E\{X_{j-1}^{(n)}|\mathcal{F}_{t_{j-1,k_n}}^{(n)}\} = 0,$$

$\{X_{j-1}^{(n)}\}$ also is a martingale difference array. By Eq. (2.2),

$$W^{(n)}(t) = \sum_{j=1}^{k_n(t)+1} X_{j-1}^{(n)},$$

so $r_n(t) = k_n(t) + 1$. As a result, the convergence of $W^{(n)}$ will follow from a more general theorem about weak convergence of processes W_n built from martingale difference arrays.

We begin with a central limit theorem for $W_n(t^*)$, where t^* is a fixed value of t, then establish weak convergence for the process W_n. For simplicity, we drop t^* from $r_n(t^*)$ and denote $W_n(t^*)$ by S_n, so $S_n = \sum_{j=1}^{r_n} X_{n,j}$.

The most widely known generalization of the classical central limit theorem for independent, identically distributed variables is to the case of independent but not identically distributed variables, and usually is referred to as Lindeberg's Theorem.

Theorem 5.2.1. Consider a doubly infinite array $\{X_{n,j} : j \geq 1, n \geq 1\}$, and assume $EX_{n,j} = 0$ and $EX_{n,j}^2 \equiv \sigma_{n,j}^2 < \infty$ for all n, j. Assume the random variables in each row $\{X_{n,j} : j \geq 1\}$ are independent, so $\mathrm{var}\left\{\sum_{j=1}^{r_n} X_{n,j}\right\} = \sum_{j=1}^{r_n} \sigma_{n,j}^2 \equiv C_n^2$. Set $Z_{n,j} = X_{n,j}C_n^{-1}$ and $Z_n = \sum_{j=1}^{r_n} Z_{n,j}$. If, for any $\epsilon > 0$,

$$E\left\{\sum_{j=1}^{r_n} Z_{n,j}^2 I_{\{|Z_{n,j}|>\epsilon\}}\right\} \to 0 \text{ as } n \to \infty, \tag{2.3}$$

then the sequence $\{Z_n\}$ converges in distribution as $n \to \infty$ to a normally distributed random variable with mean zero and variance one, (written $Z_n \xrightarrow{\mathcal{D}} Z \sim N(0, 1)$). □

When the sequence $\{C_n^2\}$ is bounded away from both 0 and ∞, Condition (2.3) is (Exercise 5.1) equivalent to

$$E\left\{\sum_{j=1}^{r_n} X_{n,j}^2 I_{\{|X_{n,j}|>\epsilon\}}\right\} \to 0 \text{ as } n \to \infty, \text{ for any } \epsilon > 0.$$

This last limit is called Lindeberg's condition, and it is not difficult to show that it implies

$$P\{|X_{n,j}| > \epsilon\} \to 0 \text{ as } n \to \infty, \text{ uniformly in } j.$$

This last condition is a version of uniform asymptotic negligibility, and essentially requires that the contributions of individual terms in the sum S_n are uniformly small, compared to S_n itself, when n is large.

One proof of the Lindeberg Theorem uses characteristic function arguments. If Z is a random variable and i is the complex number such that $i^2 = -1$, then $p(t) = E(e^{itZ}), -\infty < t < \infty$, is the characteristic function of Z. The following theorem plays an important role.

Theorem 5.2.2 (Levy's Theorem). Suppose $Z, Z_1, Z_2, \ldots$ is an infinite sequence of random variables, and let $p(t) = E(e^{itZ})$ and $p_n(t) = E(e^{itZ_n})$. A necessary and sufficient condition for $Z_n \xrightarrow{\mathcal{D}} Z$ is that $\lim_{n\to\infty} p_n(t) = p(t)$ for every t. □

We need a central limit theorem more general than Lindeberg's for sums of the form $S_n = \sum_{j=1}^{r_n} X_{n,j}$, where $\sigma_{n,j}^2$ need not be finite (a level of generality not needed when Condition (1.4) holds), and where the $\{X_{n,j} : j = 1, 2, \ldots,\}$ are not independent but are elements of a martingale difference array. A result due to McLeish provides the basis for the necessary generalization.

Theorem 5.2.3. Let $\{X_{n,j}\}$ be a martingale difference array satisfying

1. $\sup_n E\{\max_{j \le r_n} X_{n,j}^2\} < \infty$,
2. $\max_{j \le r_n} |X_{n,j}| \xrightarrow{P} 0$ as $n \to \infty$, and
3. $\sum_{j=1}^{r_n} X_{n,j}^2 \xrightarrow{P} 1$ as $n \to \infty$.

Then $S_n \xrightarrow{\mathcal{D}} Z \sim N(0,1)$ as $n \to \infty$. □

Conditions (1) and (2) in Theorem 5.2.3 are both consequences of Lindeberg's condition. Since

$$P\left\{\max_{j \le r_n} |X_{n,j}| > \epsilon\right\} = P\left\{\sum_{j=1}^{r_n} X_{n,j}^2 I_{\{|X_{n,j}|>\epsilon\}} > \epsilon^2\right\},$$

(2) is equivalent to the "weak" version of Lindeberg's condition,

$$\sum_{j=1}^{r_n} X_{n,j}^2 I_{\{|X_{n,j}|>\epsilon\}} \xrightarrow{P} 0 \text{ for any } \epsilon > 0. \tag{2.4}$$

For any $\epsilon > 0$,

$$\max_{j \le r_n} X_{n,j}^2 \le \epsilon^2 + \sum_{j=1}^{r_n} X_{n,j}^2 I_{\{|X_{n,j}|>\epsilon\}}.$$

Lindeberg's condition implies the expectation of the right-hand side converges to zero, so that (1) holds.

The assumption that $\{X_{n,j}\}$ is a martingale difference array imposes sufficient structure to replace the independence assumption in Lindeberg's Theorem.

Proof of Theorem 5.2.3. Let the array $\{Z_{n,j}\}$ be given by

$$Z_{n,j} = X_{n,j} I_{\left\{\sum_{k=1}^{j-1} X_{n,k}^2 \le 2\right\}}.$$

Since

$$\begin{aligned} E_{j-1} Z_{n,j} &= I_{\left\{\sum_{k=1}^{j-1} X_{n,k}^2 \le 2\right\}} E_{j-1} X_{n,j} \\ &= 0 \quad \text{a.s.}, \end{aligned}$$

$\{Z_{n,j}\}$ also is a martingale difference array. Assumption (3) implies that, as $n \to \infty$,

$$P\{Z_{n,j} \ne X_{n,j} \text{ for some } j \le r_n\} \le P\left\{\sum_{j=1}^{r_n} X_{n,j}^2 > 2\right\} \to 0. \tag{2.5}$$

If $\sum_{j=1}^{r_n} Z_{n,j} \xrightarrow{\mathcal{D}} N(0,1)$, then (2.5) will imply $S_n \xrightarrow{\mathcal{D}} N(0,1)$. Note that (2.5) implies that $Z_{n,j}$ satisfies the assumptions of Theorem 5.2.3.

We need only show that

$$E\,\mathrm{I}_n \to e^{-t^2/2} \text{ as } n \to \infty, \tag{2.6}$$

where $\mathrm{I}_n \equiv \exp(it \sum_{j=1}^{r_n} Z_{n,j})$. Taylor's theorem implies that

$$\log(1+ix) = ix - \frac{(ix)^2}{2} - r(x),$$

where the remainder

$$r(x) = x^3 \left\{ i\left(\frac{1}{3} - \frac{x^2}{5} + \frac{x^4}{7} - \frac{x^6}{9} + \ldots\right) + x\left(\frac{1}{4} - \frac{x^2}{6} + \frac{x^4}{8} - \frac{x^6}{10} + \ldots\right)\right\}$$

satisfies $|r(x)| \le |x|^3$ if $|x| < 1$. Consequently,

$$ix = \log(1+ix) - \frac{x^2}{2} + r(x)$$

and

$$e^{ix} = (1+ix)\exp\left\{-\frac{x^2}{2} + r(x)\right\}.$$

As a result,

$$\mathrm{I}_n = T_n e^{-t^2/2} + V_n,$$

for $T_n \equiv \prod_{j=1}^{r_n}(1 + itZ_{n,j})$ and

$$V_n = T_n \left[\exp\left\{-(t^2/2)\sum_{j=1}^{r_n} Z_{n,j}^2 + \sum_{j=1}^{r_n} r(Z_{n,j}t)\right\} - e^{-t^2/2}\right].$$

The convergence in (2.6) will hold if $E|V_n| \to 0$ and $ET_n = 1$.

To prove $E|V_n| \to 0$, it is sufficient to establish that V_n is uniformly integrable and $V_n \xrightarrow{P} 0$. We first show that $\{T_n\}$ is uniformly integrable. Define the random variables

$$J_n = \begin{cases} \min\{j \le r_n : \sum_{i=1}^{j} X_{n,i}^2 > 2\} & \text{if } \sum_{i=1}^{r_n} X_{n,i}^2 > 2 \\ r_n & \text{otherwise.} \end{cases}$$

Then

$$\begin{aligned} E|T_n|^2 &= E\prod_{j=1}^{r_n}(1 + t^2 Z_{n,j}^2) \\ &\le E\left[\left\{\exp(t^2 \sum_{j=1}^{J_n-1} X_{n,j}^2)\right\}(1 + t^2 X_{n,J_n}^2)\right] \\ &\le e^{2t^2}(1 + t^2 E X_{n,J_n}^2), \end{aligned}$$

which is uniformly bounded in n by Assumption (1). Thus, by Proposition 1.4.4, uniform integrability of $\{T_n\}$ holds. Since $E|\mathrm{I}_n|^2 = 1$ implies I_n is uniformly integrable, it follows that $V_n = \mathrm{I}_n - T_n e^{-t^2/2}$ is uniformly integrable.

Assumption (2) implies

$$\lim_{n\to\infty} P\{|\sum_{j=1}^{r_n} r(Z_{n,j}t)| \geq |t^3|\sum_{j=1}^{r_n} |Z_{n,j}|^3\} = 0,$$

while (2) and (3) imply

$$|t^3|\sum_{j=1}^{r_n} |Z_{n,j}|^3 \leq |t|^3\{\max_{j\leq r_n} |Z_{n,j}|\}\left(\sum_{j=1}^{r_n} Z_{n,j}^2\right) \xrightarrow{P} 0,$$

so

$$\lim_{n\to\infty} |\sum_{j=1}^{r_n} r(Z_{n,j}t)| = 0 \text{ in probability.}$$

Since (3) implies

$$\begin{aligned}|T_n| &= \left\{\prod_{j=1}^{r_n} (1+t^2Z_{n,j}^2)\right\}^{1/2} \\ &\leq \prod_{j=1}^{r_n} \exp\left(\frac{1}{2}t^2Z_{n,j}^2\right) \xrightarrow{P} \exp\left(\frac{1}{2}t^2\right),\end{aligned}$$

it follows that $|V_n| \xrightarrow{P} 0$.

To show $ET_n = 1$, we will appeal to the martingale property. For any k such that $1 \leq k \leq r_n$,

$$\begin{aligned}E_{k-1}\prod_{j=1}^{k}(1+itZ_{n,j}) &= \left\{\prod_{j=1}^{k-1}(1+itZ_{n,j})\right\} E_{k-1}(1+itZ_{n,k}) \\ &= \prod_{j=1}^{k-1}(1+itZ_{n,j}).\end{aligned}$$

Thus, by conditioning recursively, $ET_n = 1$. □

Theorem 5.2.3 provides conditions under which

$$W_n(t^*) \equiv \sum_{j=1}^{r_n(t^*)} X_{n,j} \xrightarrow{\mathcal{D}} Z \sim N(0,1), \quad \text{for any fixed } t^*,$$

and we now extend this result to weak convergence in $D[0,\tau]$, where τ is a fixed constant, and where $D[0,\tau]$ denotes the space of functions on $[0,\tau]$ which are right-continuous with finite left-hand limits, endowed with the Skorohod topology. Again, let W denote Brownian motion. By Corollary B.1.1 of Appendix B, this convergence result will hold if the finite-dimensional distributions converge, i.e., for any $0 \le t_1 \le t_2 \le \ldots \le t_k \le \tau$,

$$\{W_n(t_1), W_n(t_2), \ldots, W_n(t_k)\} \xrightarrow{\mathcal{D}} \{W(t_1), W(t_2), \ldots, W(t_k)\}, \qquad (2.7)$$

and if Stone's tightness condition (1.5 in Appendix B) holds here, i.e., for any $\epsilon > 0$,

$$\lim_{\delta \to 0} \limsup_{n \to \infty} P\left\{ \sup_{\substack{|s-t|<\delta \\ 0 \le s,t \le \tau}} |W_n(s) - W_n(t)| > \epsilon \right\} = 0. \qquad (2.8)$$

The next lemma is a useful tool for establishing finite-dimensional convergence. The proof is not difficult, and can be found in Billingsley (1968, p. 49).

Lemma 5.2.1 (Cramer-Wold device). Consider a sequence of random vectors $\{(Y_{n,1}, Y_{n,2}, \ldots, Y_{n,k}) : n = 1, 2, \ldots\}$ and a limit vector $(Y_1, \ldots, Y_k)$. Then, as $n \to \infty$, $(Y_{n,1}, Y_{n,2}, \ldots, Y_{n,k}) \xrightarrow{\mathcal{D}} (Y_1, Y_2, \ldots, Y_k)$ if and only if for any constants $a_1, a_2, \ldots, a_k$,

$$\sum_{i=1}^{k} a_i Y_{n,i} \xrightarrow{\mathcal{D}} \sum_{i=1}^{k} a_i Y_i. \qquad \square$$

The Cramer-Wold device is behind the conditions in Theorem 2.4 used to establish $W_n \Longrightarrow W$.

Theorem 5.2.4. Suppose $\{X_{n,j}\}$ is a martingale difference array satisfying, for any $t \in [0,\tau]$,

1. $E\left\{\max_{j \le r_n(t)} X_{n,j}^2\right\} \to 0$ as $n \to \infty$, and

2. $\sum_{j=1}^{r_n(t)} X_{n,j}^2 \xrightarrow{P} t$ as $n \to \infty$.

Then $W_n \Longrightarrow W$ on $D[0,\tau]$ as $n \to \infty$, where W denotes Brownian motion.

$\square$

Condition (2) of Theorem of 5.2.4, and Condition (3) of Theorem 5.2.3 are essentially identical. Meanwhile, Condition (1) of Theorem 5.2.4 implies Theorem 5.2.3's Conditions (1) and (2), which were also consequences of Lindeberg's condition.

Proof of Theorem 5.2.4. As in Theorem 5.2.3, the proof begins with a definition of a related martingale difference array

$$\tilde{X}_{n,j} = X_{n,j} I_{\left\{\sum_{i=1}^{j-1} X_{n,i}^2 \le \tau+1\right\}}.$$

If we define the process $\widetilde{W}_n$ at t by $\widetilde{W}_n(t) = \sum_{j=1}^{r_n(t)} \widetilde{X}_{n,j}$, it follows directly from Condition (2) that,

$$\lim_{n\to\infty} P\{W_n(t) \neq \widetilde{W}_n(t) \text{ for any } t \in [0,\tau]\} = 0.$$

By Theorem 4.1 of Billingsley (1968), $W_n \Longrightarrow W$ holds if $\widetilde{W}_n \Longrightarrow W$.

The process $\widetilde{W}_n$ is useful because

$$\max_{j\le r_n(\tau)} |\widetilde{X}_{n,j}| \xrightarrow{P} 0, \tag{2.9}$$

and

$$E\left|\sum_{j=1}^{r_n(t)} \widetilde{X}_{n,j}^2 - t\right| \to 0 \text{ for any } t. \tag{2.10}$$

The limit in (2.9) follows immediately from Condition (1) and the fact that Condition (2) implies $P\{X_{n,j} \neq \widetilde{X}_{n,j} \text{ for some } j \le r_n(\tau)\} \to 0$. To verify (2.10), observe that

$$\sum_{j=1}^{r_n(t)} \widetilde{X}_{n,j}^2 \le \tau + 1 + \max_{j\le r_n(\tau)} X_{n,j}^2.$$

By Condition (1), the right-hand side converges to $\tau + 1$ in L^1, so uniform integrability of $\sum_j \widetilde{X}_{n,j}^2$ follows. This together with Condition (2) yields (2.10).

The Cramer-Wold Device now implies that (2.9) and (2.10) are sufficient for convergence of finite-dimensional distributions. Let $0 = t_0 < t_1 < \ldots < t_m \le \tau$ and $a_1, a_2, \ldots, a_m$ be arbitrary constants. It suffices to show

$$\sum_{i=1}^{m} a_i\{\widetilde{W}_n(t_i) - \widetilde{W}_n(t_{i-1})\} \xrightarrow{\mathcal{D}} N(0,\sigma^2), \tag{2.11}$$

where $\sigma^2 \equiv \sum_{i=1}^m (a_i)^2(t_i - t_{i-1})$. For each $j \in (r_n(t_{i-1}), r_n(t_i)]$, set $Y_{n,j} = a_i\widetilde{X}_{n,j}$, $i = 1, \ldots, m$. The limit in Eq. (2.11) can be written

$$\sum_{i=1}^{m} \sum_{j=r_n(t_{i-1})+1}^{r_n(t_i)} a_i \widetilde{X}_{n,j} = \sum_{j=1}^{r_n(t_m)} Y_{n,j} \xrightarrow{\mathcal{D}} N(0,\sigma^2). \tag{2.12}$$

The convergence in Eq. (2.12) follows from Theorem 5.2.3, with unity replaced by σ^2 in (3). Assumption (2) of Theorem 5.2.3, $\max_{j\le r_n(t_m)} |Y_{n,j}| \xrightarrow{P} 0$, follows immediately from (2.9). Condition (2.10) implies $\sum_{j=1}^{r_n(t_m)} Y_{n,j}^2 \xrightarrow{L^1} \sigma^2$, from which it follows directly that

$$\sum_{j=1}^{r_n(t_m)} Y_{n,j}^2 \xrightarrow{P} \sigma^2,$$

and

$$\sup_n E\left\{\max_{j\le r_n(t_m)} Y_{n,j}^2\right\} < \infty.$$

The details for establishing the tightness condition, Eq. (2.8), are lengthy and are given in Lemmas C.1.1 and C.1.2 of Appendix C. □

We note in passing that Condition (2.9) is equivalent to Lindeberg's condition when (2.10) is true. The limit in Condition (2.9) was shown earlier to be equivalent to

$$\sum_{j=1}^{r_n(t)} \widetilde{X}_{n,j}^2 I_{\{|\widetilde{X}_{n,j}|>\epsilon\}} \xrightarrow{P} 0 \text{ for any } \epsilon > 0.$$

In addition, (2.10) implies $E\sum_j \widetilde{X}_{n,j}^2 I_{\{|\widetilde{X}_{n,j}|>\epsilon\}} < \infty$. Theorem 4.5.4 in Chung (1974) and (2.10) imply that $\sum_j \widetilde{X}_{n,j}^2$ is a uniformly integrable sequence in n, so $\sum_j \widetilde{X}_{n,j}^2 I_{\{|\widetilde{X}_{n,j}|>\epsilon\}}$ also is uniformly integrable. Thus

$$E\sum_{j=1}^{r_n(t)} \widetilde{X}_{n,j}^2 I_{\{|\widetilde{X}_{n,j}|>\epsilon\}} \to 0.$$

A corollary to Theorem 5.2.4 provides alternative conditions under which $W_n \Longrightarrow W$ in $D[0,\tau]$. We will use these in proving the weak convergence of $U^{(n)}$ in the next section.

Corollary 5.2.1. For any $t \in [0,\tau]$, let the array $\{X_{n,j}\}$ satisfy:

1. The conditional Lindeberg condition

$$\sum_{j=1}^{r_n(t)} E_{j-1}X_{n,j}^2 I_{\{|X_{n,j}|>\epsilon\}} \xrightarrow{P} 0 \text{ as } n \to \infty \text{ for any } \epsilon > 0, \tag{2.13}$$

2.

$$\sum_{j=1}^{r_n(t)} E_{j-1}X_{n,j}^2 \xrightarrow{P} t \text{ as } n \to \infty, \text{ and} \tag{2.14}$$

3.

$$\sum_{j=1}^{r_n(t)} |E_{j-1}X_{n,j}| \xrightarrow{P} 0 \text{ as } n \to \infty. \tag{2.15}$$

Then $W_n \Longrightarrow W$ on $D[0,\tau]$ as $n \to \infty$. □

When $\{X_{n,j}\}$ is a martingale difference array, Condition (3) holds automatically and Corollary 5.2.1 generalizes the central limit theorem of Dvoretzky (1972).

The Corollary's lengthy yet straightforward proof, which appears in Appendix C, has two major steps. In the first, the array $\{X_{n,j}\}$ is shown to satisfy the following conditions for all $t \in [0,\tau]$:

$$\max_{j\le r_n(t)} |X_{n,j}| \xrightarrow{P} 0, \tag{2.16}$$

$$\sum_{j=1}^{r_n(t)} X_{n,j}^2 \xrightarrow{P} t, \tag{2.17}$$

and

$$\sum_{j=1}^{r_n(t)} |E_{j-1} X_{n,j} I_{\{|X_{n,j}| \le c\}}| \xrightarrow{P} 0 \text{ for all } c \in (0, \infty). \tag{2.18}$$

In the second step, Theorem 5.2.4 is used to establish convergence from Conditions (2.16) through (2.18). Conditions (2.16), (2.13), and (1) of Theorem 5.2.4 are types of Lindeberg conditions, Conditions (2.17), (2.14), and (2) of Theorem 5.2.4 relate to second moments, while Conditions (2.18) for large c and (2.15) imply $\{X_{n,j}\}$ is nearly a martingale difference array.

The following lemma, although not necessary in our mainstream development, provides a weakening of the conditions in Theorem 5.2.4 and alternative conditions to those given in Corollary 5.2.1. In the next section, Corollary 5.2.1 will lead to conditions on the "predictable quadratic variation process" of $U^{(n)}$, while Lemma 5.2.2 will imply conditions on the "quadratic variation process."

Lemma 5.2.2. Let $\{X_{n,j}\}$ be a martingale difference array satisfying, for any $t \in [0, \tau]$:

1. $$E\left(\max_{j \le r_n(t)} |X_{n,j}|\right) \to 0 \text{ as } n \to \infty, \tag{2.19}$$

2. $$\sum_{j=1}^{r_n(t)} X_{n,j}^2 \xrightarrow{P} t \text{ as } n \to \infty. \tag{2.20}$$

Then $W_n \Longrightarrow W$ on $D[0, \tau]$ as $n \to \infty$.

Proof. It is sufficient to verify Conditions (2.16) through (2.18). Condition (2.16) follows from (2.19). Also, by Lemma 2.4 of Helland (1982), (2.19) implies

$$\sum_{j=1}^{r_n(t)} E_{j-1}\left(|X_{n,j}| I_{\{|X_{n,j}| > \delta\}}\right) \xrightarrow{P} 0 \text{ for any } \delta > 0.$$

Since $\{X_{n,j}\}$ is a martingale difference array,

$$\begin{aligned} \sum_{j=1}^{r_n(t)} |E_{j-1} X_{n,j} I_{\{|X_{n,j}| \le c\}}| &= \sum_{j=1}^{r_n(t)} |E_{j-1} X_{n,j} I_{\{|X_{n,j}| > c\}}| \\ &\le \sum_{j=1}^{r_n(t)} E_{j-1}\left(|X_{n,j}| I_{\{|X_{n,j}| > c\}}\right), \end{aligned}$$

so (2.18) holds. The proof of the lemma is completed by noting that Condition (2.20) is identical to Condition (2.17). □

We complete this section with a multivariate generalization of Corollary 5.2.1. The proof appears in Appendix C.

Theorem 5.2.5. Let $W_1, \ldots, W_r$ be r independent Brownian Motion processes and $f_1, \ldots, f_r$ be r measurable nonnegative functions such that $\alpha_\ell(t) = \int_0^t f_\ell^2(s)ds < \infty$ for all $t > 0, \ell = 1, \ldots, r$.

For each $\ell = 1, \ldots, r$, let $\{X_{\ell,n,j}\}$ be a martingale difference array with respect to $\{\mathcal{F}_{n,j}\}$. Define $W_{\ell,n}(t) = \sum_{j=1}^{r_n(t)} X_{\ell,n,j}$ and $W_{\ell,n} = \{W_{\ell,n}(t) : 0 \le t \le \tau\}$.

Assume that each of the arrays $\{X_{\ell,n,j}\}, \ell = 1, \ldots, r$, satisfies for any $t \in [0, \tau]$:

1.
$$\sum_{j=1}^{r_n(t)} E_{j-1} X_{\ell,n,j}^2 \xrightarrow{P} \alpha_\ell(t) \text{ as } n \to \infty, \tag{2.21}$$

2.
$$\sum_{j=1}^{r_n(t)} E_{j-1}\left(X_{\ell,n,j}^2 I_{\{|X_{\ell,n,j}|>\epsilon\}}\right) \xrightarrow{P} 0 \text{ as } n \to \infty \text{ for any } \epsilon > 0, \tag{2.22}$$

3.
$$\sum_{j=1}^{r_n(t)} E_{j-1}(X_{\ell,n,j} X_{\ell',n,j}) \xrightarrow{P} 0 \text{ as } n \to \infty \text{ if } \ell \neq \ell'. \tag{2.23}$$

Then $(W_{1,n}, W_{2,n}, \ldots, W_{r,n}) \Longrightarrow (\int f_1 dW_1, \int f_2 dW_2, \ldots, \int f_r dW_r)$ in $(D[0,\tau])^r$ as $n \to \infty$, where $(D[0,\tau])^r$ denotes the Cartesian product of $D[0,\tau]$ with itself r times and has the product Skorohod topology. □

By setting $X_{n,j} = X_{j-1}^{(n)}$ (or $X_{\ell,n,j} = X_{\ell,j-1}^{(n)}$, where $X_{\ell,j-1}^{(n)} \equiv U_\ell^{(n)}(t_{j,k_n}) - U_\ell^{(n)}(t_{j-1,k_n})$ for the square integrable martingale $U_\ell^{(n)}$) and $r_n(t) = k_n(t) + 1$ in Theorem 5.2.5, we obtain the multivariate weak convergence result for $(W_1^{(n)}, W_2^{(n)}, \ldots, W_r^{(n)})$, where $W_\ell^{(n)} = \sum_{j=1}^{k_n(t)+1} X_{\ell,j-1}^{(n)}$.

5.3 WEAK CONVERGENCE OF THE PROCESS, $U^{(n)}$

The results in the previous section provide the basis for the proof that $U^{(n)} \Longrightarrow \int f dW$ under Conditions (1.5) and (1.6). In fact, we will be able to remove the boundedness Condition (1.4) and allow $U^{(n)}$ to be only locally square integrable, as well as establish a multivariate version of this convergence result.

Given the conditions for convergence in Section 5.2, Condition (1.5),

$$\langle U^{(n)}, U^{(n)}\rangle(t) \xrightarrow{P} \int_0^t f^2(s)ds,$$

and Condition (1.6),

$$\langle U_\epsilon^n, U_\epsilon^n\rangle(t) \xrightarrow{P} 0,$$

are natural ones to work with. Suppose n is fixed, and $\{Q(k) : k \geq 1\}$ is an R-sequence of partitions on $[0, \infty)$. For each k, set $X_{j,k}^{(n)} = U^{(n)}(t_{j+1,k}) - U^{(n)}(t_{j,k})$, for all $j \geq 0$. The process $W_k^{(n)}$ given by $W_k^{(n)}(t) = \sum_{j=0}^{k(t)} X_{j,k}^{(n)}$, where $k(t) \equiv \max\{j : t_{j,k} \leq t\}$, approximates $U^{(n)}$ and has sample paths which are right-continuous step functions on $[0, \tau]$. Figure 5.2.1 illustrates that, on a set of sample paths of probability one, for all $t \in [0, \tau]$ and for n fixed,

$$\lim_{k \to \infty} W_k^{(n)}(t, \omega) = U^{(n)}(t, \omega). \tag{3.1}$$

Figure 5.3.1 illustrates why $U^{(n)} \Longrightarrow \int f dW$. By the convergence in Eq. (3.1), (a) in Figure 5.3.1 holds as $k \to \infty$. Let k be some function of n, denoted by k_n, such that $k_n \to \infty$ as $n \to \infty$. Then $W_k^{(n)}$ reduces to $W^{(n)}$ defined by Eq. (2.2), and Theorem 5.2.5 provides conditions under which (b) in Figure 5.3.1 holds; specifically, for any $t \in [0, \tau]$

$$\sum_{j=1}^{k_n(t)+1} E_{j-1}(X_{j-1}^{(n)})^2 \xrightarrow{P} \int_0^t f^2(s)ds \tag{3.2}$$

and

$$\sum_{j=1}^{k_n(t)+1} E_{j-1}\left\{(X_{j-1}^{(n)})^2 I_{\{|X_{j-1}^{(n)}| > \epsilon\}}\right\} \xrightarrow{P} 0 \text{ for any } \epsilon > 0, \tag{3.3}$$

where $X_{j-1}^{(n)} \equiv X_{j-1,k_n}^{(n)} \equiv U^{(n)}(t_{j,k_n}) - U^{(n)}(t_{j-1,k_n})$, and $\mathcal{F}_{n,j-1} \equiv \mathcal{F}_{t_{j-1,k_n}}^{(n)}$.

Thus we might expect that (c) holds in Figure 5.3.1 (i.e., Theorem 5.1.1) under Conditions (1.5) and (1.6) if, for each fixed n, these two conditions are approximated by Conditions (3.2) and (3.3) as $k \to \infty$. This essentially is true and follows from a result of Doléans-Dade (Meyer, 1967, p. 90–91). Her result implies that the predictable variation process for a discretized version of a square integrable martingale M converges in L^1 to the predictable variation process, $\langle M, M \rangle$, of M as the discretization becomes finer, if $\langle M, M \rangle$ is continuous.

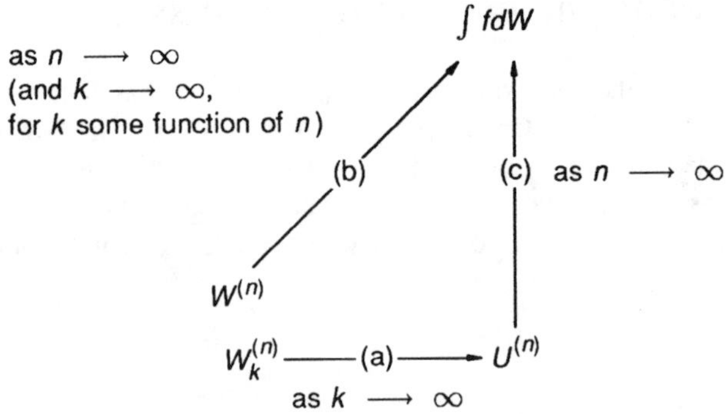

Figure 5.3.1 Schematic diagram of the convergence of $U^{(n)}$ to $\int f dW$.

Theorem 5.3.1 (Doléans-Dade). Let $\{X(s) : 0 \leq s \leq t\}$ be a submartingale adapted to some filtration $\{\mathcal{F}_s : s \geq 0\}$, and let $\{A(s) : 0 \leq s \leq t\}$ be an increasing, integrable, and continuous process, such that $X - A$ is a martingale.

For any subdivision or "skeleton" $S = (t_0, t_1, \ldots, t_m)$ of $[0, t]$, let

$$A^S(t) = \sum_{j=0}^{m-1} E\left\{A(t_{j+1}) - A(t_j)\,|\mathcal{F}_{t_j}\right\}$$

$$= \sum_{j=0}^{m-1} E\left\{X(t_{j+1}) - X(t_j)\,|\mathcal{F}_{t_j}\right\}.$$

Then $E|A^S(t) - A(t)| \to 0$ as the subdivisions become arbitrarily fine. □

Doléans-Dade also gives an immediate and important corollary.

Corollary 5.3.1. Let M be a square integrable martingale such that $\langle M, M\rangle$ is continuous. Then

$$\sum_{j=0}^{m-1} E\left[\{M(t_{j+1}) - M(t_j)\}^2 \middle| \mathcal{F}_{t_j}\right] \xrightarrow{L^1} \langle M, M\rangle(t)$$

as the partition of $[0, t]$ becomes sufficiently fine. □

Proof of Corollary 5.3.1. In Theorem 5.3.1, set $X(s) = M^2(s)$ and $A(s) = \langle M, M\rangle(s)$. Using Eq. (4.5) from Chapter 1, the corollary then follows directly from the theorem. □

Proof of Theorem 5.3.1. Suppose first that $A(t) \in L^2$. Then

$$E\{(A(t) - A^S(t))^2\}$$

$$= E\left[\left[\sum_{j=0}^{m-1}\left\{A(t_{j+1}) - A(t_j)\right\} - \sum_{j=0}^{m-1} E\left\{A(t_{j+1}) - A(t_j)\middle|\mathcal{F}_{t_j}\right\}\right]^2\right]$$

$$= E\left[\sum_{j=0}^{m-1}\left[\left\{A(t_{j+1}) - A(t_j)\right\} - E\left\{A(t_{j+1}) - A(t_j)\middle|\mathcal{F}_{t_j}\right\}\right]^2\right]$$

$$\leq 2E\left[\sum_{j=0}^{m-1}\left\{A(t_{j+1}) - A(t_j)\right\}^2\right] + 2E\left[\sum_{j=0}^{m-1}\left[E\left\{A(t_{j+1}) - A(t_j)\middle|\mathcal{F}_{t_j}\right\}\right]^2\right]$$

$$\leq 4E\left[\sum_{j=0}^{m-1}\left\{A(t_{j+1}) - A(t_j)\right\}^2\right]$$

$$\leq 4E\left[A(t)\max_j\left\{A(t_{j+1}) - A(t_j)\right\}\right],$$

where the second to the last inequality follows from Jensen's inequality for conditional expectations.

The integrability of $A(t)$ and uniform continuity on $[0,t]$ of sample paths of A imply $A(t)\max_j\{A(t_{j+1})-A(t_j)\}\to 0$ almost surely as the subdivisions become arbitrarily fine. Since $A(t)\max_j\{A(t_{j+1})-A(t_j)\}<A^2(t)$ has finite expectation, the Lebesgue Dominated Convergence Theorem implies $E[A(t)\max_j\{A(t_{j+1})-A(t_j)\}]\to 0$.

Now assume that $A(t)$ is integrable, but not necessarily in L^2. For an arbitrary η, set $B(t)=A(t)\wedge\eta$ and $C(t)=A(t)-B(t)$. Both processes B and C are increasing, continuous, and integrable with $B(t)$ bounded and hence in L^2. Since $A^S(t)=B^S(t)+C^S(t)$,

$$E|A(t)-A^S(t)|\le E|B(t)-B^S(t)|+EC(t)+EC^S(t).$$

Set $\epsilon>0$. Since $C(t)$ is integrable and $EC^S(t)=EC(t)$, we can find an η such that $EC^S(t)=EC(t)<\epsilon$. The proof is completed by observing that a sufficiently fine partition can be found such that $E|B(t)-B^S(t)|<\epsilon$. □

Since $X^{(n)}_{j-1,k}\equiv U^{(n)}(t_{j,k})-U^{(n)}(t_{j-1,k})$, it follows from Corollary 5.3.1 that

$$\sum_{j=1}^{k(t)+1} E_{j-1}(X^{(n)}_{j-1,k})^2 \xrightarrow{L^1} \langle U^{(n)},U^{(n)}\rangle(t) \text{ as } k\to\infty,$$

so, indeed, Condition (3.2) approaches (1.5) as $k\to\infty$, for fixed n.

The relationship between Conditions (1.6) and (3.3) as $k\to\infty$ is less apparent. Recall that (3.3) requires that

$$\sum_{j=1}^{k(t)+1} E_{j-1}\left\{\left(X^{(n)}_{j-1,k}\right)^2 I_{\{|X^{(n)}_{j-1,k}|>\epsilon\}}\right\} \tag{3.4}$$

$$=\sum_{j=1}^{k(t)+1} E_{j-1}\left[\left\{\sum_{i=1}^{n}\int_{t_{j-1,k}}^{t_{j,k}} H_i^{(n)}(s)dM_i^{(n)}(s)\right\}^2 I_{\left\{\left|\sum_{i=1}^{n}\int_{t_{j-1,k}}^{t_{j,k}} H_i^{(n)}(s)dM_i^{(n)}(s)\right|>\epsilon\right\}}\right]$$

converges to zero in probability, for any $\epsilon>0$. Using a simple algebraic identity, one can show that the right-hand side of Eq. (3.4) is bounded by a linear combination of the two terms which arise when $H_i^{(n)}(s)$ in that expression is replaced by $H_i^{(n)}(s)I_{\{|H_i^{(n)}(s)|\le\epsilon\}}$ and by $H_i^{(n)}(s)I_{\{|H_i^{(n)}(s)|>\epsilon\}}$ respectively. The first of these two terms becomes negligible as the discretization becomes fine since no two component processes jump at the same time. The second term is bounded by

$$\sum_{j=1}^{k(t)+1} E_{j-1}\left\{\sum_{i=1}^{n}\int_{t_{j-1,k}}^{t_{j,k}} H_i^{(n)}(s)I_{\{|H_i^{(n)}(s)|>\epsilon\}}dM_i^{(n)}(s)\right\}^2.$$

Applying Corollary 5.3.1 to the square integrable martingale

$$U_\epsilon^{(n)} = \sum_{i=1}^{n} \int H_i^{(n)}(s) I_{\{|H_i^{(n)}(s)|>\epsilon\}} dM_i^{(n)}(s),$$

it follows as the discretization becomes fine that

$$\sum_{j=1}^{k(t)+1} E_{j-1} \left[\sum_{i=1}^{n} \int_{t_{j-1,k}}^{t_{j,k}} H_i^{(n)}(s) I_{\{|H_i^{(n)}(s)|>\epsilon\}} dM_i^{(n)}(s) \right]^2$$

$$\xrightarrow{L^1} \langle U_\epsilon^{(n)}, U_\epsilon^{(n)} \rangle(t).$$

This suggests that Condition (1.6) is a sufficient condition for Condition (3.3) as $k \to \infty$, for each n.

In summary, Condition (1.5) grew out of Condition (3.2) and indicates that the predictable variation of $U^{(n)}$ is converging to that of $\int f dW$. Meanwhile, Condition (1.6), which indicates that the sample paths of $U^{(n)}$ are becoming smooth as $n \to \infty$, grew out of a condition closely related to (3.3) and therefore is a type of Lindeberg condition.

Before proving Theorem 5.1.1 and its extensions, we state a companion theorem that follows from Lemma 5.2.2. This result gives sufficient conditions for the weak convergence of $U^{(n)}$ to $\int f dW$ in terms of the quadratic variation process $[U^{(n)}, U^{(n)}]$ rather than the predictable quadratic variation process $\langle U^{(n)}, U^{(n)} \rangle$. We start with a definition.

Definition 5.3.1. For any local martingale, M, the *quadratic variation process* $[M, M]$ is that process such that as $k \to \infty$,

$$\sum_{j: t_{j,k} \leq t} \{M(t_{j+1,k} \wedge t) - M(t_{j,k})\}^2 \xrightarrow{P} [M, M](t)$$

for any $t > 0$, and any R-sequence of partitions. □

The limit in Definition 5.3.1 is well-defined. It exists and does not depend on the R–sequence used (see Doléans-Dade, 1969, and Helland, 1982).

For each fixed n, by letting $k \to \infty$, Conditions (3.5) and (3.6) of the next theorem follow immediately from Lemma 5.2.2's Conditions (2) and (1), respectively, and from Definition 5.3.1.

Theorem 5.3.2. The conclusion of Theorem 5.1.1 remains valid when Conditions (1.5) and (1.6) are replaced by

$$[U^{(n)}, U^{(n)}](t) \xrightarrow{P} \int_0^t f^2(s) ds \text{ as } n \to \infty, \tag{3.5}$$

and

$$E\{\max_{s\le t}|\Delta U^{(n)}(s)|\} \to 0 \text{ as } n \to \infty, \tag{3.6}$$

for all $t \geq 0$. □

The quadratic variation and predictable quadratic variation processes share an interesting relationship. For any $t \geq 0$,

$$[U^{(n)}, U^{(n)}](t) = \sum_{s\le t}\{\Delta U^{(n)}(s)\}^2$$
$$= \sum_{i=1}^{n}\int_0^t \{H_i^{(n)}(s)\}^2 dN_i^{(n)}(s),$$

where the first equality follows from Helland (1982, equation 4.10). Since

$$\langle U^{(n)}, U^{(n)}\rangle(t) = \sum_{i=1}^{n}\int_0^t \{H_i^{(n)}(s)\}^2 dA_i^{(n)}(s),$$

we have

$$[U^{(n)}, U^{(n)}](t) - \langle U^{(n)}, U^{(n)}\rangle(t) = \sum_{i=1}^{n}\int_0^t \{H_i^{(n)}(s)\}^2 dM_i^{(n)}(s),$$

so Conditions (1.5) and (3.5) are equivalent when

$$\sum_{i=1}^{n}\int_0^t \{H_i^{(n)}(s)\}^2 dM_i^{(n)}(s) \xrightarrow{P} 0.$$

Theorem 5.3.2 is easier to prove than Theorem 5.1.1, but Conditions (1.5) and (1.6) are easier to establish in most applications, so we give a proof of Theorem 5.1.1. We begin with three lemmas proved in Appendix C.

Lemma 5.3.1.

1. Let Y be a random variable, and $\{Y_n : n \geq 1\}$ and $\{Y_{n,k} : n, k \geq 1\}$ sequences of random variables. Suppose that

   a) $Y_{n,k} \xrightarrow{P} Y_n$ as $k \to \infty$ for each n, and

   b) $Y_n \xrightarrow{P} Y$ as $n \to \infty$.

   Then for any increasing sequence $\{k_n\}$ such that $k_n \to \infty$ fast enough, we have $\lim_{n\to\infty} Y_{n,k_n} = Y$ in probability.
2. Let $Y(\cdot)$ be a nondecreasing stochastic process, and $\{Y_n(\cdot) : n \geq 1\}$ and $\{Y_{n,k}(\cdot) : n, k \geq 1\}$ sequences of nondecreasing processes and assume that $Y(\cdot)$ is continuous. Suppose, for each $t > 0$, Conditions (a) and (b) of (1) hold for $Y_{n,k}(t), Y_n(t)$ and $Y(t)$, and let $c > 0$. Then, if $k_n \to \infty$ fast enough,

$$\lim_{n\to\infty}\sup_{0\le t\le c}|Y_{n,k_n}(t) - Y(t)| = 0 \text{ in probability.}$$

□

Lemma 5.3.2. Let $\{A_{n,k} : k \geq 1, n \geq 1\}$ be a doubly indexed family of events on a probability space $\{\Omega, \mathcal{F}, P\}$ and $\{\mathcal{F}_{n,k}; k \geq 1, n \geq 1\}$ a sequence (in n) of filtrations, i.e., for any n, $\mathcal{F}_{n,k-1} \subset \mathcal{F}_{n,k} \subset \mathcal{F}$ for all $k \geq 1$. Assume that $A_{n,k} \in \mathcal{F}_{n,k}$, and let K_n be a sequence of nonnegative integer valued random variables such that, for each n, K_n is a stopping time relative to $\{\mathcal{F}_{n,k} : k \geq 1\}$. Then

$$\lim_{n\to\infty} P\left\{\bigcup_{k=1}^{K_n} A_{n,k}\right\} = 0$$

if and only if

$$\sum_{k=1}^{K_n} E_{k-1}(I_{\{A_{n,k}\}}) \xrightarrow{P} 0 \text{ as } n \to \infty. \qquad \square$$

Lemma 5.3.3. Let $x \in D[0,\infty)$ and let $\{Q(k) : k \geq 1\} = \{\{t_{j,k} : j \geq 0\} : k \geq 1\}$ be an R-sequence of partitions. For $t > 0$, define $k(t) = \max\{j : t_{j,k} \leq t\}$ and

$$a_k = \max_{j \leq k(t)} \sup_{u,v \in [t_{j,k}, t_{j+1,k}]} |x(u) - x(v)|.$$

Then $\lim_{k\to\infty} a_k = \max_{s \leq t} |\Delta x(s)|$. $\square$

Proof of Theorem 5.1.1. We prove weak convergence in $D[0,\tau]$ for any τ, which implies weak convergence in $D[0,\infty)$. For an R-sequence of partitions $\{Q(k) : k \geq 1\} = \{\{t_{j,k} : j \geq 0\} : k \geq 1\}$, define $W_k^{(n)}(t) = \sum_{j=0}^{k(t)} X_{j,k}^{(n)}$, where $X_{j,k}^{(n)} = U^{(n)}(t_{j+1,k}) - U^{(n)}(t_{j,k})$ and where $k(t) \equiv \max\{j : t_{j,k} \leq t\}$. By Theorem 4.1 in Billingsley (1968), it is sufficient to find a sequence $\{k_n\}$ increasing so fast that, for $W^{(n)}(t) \equiv W_{k_n}^{(n)}(t)$, we have

$$\sup_{t\leq\tau} |U^{(n)}(t) - W^{(n)}(t)| \xrightarrow{P} 0 \tag{3.7}$$

and

$$W^{(n)} \Longrightarrow U^\infty \text{ on } D[0,\tau]. \tag{3.8}$$

We first establish (3.7).

By construction,

$$\sup_{t\leq\tau} |U^{(n)}(t) - W_k^{(n)}(t)| \leq \max_{j\leq k(\tau)} \sup_{u,v\in[t_{j,k},(t_{j+1,k}\wedge\tau)]} |U^{(n)}(u) - U^{(n)}(v)|.$$

For each sample path, the right-hand side is a sequence in k which decreases monotonically. By Lemma 5.3.3, as $k \to \infty$, the right-hand side converges almost surely to

$$\max_{t\leq\tau} |\Delta U^{(n)}(t)|.$$

By Lemma 5.3.1(1), (3.7) holds if we can show

$$\max_{t\leq\tau} |\Delta U^{(n)}(t)| \xrightarrow{P} 0 \text{ as } n \to \infty. \tag{3.9}$$

Note first that, because two component counting processes cannot jump at the same time, for each s,

$$\Delta U^{(n)}(s) I_{\{|\Delta U^{(n)}(s)|>\epsilon\}} = \sum_{i=1}^{n} \Delta U_i^{(n)}(s) I_{\{|\Delta U_i^{(n)}(s)|>\epsilon\}}$$

almost surely.

Because $A_i^{(n)}$ is continuous, Theorem 5.3.1 applies here with

$$X(t) = \int_0^t \{H_i^{(n)}(s)\}^2 I_{\{|H_i^{(n)}(s)|>\epsilon\}} dN_i^{(n)}(s)$$

and

$$A(t) = \int_0^t \{H_i^{(n)}(s)\}^2 I_{\{|H_i^{(n)}(s)|>\epsilon\}} dA_i^{(n)}(s),$$

and implies, as $k \to \infty$,

$$\begin{aligned} Q_k^{(n)} &\equiv \sum_{j=0}^{k(\tau)} E\left\{ \sum_{s:t_{j,k}<s\leq(t_{j+1,k}\wedge\tau)} \{\Delta U^{(n)}(s)\}^2 I_{\{|\Delta U^{(n)}(s)|>\epsilon\}} \middle| \mathcal{F}_{t_{j,k}}^{(n)} \right\} \\ &= \sum_{i=1}^{n}\sum_{j=0}^{k(\tau)} E\left\{ \sum_{s:t_{j,k}<s\leq(t_{j+1,k}\wedge\tau)} \left\{\Delta U_i^{(n)}(s)\right\}^2 I_{\{|\Delta U_i^{(n)}(s)|>\epsilon\}} \middle| \mathcal{F}_{t_{j,k}}^{(n)} \right\} \\ &\to \sum_{i=1}^{n} \langle U_{i,\epsilon}^{(n)}, U_{i,\epsilon}^{(n)} \rangle(\tau) \text{ in } L^1. \end{aligned}$$

By Condition (1.6), $\langle U_\epsilon^{(n)}, U_\epsilon^{(n)}\rangle(\tau) \xrightarrow{P} 0$ as $n \to \infty$. Therefore Lemma 5.3.1(1) implies that $\{k_n\}$ exists such that

$$Q_{k_n}^{(n)} \xrightarrow{P} 0 \text{ as } n \to \infty. \tag{3.10}$$

Now

$$\sum_{s:t_{j,k}<s\leq(t_{j+1,k}\wedge\tau)} \{\Delta U^{(n)}(s)\}^2 I_{\{|\Delta U^{(n)}(s)|>\epsilon\}} \geq \epsilon^2 I_{\left\{\max_{t_{j,k}<s\leq(t_{j+1,k}\wedge\tau)} |\Delta U^{(n)}(s)|>\epsilon\right\}}.$$

Thus (3.10) implies

$$\sum_{j=0}^{k_n(\tau)} E\left\{ I_{\left\{\max_{t_{j,k_n}<s\leq(t_{j+1,k_n}\wedge\tau)} |\Delta U^{(n)}(s)|>\epsilon\right\}} \middle| \mathcal{F}_{t_{j,k_n}}^{(n)} \right\} \xrightarrow{P} 0 \text{ as } n \to \infty$$

which, by Lemma 5.3.2, implies $P\left\{\max_{s\leq\tau}|\Delta U^{(n)}(s)|>\epsilon\right\}\rightarrow 0$. Therefore, (3.9) holds and (3.7) follows.

We now establish the limit in (3.8). Recall that $X_{n,j}\equiv X^{(n)}_{j-1,k_n}\equiv X^{(n)}_{j-1}, r_n(t)\equiv k_n(t)+1$, and $\mathcal{F}_{n,j-1}\equiv\mathcal{F}^{(n)}_{t_{j-1,k_n}}$. We must show that a sequence $\{k_n\}$ can be found so that Conditions (2.21) and (2.22) of Theorem 5.2.5 hold.

Condition (2.21),

$$\sum_{j=1}^{r_n(t)} E_{j-1}X^2_{n,j}\xrightarrow{P}\int_0^t f^2(s)ds\equiv\alpha(t),$$

can be rewritten as

$$A^{(n)}_{k_n}(t)\xrightarrow{P}\alpha(t)\text{ for any }t\in[0,\tau],$$

where

$$A^{(n)}_k(t)\equiv\sum_{j=0}^{k(t)}E\left[\left\{U^{(n)}(t_{j+1,k})-U^{(n)}(t_{j,k})\right\}^2\middle|\mathcal{F}^{(n)}_{t_{j,k}}\right].$$

Since $\langle U^{(n)},U^{(n)}\rangle$ is continuous, Corollary 5.3.1 implies

$$\lim_{k\rightarrow\infty}\sum_{j=0}^{k(t)}E\left[\left\{U^{(n)}(t_{j+1,k}\wedge t)-U^{(n)}(t_{j,k})\right\}^2\middle|\mathcal{F}^{(n)}_{t_{j,k}}\right]=\langle U^{(n)},U^{(n)}\rangle(t)$$

in probability. Also, $\lim_{k\rightarrow\infty}\{U^{(n)}(t_{k(t)+1,k})-U^{(n)}(t)\}=0$ a.s. by the right-continuity of $U^{(n)}$. Thus, as $k\rightarrow\infty$,

$$A^{(n)}_k(t)\xrightarrow{P}\langle U^{(n)},U^{(n)}\rangle(t).$$

By assumption, as $n\rightarrow\infty$,

$$\langle U^{(n)},U^{(n)}\rangle(t)\xrightarrow{P}\alpha(t)\text{ for any }t\in[0,\tau].$$

Since $A^{(n)}_k$, $\langle U^{(n)},U^{(n)}\rangle$, and $\alpha(\cdot)$ are nondecreasing and $\alpha(\cdot)$ is continuous, Lemma 5.3.1(2) implies the existence of a sequence $\{k_n\}$ such that

$$A^{(n)}_{k_n}(t)\xrightarrow{P}\alpha(t)\text{ for any }t\in[0,\tau],$$

so Condition (2.21) holds for such $\{k_n\}$.

To complete the proof, we must establish the existence of a sequence $\{k_n\}$ satisfying Condition (2.22) for all $t\in[0,\tau]$.

Condition (2.22),

$$\sum_{j=1}^{r_n(t)}E_{j-1}\left(X^2_{n,j}I_{\{|X_{n,j}|>\epsilon\}}\right)\xrightarrow{P}0\text{ for any }\epsilon>0,$$

can be rewritten (at $t = \tau$) as

$$\lim_{n\to\infty} \sum_{j=0}^{k_n(\tau)} E\left\{\left[U^{(n)}(t_{j+1,k_n}) - U^{(n)}(t_{j,k_n})\right]^2 I_{\{|U^{(n)}(t_{j+1,k_n})-U^{(n)}(t_{j,k_n})|>\epsilon\}} | \mathcal{F}^{(n)}_{t_{j,k_n}}\right\} = 0 \qquad (3.11)$$

in probability for any $\epsilon > 0$.

Fix $\delta > 0$ and write $U^{(n)} \equiv \overline{U}^{(n)}_\delta(t) + \underline{U}^{(n)}_\delta(t)$, where

$$\overline{U}^{(n)}_\delta(t) = \sum_{i=1}^n \overline{U}^{(n)}_{i,\delta}(t) = \sum_{i=1}^n \int_0^t H_i^{(n)}(s) I_{\{|H_i^{(n)}(s)|>\delta\}} dM_i^{(n)}(s),$$

and

$$\underline{U}^{(n)}_\delta(t) = \sum_{i=1}^n \underline{U}^{(n)}_{i,\delta}(t) = \sum_{i=1}^n \int_0^t H_i^{(n)}(s) I_{\{|H_i^{(n)}(s)|\le\delta\}} dM_i^{(n)}(s).$$

It is not difficult to verify that, if a_i, b_i, and c_i are real numbers satisfying $a_i = b_i + c_i$ for $i = 1, 2$, then

$$\begin{aligned}
&(a_1 - a_2)^2 I_{\{|a_1-a_2|>\epsilon\}} \\
&\le 2\{(b_1 - b_2)^2 + (c_1 - c_2)^2\}(I_{\{|b_1-b_2|>\epsilon/2\}} + I_{\{|c_1-c_2|>\epsilon/2\}}) \\
&\le 4\{(b_1 - b_2)^2 I_{(|b_1-b_2|>\epsilon/2)} + (c_1 - c_2)^2 I_{\{|c_1-c_2|>\epsilon/2\}}\}. \qquad (3.12)
\end{aligned}$$

Consequently,

$$\sum_{j=0}^{k_n(\tau)} E\left[\left\{U^{(n)}(t_{j+1,k_n}) - U^{(n)}(t_{j,k_n})\right\}^2 I_{\{|U^{(n)}(t_{j+1,k_n})-U^{(n)}(t_{j,k_n})|>\epsilon\}} \middle| \mathcal{F}^{(n)}_{t_{j,k_n}}\right]$$

$$\le 4Q^{(n)}_{1,k_n} + 4Q^{(n)}_{2,k_n}$$

where

$$Q^{(n)}_{1,k_n} \equiv \sum_{j=0}^{k_n(\tau)} E\left[\left\{\underline{U}^{(n)}_\delta(t_{j+1,k_n}) - \underline{U}^{(n)}_\delta(t_{j,k_n})\right\}^2 I_{\{|\underline{U}^{(n)}_\delta(t_{j+1,k_n})-\underline{U}^{(n)}_\delta(t_{j,k_n})|>\epsilon/2\}} \middle| \mathcal{F}^{(n)}_{t_{j,k_n}}\right]$$

and

$$Q^{(n)}_{2,k_n} \equiv \sum_{j=0}^{k_n(\tau)} E\left[\left\{\overline{U}^{(n)}_\delta(t_{j+1,k_n}) - \overline{U}^{(n)}_\delta(t_{j,k_n})\right\}^2 I_{\{|\overline{U}^{(n)}_\delta(t_{j+1,k_n})-\overline{U}^{(n)}_\delta(t_{j,k_n})|>\epsilon/2\}} \middle| \mathcal{F}^{(n)}_{t_{j,k_n}}\right].$$

We will prove (3.11) by showing $Q^{(n)}_{1,k_n} \xrightarrow{P} 0$ and $Q^{(n)}_{2,k_n} \xrightarrow{P} 0$ for some sequence $\{k_n\}$ such that $k_n \to \infty$ fast enough. By Condition (1.4), for each n there exists a constant b_n such that

$$\sup_s |\underline{U}^{(n)}_\delta(s)| \le b_n,$$

where b_n may converge to infinity as $n \to \infty$. Therefore,

$$Q^{(n)}_{1,k_n} \le (2b_n)^2 \sum_{j=0}^{k_n(\tau)} E\left\{ I_{\{|\underline{U}^{(n)}_\delta(t_{j+1,k_n}) - \underline{U}^{(n)}_\delta(t_{j,k_n})| > \epsilon/2\}} \middle| \mathcal{F}^{(n)}_{t_{j,k_n}} \right\}. \tag{3.13}$$

$\underline{U}^{(n)}_\delta$ jumps by at most δ since no two component processes jump at the same time and all have jumps no larger than δ. By setting $\delta < \epsilon/2$, we have, for fixed n,

$$\lim_{k\to\infty} P\left\{ \max_{j \le k(\tau)} \left| \underline{U}^{(n)}_\delta(t_{j+1,k}) - \underline{U}^{(n)}_\delta(t_{j,k}) \right| > \epsilon/2 \right\} = 0.$$

Hence, by Lemma 5.3.2, the right-hand side of (3.13) tends to zero in probability as $k \to \infty$.

By a direct application of Lemma 5.3.1(1), there exists a sequence $\{k_n\}$ such that $\lim_{n\to\infty} Q^{(n)}_{1,k_n} = 0$ in probability.

We now consider $Q^{(n)}_{2,k_n}$. We have

$$\begin{aligned} Q^{(n)}_{2,k} \le \sum_{j=0}^{k(\tau)} & E\left[\left\{ \overline{U}^{(n)}_\delta(\tau \wedge t_{j+1,k}) - \overline{U}^{(n)}_\delta(t_{j,k}) \right\}^2 \middle| \mathcal{F}^{(n)}_{t_{j,k}} \right] \\ & + E\left[\left\{ \overline{U}^{(n)}_\delta(t_{k(\tau)+1,k}) - \overline{U}^{(n)}_\delta(\tau) \right\}^2 \middle| \mathcal{F}^{(n)}_{t_{j,k}} \right]. \end{aligned}$$

Corollary 5.3.1 implies that the first term on the right side of this inequality converges in L^1 to $\langle \overline{U}^{(n)}_\delta, \overline{U}^{(n)}_\delta \rangle(\tau)$ as $k \to \infty$.

To see the convergence of the second term, note that the right-continuity and boundedness of $\overline{U}^{(n)}_\delta$ imply

$$|\overline{U}^{(n)}_\delta(t_{k(\tau)+1,k}) - \overline{U}^{(n)}_\delta(\tau)|^2 \to 0 \text{ in } L^1,$$

so that

$$E\left\{ \left| \overline{U}^{(n)}_\delta(t_{k(\tau)+1,k}) - \overline{U}^{(n)}_\delta(\tau) \right|^2 \middle| \mathcal{F}^{(n)}_{t_{j,k}} \right\} \to 0 \text{ in } L^1.$$

Since convergence in L^1 implies convergence in probability, the sum of terms on the right in this last inequality converges in probability to $\langle \overline{U}^{(n)}_\delta, \overline{U}^{(n)}_\delta \rangle(\tau)$ as $k \to \infty$.

By Condition (1.6), $\langle \overline{U}^{(n)}_\delta, \overline{U}^{(n)}_\delta \rangle(\tau) \xrightarrow{P} 0$ as $n \to \infty$. Thus, by Lemma 5.3.1(1), we have $Q^{(n)}_{2,k_n} \xrightarrow{P} 0$ for some $\{k_n\}$. □

It should be noted that to prove weak convergence in $D[0,\tau]$, we made use of Conditions (1.5) and (1.6) only for $t \in [0,\tau]$.

The conclusion of Theorem 5.1.1 is true even when the processes $N_i^{(n)}$, $A_i^{(n)}$, and $H_i^{(n)}$ are not necessarily uniformly bounded.

Theorem 5.3.3. Theorem 5.1.1 continues to hold when Condition (1.4) is dropped.

Proof. Let

$$\tau_{i,k}^{(n)} = \sup\left[t : \sup_{0\le s\le t}\{\max(|H_i^{(n)}(s)|, N_i^{(n)}(s), A_i^{(n)}(s))\} < k\right] \wedge k,$$

and let $\tau_k^{(n)} = \min\{\tau_{i,k}^{(n)} : i = 1, \cdots n\}$. Then $\tau_k^{(n)} \to \infty$ a.s., as $k \to \infty$.

For each n, $U^{(n)}(\tau_{k_n}^{(n)} \wedge t)$ is a square integrable martingale satisfying the conditions of Theorem 5.1.1, where all processes are replaced by versions truncated at $\tau_{k_n}^{(n)}$. By definition,

$$\langle U^{(n)}(\tau_k^{(n)} \wedge \cdot), U^{(n)}(\tau_k^{(n)} \wedge \cdot)\rangle(t) \to \langle U^{(n)}, U^{(n)}\rangle(t) \quad \text{a.s., as } k \to \infty,$$

while $\langle U^{(n)}, U^{(n)}\rangle(t) \xrightarrow{P} \alpha(t)$ as $n \to \infty$ by Condition (1.5), where $\alpha(\cdot)$ is continuous. Since a predictable quadratic variation process is monotone increasing in t, it follows from Lemma 5.3.1(2) that if k_n increases fast enough as $n \to \infty$, $\langle U^{(n)}(\tau_{k_n}^{(n)} \wedge \cdot), U^{(n)}(\tau_{k_n}^{(n)} \wedge \cdot)\rangle(t) \xrightarrow{P} \alpha(t)$ for all $t \ge 0$.

Similarly, if k_n increases fast enough,

$$\langle U_\epsilon^{(n)}(\tau_{k_n}^{(n)} \wedge \cdot), U_\epsilon^{(n)}(\tau_{k_n}^{(n)} \wedge \cdot)\rangle(t) \xrightarrow{P} 0 \text{ as } n \to \infty, \text{ for all } t \ge 0.$$

Without loss of generality, we can take $\{\tau_{k_n}^{(n)}\}$ to be increasing. Theorem 5.1.1 implies

$$U^{(n)}(\tau_{k_n}^{(n)} \wedge \cdot) \longrightarrow U^\infty \text{ on } \mathcal{D}[0,\tau] \text{ for any } \tau.$$

Since, for any τ,

$$P\{U^{(n)}(t) \ne U^{(n)}(\tau_{k_n}^{(n)} \wedge t) \quad \text{for some } t \in [0,\tau]\}$$
$$\le P\{\tau_{k_n}^{(n)} \le \tau\} \to 0 \text{ as } n \to \infty,$$
$$U^{(n)} \Longrightarrow U^\infty \text{ on } D[0,\tau] \text{ for any } \tau.$$

□

A multivariate version of Theorem 5.3.3 is provided by the next theorem.

Theorem 5.3.4. Let $W_1, \ldots, W_r$ be r independent Brownian motion processes and $f_1, \ldots, f_r$ be r measurable nonnegative functions such that $\alpha_\ell(t) = \int_0^t f_\ell^2(s)ds < \infty$ for all $t > 0$ and $\ell = 1, \ldots, r$.

Suppose Conditions (1.1) through (1.3) hold. Assume for any $t \in [0,\tau]$ and for each $\ell = 1, \ldots, r$ that, as $n \to \infty$,

$$\langle U_\ell^{(n)}, U_\ell^{(n)}\rangle(t) \xrightarrow{P} \int_0^t f_\ell^2(s)ds, \tag{3.14}$$

and

$$\langle U_{\ell,\epsilon}^{(n)}, U_{\ell,\epsilon}^{(n)}\rangle(t) \xrightarrow{P} 0 \text{ for any } \epsilon > 0. \tag{3.15}$$

Then

$$(U_1^{(n)}, \ldots, U_r^{(n)}) \Longrightarrow \left(\int f_1 dW_1, \ldots, \int f_r dW_r\right) \text{ in } (D[0,\tau])^r \text{ as } n \to \infty.$$

□

The proof of Theorem 5.3.4 involves an extension of the proofs of Theorems 5.1.1 and 5.3.3 and appears in Appendix C. In addition to Conditions (3.14) and (3.15), the proof requires showing that, for any $t \in [0, \tau]$,

$$\langle U_\ell^{(n)}, U_{\ell'}^{(n)}\rangle(t) \xrightarrow{P} 0 \text{ as } n \to \infty \text{ for any } \ell \neq \ell'. \tag{3.16}$$

Condition (3.16) can be motivated by Condition (2.23) in the same manner that Condition (3.14) was motivated by Condition (2.21), by using Lemma C.7.1 in Appendix C. When Conditions (1.1) through (1.3) hold, $\langle U_\ell^{(n)}, U_{\ell'}^{(n)}\rangle(t) = 0$ for any $\ell \neq \ell'$ and $t \in [0, \tau]$, and so Condition (3.16) is satisfied. Exercise 5.3 provides an illustration of the application of this multivariate result.

We complete this chapter with a second multivariate extension of Theorem 5.3.3. This multivariate version will be used in Chapter 8 in the development of large sample properties of partial likelihood statistics arising from the multiplicative intensity model introduced in Chapter 4. The proof of this multivariate result follows directly from Lemma C.3.1 in Appendix C, and is left as Exercise 5.2.

Theorem 5.3.5. Let $W_1^*, \ldots, W_r^*$ be r dependent time-transformed Brownian motion processes. Specifically, $(W_1^*, \ldots, W_r^*)$ is an r-variate Gaussian process having components with independent increments, $W_\ell^*(0) = 0$ a.s. and, for all $0 \leq s \leq t$, $EW_\ell^*(t) = 0$ and $E\{W_\ell^*(s)W_{\ell'}^*(t)\} = C_{\ell\ell'}(s)$, where $C_{\ell\ell'}$ is a continuous function for all $\ell, \ell' \in \{1, \ldots, r\}$.

Suppose $\{N_i^{(n)} : i = 1, \ldots, n\}$ satisfies Conditions (1.1) and (1.2), and $\{H_{i,\ell}^{(n)} : i = 1, \ldots, n, \ell = 1, \ldots, r\}$ satisfies (1.3). Let $A_i^{(n)}$ denote the compensator of $N_i^{(n)}$ identified in (1.2), and consider the vector of local square integrable martingales $(U_1^{(n)}, \ldots, U_r^{(n)})$ where, for $\ell = 1, \ldots, r$ and for $t \geq 0$,

$$U_\ell^{(n)}(t) = \sum_{i=1}^{n} \int_0^t H_{i,\ell}^{(n)}(s) d\{N_i^{(n)}(s) - A_i^{(n)}(s)\}.$$

Suppose for each $\ell, \ell' \in \{1, \ldots, r\}$ and for all $t > 0$:

$$\begin{aligned} \langle U_\ell^{(n)}, U_{\ell'}^{(n)}\rangle(t) &= \sum_{i=1}^{n} \int_0^t H_{i,\ell}^{(n)}(s) H_{i,\ell'}^{(n)}(s) dA_i^{(n)}(s) \\ &\xrightarrow{P} C_{\ell\ell'}(t) \text{ as } n \to \infty, \end{aligned} \tag{3.17}$$

and

$$\langle U_{\ell,\epsilon}^{(n)}, U_{\ell,\epsilon}^{(n)}\rangle(t) = \sum_{i=1}^{n} \int_0^t \{H_{i,\ell}^{(n)}(s)\}^2 I_{\{|H_{i,\ell}^{(n)}(s)|>\epsilon\}} dA_i^{(n)}(s) \tag{3.18}$$

$$\xrightarrow{P} 0 \text{ as } n \to \infty \text{ for any } \epsilon > 0.$$

Then

$$(U_1^{(n)}, \ldots, U_r^{(n)}) \Longrightarrow (W_1^*, \ldots, W_r^*) \text{ in } (D[0,\tau])^r \text{ as } n \to \infty. \qquad \square$$

5.4 BIBLIOGRAPHIC NOTES

During the latter half of the 1970s, general central limit theorems for sequences of martingales were proved by Aalen, Rebolledo, and Liptser and Shiryayev. Their work relied on the modern theory of martingales and stochastic integrals summarized in Meyer (1976). They used the development earlier in the decade in the theory of point processes, as well as central limit theorems for arrays of dependent random variables.

Section 5.2

Central limit theorems for discretized martingale processes were established by Brown (1971), Dvoretsky (1972) and McLeish (1974). These authors extended the classical central limit theorem for arrays of independent variables to arrays of dependent random variables. The classical conditions, which required convergence of sums of expectations and of variances and validity of the Lindeberg condition, were replaced by analogous conditions on conditional expectations given the past. Section 2 gives particular attention to the proof of McLeish which is based on simple manipulations of characteristic functions.

Section 5.3

Bremaud (1972) began the development of point processes based on the modern theory of martingales and stochastic integrals. This work was continued by Dolivo (1974), Bremaud (1974), Boel, Varaiya and Wong (1975a,b) and Segall and Kailath (1975a,b), while Jacod (1973, 1975) took a different approach. Additional references and summaries are provided by Bremaud and Jacod (1977), Jacod (1979), Dellacherie (1980), Bremaud (1981) and Shiryayev (1981).

Working from this theory, Aalen (1975, 1977) and Rebolledo (1977, 1979, 1980) estabilished weak convergence results for continuous time martingales and local martingales. Conditions for convergence were formulated in terms of convergence of predictable covariation processes. Helland (1982) provided more elementary proofs of the more general results of Rebolledo, and we largely follow his development in this section of the book.

Central limit theorems more general than those of Rebolledo were provided by Liptser and Shiryayev (1980). However, the results are not stronger in the setting we consider in this book. A survey was provided by Shiryayev (1981).

CHAPTER 6

Large Sample Results for the Kaplan-Meier Estimator

6.1 INTRODUCTION

Large sample properties for estimators and for confidence bands of the survival function S follow from the theorems developed in the preceding chapter. Our attention will be focused on the Kaplan-Meier estimator $\hat{S}$ defined in Section 3.2 and on the general Random Censorship Model in Definition 3.1.1. Before exploring the asymptotic properties of $\hat{S}$, we prove in Section 6.2 a theorem from Gill (1980) which will be useful not only in the $r = 1$ sample setting of this chapter, but also in the $r = 2$ sample setting of Chapter 7, where weighted logrank statistics are discussed. This theorem will follow directly from the multivariate weak convergence result, Theorem 5.3.4. Later, in Section 6.3, a second theorem of Gill's (1983) will be established which allows the calculation of confidence bands for S over the largest possible interval. We use the notation of Section 3.1 throughout the chapter.

6.2 A LARGE SAMPLE RESULT FOR KAPLAN-MEIER AND WEIGHTED LOGRANK STATISTICS

For each n, let (T_{ij}^n, U_{ij}^n), $j = 1, \ldots, n_i, i = 1, \ldots, r$, be n $(n = n_1 + \cdots + n_r)$ independent pairs of positive failure time and censoring random variables, such that T_{ij}^n is almost surely finite and Condition (3.7) in Chapter 1 holds.

Let $T_{ij}^n \sim F_i^n = 1 - S_i^n, U_{ij}^n \sim L_{ij}^n = 1 - C_{ij}^n$, and $\Lambda_i^n = \int_0^{\cdot}\{1 - F_i^n(s-)\}^{-1}dF_i^n(s)$. The observable random variables are $X_{ij}^n = T_{ij}^n \wedge U_{ij}^n$ and $\delta_{ij}^n = I_{\{X_{ij}^n = T_{ij}^n\}}$. Set $\pi_{ij}^n = P\{X_{ij}^n \geq t\}$. The dependence on n of S_i^n is used to study a contiguous sequence of alternative hypotheses in the discussion of efficiencies in Chapter 7. The dependence on j of L_{ij}^n allows consideration of the fixed censorship model in which each individual has a deterministic censorship time. In the special case of the "usual" random censorship model, $L_{ij}^n = L_i$ for all i, j, and n.

The next assumption is used in Chapters 6 and 7.

Assumption 6.2.1. For each $i = 1, \ldots, r$, there exists a function π_i and a constant $a_i \in [0, 1]$ such that, as $n \to \infty$,

$$\sup_{0 \leq t < \infty} \left| \frac{\overline{Y}_i(t)}{n_i} - \pi_i(t) \right| \xrightarrow{P} 0 \tag{2.1}$$

and

$$\frac{n_i}{n} \to a_i . \tag{2.2}$$

□

If T_{ij}^n and U_{ij}^n are independent (rather than simply satisfying Condition (3.7) of Chapter 1), and if $\sup_{0 \leq t < \infty} |F_i^n(t) - F_i(t)| \to 0$ as $n \to \infty$ for some fixed distribution function F_i, then the Glivenko-Cantelli Theorem for independent but not identically distributed random variables (van Zuijlen, 1978) implies that Condition (2.1) is equivalent to

$$\sup_{0 \leq t < \infty} \left| \frac{1}{n_i} \sum_{j=1}^{n_i} L_{ij}^n(t) - L_i(t) \right| \to 0 \tag{2.3}$$

as $n \to \infty$, for some right-continuous function L_i. In this case, $\pi_i(t) = \{1 - F_i(t-)\}\{1 - L_i(t-)\}$.

In Chapters 6 and 7, we examine statistics

$$U_i^{(n)} = \sum_{j=1}^{n_i} \int H_{ij}^{(n)}(s) dM_{ij}^{(n)}(s)$$

in which $H_{ij}^{(n)}(s) = H_i^{(n)}(s)$ for each i, j, n, and s. Hence

$$U_i^{(n)} = \int H_i^{(n)}(s) dM_i^{(n)}(s),$$

where $M_i^{(n)}(s) = \overline{N}_i(s) - \int_0^s \overline{Y}_i(x) d\Lambda_i^n(x)$. It follows that

$$\langle U_i^{(n)}, U_i^{(n)} \rangle(t) = \int_0^t \{H_i^{(n)}(s)\}^2 \overline{Y}_i(s) d\Lambda_i^n(s),$$

$$U_{i,\epsilon}^{(n)} = \int H_i^{(n)}(s) I_{\{|H_i^{(n)}(s)| > \epsilon\}} dM_i^{(n)}(s),$$

and

$$\langle U_{i,\epsilon}^{(n)}, U_{i,\epsilon}^{(n)} \rangle(t) = \int_0^t \{H_i^{(n)}(s)\}^2 I_{\{|H_i^{(n)}(s)| > \epsilon\}} \overline{Y}_i(s) d\Lambda_i^n(s).$$

The next theorem, a modification of Theorem 4.2.1 of Gill (1980), provides conditions for the weak convergence of $\{U_i^{(n)}, i = 1, \ldots, r\}$ and is an important tool in the development of large sample properties of one- and two-sample statistics. The theorem specifies three situations under which the processes $U_i^{(n)}$ converge weakly to Gaussian processes, ranging from relatively weak conditions (Condition (1) alone) to stronger conditions (Conditions (1) and (2), or Conditions (1), (2) and (3)) that are more difficult to check but extend convergence to larger intervals of the real line.

Definition 6.2.1. Set $\mathcal{I} = \{t : \prod_{i=1}^{r} \pi_i(t) > 0\}$ and $u = \sup \mathcal{I}$, so $u \in (0, \infty]$. Note $u \notin \mathcal{I}$ when $u = \infty$. □

Theorem 6.2.1. For each $i = 1, \ldots, r$ and $n = 1, 2, \ldots$, let

$$U_i^{(n)} = \int H_i^{(n)}(s) dM_i^{(n)}(s),$$

where $H_i^{(n)}$ is a locally bounded predictable process and

$$M_i^{(n)}(t) = \overline{N}_i(t) - \int_0^t \overline{Y}_i(s) d\Lambda_i^n(s).$$

Suppose Assumption 6.2.1 holds and, for $i = 1, \ldots, r$, let Conditions (1), (2) and (3) be given by:

1. For any $t \in \mathcal{I}$,
   a) There exists a distribution function F_i satisfying
   $$\Lambda_i(t) = \int_0^t \{1 - F_i(s-)\}^{-1} dF_i(s) < \infty$$
   and
   $$\sup_{0 \leq s \leq t} |F_i^n(s) - F_i(s)| \to 0 \text{ as } n \to \infty.$$
   b) There exists a nonnegative function h_i such that, for any $t \in \mathcal{I}$,
   $$\sup_{0 \leq s \leq t} |\{H_i^{(n)}(s)\}^2 \overline{Y}_i(s) - h_i(s)| \xrightarrow{P} 0 \text{ as } n \to \infty.$$
   Also, h_i is left-continuous and has right-hand limits on $\mathcal{I}$, with a right-continuous adaptation h_i^+ of bounded variation on each subinterval of $\mathcal{I}$. (When $F_i^n \equiv F_i$, the conditions in the previous sentence on the nonnegative function h_i can be weakened to boundedness on each closed subinterval of $\mathcal{I}$). The function h_i also is assumed to satisfy $h_i(t) < \infty$ for any $t \in \mathcal{I}$, and $h_i(t) = 0$ for any $t \notin \mathcal{I}$.
2. When $u \notin \mathcal{I}$,
   a) $\int_{\mathcal{I}} h_i(1 - \Delta\Lambda_i) d\Lambda_i < \infty$, and
   b)
   $$\lim_{t \uparrow u} \limsup_{n \to \infty} P\left\{ \int_t^u \{H_i^{(n)}\}^2 \overline{Y}_i d\Lambda_i^n > \epsilon \right\} = 0$$
   for any $\epsilon > 0$.

3. When $u < \infty$,

$$\int_u^\infty \{H_i^{(n)}\}^2 \overline{Y}_i d\Lambda_i^n \xrightarrow{P} 0 \text{ as } n \to \infty.$$

Suppose $Z_1^\infty, \ldots, Z_r^\infty$ are independent zero-mean Gaussian processes with independent increments, continuous sample paths, and variance functions

$$\int h_i(1 - \Delta\Lambda_i)d\Lambda_i. \tag{2.4}$$

Under (1),

$$(U_1^{(n)}, \ldots, U_r^{(n)}) \Longrightarrow (Z_1^\infty, \ldots, Z_r^\infty) \text{ in } (D[0,t])^r, t \in \mathcal{I}, \tag{2.5}$$

as $n \to \infty$. Under (1) and (2), the convergence in Eq. (2.5) holds in $(D[0,u])^r$. Under (1), (2) and (3), convergence holds in $D([0,\infty])^r$, where $D[0,\infty]$ is the space of functions on $[0,\infty]$ which are right-continuous with finite left-hand limits. □

Condition (1) gives relatively weak conditions for convergence of $U_i^{(n)}$ on $[0,t]$ for any t with $\pi_i(t) > 0$. Often, however, greatest interest is in weak convergence over $[0,u]$, even when $\pi_i(u) = 0$. Theorem 6.2.1 provides this stronger result for $U_i^{(n)}$ if $U_i^{(n)}$ has finite variance at u (i.e., (2a) holds) and the process $U_i^{(n)}$ is "tight" at u (i.e., (2b) holds). When $F_i^{(n)} \equiv F_i$ and $L_{ij}^n \equiv L_i$ for any i, j, and n, Condition (3) always holds.

We prove Theorem 6.2.1 under the following assumption.

Assumption 6.2.2. For each i and n, F_i^n is absolutely continuous. □

Under Assumption 6.2.2, the term $(1 - \Delta\Lambda_i)$ in Expression (2.4) and in Condition (2a) is unity. Gill (1980) provides a more general proof that does not require Assumption 6.2.2.

In the proof of Theorem 6.2.1, convergence over any finite interval will follow directly from the multivariate weak convergence result, Theorem 5.3.4, since Assumption 6.2.2 implies Conditions (1.1) through (1.3) in Chapter 5 hold. Extension to convergence on $[0,\infty]$ is possible by using Theorem 4.2 of Billingsley (1968). The idea underlying this theorem is conceptually simple. Suppose, for some $t_0 \in (0,\infty]$, that a sequence of processes $V^{(n)}$ defined on $[0,t_0]$ satisfies $V^{(n)} \Rightarrow V^{(\infty)}$ on $D[0,t]$ for every $t < t_0$. If $V^{(\infty)}(t)$ approaches $V^{(\infty)}(t_0)$ as $t \to t_0$ (usually satisfied when $V^{(\infty)}(t_0)$ has finite variance) and if $V^{(n)}(t)$ is "close" to $V^{(n)}(t_0)$ for sufficiently large n (more formally, $V^{(n)}$ is tight at t_0), then $V^{(n)} \Rightarrow V^{(\infty)}$ on $D[0,t_0]$.

Proof of Theorem 6.2.1. To establish that Eq. (2.5) holds under (1), we must verify that Conditions (3.14) and (3.15) in Chapter 5 hold for any $t \in \mathcal{I}$, where $\int f_i^2(s)ds \equiv \int h_i(s)d\Lambda_i(s)$. For any t,

$$\begin{aligned}\langle U_i^{(n)}, U_i^{(n)}\rangle(t) &= \int_0^t \{H_i^{(n)}(s)\}^2 \overline{Y}_i(s) d\Lambda_i^n(s) \\ &= \int_0^t h_i(s) d\Lambda_i(s) + \int_0^t \left[\{H_i^{(n)}(s)\}^2 \overline{Y}_i(s) - h_i(s)\right] d\Lambda_i^n(s) \\ &\quad + \int_0^t h_i(s) d\{\Lambda_i^n(s) - \Lambda_i(s)\}.\end{aligned}$$

Conditions (1a) and (1b) imply that

$$\int_0^t \left[\{H_i^{(n)}(s)\}^2 \overline{Y}_i(s) - h_i(s)\right] d\Lambda_i^n(s) \xrightarrow{P} 0$$

for any $t \in \mathcal{I}$. Since the left-continuous function h_i has right-hand limits (so that h_i^+ is well defined), integration by parts, Eq. (2.8) of Chapter 3, yields

$$\int_0^t h_i(s) d\{\Lambda_i^n(s) - \Lambda_i(s)\} = \{\Lambda_i^n(t) - \Lambda_i(t)\} h_i^+(t) - \int_0^t \{\Lambda_i^n(s) - \Lambda_i(s)\} dh_i^+(s).$$

The right-hand side converges to 0 as $n \to \infty$ because $\sup_{0 \le s \le t} |\Lambda_i^n(s) - \Lambda_i(s)| \to 0$ and h_i^+ is of bounded variation over $[0, t]$. Thus, Condition (3.14) of Chapter 5 holds for any $t \in \mathcal{I}$.

To establish Condition (3.15) of Chapter 5, fix $\epsilon > 0$ and $t \in \mathcal{I}$. By (1b),

$$\sup |\{H_i^{(n)}(s)\}^2 \overline{Y}_i(s) - h_i(s)| \xrightarrow{P} 0 \text{ over } [0, t], \tag{2.6}$$

where h_i is bounded on $[0, t]$. Assumption 6.2.1 implies $\overline{Y}_i(t) \xrightarrow{P} \infty$ for any $t \in \mathcal{I}$. In turn the monotonicity of $\overline{Y}_i$ yields $\inf_{0 \le s \le t} \overline{Y}_i(s) \xrightarrow{P} \infty$. This, together with Eq. (2.6), establishes $\sup_{0 \le s \le t} |H_i^{(n)}(s)| \xrightarrow{P} 0$, and consequently

$$\int_0^t \{H_i^{(n)}(s)\}^2 \overline{Y}_i(s) I_{\{|H_i^{(n)}(s)| > \epsilon\}} d\Lambda_i^n(s) \xrightarrow{P} 0. \tag{2.7}$$

We must show next that adding Condition (2) implies that Eq. (2.5) holds over $(D[0, u])^r$. To begin, assume $u < \infty$. Then it suffices to show that Conditions (3.14) and (3.15) of Chapter 5 hold at $t = u$. Since Eq. (2.7) holds for any $t < u$, Condition (2b) establishes Condition (3.15) at $t = u$. To show Condition (3.14) holds at $t = u$, observe that, for any $t \le u$,

$$\begin{aligned}&\left|\int_0^u \{H_i^{(n)}(s)\}^2 \overline{Y}_i(s) d\Lambda_i^n(s) - \int_0^u h_i(s) d\Lambda_i(s)\right| \\ &\quad \le \left|\int_0^t \{H_i^{(n)}(s)\}^2 \overline{Y}_i(s) d\Lambda_i^n(s) - \int_0^t h_i(s) d\Lambda_i(s)\right| \\ &\quad\quad + \int_t^u \{H_i^{(n)}(s)\}^2 \overline{Y}_i(s) d\Lambda_i^n(s) + \int_t^u h_i(s) d\Lambda_i(s).\end{aligned} \tag{2.8}$$

By (2a), there exists t_0 sufficiently close to u such that

$$\int_t^u h_i(s)d\Lambda_i(s) < \epsilon/3 \text{ for any } t \in [t_0, u].$$

By (2b), there exists $t^* \in [t_0, u]$ and n_0 such that, for any $n \geq n_0$,

$$P\left\{\int_{t^*}^u \{H_i^{(n)}(s)\}^2 \overline{Y}_i(s) d\Lambda_i^n(s) > \epsilon/3\right\} < \epsilon/2.$$

Since Condition (3.14) of Chapter 5 holds for any $t \in \mathcal{I}$, there exists n_1 such that for any $n \geq n_1$,

$$P\left\{\left|\int_0^{t^*} \{H_i^{(n)}(s)\}^2 \overline{Y}_i(s) d\Lambda_i^n(s) - \int_0^{t^*} h_i(s)d\Lambda_i(s)\right| > \epsilon/3\right\} < \epsilon/2.$$

Thus, by Inequality (2.8), for any $n > \max(n_0, n_1)$,

$$P\left\{\left|\int_0^u \{H_i^{(n)}(s)\}^2 \overline{Y}_i(s) d\Lambda_i^n(s) - \int_0^u h_i(s)d\Lambda_i(s)\right| > \epsilon\right\} < \epsilon.$$

Next, we suppose $u = \infty$ and show that adding Condition (2) implies that Eq. (2.5) holds over $(D[0,\infty])^r$. We can assume Eq. (2.5) holds for any $t \in \mathcal{I} = (0, \infty)$. By Theorem 4.2 of Billingsley (1968) and Condition (2a), it is sufficient to show $(U_1^{(n)}, \ldots, U_r^{(n)})$ is "tight at ∞", i.e., for each $i = 1, \ldots, r$,

$$\lim_{t\uparrow\infty} \limsup_{n\to\infty} P\left\{\sup_{t\leq s\leq\infty} \left|\int_t^s H_i^{(n)}(s)dM_i^{(n)}(s)\right| > \epsilon\right\} = 0, \tag{2.9}$$

for any $\epsilon > 0$.

Set $t' > t$. Applying Lenglart's Inequality (i.e., Corollary 3.4.1), for any ϵ, $\eta > 0$,

$$P\left\{\sup_{t\leq s\leq t'} \left|\int_t^s H_i^{(n)}(s)dM_i^{(n)}(s)\right| > \epsilon\right\}$$
$$\leq \frac{\eta}{\epsilon^2} + P\left\{\int_t^{t'} \{H_i^{(n)}(s)\}^2 \overline{Y}_i(s) d\Lambda_i^n(s) \geq \eta\right\}. \tag{2.10}$$

Taking limits as $t' \uparrow \infty$, Eq. (2.10) holds with t' replaced by ∞. Thus Eq. (2.9) follows from (2b).

Last, we must show that (3) implies weak convergence on $(D[0,\infty])^r$. By similar arguments, this is immediate since $h_i(t) \equiv 0$ for any $t > u$. □

6.3 CONFIDENCE BANDS FOR THE SURVIVAL DISTRIBUTION

Theorem 6.2.1 will be used to establish the large-sample distribution of the Kaplan-Meier estimator $\hat{S} \equiv 1 - \hat{F}$, and confidence intervals and confidence bands for $S \equiv 1 - F$ on intervals $[0, t]$, for $t \in \mathcal{I}$.

Throughout this section, $r = 1$ and $F_i^n \equiv F$, so we will suppress i and n. We also presume Assumptions 6.2.1 and 6.2.2.

Theorem 6.3.1. Consider the general Random Censorship Model in Definition 3.1.1, with $r = 1$ and with F_i^n, L_{ij}^n, and π_{ij}^n independent of n. Suppose Assumption 6.2.1 and 6.2.2 hold, and that W is Brownian Motion.

Then, as $n \longrightarrow \infty$, for any $t \in \mathcal{I}$:

1. $$\sqrt{n}\,(\hat{F}(\cdot) - F(\cdot)) \Longrightarrow (1 - F(\cdot))W(v(\cdot)) \text{ on } D[0, t], \tag{3.1}$$
   where $v(t) \equiv \int_0^t \pi^{-1}(s)d\Lambda(s)$. For any $0 \leq s \leq t$,
   $$\text{cov}[\{1-F(s)\}W\{v(s)\},\{1-F(t)\}W\{v(t)\}] = \{1-F(s)\}\{1-F(t)\}\int_0^s \frac{d\Lambda(x)}{\pi(x)}.$$

2. If $\hat{v}(t) \equiv n \int_0^t [\{\overline{Y}(s) - \Delta\overline{N}(s)\}\overline{Y}(s)]^{-1} d\overline{N}(s)$, then
   $$\sup_{0 \leq s \leq t} |\hat{v}(s) - v(s)| \xrightarrow{P} 0. \tag{3.2}$$

3. $$\sqrt{n}\left\{\frac{\hat{F}(\cdot) - F(\cdot)}{1 - \hat{F}(\cdot)}\right\} \Longrightarrow W(v(\cdot)) \text{ on } D[0, t]. \tag{3.3}$$

4. $$\sup_{0 \leq s \leq t} \left\{\frac{n}{\hat{v}(t)}\right\}^{1/2} \left\{\frac{\hat{F}(s) - F(s)}{1 - \hat{F}(s)}\right\} \xrightarrow{\mathcal{D}} \sup_{0 \leq x \leq 1} W(x), \tag{3.4}$$
   and
   $$\sup_{0 \leq s \leq t} \left\{\frac{n}{\hat{v}(t)}\right\}^{1/2} \frac{|\hat{F}(s) - F(s)|}{1 - \hat{F}(s)} \xrightarrow{\mathcal{D}} \sup_{0 \leq x \leq 1} |W(x)|. \tag{3.5}$$

□

Result (1) provides a slight generalization of a well-known result by Breslow and Crowley (1974). Their proof used results on weak convergence of empirical distribution functions.

Proof of Theorem 6.3.1.

1. For any $t \in \mathcal{I}$, $S(t) > 0$. Fix $t \in \mathcal{I}$ and suppose $0 \leq s \leq t$. By Eq. (2.11) in Chapter 3,
   $$\sqrt{n}\,\{\hat{F}(s) - F(s)\} =$$
   $$S(s)\int_0^s \frac{\hat{S}(x-)}{S(x)}\sqrt{n}\left\{\frac{d\overline{N}(x)}{\overline{Y}(x)} - I_{\{\overline{Y}(x)>0\}}d\Lambda(x)\right\} + \sqrt{n}\,B(s),$$

where

$$B(s) = I_{\{T<s\}} \frac{\hat{S}(T)\{S(T) - S(s)\}}{S(T)}$$

and $T = \inf\{x : \overline{Y}(x) = 0\}$. Since $0 \leq B(s) \leq \{1 - S(s)\} I_{\{T<s\}}$,

$$\sup_{0 \leq s \leq t} \left|\sqrt{n}\, B(s)\right| \leq \sqrt{n}\,\{1 - S(t)\} I_{\{\overline{Y}(t)=0\}} \xrightarrow{P} 0 \text{ as } n \to \infty.$$

Hence, it suffices to obtain the asymptotic distribution of the process $\{\int_0^s H(x)dM(x) : 0 \leq s \leq t\}$, where

$$H(s) = \sqrt{n}\,\frac{\hat{S}(s-)}{S(s)} \frac{I_{\{\overline{Y}(s)>0\}}}{\overline{Y}(s)}$$

and

$$M(s) = \overline{N}(s) - \int_0^s \overline{Y}(x)d\Lambda(x).$$

Theorem 6.2.1 yields the desired result once Condition (1) has been verified for $h(s) = \{\pi(s)\}^{-1}$. The only non-trivial component of Condition (1) requires verification of

$$\sup_{0 \leq s \leq t} \left| H^2(s)\overline{Y}(s) - \frac{1}{\pi(s)} \right| \xrightarrow{P} 0, \text{ as } n \to \infty. \tag{3.6}$$

By adding and subtracting $\{S(s)\}^{-2}\hat{S}^2(s-)\{\pi(s)\}^{-1}$,

$$\sup_{0 \leq s \leq t} \left| H^2(s)\overline{Y}(s) - \frac{1}{\pi(s)} \right| \leq \{S^2(t)\pi(t)\}^{-1} \sup_{0 \leq s \leq t} |\hat{S}^2(s-) - S^2(s)|$$
$$+ \{S(t)\}^{-2} \sup_{0 \leq s \leq t} \left| \frac{nI_{\{\overline{Y}(s)>0\}}}{\overline{Y}(s)} - \frac{1}{\pi(s)} \right|.$$

Thus, Condition (3.6) follows by Assumption 6.2.1 and by the consistency of the Kaplan-Meier estimator established in Theorem 3.4.2.

2. By the continuity of F,

$$\hat{v}(s) - v(s) = \int_0^s \left\{ \frac{n}{\overline{Y}(x) - 1} \right\} \frac{d\overline{N}(x)}{\overline{Y}(x)} - \int_0^s \frac{d\Lambda(x)}{\pi(x)}.$$

Part (2) now follows from Assumption 6.2.1 and from the consistency of the Nelson estimator in Eq. (4.3) of Chapter 3.

3. The convergence in Part (3) follows from Part (1) and from the uniform consistency of the Kaplan-Meier estimator in Eq. (4.4) in Chapter 3.
4. Results in Part (4) are immediate from Part (3), from the consistency of $\hat{v}$ established in (2), and from the Continuous Mapping Theorem (Theorem B.1.1 of Appendix B). □

Example 6.3.1. In a clinical trial with entry uniformly distributed over $[0, a]$ and analysis occurring at time $b, (b > a)$ (Example 3.2.1), the censoring survivor function C has the form

$$C(t) = \begin{cases} 1 & 0 < t \le b - a \\ a^{-1}(b - t) & b - a < t \le b \\ 0 & b < t < \infty. \end{cases}$$

Theorem 6.3.1 (1) provides the large sample approximation to the variance of the Kaplan-Meier estimator. When $0 \le s \le b - a$, the asymptotic variance of $\sqrt{n}\{\hat{F}(s) - F(s)\}$ has the form

$$\begin{aligned} S^2(s) \int_0^s \frac{d\Lambda(x)}{S(x)} &= -S^2(s) \int_0^s \frac{dS(x)}{S^2(x)} \\ &= S(s)\{1 - S(s)\}, \end{aligned}$$

which is the variance of the empirical cumulative distribution function. When $b - a < s < b$ and $S(b) > 0$, the asymptotic variance will be

$$\begin{aligned} S^2(s) \int_0^{b-a} \frac{d\Lambda(x)}{S(x)} &+ S^2(s) \int_{b-a}^s \frac{a}{b - x} \frac{1}{S(x)} d\Lambda(x) \\ &= S^2(s) \left\{ \frac{1 - S(b - a)}{S(b - a)} \right\} + S^2(s) \int_{b-a}^s \frac{a}{b - x} \frac{1}{S(x)} d\Lambda(x). \end{aligned}$$

For an exponential survival distribution with rate λ, this becomes

$$e^{-2\lambda s} \left\{ e^{\lambda(b-a)} - 1 + \lambda a \int_{b-a}^s \frac{e^{\lambda x}}{b - x} dx \right\}.$$

Figure 6.3.1a shows a plot of this variance function over $0 \le s \le 4$ when $a = 4, b = 5$, and $\lambda = 1/2$.

Figure 6.3.1b shows the ratio of the above variance function under uniform censoring to the variance of the empirical cumulative distribution function if there had been no censoring for $0 \le s \le 4$, and thus indicates the amount of variability in $\hat{S}$ induced by censoring. □

Theorem 6.3.1 provides approximate confidence limits for $F(t)$ when $\pi(t) > 0$; two-sided $(1 - \alpha)$-level confidence limits for $F(t)$ are

$$\hat{F}(t) \pm \left\{ \frac{\hat{v}(t)}{n} \right\}^{1/2} \{1 - \hat{F}(t)\} z_{1-\alpha/2},$$

where z_γ is the γ-quantile of the standard normal distribution.

Two functions, ℓ and u, provide a $(1 - \alpha)$-level confidence band for F on an interval $[0, t]$ if $\ell(s) \le u(s)$ for $0 \le s \le t$ and

$$P\{\ell(s) \le F(s) \le u(s) : 0 \le s \le t\} \ge 1 - \alpha.$$

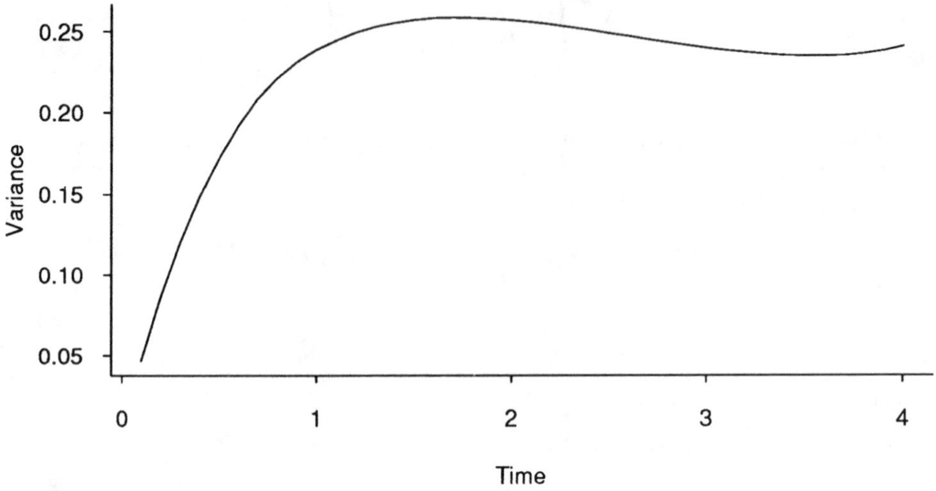

Figure 6.3.1a Asymptotic variance of $\sqrt{n}\{\hat{F}(s) - F(s)\}$ when $F(s) = 1 - e^{-s/2}$ and censoring is uniformly distributed over $[1, 5]$.

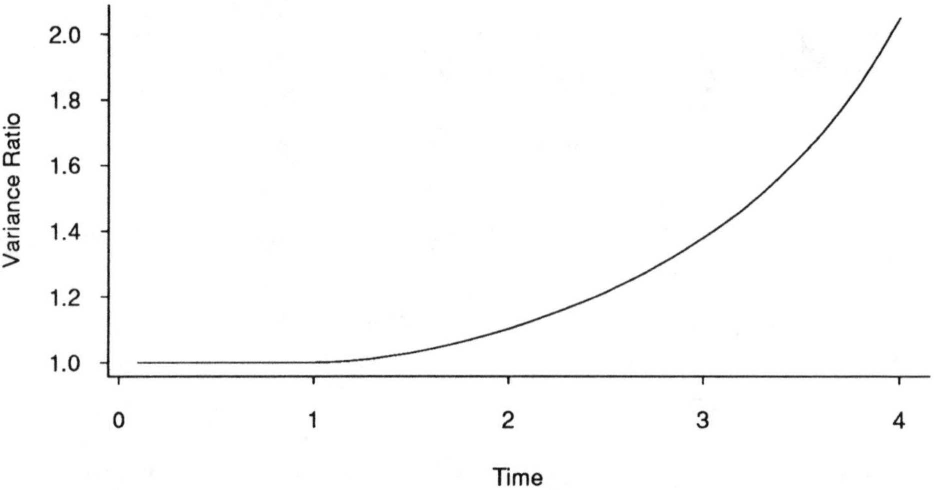

Figure 6.3.1b Ratio of asymptotic variance of $\sqrt{n}\{\hat{F}(s) - F(s)\}$ when $F(s) = 1 - e^{-s/2}$ and censoring is uniformly distributed over $[1, 5]$ vs. no censoring.

A collection of upper- and lower- $(1 - \alpha)$-level confidence limits on $[0, t]$ do not form a $(1 - \alpha)$-level confidence band, since the collection will not in general have the right coverage probability. Confidence bands are more difficult to construct than pointwise confidence intervals. The weak convergence of processes of the form $g_n(\hat{F} - F)$ for nonnegative functions g_n, however, does provide a general method for calculating confidence bands for F. When $g_n(\hat{F} - F)$ converges on some interval $[0,t]$ to a limit process Q, then the Continuous Mapping Theorem implies that

$$\sup_{0\le s\le t} g_n(s)|\hat{F}(s) - F(s)| \xrightarrow{\mathcal{D}} \sup_{0\le s\le t} |Q(s)|.$$

If $q_\alpha(t)$ satisfies

$$P\left\{\sup_{0\le s\le t} |Q(s)| \le q_\alpha(t)\right\} \ge 1-\alpha,$$

then, asymptotically,

$$P\left\{\sup_{0\le s\le t} g_n(s)|\hat{F}(s) - F(s)| \le q_\alpha(t)\right\} \ge 1-\alpha.$$

Consequently,

$$P\left\{\hat{F}(s) - \frac{q_\alpha(t)}{g_n(s)} \le F(s) \le \hat{F}(s) + \frac{q_\alpha(t)}{g_n(s)} : 0 \le s \le t\right\} \ge 1-\alpha,$$

and $F_n(s) \pm \{g_n(s)\}^{-1} q_\alpha(t), 0 \le s \le t$, is a $(1-\alpha)$-level band for F on $[0,t]$.

Confidence bands based on $\sup_{0\le s\le t} g_n(s)|\hat{F}(s) - F(s)|$ will be useful only when sufficient conditions for the convergence of $g_n(\hat{F} - F)$ on reasonable intervals $[0,t]$ are not too restrictive, when $q_\alpha(t)$ is easy to calculate, and when the resulting bands have appealing properties. The simplest approach uses Theorem 6.3.1 directly. While easy to describe, this approach does not always produce confidence bands on sufficiently wide intervals of the real line, and in some cases relies on processes whose asymptotic distributions are difficult to calculate. Theorem 6.3.1 can be extended to provide alternative confidence bands that are easier to work with while still having appealing properties.

We begin with methods based on Theorem 6.3.1. When $\pi(t) > 0$, Eq. (3.1) implies that asymptotic two-sided confidence bands based on $\sup_{0\le s\le t} \sqrt{n}|\hat{F}(s) - F(s)|$ can be constructed from the distribution of $\sup_{0\le s\le t}\{1 - \bar{F}(s)\}|W\{v(s)\}|$. Quantiles of this distribution are not difficult to calculate in uncensored data, where the Kaplan-Meier estimator and the empirical cumulative distribution function coincide, and there already is a large body of literature on nonparametric confidence bands for the distribution function of independent, identically distributed, uncensored observations. The distribution of $\sup_{0\le s\le t}\{1-F(s)\}|W\{v(s)\}|$, for $\pi(t) > 0$, is considerably more difficult to calculate in censored data, so confidence bands using this approach are not practical.

Part (4) of Theorem 6.3.1 does lead to accessible confidence bands for F on intervals $[0,t]$, when $\pi(t) > 0$. For any $y > 0$, Eq. (3.5) implies that, as $n \to \infty$,

$$P\left\{\sup_{0\le s\le t} \left\{\frac{n}{\hat{v}(t)}\right\}^{1/2} \left|\frac{\hat{F}(s) - F(s)}{1-\hat{F}(s)}\right| \le y\right\} \to P\{\sup_{0\le x\le 1} |W(x)| \le y\}.$$

Feller (1966, p. 330) has derived the distribution of $\sup_{0\le x\le 1} |W(x)|$. A formula is given in Eq. (5.13) in the next chapter, and tabulated values appear in several places (e.g., Schumacher, 1984). If $\psi_{1-\alpha}$ denotes the $(1-\alpha)$-quantile of this distribution, the resulting bands are

$$\hat{F}(s) \pm \psi_{1-\alpha}\{1 - \hat{F}(s)\}\left\{\frac{\hat{v}(t)}{n}\right\}^{1/2}.$$

These bands were formally proposed by Gill (1980), with closely related bands proposed by Gillespie and Fisher (1979). One sided Gill bands using Eq. (3.4) are developed in Exercise 6.2.

The Gill bands arise by taking the supremum of the process $g_n|\hat{F} - F|$, where the weighting process at s, $g_n(s)$, is proportional to $\{1 - \hat{F}(s)\}^{-1}$. Because g_n gives much greater weight at later times, the Gill bands can be expected to be wide for small values of t, as illustrated in the next example.

Example 6.3.2. Chapters 0 and 4 provide the background and an analysis of the Mayo Clinic data set for patients having PBC disease. The 418 patients form the complete collection of all PBC patients with follow-up data, referred to Mayo between January 1974 and May 1984, who met eligibility criteria for a randomized, placebo-controlled trial of the drug D-penicillamine. The Kaplan-Meier plot in Figure 4.4.1 and the Cox regression analysis in Section 4.4 reveal that the drug fails to provide a detectable effect on patient survival. Hence, the natural history survival experience for PBC patients meeting protocol eligibility criteria can be estimated by a Kaplan-Meier survival curve based on data from the 418 patients. This curve extending over 12 years of follow-up appears as the solid line at the center of Figure 6.3.2a. The collection of pointwise upper and lower 95% confidence limits appears in the figure, along with valid confidence bands over a 10 year follow-up period which were generated using Gill's method. As anticipated, the pointwise confidence limits at each t are narrower than the Gill confidence bands. The Kaplan-Meier estimate for five-year survival is 0.7029. The 95% confidence limits are (.6565, .7492), while five-year limits from the Gill 10-year bands are (.5625, .8432).

It is clear from their structure and from Figure 6.3.2a that Gill bands are wide early on, and then approach the pointwise confidence limits near the right-hand tail of the bands. If the emphasis in a data analysis is on predicting shorter term survival, say over a 5- to 6-year period, considerably narrower bands can be obtained by generating 6-year rather than 10-year bands, as displayed in Figure 6.3.2(b). The five-year limits from the Gill 6-year bands are (.6428, .7629). □

Confidence bands for F on an interval $[0, t]$ that use Theorem 6.3.1 all require $\pi(t) > 0$. This assumption might be reasonable in some historical record studies where the likelihoods of both long-term survival and long-term follow up are substantial, but it is less tenable in the late stages of follow up of a randomized clinical trial, where censoring is the result of random registration times. In the latter case, confidence bands over $[0, u]$ or over the random interval of observed data, $[0, T]$, where $T = \inf\{x : \overline{Y}(x) = 0\}$, are more appealing. Suppose $L_j \equiv L$

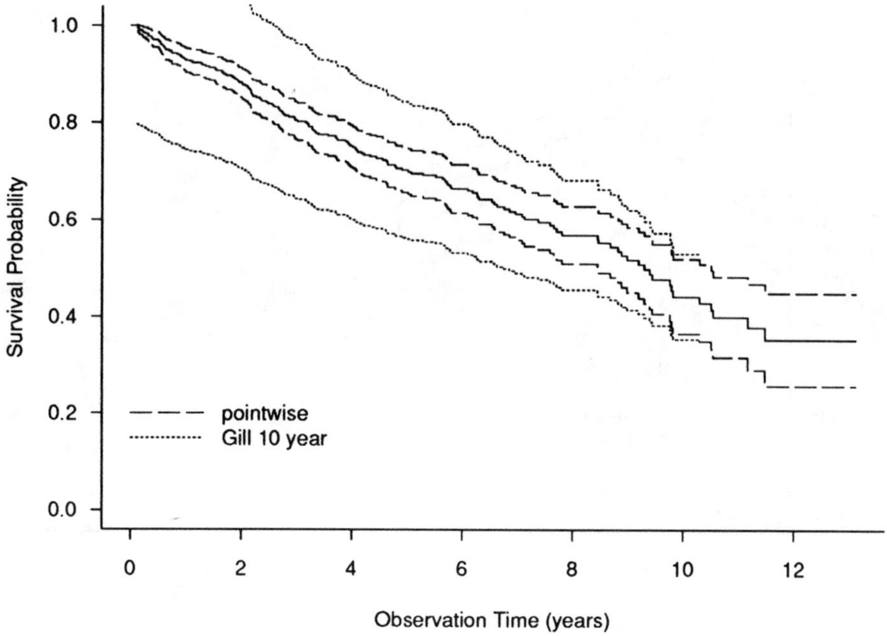

Figure 6.3.2a Pointwise confidence intervals and confidence bands (using the Gill ten-year bands) for survival probability using all 418 patients in the PBC data set.

for all j, so that $T \leq u$ a.s. If $u \notin \mathcal{I}$, as is the case in most applications, the results in Theorem 6.3.1 need to be strengthened before they can be used for bands on $[0, u]$. The convergence in (3.1) can be extended to $D[0, u]$ when Condition (2) of Theorem 6.2.1 holds for

$$H(s) = \sqrt{n}\frac{\hat{S}(s-)}{S(s)}\frac{I_{\{\overline{Y}(s)>0\}}}{\overline{Y}(s)}$$

and $h(s) = \{\pi(s)\}^{-1}$. In most applications, however, Condition (2a) of Theorem 6.2.1, requiring finiteness of the variance of $W\{v(u)\}$, i.e., requiring

$$v(u) = \int_0^u \frac{d\Lambda(s)}{\pi(s)} < \infty, \tag{3.7}$$

fails to hold. (Figure 6.3.3 provides one example in which Eq. (3.7) is true.)

The variance function for $\sqrt{n}(\hat{F} - F)$ is $(1 - F)^2 \int \pi^{-1} d\Lambda$, and, even when $W\{v(u)\}$ has infinite variance, the asymptotic variance function for $\sqrt{n}(\hat{F} - F)$ can satisfy

$$\lim_{t \uparrow u}\{1 - F(t)\}^2 \int_0^t \frac{d\Lambda(s)}{\pi(s)} < \infty, \tag{3.8}$$

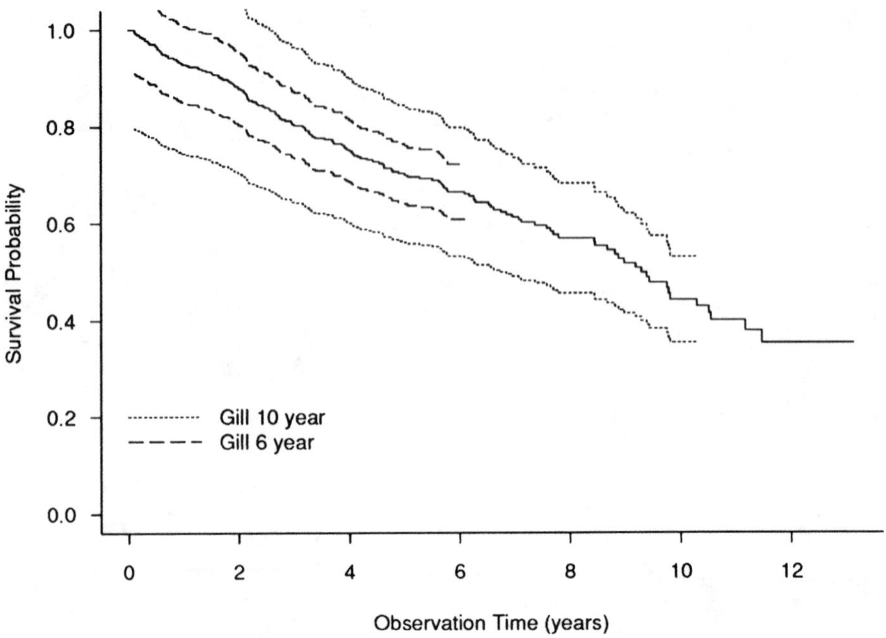

Figure 6.3.2b Gill six- and ten-year confidence bands for survival probability using all 418 patients in the PBC data set.

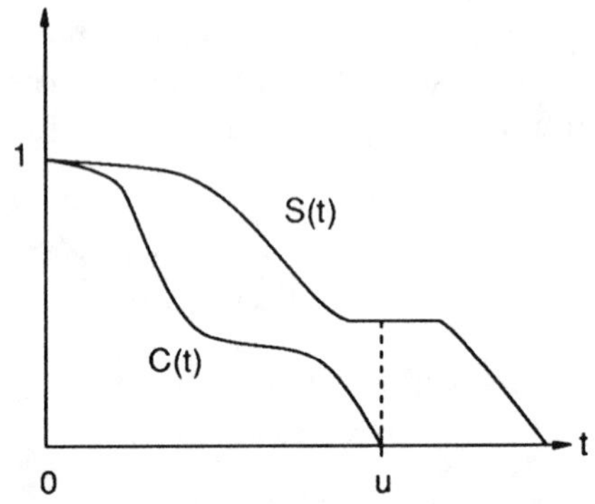

Figure 6.3.3 A situation in which $v(u) = \int_0^u \{\pi(s)\}^{-1} d\Lambda(s) < \infty$. In general, $v(u) = \infty$ in clinical trial applications.

suggesting that $\sqrt{n}(\hat{F} - F)$ can converge to a limiting process in $D[0, u]$, even when $\sqrt{n}(\hat{F} - F)/(1 - F)$ does not. In fact, it is always possible to find a weight function K so that the asymptotic variance for $\sqrt{n}K(\hat{F} - F)$ has a finite limit from the left at $t = u$. If $\sqrt{n}K(\hat{F} - F)$ then converges in $D[0, u]$, the limit distribution can be used to construct confidence bands as in earlier examples. Of course, the function K must be a natural one in this setting, and we will return to this point later.

The convergence of weighted Kaplan-Meier processes $\sqrt{n}K(\hat{F}-F)$ on $D[0,u]$ is provided in Theorem 6.3.2. The proof given by Gill (1983) uses Lenglart's Inequality and Theorem 4.2 of Billingsley (1968). A simple corollary to Theorem 6.3.2 then establishes convergence of $\sqrt{n}(\hat{F}-F)$ on $D[0,u]$ under a hypothesis slightly stronger than Inequality (3.8).

We introduce one more assumption before proving Gill's theorem.

Assumption 6.3.1.

1. T_j and U_j are independent for $j = 1, \ldots, n$.
2. $U_j \sim L_j \equiv L = 1 - C$ for all $j = 1, \ldots, n$, where L is a right-continuous distribution function.

□

Under Assumption 6.3.1, $\pi(t) = \{1-F(t)\}\{1-L(t-)\} = S(t)C(t-)$. The point $u \notin \mathcal{I}$ if and only if $\pi(u) = 0$, i.e., if and only if either $S(u) = 0$ or $C(u-) = 0$. Also, Assumption 6.2.1 automatically holds.

Theorem 6.3.2. Set $Z = \sqrt{n}\,(\hat{F}-F)\,/\,(1-F)$ and $v = \int \pi^{-1}d\Lambda$, where $\pi(s) = S(s)C(s-)$. Let $X^T(t) \equiv X(t \wedge T)$ for any process X, where $T = \inf\{x : \overline{Y}(x) = 0\}$.

Suppose h is any nonnegative continuous nonincreasing function on $[0,u]$ where $u = \sup\{t : \pi(t) > 0\}$, and assume

$$\int_0^u \{h(t)\}^2 dv(t) < \infty. \tag{3.9}$$

Then, under Assumption 6.3.1 and the hypotheses of Theorem 6.3.1,

$$(hZ)^T \Longrightarrow hW(v) \text{ on } D[0,u], \tag{3.10}$$

$$\left(\int hdZ\right)^T \Longrightarrow \int hW(v) \text{ on } D[0,u], \tag{3.11}$$

and

$$\left(\int Zdh\right)^T \Longrightarrow \int W(v)dh \text{ on } D[0,u]. \tag{3.12}$$

□

Inequality (3.9) is slightly stronger than the assumption that var $(hZ)^T(u) = h^2(u)v(u) < \infty$.

Proof. By a straightforward application of Theorem 6.2.1, Eqs. (3.10) through (3.12) hold on $D[0,t]$ for any $t \in \mathcal{I}$. Each limiting process exists at $t = u$ and can be defined as a pathwise limit as $t \to u$, even when $v(u) = \infty$ (see Gill, 1983). Thus, by Theorem 4.2 of Billingsley (1968), Eq. (3.10) holds if, for any $\epsilon > 0$,

$$\lim_{t\uparrow u} \limsup_{n\to\infty} P\left\{ \sup_{t\le s\le T} \left| h(s)Z(s) - h(t)Z(t) \right| > \epsilon \right\} = 0, \tag{3.13}$$

while Eq. (3.11) holds if, for any $\epsilon > 0$,

$$\lim_{t\uparrow u} \limsup_{n\to\infty} P\left\{ \sup_{t\le s\le T} \left| \int_t^s h(x)dZ(x) \right| > \epsilon \right\} = 0. \tag{3.14}$$

In turn, Eqs. (3.10) and (3.11) imply Eq. (3.12), since $\int Zdh = hZ - \int hdZ$.
For any $t < u$, Eq. (2.7) in Chapter 3 implies

$$\{Z(t)\}^T = \sqrt{n} \int_0^t \frac{\hat{S}(s-)}{S(s)\overline{Y}(s)} \left\{ d\overline{N}(s) - \overline{Y}(s)d\Lambda(s) \right\}.$$

Thus, for $t < t' < u$,

$$\int_{t\wedge T}^{t'\wedge T} h(s)dZ(s) = \sqrt{n} \int_t^{t'} \frac{h(s)\hat{S}(s-)}{S(s)\overline{Y}(s)} \left\{ d\overline{N}(s) - \overline{Y}(s)d\Lambda(s) \right\}.$$

Applying Lenglart's Inequality (i.e., Corollary 3.4.1), for any ϵ, $\eta > 0$ and $\beta \in (0,1)$,

$$\begin{aligned} P&\left\{ \sup_{t\le s\le t'\wedge T} \left| \int_t^s h(x)dZ(x) \right| > \epsilon \right\} \\ &\le \frac{\eta}{\epsilon^2} + P\left\{ \int_t^{t'} \left\{ \frac{h(s)\hat{S}(s-)}{S(s)} \right\}^2 \frac{n}{\overline{Y}(s)} I_{\{\overline{Y}(s)>0\}} d\Lambda(s) \ge \eta \right\} \\ &\le \frac{\eta}{\epsilon^2} + \beta + (e/\beta)e^{-1/\beta} + P\left\{ \int_t^{t'} \frac{\{h(s)\}^2 d\Lambda(s)}{\beta^3 \pi(s)} \ge \eta \right\} \end{aligned}$$

by Theorem 3.2.4, and by the inequality

$$P\left\{ \frac{n}{\overline{Y}(t)} \le \frac{1}{\beta\pi(t)} \text{ for all } t \le T \right\} \ge 1 - (e/\beta)e^{-1/\beta},$$

from Shorack and Wellner (1986, Inequality (10.3.1)). Let $t' \uparrow u$ and set

$$\eta = \int_t^u \frac{\{h(s)\}^2 d\Lambda(s)}{\beta^3 \pi(s)}.$$

Then,

$$P\left\{\sup_{t\le s\le T}\left|\int_t^s h(x)dZ(x)\right| > \epsilon\right\} \le \frac{1}{\beta^3\epsilon^2}\int_t^u h^2(s)dv(s) + \beta + \frac{e}{\beta}e^{-1/\beta}.$$

By Condition (3.9), Eq. (3.14) holds.

To complete the proof, we show Eq. (3.14) implies Eq. (3.13). Observe that

$$\sup_{t\le s\le T} |h(s)Z(s) - h(t)Z(t)| \le |\{h(t) - h(u)\}Z(t)| + \sup_{t\le s\le T} |h(s)\{Z(s) - Z(t)\}|.$$

Since Theorem 6.3.1 yields $Z \Longrightarrow W(v)$ as $n \to \infty$, Chebyshev's Inequality implies that

$$\limsup_{n\to\infty} P\{|\{h(t) - h(u)\}Z(t)| > \epsilon\} \le \frac{\{h(t) - h(u)\}^2 v(t)}{\epsilon^2}.$$

This upper bound converges to zero as $t \uparrow u$ if $v(u) < \infty$. When $v(u) = \infty$, $\lim_{t\uparrow u} h(t) = h(u) = 0$ by Inequality (3.9), so the Dominated Convergence Theorem implies that, as $t \uparrow u$,

$$h^2(t)v(t) = \int_0^u h^2(t)I_{[0,t]}(s)dv(s) \to 0.$$

It remains to establish that Eq. (3.14) implies

$$\lim_{t\uparrow u}\limsup_{n\to\infty} P\{\sup_{t\le s\le T} |h(s)\{Z(s) - Z(t)\}| > \epsilon\} = 0.$$

Define $U(s) \equiv \int_t^s h(x)dZ(x)$ where $0 \le t \le s$. When $h(s) > 0$, integration by parts yields

$$\begin{aligned} Z(s) - Z(t) &= \int_t^s \frac{dU(x)}{h(x)} \\ &= \frac{U(s)}{h(s)} - \int_t^s U(x-)d\left\{\frac{1}{h(x)}\right\} \\ &= \int_t^s \{U(s) - U(x-)\}d\left\{\frac{1}{h(x)}\right\} + \frac{U(s)}{h(t)}. \end{aligned}$$

Thus,

$$\begin{aligned} |h(s)\{Z(s) - Z(t)\}| &= \left|\int_t^s \{U(s) - U(x-)\}d\left\{\frac{h(s)}{h(x)}\right\} + U(s)\frac{h(s)}{h(t)}\right| \\ &\le 2\sup_{t\le x\le s} |U(x)| \left\{1 - \frac{h(s)}{h(t)} + \frac{h(s)}{h(t)}\right\}, \end{aligned}$$

so

$$\sup_{t\le s\le T} |h(s)\{Z(s) - Z(t)\}| \le 2\sup_{t\le s\le T}\left|\int_t^s h(x)dZ(x)\right|. \qquad \square$$

Theorem 6.3.1(1) can now be extended by setting $h(x) = 1 - F(x)$.

Corollary 6.3.1. Under Assumption 6.3.1 and the hypotheses of Theorem 6.3.1,

$$\sqrt{n}\,(\hat{F} - F)^T \Longrightarrow (1 - F)W(v) \text{ in } D[0, u]$$

whenever

$$\int_0^u \frac{S(s)d\Lambda(s)}{C(s-)} < \infty. \tag{3.15}$$

□

Inequality (3.15) is slightly stronger than the assumption of finite variance of $\sqrt{n}\,\{\hat{F}(u) - F(u)\}$, since

$$\begin{aligned}\lim_{t\to u}\{S(t)\}^2 \int_0^t \frac{d\Lambda(s)}{S(s)C(s-)} &= \lim_{t\to u}\int_0^t \left\{\frac{S(t)}{S(s)}\right\}^2 \frac{S(s)d\Lambda(s)}{C(s-)} \\ &\le \int_0^u \frac{S(s)d\Lambda(s)}{C(s-)}.\end{aligned}$$

There is currently no proof for Corollary 6.3.1 using only the weaker Inequality (3.8).

Some examples in Figure 6.3.4 provide insight into the restrictions in Conditions (3.7), (3.8) and (3.15). Condition (3.7), (i.e., var $W\{v(u)\} < \infty$), fails to hold in all situations shown. Condition (3.8) (i.e., finite asymptotic variance of $\sqrt{n}\,\{\hat{F}(u) - F(u)\}$) holds in Case 1 and in Case 3 if $0 \le \gamma \le 1$. Condition (3.15) holds in Case 1 and Case 3 if $0 \le \gamma < 1$. These examples suggest that (3.7) rarely holds in applications in which $\pi(u) = 0$, and that Gill confidence bands using Eqs. (3.4) or (3.5) can rarely be extended to $[0, u]$.

When Inequality (3.8) holds in the examples shown, Gill's Condition (3.15) fails only when $S = C$. In these examples, then, Corollary 6.3.1 is nearly as strong as could be hoped.

Unfortunately, even Condition (3.8) fails in some important situations, such as Case 2. This case represents the prospective clinical trial in which patients enter during an accrual period and have positive probability of long term survival. One-sided and two-sided confidence bands over $[0, u]$ using Corollary 6.3.1 are restricted by the need to satisfy Conditions (3.15) or (3.8), and their usefulness is hindered by the complexity of the distributions of $\sup_{0\le t\le u}\{1 - F(t)\}W\{v(t)\}$ and $\sup_{0\le t\le u}\{1 - F(t)\}|W\{v(t)\}|$ in censored data. To deal with both of these problems, we consider confidence bands based on a "weighted" Kaplan-Meier process.

In uncensored data, Corollary 6.3.1 suggests one way to choose the weight K. When there is no censoring, $\hat{F}$ is the empirical distribution function, Inequality (3.15) is true, and $v = F/(1 - F)$, so

$$\sqrt{n}\,(\hat{F} - F)^T \Longrightarrow (1 - F)W\left(\frac{F}{1 - F}\right) \quad \text{on } D[0, u],$$

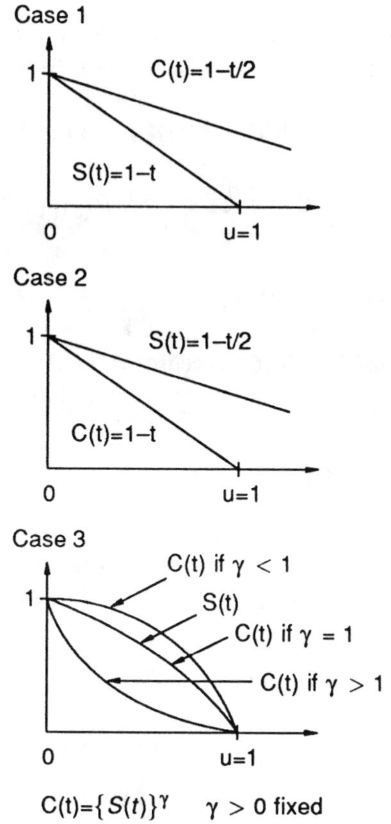

Figure 6.3.4 $\text{var}[W\{v(u)\}] = \infty$ (i.e., Condition (3.7) fails to hold) in Cases 1, 2, and 3. Condition (3.8) is true in only Case 1 and Case 3 when $0 \le \gamma \le 1$. Condition (3.15) is true in only Case 1 and Case 3 when $0 \le \gamma < 1$.

where W is a standard Brownian motion process. The Gaussian process W^o defined by $W^o(t) = W(t) - tW(1), 0 \le t \le 1$, satisfies

$$\text{cov}\{W^o(s), W^o(t)\} = s(1-t), \quad 0 \le s \le t \le 1,$$

and is called a Brownian Bridge since $W^o(0) = W^o(1) = 0$. It is not difficult to show that the processes given by $(1-t)W\{t/(1-t)\}$ and $W^o(t), 0 \le t \le 1$, have the same distribution, so $\{1-F(t)\}W\{F(t)/(1-F(t))\}$ and $W^o\{F(t)\}, 0 \le t \le u$, have the same distributions, and thus

$$\sqrt{n}(\hat{F} - F)^T \Longrightarrow W^o(F) \text{ in } D[0, u]. \tag{3.16}$$

The transformation $(1-t)W\{t/(1-t)\}$ is the inverse of the "Doob transformation" (Doob, 1949) and is discussed in Shorack and Wellner (1986, page 30).

In uncensored data, $F = v/(1+v)$, so, in censored data, we take $K = v/(1+v)$. The function K reduces to F in uncensored data and, for any fixed s, $K(s)$ increases as the probability of censoring increases, so $0 \le F(s) \le K(s) \le 1$ for any s. Since $v(s) = K(s)\{1 - K(s)\}^{-1}$, Corollary 6.3.1 implies, for any $0 \le s \le t$,

$$\text{cov}[\sqrt{n}\,\{\hat{F}(s) - F(s)\}^T, \sqrt{n}\,\{\hat{F}(t) - F(t)\}^T] \to \{1 - F(s)\}\{1 - F(t)\}\frac{K(s)}{1 - K(s)}$$

$$= \left\{\frac{1 - F(s)}{1 - K(s)}\right\}\left\{\frac{1 - F(t)}{1 - K(t)}\right\} K(s)\{1 - K(t)\}.$$

In turn,

$$\sqrt{n}\left\{\left(\frac{1 - K}{1 - F}\right)(\hat{F} - F)\right\}^T \Longrightarrow W^o(K), \tag{3.17}$$

so Eq. (3.17) generalizes Eq. (3.16) to censored data.

Define $\hat{K} = 1 - \{1 + \hat{v}(s)\}^{-1}$. Because $W^o(K)$ has variance bounded by 1/4, we conjecture that

$$\sqrt{n}\left\{\left(\frac{1 - \hat{K}}{1 - \hat{F}}\right)(\hat{F} - F)\right\}^T \Longrightarrow W^o(K) \quad \text{on } D[0, u].$$

The next corollary establishes the conjecture when (3.15) holds. It is unclear whether Condition (3.15) can be relaxed. In the statement of the corollary, we take $\{1 - K(u)\}/\{1 - F(u)\}$ to be

$$\lim_{t \uparrow u} \frac{1 - K(t)}{1 - F(t)}$$

when $F(u) = 1$, and $\{1 - \hat{K}(T)\}/\{1 - \hat{F}(T)\}$ to be

$$\frac{1 - \hat{K}(T-)}{1 - \hat{F}(T-)}$$

when $\hat{F}(T) = 1$.

Corollary 6.3.2. Under Assumption 6.3.1, the hypotheses of Theorem 6.3.1, and Inequality (3.15),

$$\sqrt{n}\left\{\left(\frac{1 - \hat{K}}{1 - \hat{F}}\right)(\hat{F} - F)\right\}^T \Longrightarrow W^o(K) \text{ on } D[0, u].$$

Proof. The uniform consistency of $\hat{v}$ over $[0, t], t < u$, established in the proof of Theorem 6.3.1 yields

$$\sup_{0 \le s \le t} |\hat{K}(s) - K(s)| \xrightarrow{P} 0 \quad \text{as } n \to \infty.$$

This, together with uniform consistency of the Kaplan-Meier estimator in (4.4) of Chapter 3 and with Theorem 6.3.1(1), yields

$$\sqrt{n}\left\{\left(\frac{1 - \hat{K}}{1 - \hat{F}}\right)(\hat{F} - F)\right\}^T \Longrightarrow W^o(K) \quad \text{on } D[0, t].$$

As in the proof of Theorem 6.3.2, Billingsley's Theorem 4.2 yields the desired result if we can show "tightness at u"; i.e., for any $\epsilon > 0$,

$$\lim_{t \uparrow u} \limsup_{n \to \infty} P\left\{ \sup_{t \le s \le T} |f(s) - f(t)| > \epsilon \right\} = 0, \tag{3.18}$$

where

$$f \equiv \sqrt{n}\left\{ \left(\frac{1-\hat{K}}{1-\hat{F}} \right) (\hat{F} - F) \right\}^T .$$

Since it can be shown that $\{(1-\hat{K})/(1-\hat{F})\}^T$ is decreasing,

$$|f(s) - f(t)| \le \left| \left(\frac{1-\hat{K}}{1-\hat{F}} \right)^T (s) \{ \sqrt{n}(\hat{F} - F)^T(s) - \sqrt{n}(\hat{F} - F)^T(t) \} \right|$$
$$+ \left| \left\{ \left(\frac{1-\hat{K}}{1-\hat{F}} \right)^T (u) - \left(\frac{1-\hat{K}}{1-\hat{F}} \right)^T (t) \right\} \sqrt{n}(\hat{F} - F)^T(t) \right| .$$

Tightness now follows from Corollary 6.3.1, and by observing that

$$0 \le \left(\frac{1-\hat{K}}{1-\hat{F}} \right)^T \le 1,$$

that $(1-K)(1-F)^{-1}$ is left-continuous at u, and that

$$\left| \left(\frac{1-\hat{K}}{1-\hat{F}} \right)^T - \left(\frac{1-K}{1-F} \right) \right| \xrightarrow{P} 0$$

for any $t \in [0, u]$. □

Corollary 6.3.2 can be used to construct two-sided confidence bands for F over $[0, u]$. For any $k \ge 0$,

$$P\left\{ \sup_{0 \le t \le u} \sqrt{n} \left\{ \frac{1-\hat{K}(t)}{1-\hat{F}(t)} \right\}^T |\{\hat{F}(t) - F(t)\}^T| \ge k \right\}$$
$$\longrightarrow P\left\{ \sup_{0 \le t \le u} |W^o\{K(t)\}| \ge k \right\}. \tag{3.19}$$

When Condition (3.7) fails, as almost always occurs, $K(u) = 1$ and the right side of Eq. (3.19) becomes

$$P\left\{ \sup_{0 \le t \le 1} |W^o(t)| \ge k \right\}.$$

The distribution of $\sup_{0\leq t\leq 1}|W^o(t)|$ is given by

$$P\left\{\sup_{0\leq t\leq 1}|W^o(t)|\geq k\right\}=2\sum_{j=1}^{\infty}(-1)^{j+1}e^{-2j^2k^2},$$

(see, for instance, Billingsley, 1968, equation 11.39). Confidence bands using weighted Kaplan-Meier processes and $\hat{K}=\hat{v}/(1+\hat{v})$ originally were proposed by Hall and Wellner (1980), who also provided formulas for the right side of Eq. (3.19) when $K(u)<1$ and for one-sided confidence bands for F. In uncensored data, $K(s)=F(s)$ and $\hat{K}(s)=\hat{F}(s)$, so the Hall-Wellner bands reduce to the Kolmogorov-Smirnov bands.

By its construction, a Hall-Wellner confidence band will be at its narrowest (Exercise 6.4) relative to the estimated standard error of $\hat{F}$ when $K\approx 1/2$. Since K increases monotonically from zero to one at a rate which depends on $\{C(t):0\leq t\leq u\}$, the time t_0 such that $K(t_0)=1/2$ depends on the censoring distribution. More precisely, for any $t\in[0,u]$, if we set $p\equiv F(t)\in[0,F(u)]$ and $y(p)\equiv K\{F^{-1}(p)\}$, then the ratio of the Hall-Wellner band width to the estimated standard error of $\hat{F}(t)$ is proportional to

$$R\equiv\left[K(t)\{1-K(t)\}\right]^{-1/2}=\left[y(p)\{1-y(p)\}\right]^{-1/2}.$$

In uncensored data, $y(p)\equiv p$, so the ratio, R, is minimized in the vicinity of $p=F(t)=1/2$. On the other hand, in censored data, $y(p)=p$ if $C(s)\equiv 1$ for all $s\in[0,F^{-1}(p)]$; otherwise $p<y(p)<K\{F^{-1}(F(u))\}=K(u)$, where $K(u)=1$ whenever Condition (3.7) fails. The effect of $y(1/2)\gg 1/2$ is that the place where R is minimized is shifted to an earlier point in the distribution of F.

Fleming et al. (1980) used another method for obtaining confidence bands for F. In their approach, $y(p)\equiv p$, but their bands do not reduce in uncensored data to the usual Kolmogorov-Smirnov bands. The bands in that paper are based on the result

$$\sqrt{n}(1-\hat{F})\int\{\hat{C}^-\}^{1/2}I_{\{\bar{Y}>0\}}d(\Lambda-\hat{\Lambda})\Longrightarrow W^o(F).$$

The Gill bands obtained from Eqs. (3.4) and (3.5) are valid over $[0,u]$ only when $\pi(u)>0$. These bands over any interval $[0,t_0]$ are narrowest (Exercise 6.4), relative to the estimated standard error of $\hat{F}$, near t_0. However, unlike the previously discussed bands, Gill bands over $[0,t_0]$ have width at any time $t<t_0$,

$$2\psi_{1-\alpha}\{1-\hat{F}(t)\}\{\hat{v}(t_0)/n\}^{1/2},$$

which depends on the degree of censorship after t. This point is discussed further in Fleming and Harrington (1981, §3.3).

Example 6.3.2 (continued). In the PBC data set, follow up on some patients extended beyond 12 years. The Hall-Wellner 95% confidence bands in Figure 6.3.5a extend to the end of the data and are close to the pointwise confidence intervals, especially over the middle region of the data where the Kaplan-Meier estimator

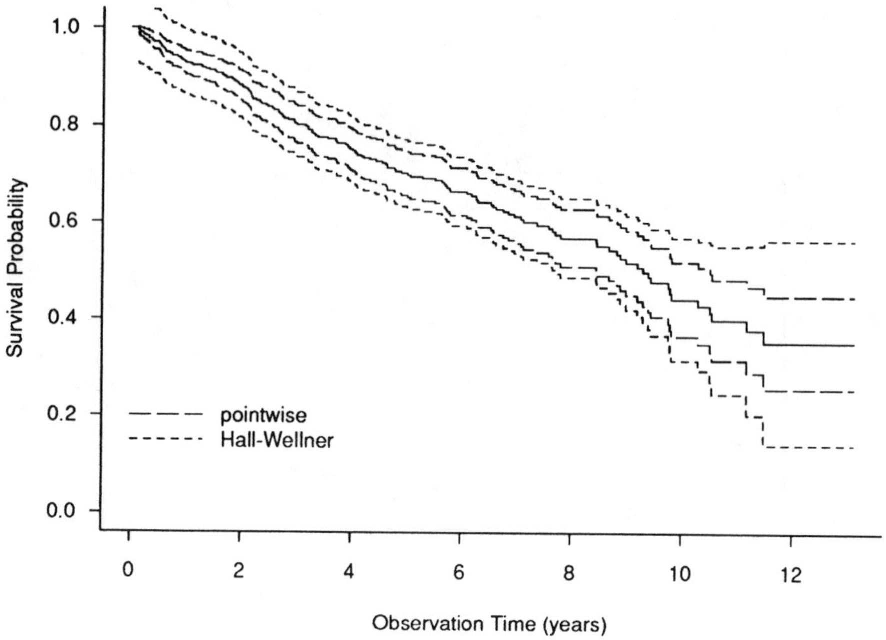

Figure 6.3.5a Pointwise confidence bands and Hall-Wellner confidence bands for survival probability, using all 418 patients in the PBC data set.

$\hat{S}$ takes values near 0.6. The 95% confidence limits for five-year survival are (.6565, .7492), while five-year limits from the Hall-Wellner bands are (.6341, .7716).

The Gill ten-year bands in Figure 6.3.5b are wider than Hall-Wellner bands until the interval from nine to ten years.

Figure 6.3.6 illustrates the sensitivity of the width of Gill bands at any time t to the degree of censorship after t. For this illustration, an additional independent censoring mechanism was artificially imposed on the original data, termed the "full" data set. For each patient, a variable V was randomly generated from a distribution which was uniform over five to eleven years. The follow up data on the ith patient was censored at V_i if the patient had been alive and uncensored beyond V_i in the original data set. The Kaplan-Meier curves on the original "full" data set and on the additionally "censored" data set are displayed in Figure 6.3.6, along with the Gill ten-year bands computed using each data set.

The Gill bands from the "censored" data set are wider than those from the original "full" data, not only over the period from five to ten years where additional censoring occurred, but also over the period from zero to five years where there were no changes in the data. Increased censorship after five years widened the five-year limits of the Gill bands in Figure 6.3.6 from (.5625, .8432) to (.4826, .9231). □

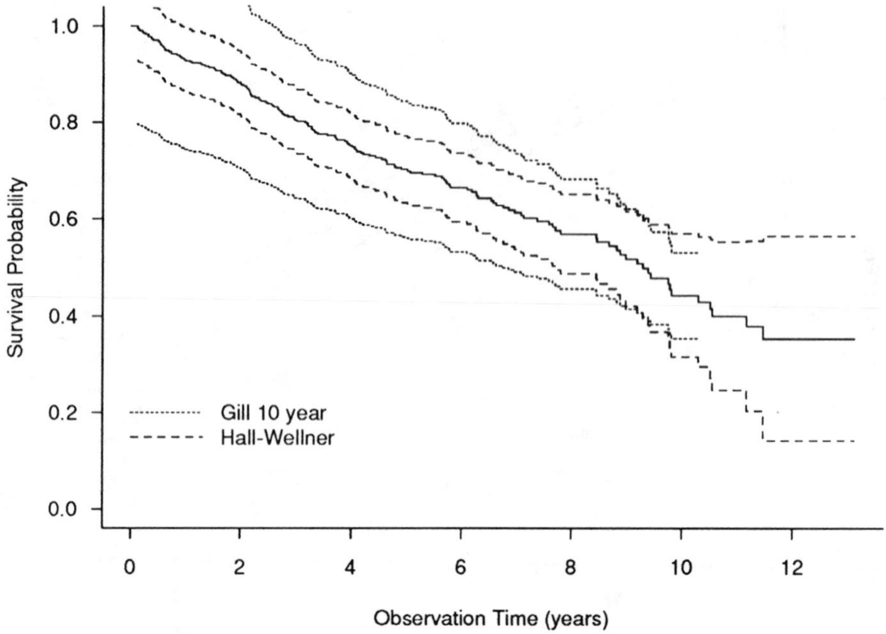

Figure 6.3.5b Gill 10-year and Hall-Wellner confidence bands in the PBC data.

6.4 BIBLIOGRAPHIC NOTES

Section 6.3

Many authors, beginning with Kaplan and Meier (1958) and Efron (1967), have considered the large sample properties of the Kaplan-Meier estimator. Breslow and Crowley (1974) established Theorem 6.3.1 (1) under the general Random Censorship Model for the case in which the distribution functions F and L are continuous. Meier (1975) obtained large sample results by conditioning on the times of censorship. Gill (1980) established the weak convergence results in Theorem 6.3.1 in a general setting which allows F and L to have jumps. By restricting to continuous F, Gill (1983) then established Theorem 6.3.2 providing weak convergence results over the whole real line. Wellner (1985) considered a model for heavy censoring to obtain Poisson type limit theorems under heavy censoring for the Kaplan-Meier estimator. Using equality (2.7) in Chapter 3, he then derived new approximate variance formulas.

Gill (1989) obtained Breslow and Crowley's large sample result for the Kaplan-Meier estimator through an application of the delta-method and the chain rule for Hadamard derivatives, after observing that the product integral $x \to \prod_0^{(\cdot)}(1 + dx)$ (see Gill and Johansen, 1989) is a Hadamard-differentiable mapping.

As noted in Chapter 3, the asymptotic optimality of the Kaplan-Meier estimator was established by Wellner (1982). Miller (1983) evaluated its asymptotic efficiency relative

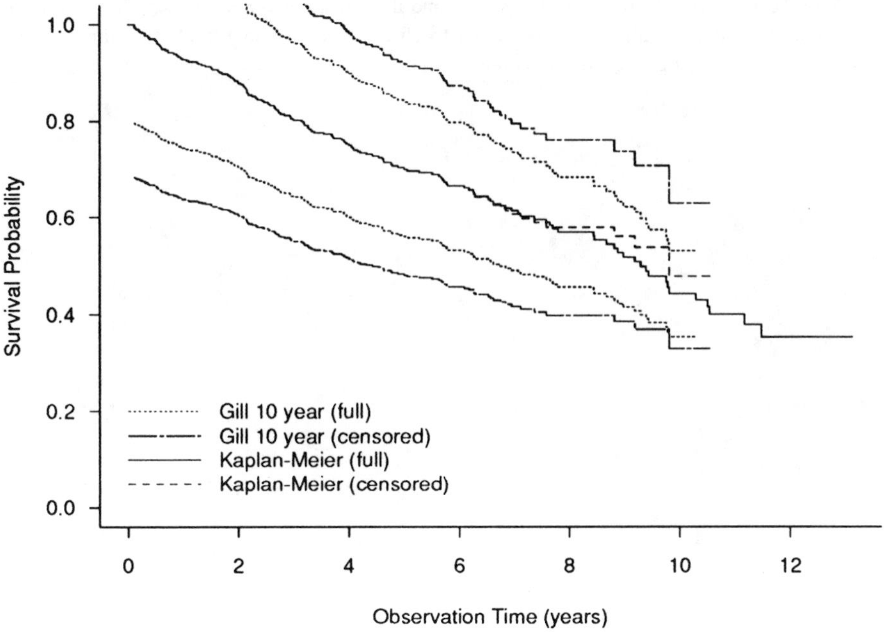

Figure 6.3.6 Gill ten-year confidence bands for the original PBC dats set of 418 cases (full), and for the same cases with additional censoring imposed (censored).

to the maximum likelihood estimator of a parametric survival function, while Hollander, Proschan, and Sconing (1985) studied the loss of efficiency relative to the empirical distribution function under the assumption that $S(t) = \{C(t)\}^{\alpha}$. Recently, Gill (1989) has shown that the generalized likelihood approach can be used to obtain efficiencies of the Kaplan-Meier estimator.

Aalen and Johansen (1978) and Fleming (1978) established asymptotic properties of nonparametric estimators for non-time-homogeneous Markov chains. These estimators reduce to the Kaplan-Meier in the special case of a single transient state. Fleming and Harrington (1978) investigated the analogous approach in the discrete time setting.

Nair (1984) proposed a confidence band through an adaption of the usual pointwise confidence intervals based on Greenwood's (1926) variance formula, and compared Hall-Wellner bands, Gill bands, and his proposed "equal precision" bands. Further comparisons are provided by Csörgő and Horváth (1986) and Hollander and Peña (1989). Hollander and Peña also proposed a family of confidence bands which include the Hall-Wellner bands and Gill bands, the latter referred to as "Gillespie-Fisher" bands. In evaluating relative properties of the two bands, both Nair and Hollander-Peña used the notion of asymptotic relative width efficiency, defined as the limiting ratio of the sample sizes needed in order for the bands to have equal asymptotic widths.

Lin and Fleming (1991) provide confidence bands for the survival curve estimates, $\hat{S}(t|\mathbf{Z})$, formulated in Equation (4.3) of Chapter 4 under the proportional hazards model. These reduce to Hall-Wellner bands when there are no covariates.

Ying (1989) strengthened Corollary 6.3.1 and the weak convergence result in Eq. (3.17), obtained under the assumption in Inequality (3.15), by showing that the basic process itself converges and the stopping time T is not needed.

CHAPTER 7

Weighted Logrank Statistics

7.1 INTRODUCTION

In this chapter we consider statistics used to test the hypothesis $H_0 : F_1 = F_2$, for two survival distributions F_1 and F_2, in the general Random Censorship Model of Definition 3.1.1 and Section 6.2. A number of approaches have been used to formulate two-sample statistics for censored data. There is a large literature on parametric methods which lean heavily on likelihood methods for one- and two-parameter exponential distributions, for two- and three-parameter Weibull models, and for other distributions such as the log-logistic, log-normal and gamma. These approaches are summarized in Lawless (1982) and in Kalbfleisch and Prentice (1980).

We concentrate here on nonparametric methods, which in themselves have been studied from varying perspectives. The first nonparametric statistics to gain widespread use in censored data were those proposed by Gehan (1965) and Mantel (1966). Gehan's generalization of the Wilcoxon rank-sum statistic was based on a natural adaptation of the Mann-Whitney form of the statistic. Mantel's statistic was based on the Mantel-Haenszel statistic for stratified contingency tables, as discussed in Chapter 0. This statistic subsequently became known as the logrank statistic and is probably the most commonly used two-sample test statistic for censored data.

In 1978, Prentice proposed an adaptation of the efficient score function used to construct efficient rank tests, so that these tests could be used with censored data. He showed that the logrank test was the adapted version of the Savage exponential scores test, and suggested that another generalization of the Wilcoxon test might be more appropriate than that of Gehan (1965). Mehrotra, Michalek, and Mihalko (1982), Cuzick (1985) and Struthers (1984) have shown under regularity conditions that Prentice's adapted rank tests remain efficient in the presence of censoring.

Aalen (1978b) showed that nearly all of the previously proposed two-sample test statistics for censored data could be studied in the martingale framework. Gill

(1980) provided a thorough examination of the two-sample problem by studying properties of various statistics under both null and alternative hypotheses.

In Section 7.2, we discuss the asymptotic null distribution of a class of linear rank statistics introduced earlier, named "tests of the class $\mathcal{K}$" by Gill (1980). Section 7.3 explores the consistency of these tests against ordered hazards and stochastic ordering alternatives. Large-sample distributions under contiguous alternatives are derived in Section 7.4. That section contains a characterization of the alternative hypotheses against which tests of the class $\mathcal{K}$ are efficient. The large sample joint null distribution of vectors of statistics of the class $\mathcal{K}$ is derived in Section 7.5, allowing the study of procedures based on clusters of such statistics. Supremum versions of statistics of the class $\mathcal{K}$ also are explored in Section 7.5. Finally, the joint distribution of the linear rank statistics and of their supremum analogues is developed.

7.2 LARGE SAMPLE NULL DISTRIBUTION

Mantel's derivation of the logrank statistic is discussed in Chapter 0. There, we define $\{T_j^o : j = 1, \ldots, d\}$ to be the set of d distinct ordered observed failure times in the pooled sample, $\overline{Y}_{ij}$ to be the risk set size in sample i at T_j^o, and D_{ij} to be number of failures in sample i at T_j^o. Mantel's logrank statistic is

$$G^0 = c\sum_{j=1}^{d}(\mathcal{O}_{1j} - \mathcal{E}_{1j}), \tag{2.1}$$

where $\mathcal{O}_{1j} = D_{1j}$, the observed number of failures in sample 1 at T_j^o, and

$$\mathcal{E}_{1j} = (D_{1j} + D_{2j})\left(\frac{\overline{Y}_{1j}}{\overline{Y}_{1j} + \overline{Y}_{2j}}\right),$$

the conditional expected number of failures in sample 1 at T_j^o, given $\overline{Y}_{1j}, \overline{Y}_{2j}$, and $D_{1j} + D_{2j}$. The constant c is $\{(n_1 + n_2)/n_1 n_2\}^{1/2}$, where n_i is the initial sample size in group i.

To weight early and late failure times differently, (2.1) can be changed to

$$W_K = c\sum_{j=1}^{d} W(T_j^o)(\mathcal{O}_{1j} - \mathcal{E}_{1j}), \tag{2.2}$$

where we assume W is an adapted bounded nonnegative predictable process. Equation (2.1) is equivalent to

$$G^0 = c\int_0^\infty \frac{\overline{Y}_1(s)\overline{Y}_2(s)}{\overline{Y}_1(s) + \overline{Y}_2(s)}\left\{\frac{d\overline{N}_1(s)}{\overline{Y}_1(s)} - \frac{d\overline{N}_2(s)}{\overline{Y}_2(s)}\right\},$$

and (2.2) can be written

$$W_K = c\int_0^\infty W(s)\frac{\overline{Y}_1(s)\overline{Y}_2(s)}{\overline{Y}_1(s)+\overline{Y}_2(s)}\left\{\frac{d\overline{N}_1(s)}{\overline{Y}_1(s)} - \frac{d\overline{N}_2(s)}{\overline{Y}_2(s)}\right\}. \tag{2.3}$$

In Chapter 3, we saw that $\int_0^t (\overline{Y}_i)^{-1} d\overline{N}_i$ is a useful estimator of $\Lambda_i(t) = \int_0^t \lambda_i(s)ds$. Thus W_K is a "weighted logrank statistic" and is essentially a sum, over all times, of weighted differences in the estimated hazards.

This chapter explores properties of these weighted logrank statistics referred to in Section 3.3 as statistics of the class $\mathcal{K}$. They are of the form

$$W_K = \int K\frac{d\overline{N}_1}{\overline{Y}_1} - \int K\frac{d\overline{N}_2}{\overline{Y}_2},$$

where

$$K(s) = \left(\frac{n_1+n_2}{n_1 n_2}\right)^{1/2} W(s)\frac{\overline{Y}_1(s)\overline{Y}_2(s)}{\overline{Y}_1(s)+\overline{Y}_2(s)},$$

and so K is an adapted bounded nonnegative predictable process such that

$$Y_1(t)\wedge Y_2(t) = 0 \text{ implies } K(t) = 0.$$

The applications and examples here focus on some subsets of the class $\mathcal{K}$. Let f be a nonnegative bounded continuous function with bounded variation on $[0,1]$. These subsets arise in Eq. (2.3) by setting

$$W(s) = f\{\hat{S}(s-)\}$$

or

$$W(s) = f\{\hat{\pi}(s)\},$$

where $\hat{S}$ is the Kaplan-Meier estimator in the pooled sample and $\hat{\pi}(s) = (n_1+n_2)^{-1}(\overline{Y}_1(s)+\overline{Y}_2(s))$ is the pooled sample estimator of $a_1\pi_1(s)+a_2\pi_2(s)$.

Definition 7.2.1. Consider the weighted logrank statistics given by Eq. (2.3). Statistics with weight function $W(s) = \{\hat{S}(s-)\}^\rho\{1-\hat{S}(s-)\}^\gamma$ for $\rho \geq 0, \gamma \geq 0$ will be called $G^{\rho,\gamma}$ statistics. □

When $\gamma = 0$, this $G^{\rho,\gamma}$ class reduces to the G^ρ class introduced by Harrington and Fleming (1982), and $\rho = 0$ and $\rho = 1$ correspond to the logrank and Prentice-Wilcoxon statistics, respectively. Figure 7.2.1 displays the range of weight functions used in the $G^{\rho,\gamma}$ statistics. The Tarone-Ware (1977) class of statistics arises when $W(s) = \{\hat{\pi}(s)\}^\rho$ for $\rho \geq 0$. The values $\rho = 0$ and $\rho = 1$ correspond to the logrank and Gehan-Wilcoxon (Gehan, 1965) statistics respectively.

By (2.3),

$$K = \left(\frac{n_1 n_2}{n_1+n_2}\right)^{1/2}(\hat{S}^-)^\rho(1-\hat{S}^-)^\gamma\frac{\overline{Y}_1}{n_1}\frac{\overline{Y}_2}{n_2}\frac{n_1+n_2}{\overline{Y}_1+\overline{Y}_2} \tag{2.4}$$

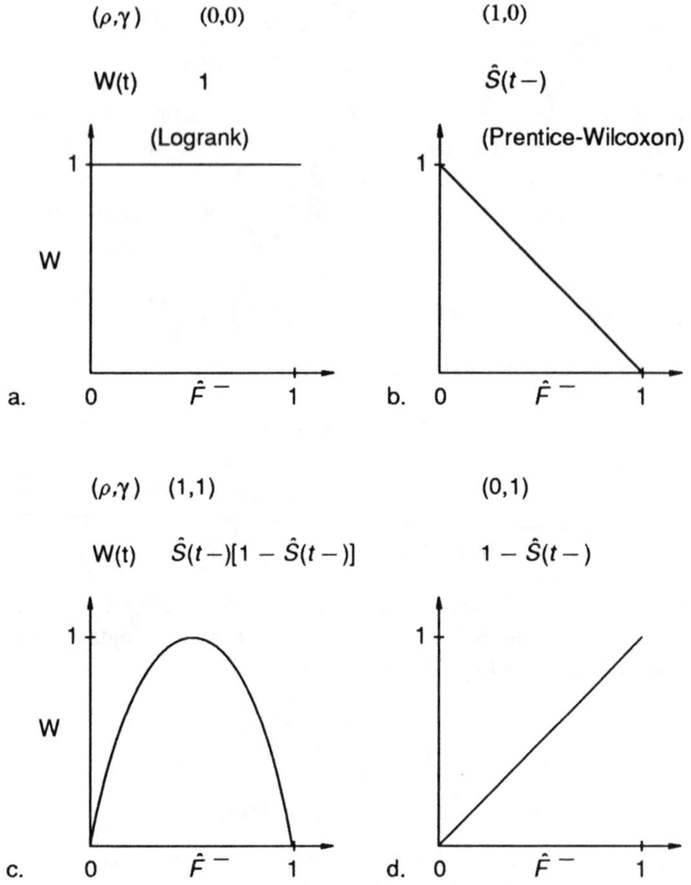

Figure 7.2.1 Weight functions W in the $G^{\rho,\gamma}$ class of statistics. W provides evenly distributed weights in (a) for the logrank statistics, emphasizes early differences in (b) for the Wilcoxon statistic, and emphasizes middle and late differences, respectively, in (c) and (d).

for the $G^{\rho,\gamma}$ statistic, while

$$K = \left(\frac{n_1 n_2}{n_1 + n_2}\right)^{1/2} \frac{\overline{Y}_1}{n_1} \frac{\overline{Y}_2}{n_2}$$

for the Gehan-Wilcoxon statistic.

For any statistic of the class $\mathcal{K}$,

$$W_K = \int \frac{K}{\overline{\overline{Y}}_1} dM_1 - \int \frac{K}{\overline{\overline{Y}}_2} dM_2 + \int K(d\Lambda_1 - d\Lambda_2), \tag{2.5}$$

where $M_i = \overline{N}_i - \int \overline{Y}_i d\Lambda_i$. Thus, under $H_0 : F_1 = F_2$,

$$W_K = \int H_1 dM_1 - \int H_2 dM_2,$$

where $H_i = K(\overline{Y}_i)^{-1}$. Weak convergence of W_K follows from Theorem 6.2.1 as long as Conditions (1), (2) and (3) hold. If h_i denotes the limit of $H_i^2 \overline{Y}_i = K^2(\overline{Y}_i)^{-1}$ then, under H_0, Theorem 6.2.1 implies that

$$W_K \xrightarrow{\mathcal{D}} N(0, \sigma^2),$$

where

$$\sigma^2 = \int (h_1 + h_2)(1 - \Delta\Lambda) d\Lambda.$$

An appealing estimator for σ^2 is

$$\hat{\sigma}^2 = \int \left(\frac{K^2}{\overline{Y}_1} + \frac{K^2}{\overline{Y}_2} \right) \left(1 - \frac{\Delta \overline{N}_1 + \Delta \overline{N}_2 - 1}{\overline{Y}_1 + \overline{Y}_2 - 1} \right) \frac{d(\overline{N}_1 + \overline{N}_2)}{\overline{Y}_1 + \overline{Y}_2}. \tag{2.6}$$

Theorem 3.3.2 establishes that $\hat{\sigma}^2$ is an unbiased estimate of the variance of W_K. When the weight function $W(s) \equiv 1$, $\hat{\sigma}^2$ reduces to the variance estimator proposed by Mantel (1966).

We begin with a lemma from Gill (1980) which establishes that $\hat{\sigma}^2$ is a uniformly consistent estimator for σ^2. We do not give the proof, which uses an approach similar to that used for the uniform consistency of $\hat{F}$ and $\hat{\Lambda}$ in Theorem 3.4.2. We state the lemma in its full generality, allowing F_i^n to vary with n, and to have both continuous and discontinuous components. When F_i^n does vary with n, we assume for all i and n that F_i^n is absolutely continuous with respect to a common limiting distribution function F. (When F_i^n and F have hazard functions λ_i^n and λ, respectively, this absolute continuity holds if $\lambda_i^n(t) = 0$ whenever $\lambda(t) = 0$ or $F(t) = 1$.)

In this lemma and other theorems and lemmas in this chapter, we use the general Random Censorship Model in Section 6.2 (for $r = 2$), implying Assumption 2.6.1 does hold, and we presume Assumption 6.2.1. In this chapter, Assumption 6.2.2 on absolute continuity of F is presumed only in Section 7.3. As in Chapter 6, let $\mathcal{I} = \{t : \pi_1(t)\pi_2(t) > 0\}$ and $u = \sup \mathcal{I}$.

Lemma 7.2.1. Suppose Assumption 6.2.1 holds. In Condition (1) of Theorem 6.2.1, assume that the functions h_i are left-continuous, have right-hand limits, and that each h_i^+ has bounded variation on closed subintervals of $\mathcal{I}$, even when F_i^n does not depend on n.

Assume the limiting distribution functions F_1 and F_2 equal a common distribution function F and, when F_i^n does vary with n, assume for all n and i that F_i^n is absolutely continuous with respect to F.

For $i \neq i'$, let Conditions (2.7)–(2.9) be given by

$$\lim_{n \to \infty} \int_0^t |d\Lambda_i^n - d\Lambda| = 0 \quad \text{for } t \in \mathcal{I}; \tag{2.7}$$

$$\lim_{t\uparrow u}\limsup_{n\to\infty}\sup_{s\in(t,u]}\left|\frac{d\Lambda_i^n}{d\Lambda_{i'}^n}(s)\right| < \infty \text{ whenever } u \notin \mathcal{I}; \tag{2.8}$$

$$\limsup_{n\to\infty}\sup_{s\in(u,\infty)}\left|\frac{d\Lambda_i^n}{d\Lambda_{i'}^n}(s)\right| < \infty \text{ whenever } u < \infty, \tag{2.9}$$

where $d\Lambda_i^n/d\Lambda_{i'}^n$ is the Radon-Nikodym derivative of Λ_i^n with respect to $\Lambda_{i'}^n$, which reduces to $\lambda_i^n(\cdot)/\lambda_{i'}^n(\cdot)$ when F_i^n and $F_{i'}^n$ have hazard functions λ_i^n and $\lambda_{i'}^n$.

Under (2.7) and the modified (1) of Theorem 6.2.1,

$$\sup_{s\in[0,t]}\left|\hat{\sigma}^2(s) - \int_0^s (h_1 + h_2)(1 - \Delta\Lambda)d\Lambda\right| \xrightarrow{P} 0 \tag{2.10}$$

as $n \to \infty$ for any $t \in \mathcal{I}$.

If in addition, (2.8) and (2) of Theorem 6.2.1 hold, (2.10) is true for $t = u$. , If (2.9) and (3) also hold, then (2.10) is true for $t \in [0, \infty]$. □

When $F_1^n = F_2^n = F$ for all n, Conditions (2.7)–(2.9) are empty. Convergence in distribution under the null hypothesis for statistics in $\mathcal{K}$ is now a simple consequence of Theorems 6.2.1 and Lemma 7.2.1. The conditions sufficient for convergence given in Corollary 7.2.1 below are the specific form of Conditions (1), (2), and (3) in Theorem 6.2.1 for the two-sample problem.

Corollary 7.2.1. Suppose W_K is in $\mathcal{K}$ and $F_1^n = F_2^n = F$ for all n. Suppose **(1)** that as $n \to \infty$

$$\frac{K^2(s)}{\overline{Y}_i(s)} \xrightarrow{P} h_i(s)$$

uniformly on $[0, t]$ for any $t \in \mathcal{I}$, where h_i is a nonnegative, left-continuous function with right-hand limits such that $h_i(t) < \infty$ and h_i^+ has bounded variation on each closed subinterval of $\mathcal{I}$, and $h_i(t) = 0$ for any $t \notin \mathcal{I}$.

If $u \notin \mathcal{I}$, assume **(2a)**,

$$\sigma^2(u) = \int_0^u (h_1 + h_2)(1 - \Delta\Lambda)d\Lambda < \infty$$

and, for any $\epsilon > 0$, **(2b)**,

$$\lim_{t\uparrow u}\limsup_{n\to\infty} P\left\{\int_t^u K^2\frac{\overline{Y}_1 + \overline{Y}_2}{\overline{Y}_1\overline{Y}_2}d\Lambda > \epsilon\right\} = 0.$$

If $u < \infty$, assume **(3)** that for any $\epsilon > 0$,

$$\lim_{n\to\infty} P\left\{\int_u^\infty K^2\frac{\overline{Y}_1 + \overline{Y}_2}{\overline{Y}_1\overline{Y}_2}d\Lambda > \epsilon\right\} = 0.$$

Let

$$\hat{\sigma}^2(t) = \int_0^t K^2 \left(\frac{\overline{Y}_1 + \overline{Y}_2}{\overline{Y}_1 \overline{Y}_2} \right) \left(1 - \frac{\Delta \overline{N}_1 + \Delta \overline{N}_2 - 1}{\overline{Y}_1 + \overline{Y}_2 - 1} \right) \frac{d(\overline{N}_1 + \overline{N}_2)}{\overline{Y}_1 + \overline{Y}_2}.$$

Then,

$$W_K \Longrightarrow W_0(\sigma^2) \text{ on } D[0, t],$$

and

$$\hat{\sigma}^2(t) \xrightarrow{P} \sigma^2(t)$$

as $n \to \infty$, for any $t \in [0, \infty]$, where W_0 denotes Brownian motion.

If $T \equiv \inf\{s : \overline{Y}_1(s) \wedge \overline{Y}_2(s) = 0\}$ then, as $n \to \infty$,

$$\frac{W_K(T)}{\{\hat{\sigma}^2(T)\}^{1/2}} \xrightarrow{\mathcal{D}} N(0, 1). \tag{2.11}$$

□

The limit in Eq. (2.11) provides the large sample distribution for statistics in $\mathcal{K}$ as they are used in practice. The next theorem illustrates the use of Corollary 7.2.1 for a particular subset of statistics of the class $\mathcal{K}$.

Theorem 7.2.1. Let $\hat{S}$ be the Kaplan-Meier estimator in the pooled sample and let $\hat{\pi}$ be the pooled sample estimate of π, i.e.

$$\hat{\pi}(t) = \frac{\overline{Y}_1(t) + \overline{Y}_2(t)}{n_1 + n_2}.$$

Let f be a nonnegative bounded continuous function with bounded variation on $[0, 1]$. Suppose W_K is a statistic of the class $\mathcal{K}$, where

$$K(t) = \left(\frac{n_1 n_2}{n_1 + n_2} \right)^{1/2} W(t) \frac{\overline{Y}_1(t)}{n_1} \frac{\overline{Y}_2(t)}{n_2} \frac{n_1 + n_2}{\overline{Y}_1(t) + \overline{Y}_2(t)},$$

and where $W(t) = f(\hat{S}(t-))$ or $W(t) = f(\hat{\pi}(t))$.

Suppose $a_i \in (0, 1)$ for $i = 1, 2$. When $F_1^n = F_2^n = F$ for all n,

$$\frac{W_K(T)}{\{\hat{\sigma}^2(T)\}^{1/2}} \xrightarrow{\mathcal{D}} N(0, 1) \text{ as } n \to \infty,$$

where $T = \inf\{s : \overline{Y}_1(t) \wedge \overline{Y}_2(s) = 0\}$ and $\hat{\sigma}^2$ is given by Eq. (2.6). □

The next lemma is needed in the proof of Theorem 7.2.1.

Lemma 7.2.2. Assume $F_1^n = F_2^n = F$, and that Assumption 6.2.1 holds. Then

$$\int_0^\infty \pi_i(s) d\Lambda(s) < \infty \tag{2.12}$$

and

$$\sup_{t\in[0,\infty)}\left|\int_0^t \frac{\overline{Y}_i}{n_i}d\Lambda - \int_0^t \pi_i d\Lambda\right| \xrightarrow{P} 0 \text{ as } n \to \infty. \tag{2.13}$$

Proof. Inequality (2.12) holds since $\pi_i \leq (1-F_i^-)$. Equation (2.13) also is true when $[0,\infty)$ is replaced by $[0,s]$ when $\Lambda(s) < \infty$. Define $\tau = \sup\{t : F(t) < 1\}$, and suppose $\Lambda(\tau) = \infty$. Since $\overline{Y}_i(t) = 0$ a.s. for any $t > \tau$ and for any n, and $\pi_i(t) = 0$ for any $t > \tau$, it remains to establish Eq. (2.13) for $t \in [0,\tau]$.

Now $\Lambda(\tau) = \infty$ implies $\Delta F(\tau) = 0$, so

$$\begin{aligned} E\int_t^\tau \frac{\overline{Y}_i}{n_i}d\Lambda &= E\int_t^\tau \frac{d\overline{N}_i}{n_i} \\ &\leq \{F(\tau) - F(t)\} \downarrow 0 \text{ as } t \uparrow \tau. \end{aligned} \tag{2.14}$$

The convergence in Eq. (2.14) is uniform in n. Therefore, by Chebyshev's inequality, we have, for any $\epsilon, \eta > 0$, that t_0 exists such that

$$P\left\{\int_{t_0}^\tau \frac{\overline{Y}_i}{n_i}d\Lambda > \epsilon\right\} \leq \eta \text{ for all } n. \tag{2.15}$$

By the fact that Eq. (2.13) holds when $[0,\infty)$ is replaced by $[0,s]$ for $s < \tau$, and by Inequalities (2.12) and (2.15), Theorem 4.2 of Billingsley (1968) implies that (2.13) holds for $t \in [0,\tau]$. □

Proof of Theorem 7.2.1. Suppose $W = f(\hat{S}^-)$. We begin by verifying Condition (1) of Corollary 7.2.1. Note

$$\begin{aligned} (H_i)^2\overline{Y}_i &= K^2/\overline{Y}_i \\ &= f^2(\hat{S}^-)\left(\frac{n_{i'}}{n_1+n_2}\right)\frac{n_i}{\overline{Y}_i}\left(\frac{\overline{Y}_1}{n_1}\right)^2\left(\frac{\overline{Y}_2}{n_2}\right)^2\left(\frac{n_1+n_2}{\overline{Y}_1+\overline{Y}_2}\right)^2, \end{aligned}$$

for $i' \neq i$. If

$$h_i = f^2(S^-)\frac{a_{i'}}{\pi_i}\frac{\pi_1^2\pi_2^2}{(a_1\pi_1+a_2\pi_2)^2},$$

then h_i is a nonnegative left-continuous function having right-hand limits, with $h_i(t) < \infty$ for $t \in \mathcal{I}$ and $h_i(t) = 0$ for any $t \notin \mathcal{I}$. Also h^+ is of bounded variation on each closed subinterval of $\mathcal{I}$. The consistency of $\hat{S}$ (Eq. (4.4) of Chapter 3) and Assumption 6.2.1 imply

$$\sup_{0\leq s\leq t}|H_i^2(s)\overline{Y}_i(s) - h_i(s)| \xrightarrow{P} 0 \text{ as } n \to \infty \text{ for any } t \in \mathcal{I}.$$

We now verify Condition (2a) that $\sigma^2(u) < \infty$. Without loss of generality, we can assume $|f| \le 1$. If $i \ne i'$,

$$\begin{aligned}\int_0^u h_i(1-\Delta\Lambda)d\Lambda &\le \int_0^u (h_1+h_2)d\Lambda \\ &= \int_0^u f^2(S^-)\left(\frac{\pi_1\pi_2}{a_1\pi_1+a_2\pi_2}\right)d\Lambda \\ &\le \int_0^u \frac{\pi_i}{a_{i'}}d\Lambda \\ &\le \frac{1}{a_{i'}}\int_0^u S^- d\Lambda \\ &< \frac{1}{a_{i'}} \\ &< \infty.\end{aligned}$$

Consequently $\sigma^2(u) < \infty$.

To establish Condition (2b), observe that

$$\begin{aligned}H_1^2\overline{Y}_1 + H_2^2\overline{Y}_2 &= K^2\frac{\overline{Y}_1+\overline{Y}_2}{\overline{Y}_1\overline{Y}_2} \\ &= f^2(\hat{S}^-)\frac{n_1+n_2}{n_1n_2}\frac{\overline{Y}_1\overline{Y}_2}{\overline{Y}_1+\overline{Y}_2} \\ &\le \frac{n_1+n_2}{n_i'}\frac{\overline{Y}_i}{n_i},\end{aligned}$$

where $i' \ne i$. Since $a_{i'} > 0$, Lemma 7.2.2 implies

$$\int_t^u \frac{n_1+n_2}{n_{i'}}\frac{\overline{Y}_i}{n_i}d\Lambda \xrightarrow{P} \frac{1}{a_{i'}}\int_t^u \pi_i d\Lambda.$$

Condition (2b) follows immediately if $\Delta\Lambda(u) = 0$. Since $u \notin \mathcal{I}$, we can assume $\pi_i(u) = 0$. If $\Delta\Lambda(u) > 0$, the condition is satisfied, since $a_{i'} > 0$ and $\pi_i(u) = 0$ imply

$$\int_{\{u\}} \frac{n_1+n_2}{n_{i'}}\frac{\overline{Y}_i}{n_i}d\Lambda \xrightarrow{P} 0.$$

Condition (3) follows easily since Lemma 7.2.2 implies

$$\int_u^\infty \frac{n_1+n_2}{n_{i'}}\frac{\overline{Y}_i}{n_i}d\Lambda \xrightarrow{P} a_{i'}^{-1}\int_u^\infty \pi_i d\Lambda = 0.$$

The proof for $W = f(\hat{\pi})$ proceeds exactly as above. □

Conditions (1), (2), and (3) in Corollary 7.2.1 play important roles in the asymptotic distribution of the weighted logrank statistic $W_K(T)/\{\hat{\sigma}^2(T)\}^{1/2}$. For independent and identically distributed failure and censoring variables, and when

$F_i^n = F$ and $L_{ij}^n = L$ for all i, j, and n, Condition (3) is empty since $T \leq u \equiv \sup\{s : \pi_1(s) \wedge \pi_2(s) > 0\}$ a.s. On the other hand, it is easy to construct examples where Condition (3) is needed. Suppose $F_1^n = F_2^n = F$, where $F(t) = 1 - t/2$ for $0 \leq t \leq 2$. Let $1 - L_{ij}^n \equiv C_{ij}^n = C^n$, where $C^n(t) = 1 - n(n-1)^{-1}t$ for $0 \leq t \leq n(n-1)^{-1}$ (or for $0 \leq t < \infty$ when $n = 1$), and suppose X_{ij}^n and U_{ij}^n are independent. Then $u = 1$ and yet, for any n, $P\{T > 1\} > 0$.

To explore the role for Condition (2), consider the situation in which

$$F_1^n = F_2^n = F \quad \text{for all } n, \tag{2.16}$$

$$L_{1j}^n = L_{2j}^n = L \quad \text{for all } j, n, \tag{2.17}$$

and

$$F \text{ is absolutely continuous.} \tag{2.18}$$

When L is continuous at u, it follows that $\mathcal{I} = [0, u)$, i.e., $u \notin \mathcal{I}$, and that $T \xrightarrow{P} u$ as $n \to \infty$. Clearly, then, this implies we need a stronger result than convergence of $W_K(t)/\{\hat{\sigma}^2(t)\}^{1/2}$ for any fixed $t \in \mathcal{I}$. However, since $T \in \mathcal{I}$ a.s. for all n, one might speculate that under (2.16)–(2.18) the convergence result

$$W_K(T) \to N\left(0, \int_0^\infty (h_1 + h_2)d\Lambda\right) \tag{2.19}$$

would follow from Condition (1) alone. The next example illustrates that Condition (1) is not sufficient to establish Eq. (2.19), even when Eqs. (2.16), (2.17), and (2.18) all hold.

Example 7.2.1. Suppose $r = 2$, $n_1 = n_2$ $(= n/2) \equiv n_0$, $L_{ij}^n \equiv 0$ (i.e., data are uncensored), and $F_1^n = F_2^n = F$, where $F(t) = t$ for $t \in [0, 1]$. Hence, (2.16)–(2.18) all hold. For $j = 0, 1$, let $t_{(j)}^* = \inf\{t : \overline{Y}_1(t) = j\}$ and $t_{(j)}^{**} = \inf\{t : \overline{Y}_2(t) = j\}$, and observe that $T = \min\{t_{(0)}^*, t_{(0)}^{**}\}$.

If $K(t) = I_{\{\overline{Y}_2(t)=1; \overline{Y}_1(t)>1\}}$, then

$$W_K = \int K\left(\frac{d\overline{N}_1}{\overline{Y}_1} - \frac{d\overline{N}_2}{\overline{Y}_2}\right)$$

is a statistic of the class $\mathcal{K}$. W_K satifies

$$W_K(T) = \begin{cases} 0 & \text{if } t_{(1)}^* \leq t_{(1)}^{**} \quad \text{Case } (a) \\ \int_{t_{(1)}^{**}}^{t_{(1)}^*} \frac{d\overline{N}_1}{\overline{Y}_1} & \text{if } t_{(1)}^{**} < t_{(1)}^* < t_{(0)}^{**} \quad \text{Case } (b) \\ \left(\int_{t_{(1)}^{**}}^{t_{(0)}^{**}} \frac{d\overline{N}_1}{\overline{Y}_1}\right) - 1 & \text{if } t_{(0)}^{**} \leq t_{(1)}^* \quad \text{Case } (c). \end{cases}$$

Now $P\{t_{(1)}^* \leq t_{(1)}^{**}\} = 1/2$, and Figure 7.2.2 implies that

$$P\{t_{(1)}^{**} < t_{(1)}^* < t_{(0)}^{**}\} = \frac{n_0}{2(2n_0 - 1)} \in (\frac{1}{4}, \frac{1}{2}].$$

	TYPE							
	#1	#2	#3	#4	#5	#6	#7	#8
Sample for:								
last observation	1	1	1	1	2	2	2	2
2nd last observation	1	1	2	2	1	1	2	2
3rd last observation	1	2	1	2	1	2	1	2
	⋮	⋮	⋮	⋮	⋮	⋮	⋮	⋮

Figure 7.2.2 Possible orderings of final three observations in Example 7.2.1.

Specifically, Case (a) holds for types #4, #6, #7 and #8 in Figure 7.2.2, which occur with probability 1/2. Case (b) holds for types #3 and #5, while Case (c) holds for types #1 and #2. Now Pr(#1 or #2) / Pr(#3 or #4) = $(n_0 - 1)/n_0$ and Pr(#1 or #2 or #3 or #4) = 1/2, so Pr(#3 or #4) = $n_0/[2(2n_0 - 1)]$. Also Pr(#4) = Pr(#5) and, by disjointness, P(#3 or #4) = P(#3 or #5).

Since $\int_{t_{(1)}^{**}}^{t_{(1)}^{*}} \frac{dN_1}{\overline{Y}_1} \geq 1/2$ in Case (b), $P\{W_K(T) \geq 1/2\} > 1/4$, so $W_K(T) \overset{P}{\nrightarrow} 0$. However, $P\{K(t) = 0\} \to 1$ for any $t \in (0, 1)$, implying $h_i(t) = 0$ for any $t \in (0, 1)$, so $\sigma^2 = \int_0^1 (h_1 + h_2) d\Lambda = 0$. Condition (1) of Corollary 7.2.1 is easily verified. Yet, if Eq. (2.19) were true, then $W_K(T) \xrightarrow{P} 0$. In Exercise 7.2, it is established that Condition (2b) of Corollary 7.2.1 fails to hold. □

One additional distinction should be made between results obtained from verification of Condition (1) *vis-a-vis* the stronger results which follow from verification of Conditions (1) and (2). Under Condition (1), weak convergence over $[0, t^*]$ requires $\Lambda_i(t^*) < \infty$. Under (1) and (2) however, weak convergence over $[0, t^*]$ instead requires that $\int_0^{t^*} h_i(1 - \Delta\Lambda_i) d\Lambda_i < \infty$, that is, the limiting process has finite variance at t^*.

7.3 CONSISTENCY OF TESTS OF THE CLASS $\mathcal{K}$

To investigate the power of a test of the class $\mathcal{K}$, we remove the assumption that $F_1^n = F_2^n = F$. When $H_i = K/\overline{Y}_i$,

$$W_K = \int H_1 dM_1 - \int H_2 dM_2 + \int K(d\Lambda_1 - d\Lambda_2), \tag{3.1}$$

so $EW_K = \int (EK)(d\Lambda_1 - d\Lambda_2)$. It is clear that $\int K(d\Lambda_1 - d\Lambda_2)$ plays a central role in the power properties of W_K.

We begin by assuming $F_i^n = F_i$ for all i and n, then investigating consistency of W_K under some alternatives to $H_0 : F_1 = F_2$. We assume throughout the section that each F_i is absolutely continuous.

Definition 7.3.1. Let $Q_n, n = 1, 2, \ldots$, be a sequence of test statistics used to test a hypothesis H, and $R_n, n = 1, 2, \ldots$, an associated set of level α rejection regions. The sequence Q_n is said to be *consistent* under an alternative hypothesis H_A if

$$\lim_{n \to \infty} P(Q_n \in R_n) = 1$$

when H_A true. □

We establish the consistency of tests based on W_K under two types of alternative hypotheses.

Definition 7.3.2. For $i = 1, 2$, let $1 - F_i = S_i$, and $\lambda_i(t) = \frac{d}{dt}\Lambda_i(t)$.

1. The alternative $H_1 : \lambda_1(t) \geq \lambda_2(t)$ for all t is called the *ordered hazards alternative.*
2. The alternative $H_2 : S_2(t) \geq S_1(t)$ for all t is called the *alternative of stochastic ordering.*

□

It is clear that H_1 implies H_2.

Equation (3.1) suggests that W_K will be consistent under an alternative H_A when

$$\int_0^\infty K(d\Lambda_1 - d\Lambda_2) \xrightarrow{P} \infty \text{ as } n \to \infty. \tag{3.2}$$

The limit in (3.2) is used in establishing consistency against ordered hazards or stochastic ordering for a one-sided statistic, W_K, which rejects $F_1 = F_2$ whenever $W_K > z_{1-\alpha}$. We summarize these results for a subset of the statistics in the class $\mathcal{K}$, with details of the proofs left as exercises.

One would expect Eq. (3.2) to hold and consistency of W_K to follow under an ordered hazards alternative as long as there exists t_0 such that $\Lambda_1(t_0) > \Lambda_2(t_0)$. The next theorem makes this precise.

Theorem 7.3.1. Suppose in an ordered hazards alternative there exists a $t_0 > 0$ such that, for $i = 1, 2$, $\Lambda_1(t_0) > \Lambda_2(t_0)$ and $\pi_i(t_0) > 0$. Assume $0 < a_i = \lim_{n\to\infty} n_i/(n_1 + n_2) < 1$.

Let

$$W_K(\infty) = \int_0^\infty K\left(\frac{d\overline{N}_1}{\overline{Y}_1} - \frac{d\overline{N}_2}{\overline{Y}_2}\right),$$

where K is one of the weight functions defined in Theorem 7.2.1, so $W(t)$ is $f(\hat{S}(t-))$ or $f(\hat{\pi}(t))$. Suppose $f(t) > 0$ for $0 < t < 1$. Let $\hat{\sigma}^2(\infty)$ be the variance estimator defined in Eq. (2.6). Then $W_K(\infty)/\{\hat{\sigma}^2(\infty)\}^{1/2}$ is consistent. □

For a statistic W_K,

$$K(s) = \left(\frac{n_1 n_2}{n_1 + n_2}\right)^{1/2} W(s) \frac{\overline{Y}_1(s)}{n_1} \frac{\overline{Y}_2(s)}{n_2} \frac{n_1 + n_2}{\overline{Y}_1(s) + \overline{Y}_2(s)}.$$

If $W(s) \xrightarrow{P} w(s)$, then

$$\left(\frac{n_1+n_2}{n_1 n_2}\right)^{1/2} K(s) \xrightarrow{P} k(s) = w(s)\frac{\pi_1(s)\pi_2(s)}{a_1\pi_1(s)+a_2\pi_2(s)}. \tag{3.3}$$

Hence, W_K should be consistent against stochastic ordering if

$$\int_0^\infty w(s)\frac{\pi_1(s)\pi_2(s)}{a_1\pi_1(s)+a_2\pi_2(s)}\{d\Lambda_1(s)-d\Lambda_2(s)\} > 0.$$

The next result makes this conjecture precise.

Theorem 7.3.2. Consider a stochastic ordering alternative, and let W_K and K be defined as in Theorem 7.2.1. When $W(t)$ is $f(\hat{S}(t-))$ or $f(\hat{\pi}(t))$, then set $w(t)$ to be $f(S^*(t-))$ or $f(a_1\pi_1(t)+a_2\pi_2(t))$, respectively, where S^* denotes the limit to which $\hat{S}$ converges in probability.

Suppose $a_i \in (0,1)$ and

$$\int_0^\infty w(s)\frac{\pi_1(s)\pi_2(s)}{a_1\pi_1(s)+a_2\pi_2(s)}\{d\Lambda_1(s)-d\Lambda_2(s)\} > 0. \tag{3.4}$$

Then $W_K(\infty)/\{\hat{\sigma}^2(\infty)\}^{1/2}$ is consistent, where $\hat{\sigma}^2(\infty)$ is defined in (2.6). □

Suppose w is strictly decreasing. By integration of parts, Inequality (3.4) is equivalent to

$$-\int_0^\infty \{\Lambda_1(s)-\Lambda_2(s)\}d\left\{w(s)\frac{\pi_1(s)\pi_2(s)}{a_1\pi_1(s)+a_2\pi_2(s)}\right\} > 0. \tag{3.5}$$

By Inequality (3.5), it easily follows that statistics W_K, with weight functions $W(t) = f\{\hat{S}(t-)\}$ or $W(t) = f\{\hat{\pi}(t)\}$ with f strictly increasing on $[0,1]$, will be consistent against stochastic ordering as well as ordered hazards alternatives under the weak conditions of Theorem 7.3.1. In the G^ρ statistic, $f(t) = t^\rho$ so these statistics produce tests which are nearly always consistent against stochastic ordering. On the other hand, if $f(t) = t^\rho(1-t)^\gamma$, which includes the $G^{\rho,\gamma}$ statistics, it is easy to construct examples in which consistency under stochastic ordering fails when $\gamma > 0$.

7.4 EFFICIENCIES OF TESTS OF THE CLASS $\mathcal{K}$

Tests based on statistics in $\mathcal{K}$ which are consistent against ordered hazards and stochastic ordering alternatives all have power converging to unity as $n \to \infty$ for any fixed alternative of these two types. As a result, a more refined measure of the asymptotic operating characteristics must be used to discriminate among these tests. The concept of asymptotic relative efficiency plays a central role in the asymptotic theory of hypothesis testing, and is discussed in the context of nonparametric tests by Hájek and Šidák (1967). A careful study of asymptotic efficiencies requires more time and mathematical rigor than is appropriate here, and we only describe briefly the use of asymptotic efficiencies in studying censored data rank tests.

The underlying idea is not difficult. When a test has a power function increasing to one for any fixed alternative as $n \to \infty$, the distribution of an approximately normally distributed statistic must converge to point mass at ∞ under the alternative. It is sometimes possible to show, however, that under a sequence of alternative hypotheses converging to the null hypothesis at the right rate as $n \to \infty$, the distribution of a statistic has, asymptotically, a finite mean and positive variance. The asymptotic distribution under the sequence of alternatives then provides approximate operating characteristics of a test in large samples and for alternatives "close" to the null hypothesis. Theorem 6.2.1 can be used to obtain the asymptotic distribution of censored data rank test statistics under sequences of alternative hypotheses.

Equation (3.1) helps motivate the alternatives used to study asymptotic relative efficiencies in censored data. Suppose F_i^n varies with both i and n, and assume

$$\sup_{0 \le t < \infty} |F_i^n(t) - F(t)| \to 0 \quad \text{as } n \to \infty,$$

for a distribution function F with respect to which F_i^n is absolutely continuous. Equation (3.1) now becomes

$$\begin{aligned} W_K &= \int \frac{K}{\overline{Y}_1} dM_1 - \frac{K}{\overline{Y}_2} dM_2 + \int K(d\Lambda_1^n - d\Lambda) - \int K(d\Lambda_2^n - d\Lambda) \\ &= \int \frac{K}{\overline{Y}_1} dM_1 - \frac{K}{\overline{Y}_2} dM_2 + \int K\left(\frac{d\Lambda_1^n}{d\Lambda} - 1\right) d\Lambda \\ &\quad - \int K\left(\frac{d\Lambda_2^n}{d\Lambda} - 1\right) d\Lambda. \end{aligned} \tag{4.1}$$

When F_i^n and F have hazard functions λ_i^n and λ,

$$\frac{d\Lambda_i^n}{d\Lambda}(t) = \frac{\lambda_i^n(t)}{\lambda(t)}.$$

Under the hypotheses of Theorem 6.2.1, with $K^2/\overline{Y}_i = H_i^2 \overline{Y}_i \to h_i$,

$$\int \frac{K}{\overline{Y}_1} dM_1 - \int \frac{K}{\overline{Y}_2} dM_2 \Longrightarrow Z^\infty, \tag{4.2}$$

where $EZ^\infty(t) = 0$ and

$$\text{var}\{Z^\infty(t)\} = \sigma^2(t) = \int_0^t (h_1 + h_2)(1 - \Delta\Lambda) d\Lambda.$$

Suppose for a sequence of alternatives $\{F_1^n, F_2^n : n = 1, 2, \ldots\}$ that, for $i = 1, 2$,

$$\int_0^t K\left(\frac{d\Lambda_i^n}{d\Lambda} - 1\right) d\Lambda \to r_i(t) \tag{4.3}$$

for some function r_i, where the convergence is uniform in probability over $[0,t]$ for any $t \in \mathcal{I}$. It then follows from Eqs. (4.1), (4.2), and (4.3) that, under $\{F_1^n, F_2^n\}$,

$$W_K(\cdot) \xrightarrow{\mathcal{D}} N(r_1(\cdot) - r_2(\cdot), \sigma^2(\cdot)).$$

This result is exploited later in comparing the relative asymptotic power under $\{F_1^n, F_2^n\}$ for any two statistics.

A study of asymptotic relative efficiency in this setting, then, requires a sequence $\{F_1^n, F_2^n\}$ of alternatives in which Eq. (4.3) is true. In Eq. (3.3),

$$\left(\frac{n_1+n_2}{n_1 n_2}\right)^{1/2} K \xrightarrow{P} k,$$

and the convergence is uniform whenever $W \to w$ uniformly in probability. Hence, if there exist real valued functions γ_i such that

$$\left(\frac{n_1 n_2}{n_1+n_2}\right)^{1/2} \left(\frac{d\Lambda_i^n}{d\Lambda} - 1\right) \to \gamma_i \tag{4.4}$$

uniformly in probability, Eq. (4.3) should hold with $r_i(t) = \int_0^t k\gamma_i \, d\Lambda$.

Theorem 7.4.1. Consider a test statistic in $\mathcal{K}$,

$$W_K = \int K \left(\frac{d\overline{N}_1}{\overline{Y}_1} - \frac{d\overline{N}_2}{\overline{Y}_2}\right).$$

In addition to Assumption 6.2.1, suppose that each of the following conditions is satisfied:

$$\sup_{0 \le t < \infty} |F_i^n(t) - F(t)| \to 0 \quad \text{as } n \to \infty \tag{4.5}$$

for a distribution function F with respect to which each F_i^n is absolutely continuous;

$$\left(\frac{n_1 n_2}{n_1+n_2}\right)^{1/2} \left(\frac{d\Lambda_i^n}{d\Lambda}(t) - 1\right) \to \gamma_i(t) \quad \text{as } n \to \infty \tag{4.6}$$

uniformly on each closed subinterval of $\{t : F(t-) < 1\}$, where γ_i is a real valued function;

$$\left(\frac{n_1+n_2}{n_1 n_2}\right)^{1/2} K(t) \to k(t) \quad \text{as } n \to \infty \tag{4.7}$$

uniformly in probability on closed subintervals of $\mathcal{I}$, where k is left-continuous with right-hand limits, k^+ is of bounded variation on closed subintervals of $\mathcal{I}$, and $k = 0$ outside $\mathcal{I}$; and

$$\int_0^t |\gamma_i| \, d\Lambda < \infty \quad \text{for all } t \in \mathcal{I}, i = 1, 2. \tag{4.8}$$

Then, as $n \to \infty$, and for all $t \in \mathcal{I}$,

$$W_K(t) \xrightarrow{\mathcal{D}} N\left(\int_0^t k\gamma d\Lambda, \int_0^t \frac{a_1\pi_1 + a_2\pi_2}{\pi_1\pi_2} k^2(1 - \Delta\Lambda)d\Lambda\right), \tag{4.9}$$

where $\gamma = \gamma_1 - \gamma_2$. If $\hat{\sigma}^2(t)$ is defined by (2.6),

$$\hat{\sigma}^2(t) \xrightarrow{P} \sigma^2(t) = \int_0^t \frac{a_1\pi_1 + a_2\pi_2}{\pi_1\pi_2} k^2(1 - \Delta\Lambda)d\Lambda. \tag{4.10}$$

If $u \notin \mathcal{I}$, then, for $i = 1, 2$, assume in addition Inequality (2.8), Condition (2) of Theorem 6.2.1, and:

$$\int_{\mathcal{I}} |k\gamma_i| d\Lambda < \infty,$$

and

$$\lim_{t \uparrow u} \limsup_{n \to \infty} P\left\{\int_{t-}^{u} |K|\,|d\Lambda_i^n - d\Lambda| > \epsilon\right\} = 0$$

for any $\epsilon > 0$. Then, as $n \to \infty$, Eqs. (4.9) and (4.10) also hold at $t = u$.

If $u < \infty$, then for $i = 1, 2$, assume in addition Inequality (2.9), Condition (3) of Theorem 6.2.1, and:

$$\int_u^\infty |K|\,|d\Lambda_i^n - d\Lambda| \xrightarrow{P} 0 \quad \text{as } n \to \infty.$$

Then, as $n \to \infty$, Eqs. (4.9) and (4.10) hold for any $t \in [0, \infty]$. □

Theorem 7.4.1 follows from Theorem 6.2.1 and Lemma 7.2.1.

Consider a sequence of alternatives $\{F_1^n, F_2^n\}$ satisfying the conditions in Theorem 7.4.1 and two statistics W_{K_1} and W_{K_2}. The statistic with the higher power against alternatives in this sequence should have the larger value (for $\ell = 1, 2$) of the noncentrality parameter

$$\frac{|\int k_\ell \gamma d\Lambda|}{\left\{\int (a_1\pi_1 + a_2\pi_2)(\pi_1\pi_2)^{-1} k_\ell^2 (1 - \Delta\Lambda) d\Lambda\right\}^{1/2}}.$$

For such a sequence of alternatives, the asymptotic efficacy of $W_K(\infty)/\{\hat{\sigma}^2(\infty)\}^{1/2}$ is defined to be

$$e(k, \infty) = \frac{\left(\int_0^\infty k\gamma\ d\Lambda\right)^2}{\int_0^\infty (a_1\pi_1 + a_2\pi_2)(\pi_1\pi_2)^{-1} k^2(1 - \Delta\Lambda)d\Lambda}. \tag{4.11}$$

Under appropriate conditions, the Pitman asymptotic relative efficiency (ARE) of W_{K_1} with respect to W_{K_2} is $e(k_1, \infty)/e(k_2, \infty)$.

The remainder of this section outlines how statistics from the class $\mathcal{K}$ are selected to maximize efficacy and discusses the ARE of this "best test" of the class $\mathcal{K}$ with respect to the likelihood ratio test.

The two important components in $e(k, \infty)$ are γ and k. The function γ depends only on the sequence $\{F_1^n, F_2^n\}$, while k depends only on the statistic W_K. Gill (1980) has shown that the statistic in $\mathcal{K}$ maximizing $e(k, \infty)$ for a sequence $\{F_1^n, F_2^n\}$ can be found by "solving for" k as a function of γ. Specifically the function k which maximizes $e(k, \infty)$ over all possible k must be proportional to

$$\frac{\pi_1 \pi_2}{a_1 \pi_1 + a_2 \pi_2} \frac{\gamma}{1 - \Delta\Lambda} \tag{4.12}$$

on $[0, \infty]$. For this choice of k (call it k^*), Eq. (4.11) reduces to

$$e(k^*, \infty) = \int_0^\infty \frac{\pi_1 \pi_2}{a_1 \pi_1 + a_2 \pi_2} \frac{\gamma^2}{(1 - \Delta\Lambda)} d\Lambda.$$

In some cases, there are simple expressions for γ for sequences of parametric alternatives. Let $\{F_\theta : \theta \in \Theta\}$ be a family of continuous distribution functions on $[0, \infty)$ indexed by a parameter $\theta \in \Theta$, where Θ is an interval of the real line. Let $\Lambda_\theta = -\log S_\theta$ and $\lambda_\theta(t) = d\Lambda_\theta(t)/dt$. Let $F_i^n = F_{\theta_i^n}$ for any i, n, and $F = F_{\theta_0}$, where $\theta_i^n, \theta_0 \in \Theta$. If $F_i^n \to F$, then $\theta_i^n \to \theta_0$. Suppose λ is differentiable with respect to θ at $\theta = \theta_0$. By Taylor's Theorem with remainder,

$$\begin{aligned} \lambda_{\theta_i^n} &= \lambda_{\theta_0} + \frac{\partial}{\partial\theta}\lambda_\theta \bigg|_{\theta=\theta^*} (\theta_i^n - \theta_0) \\ &\approx \lambda_{\theta_0} + (\theta_i^n - \theta_0) \frac{\partial}{\partial\theta}\lambda_\theta \bigg|_{\theta=\theta_0} . \end{aligned} \tag{4.13}$$

By Eqs. (4.6) and (4.13),

$$\begin{aligned} \gamma_i(t) &= \lim_{n\to\infty} \left(\frac{n_1 n_2}{n_1 + n_2}\right)^{1/2} \frac{\lambda_{\theta_i^n}(t) - \lambda_{\theta_0}(t)}{\lambda_{\theta_0}(t)} \\ &= \frac{1}{\lambda_{\theta_0}(t)} \left(\frac{\partial}{\partial\theta} \lambda_\theta(t) \bigg|_{\theta=\theta_0} \right) \lim_{n\to\infty} \left(\frac{n_1 n_2}{n_1 + n_2}\right)^{1/2} (\theta_i^n - \theta_0). \end{aligned}$$

Assume for some constant c,

$$\theta_i^n = \theta_0 + (-1)^{i+1} c \left\{ \frac{n_{i'}}{n_i(n_1 + n_2)} \right\}^{1/2}, \quad i' \neq i. \tag{4.14}$$

Then,

$$\gamma_i(t) = \frac{1}{\lambda_{\theta_0}(t)} \left(\frac{\partial}{\partial\theta} \lambda_\theta(t) \bigg|_{\theta=\theta_0} \right) c(-1)^{i+1} \lim_{n\to\infty} \frac{n_{i'}}{(n_1 + n_2)},$$

so

$$\gamma(t) = \gamma_1(t) - \gamma_2(t) = c\frac{\partial}{\partial\theta}\log\lambda_\theta(t)\bigg|_{\theta=\theta_0}. \quad (4.15)$$

Equations (4.12) and (4.15) imply that statistics with maximal asymptotic efficacy against parametric alternatives will have limiting weight functions k proportional to

$$\left(\frac{\partial}{\partial\theta}\log\lambda_\theta\bigg|_{\theta=\theta_0}\right)\left(\frac{\pi_1\pi_2}{a_1\pi_1 + a_2\pi_2}\right). \quad (4.16)$$

We will give the exact expression for these statistics under "time transformed location alternatives" to specified error distributions.

Before doing this, we can now display the Pitman ARE of the test of the class $\mathcal{K}$ satisfying (4.16) with respect to the most powerful test against the contiguous alternatives $\{F_{\theta_i^n}\}$ defined by Eq. (4.14). Specifically, the ARE of the optimal test in $\mathcal{K}$, whose efficacy is given by

$$\int_0^\infty \left(\frac{\partial}{\partial\theta}\log\lambda_\theta\bigg|_{\theta=\theta_0}\right)^2 \frac{\pi_1\pi_2}{a_1\pi_1 + a_2\pi_2}d\Lambda,$$

relative to the most powerful test (i.e., the likelihood ratio test) against the contiguous alternatives defined by Eq. (4.14), is

$$\frac{\int_0^\infty \left(\frac{\partial}{\partial\theta}\log\lambda_\theta\,|_{\theta=\theta_0}\right)^2 \frac{\pi_1\pi_2}{a_1\pi_1+a_2\pi_2}d\Lambda,}{\int_0^\infty \left(\frac{\partial}{\partial\theta}\log\lambda_\theta\,|_{\theta=\theta_0}\right)^2 (a_1\pi_1 + a_2\pi_2)d\Lambda} \leq 1.$$

The test of the class $\mathcal{K}$ with limiting weight function proportional to (4.16) will be fully efficient if and only if $\pi_1 = \pi_2$.

A test of the class $\mathcal{K}$ satisfying (4.16), derived to provide an efficient test relative to the sequence $(\lambda_{\theta_1^n}, \lambda_{\theta_2^n})$ for θ_i^n converging to θ_0 as specified by Eq. (4.14), can also be obtained as a partial likelihood score statistic. The derivation is explored in Exercise 7.5.

We now construct statistics W_K with limiting weight functions proportional to (4.16) when

$$F_\theta(t) = \Psi\{g(t) + \theta\}, \qquad t \in [0, \infty), \quad (4.17)$$

$\theta \in \Theta = (-\infty, \infty)$, where g is a differentiable nondecreasing function from $[0, \infty]$ to $[-\infty, \infty]$ and Ψ is a continuous distribution function with positive density ψ, having a derivative ψ' continuous at all but finitely many points. The family $\{F_\theta(t) : -\infty < \theta < \infty\}$ consists of location alternatives to the time-transformed distribution Ψ. Typically, Ψ may be a distribution such as the extreme value or logistic, and the time transformation g is often the log function. It is not difficult to find the test statistics in class $\mathcal{K}$ with limiting weight functions proportional to (4.16), and hence with an optimal limiting weight function.

Let $\lambda = \psi/(1-\Psi)$ and $\ell = \log\lambda$. Since $\lambda_\theta(t) = \lambda\{g(t)+\theta\}g'(t)$,

$$\begin{aligned}
\frac{\partial}{\partial\theta}\log\lambda_\theta(t)\bigg|_{\theta=\theta_0} &= \frac{\partial}{\partial\theta}\ell\{g(t)+\theta\}\bigg|_{\theta=\theta_0} \\
&= \ell'\{g(t)+\theta_0\} \\
&= \ell'[\Psi^{-1}\{F_{\theta_0}(t)\}]. \qquad (4.18)
\end{aligned}$$

If the weight function K is

$$K = K_{\text{opt}} = \left(\frac{n_1 n_2}{n_1+n_2}\right)^{1/2} \ell'\{\Psi^{-1}(\hat{F}^-)\}\frac{\overline{Y}_1}{n_1}\frac{\overline{Y}_2}{n_2}\frac{n_1+n_2}{\overline{Y}_1+\overline{Y}_2}, \qquad (4.19)$$

then, under H_0

$$\begin{aligned}
\left(\frac{n_1+n_2}{n_1 n_2}\right)^{1/2} K_{\text{opt}}(t) &\xrightarrow{P} \ell'[\Psi^{-1}\{F_{\theta_0}(t)\}]\frac{\pi_1(t)\pi_2(t)}{a_1\pi_1(t)+a_2\pi_2(t)} \\
&= \frac{\partial}{\partial\theta}\log\lambda_\theta(t)\bigg|_{\theta=\theta_0}\frac{\pi_1(t)\pi_2(t)}{a_1\pi_1(t)+a_2\pi_2(t)}.
\end{aligned}$$

Consequently, W_K has maximal efficacy.

Example 7.4.1. Consider alternatives indexed by location parameters θ_i^n satisfying Eq. (4.14), and let $F_i^n(t) = \Psi\{g(t)+\theta_i^n\}$. We solve for K_{opt} in Eq. (4.19) for various "error" distributions, Ψ.

1. Suppose $\Psi(t) = 1-e^{-e^t}$, the extreme value distribution. Then $\psi(t) = e^t e^{-e^t}$, $\lambda(t) = e^t$, and $\ell(t) = \log\lambda(t) = t$. Since $\ell'(t) = 1$,

$$K = K_{\text{opt}} = \left(\frac{n_1 n_2}{n_1+n_2}\right)^{1/2}\frac{\overline{Y}_1}{n_1}\frac{\overline{Y}_2}{n_2}\frac{n_1+n_2}{\overline{Y}_1+\overline{Y}_2}.$$

   Therefore, by Eq. (2.4), the logrank weight function provides a fully-efficient statistic.

2. Suppose $\Psi(t) = (1+e^{-t})^{-1}$, the logistic distribution. Then $\psi(t) = (1+e^{-t})^{-2}e^{-t}$, so $\lambda(t) = (1+e^{-t})^{-1}$ and $\ell(t) = -\log(1+e^{-t})$. Thus $\ell'(t) = e^{-t}(1+e^{-t})^{-1} = 1-\Psi(t)$, and

$$\ell'\{\Psi^{-1}(\hat{F}^-)\} = 1-\Psi\{\Psi^{-1}(\hat{F}^-)\} = 1-\hat{F}^- = \hat{S}^-,$$

   so

$$K = K_{\text{opt}} = \left(\frac{n_1 n_2}{n_1+n_2}\right)^{1/2}\hat{S}^-\frac{\overline{Y}_1}{n_1}\frac{\overline{Y}_2}{n_2}\frac{n_1+n_2}{\overline{Y}_1+\overline{Y}_2}.$$

Therefore, by Eq. (2.4), the Prentice-Wilcoxon weight function provides a fully-efficient statistic. □

It is possible to work in the reverse direction from this example, by beginning with a test statistic W_K, and computing the error distribution Ψ for which W_K is fully efficient.

Example 7.4.2. Consider a G^ρ statistic with

$$K^{\rho,0} = \left(\frac{n_1 n_2}{n_1+n_2}\right)^{1/2} (\hat{S}^-)^\rho \frac{\overline{Y}_1}{n_1}\frac{\overline{Y}_2}{n_2}\frac{n_1+n_2}{\overline{Y}_1+\overline{Y}_2}.$$

Equation (4.19) suggests Ψ should satisfy

$$\ell'\{\Psi^{-1}(\hat{F}^-)\} = (1-\hat{F}^-)^\rho.$$

A sufficient condition ensuring this is

$$\ell'(t) = \{1-\Psi(t)\}^\rho. \tag{4.20}$$

Since $\ell = \log\lambda = \log\psi - \log(1-\Psi)$,

$$\begin{aligned}\ell'(t) &= \frac{\psi'(t)}{\psi(t)} + \lambda(t)\\ &= \frac{\Psi''(t)}{\Psi'(t)} + \frac{\Psi'(t)}{1-\Psi(t)}\\ &= \frac{H''(t)}{H'(t)} - \frac{H'(t)}{H(t)},\end{aligned}$$

where $H(t) = 1-\Psi(t)$. Substitution into Eq. (4.20) yields the differential equation

$$\frac{H''(t)}{H'(t)} - \frac{H'(t)}{H(t)} = \{H(t)\}^\rho. \tag{4.21}$$

Solving Eq. (4.21), Harrington and Fleming (1982) obtain

$$H(t) \equiv H_\rho(t) = \begin{cases} (1+\rho e^t)^{-1/\rho} & \text{if } \rho > 0\\ \exp(-e^t) & \text{if } \rho = 0\,. \end{cases}$$

Interestingly, $\lim_{\rho\to 0} H_\rho(t) = H_0(t)$. □

The results in Example 7.4.2 generalize those in Example 7.4.1. The G^ρ statistics and the sequence of alternatives $\{F_1^n, F_2^n\}$ in these two examples do satisfy the assumptions of Theorem 7.4.1. The next theorem provides a precise statement.

Theorem 7.4.2. Let $-\infty < t < \infty$ and let $H_\rho(t)$ be the family of survival functions given by

$$\begin{aligned}H_0(t) &= \exp(-e^t) \text{ if } \rho = 0,\\ H_\rho(t) &= (1+\rho e^t)^{-1/\rho} \text{ if } \rho > 0.\end{aligned}$$

Let $S^{\rho}(t,\theta) = H_{\rho}\{g(t)+\theta\}$ and let $S^{\rho}(t,\theta_i^n)$ $(i = 1,2)$ be a sequence of location alternatives, with θ_i^n defined by (4.14). Let $\rho \geq 0$ be fixed and known, and z_{α} be the α quantile of a standard normal distribution.

The level-α test which rejects $H_0^n : \theta_1^n = \theta_2^n$ in favor of $H_1^n : \theta_1^n \neq \theta_2^n$ whenever

$$\frac{\left|\int_0^{\infty}(\hat{S}-)^{\rho}\frac{\overline{Y}_1\overline{Y}_2}{\overline{Y}_1+\overline{Y}_2}\left(\frac{d\overline{N}_1}{\overline{Y}_1}-\frac{d\overline{N}_2}{\overline{Y}_2}\right)\right|}{\left\{\int_0^{\infty}(\hat{S}-)^{2\rho}\frac{\overline{Y}_1\overline{Y}_2}{\overline{Y}_1+\overline{Y}_2}\left(1-\frac{\Delta\overline{N}_1+\Delta\overline{N}_2-1}{\overline{Y}_1+\overline{Y}_2-1}\right)\frac{d(\overline{N}_1+\overline{N}_2)}{\overline{Y}_1+\overline{Y}_2}\right\}^{1/2}} > z_{1-\alpha/2}$$

is a fully-efficient test against time-transformed location alternatives to $H_{\rho}(t)$ if and only if $\pi_1 = \pi_2$ a.s. with respect to the probability measure specified by H_{ρ}.

□

The transformation g plays an important role in the generality of Theorem 7.4.2. Consider the logrank statistic, G^0, and suppose $g(t) \equiv t$. Then Theorem 7.4.2 implies that G^0 is efficient in detecting the alternative

$$\bar{S}_1(t) \equiv H_0(t+\theta_1) = \exp(-e^{t+\theta_1}), \quad \bar{S}_2(t) \equiv H_0(t+\theta_2) = \exp(-e^{t+\theta_2}), \quad (4.22)$$

where $\theta_1 > \theta_2$. However the monotonic differentiable transformation $g(t) = \ln\{-\ln S_1(t)\} - \theta_1$ from $[0,\infty]$ to $[-\infty,\infty]$ maps $\bar{S}_1(t)$ into any survival distribution $S_1(t)$. Then if $\Delta \equiv \theta_2 - \theta_1$ (so $\Delta < 0$),

$$\begin{aligned} S_2(t) \equiv \bar{S}_2(g(t)) &= \exp[-e^{\{g(t)+\theta_2\}}] = \exp[-\{-\log S_1(t)\}e^{\Delta}] \\ &= \{S_1(t)\}^{e^{\Delta}}. \end{aligned}$$

Since $S_i(t) = e^{-\Lambda_i(t)}$, this implies

$$\lambda_2(t) = \lambda_1(t)e^{\Delta}. \quad (4.23)$$

The logrank statistic G^0 is thus efficient not only against shifts in the extreme value distribution but against proportional hazards alternatives in general.

Now let $\rho > 0$. By a similar argument (Exercise 7.4), G^{ρ} is efficient against

$$\lambda_2(t) = \lambda_1(t)e^{\Delta}[\{S_1(t)\}^{\rho} + [1-\{S_1(t)\}^{\rho}]e^{\Delta}]^{-1}, \quad (4.24)$$

for arbitrary λ_1. Equation (4.24) reduces to Eq. (4.23) when $\rho = 0$. A G^{ρ} statistic then (for $\rho > 0$) is efficient against alternatives in which the hazard ratio decreases monotonically from e^{Δ} to unity. This is expected since the G^{ρ} weight function $W(s) = \{\hat{S}(s-)\}^{\rho}$ decreases monotonically. The type of alternative characterized by Eq. (4.24) is biologically plausible in many applications since the effect of an intervention or of a covariate on the failure rate often deteriorates or becomes diluted as time passes. The parameter Δ indicates the strength of the effect while the parameter ρ is related to the rate at which the effect diminishes.

Simulation studies show that the asymptotic relative efficiencies of the G^ρ statistics are consistent with small sample power properties. For simplicity, we present here some simulation results for uncensored data and shift alternatives of the form $H_\rho(\log t + \theta_i), i = 1, 2, \theta_1 \neq \theta_2$. For a fixed value ρ^*, the alternative to equal survival distributions for two groups is

$$S_1^{\rho^*}(t) = H_{\rho^*}(\log t + \theta_1), \quad S_2^{\rho^*}(t) = H_{\rho^*}(\log t + \theta_2). \tag{4.25}$$

Table 7.4.1 shows results for four such pairs corresponding to $\rho^* = 0, 1/2, 1$, and 2. Each ρ^* has a different pair of constants, θ_1, θ_2. For each of these four alternatives, we have simulated the small- and moderate-sized sample power of the G^ρ statistics for $\rho = 0$ (logrank), $\rho = 1/2$, $\rho = 1$ (Wilcoxon), and $\rho = 2$.

As noted at the bottom of Table 7.4.1, the Pitman ARE of G^ρ with respect to G^{ρ^*} for the alternatives against which G^{ρ^*} is fully efficient is $(2\rho + 1)(2\rho^* + 1)/(\rho + \rho^* + 1)^2$. The table shows that small sample relative power is consistent with ARE results.

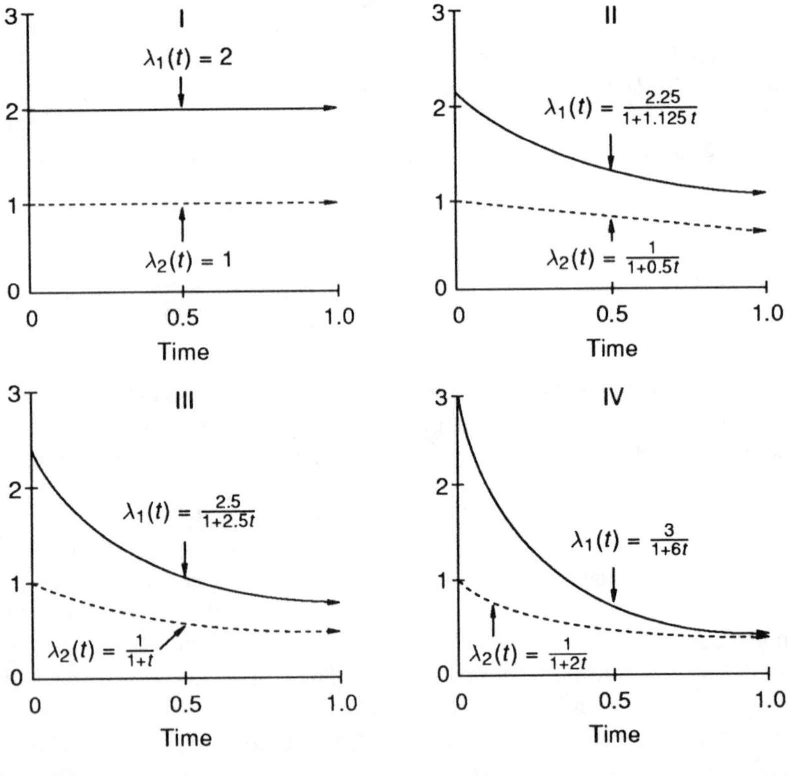

Figure 7.4.1 Hazard functions used in the simulations presented in Table 7.4.1

Table 7.4.1 Monte Carlo Estimates of Power for G^{ρ} Test Statistics

$G^{\rho}(\rho = 0, \frac{1}{2}, 1, 2,)$ $\alpha = .05$ level one-sided test procedures of $H_0 : S_1 = S_2$ vs. $H_A : S_2 > S_1$; $n_1 = n_2$ (Uncensored data)

		Proportional hazards $\rho^* = 0$	$\rho^* = \frac{1}{2}$	Time-transformed logistic shift $\rho^* = 1$	$\rho^* = 2$
$(e^{\theta_1}, e^{\theta_2})$:		I:(2, 1)	II:(2.25, 1)	III:(2.50, 1)	IV:(3, 1)
	$n_1 = n_2$				
$\rho = 0$	20	.668	.548	.444	.336
logrank	50	.954	.844	.754	.534
		(1.00)	(.889)	(.750)	(.556)
$\rho = \frac{1}{2}$	20	.620	.576	.470	.402
	50	.938	.878	.834	.662
		(.889)	(1.00)	(.960)	(.816)
$\rho = 1$	20	.578	.564	.488	.416
Wilcoxon	50	.894	.868	.864	.722
		(.750)	(.960)	(1.00)	(.938)
$\rho = 2$	20	.456	.516	.470	.426
	50	.812	.830	.828	.742
		(.556)	(.816)	(.938)	(1.00)

Numbers in parenthesis give Pitman ARE: $(2\rho + 1)(2\rho^* + 1)/(\rho + \rho^* + 1)^2$ for G^{ρ} relative to G^{ρ^*} under time-transformed location alternatives for $H_{\rho*}$.

7.5 SOME VERSATILE TEST PROCEDURES

In many applied situations, investigators are not able to specify in advance the type of the survival differences that may exist between two groups. Versatile test procedures sensitive to a range of types of survival differences can be useful in such situations. The previous section describes efficient tests for detecting specific local alternatives. However, as seen from efficiency calculations, simulations in Table 7.4.1, and other investigations such as those of Fleming, Harrington and O'Sullivan (1987), the power of these tests may not be robust against different types of alternative hypotheses.

This section discusses two approaches to obtaining procedures with robust power, beginning with procedures using clusters of linear rank statistics, then examining statistics which are supremum versions of the statistics in $\mathcal{K}$.

The next theorem provides the large sample joint distribution of L statistics of the class $\mathcal{K}$. The result is used in procedures based upon the simultaneous use of several statistics.

Theorem 7.5.1. Consider L statistics of the class $\mathcal{K}$,

$$W_{K_\ell} = \int_0^\infty K_\ell(s)\left\{\frac{d\overline{N}_1(s)}{\overline{Y}_1(s)} - \frac{d\overline{N}_2(s)}{\overline{Y}_2(s)}\right\}, \quad \ell = 1, \ldots . L.$$

Suppose each satisfies all the conditions of Corollary 7.2.1. Then under $H_0 : F_1^n = F_2^n = F$ for all n,

$$(W_{K_1}, \ldots, W_{K_L}) \xrightarrow{\mathcal{D}} (Z_1^\infty, \ldots, Z_L^\infty),$$

where $(Z_1^\infty, \ldots, Z_L^\infty)$ is a mean zero L-variate Gaussian random vector. If k_ℓ denotes the limit in probability of $\{(n_1 + n_2)/(n_1 n_2)\}^{1/2} K_\ell$, then

$$\begin{aligned}\sigma_{k,\ell} &\equiv \text{cov}(Z_k^\infty, Z_\ell^\infty)\\ &= \int_0^\infty k_k(s) k_\ell(s)\left\{\frac{a_1\pi_1(s) + a_2\pi_2(s)}{\pi_1(s)\pi_2(s)}\right\}\{1 - \Delta\Lambda(s)\} d\Lambda(s). \quad (5.1)\end{aligned}$$

If

$$\hat{\sigma}_{k,\ell} = \int_0^\infty K_k K_\ell \left(\frac{\overline{Y}_1 + \overline{Y}_2}{\overline{Y}_1 \overline{Y}_2}\right)\left(1 - \frac{\Delta\overline{N}_1 + \Delta\overline{N}_2 - 1}{\overline{Y}_1 + \overline{Y}_2 - 1}\right)\frac{d(\overline{N}_1 + \overline{N}_2)}{\overline{Y}_1 + \overline{Y}_2}, \quad (5.2)$$

then

$$|\hat{\sigma}_{k,\ell} - \sigma_{k,\ell}| \xrightarrow{P} 0 \text{ as } n \to \infty.$$

□

When

$$K_\ell = \left(\frac{n_1 n_2}{n_1 + n_2}\right)^{1/2} W_\ell \frac{\overline{Y}_1}{n_1} \frac{\overline{Y}_2}{n_2} \frac{n_1 + n_2}{\overline{Y}_1 + \overline{Y}_2},$$

where W_ℓ converges in probability to w_ℓ,

$$k_\ell = w_\ell \frac{\pi_1 \pi_2}{a_1\pi_1 + a_2\pi_2}.$$

Thus, Eq. (5.1) reduces to

$$\sigma_{k,\ell} = \int_0^\infty w_k(s) w_\ell(s)\left\{\frac{\pi_1(s)\pi_2(s)}{a_1\pi_1(s) + a_2\pi_2(s)}\right\}\{1 - \Delta\Lambda(s)\} d\Lambda(s), \quad (5.3)$$

and Eq. (5.2) to

$$\begin{aligned}\hat{\sigma}_{k,\ell} = \int_0^\infty W_k W_\ell &\frac{\overline{Y}_1}{n_1} \frac{\overline{Y}_2}{n_2} \frac{n_1 + n_2}{\overline{Y}_1 + \overline{Y}_2}\\ &\left(1 - \frac{\Delta\overline{N}_1 + \Delta\overline{N}_2 - 1}{\overline{Y}_1 + \overline{Y}_2 - 1}\right)\frac{d(\overline{N}_1 + \overline{N}_2)}{\overline{Y}_1 + \overline{Y}_2}. \quad (5.4)\end{aligned}$$

For the statistics in Theorem 7.2.1, $W(t) = f\{\hat{S}(t-)\}$ or $W(t) = f\{\hat{\pi}(t)\}$, so $w(t) = f\{S(t-)\}$ or $w(t) = f\{a_1\pi_1(t) + a_2\pi_2(t)\}$, respectively. Theorem 7.5.1 implies, for example, that the asymptotic joint distribution of the Prentice-Wilcoxon statistic W_{K_1} (with $W_1 = \hat{S}(t-)$) and the Gehan-Wilcoxon statistic W_{K_2} (with $W_2 = \hat{\pi}(t)$) is mean-zero bivariate normal with covariance

$$\sigma_{12} = \int_0^\infty S(s-)\pi_1(s)\pi_2(s)\{1 - \Delta\Lambda(s)\}d\Lambda(s).$$

Proof of Theorem 7.5.1. To establish the multivariate convergence, it is sufficient by the Cramer-Wold device to prove

$$\sum_{\ell=1}^{L} b_\ell W_{K_\ell} \xrightarrow{\mathcal{D}} \sum_{\ell=1}^{L} b_\ell Z_\ell^\infty, \tag{5.5}$$

for any set of L constants $(b_1, \ldots, b_L)$. Observe that

$$\sum_{\ell=1}^{L} b_\ell W_{K_\ell} = \int_0^\infty K_*(s)\left\{\frac{d\overline{N}_1(s)}{\overline{Y}_1(s)} - \frac{d\overline{N}_2(s)}{\overline{Y}_2(s)}\right\},$$

where $K_*(s) = \sum_{\ell=1}^{L} b_\ell K_\ell(s)$.

If $k_*(s) = \sum_{\ell=1}^{L} b_\ell k_\ell(s)$, the hypotheses of Corollary 7.2.1 are true when K_ℓ and k_ℓ are replaced by K_* and k_*. It follows immediately from that corollary that

$$\sum_{\ell=1}^{L} b_\ell W_{K_\ell} \xrightarrow{\mathcal{D}} Z_*^\infty \sim N(0, \sigma_*^2), \tag{5.6}$$

where

$$\begin{aligned}\sigma_*^2 &= \int_0^\infty \{k_*(s)\}^2 \left\{\frac{a_1\pi_1(s) + a_2\pi_2(s)}{\pi_1(s)\pi_2(s)}\right\}\{1 - \Delta\Lambda(s)\}d\Lambda(s),\\ &= \int_0^\infty \left\{\sum_{\ell=1}^{L} b_\ell k_\ell(s)\right\}^2 \frac{a_1\pi_1(s) + a_2\pi_2(s)}{\pi_1(s)\pi_2(s)}\{1 - \Delta\Lambda(s)\}d\Lambda(s).\end{aligned} \tag{5.7}$$

The limit (5.5) follows from Eqs. (5.6) and (5.7) if

$$\text{var}\left(\sum_{\ell=1}^{L} b_\ell Z_\ell^\infty\right) = \sigma_*^2.$$

To verify this, note

$$\begin{aligned}
\operatorname{var}\left(\sum_{\ell=1}^{L} b_\ell Z_\ell^\infty\right) &= \sum_{\ell=1}^{L}\sum_{k=1}^{L} b_\ell b_k \operatorname{cov}(Z_\ell^\infty, Z_k^\infty) \\
&= \int_0^\infty \sum_{\ell=1}^{L}\sum_{k=1}^{L} b_\ell b_k k_\ell k_k \left(\frac{a_1\pi_1 + a_2\pi_2}{\pi_1\pi_2}\right)(1-\Delta\Lambda)d\Lambda \\
&= \int_0^\infty \left(\sum_{\ell=1}^{L} b_\ell k_\ell\right)^2 \left(\frac{a_1\pi_1 + a_2\pi_2}{\pi_1\pi_2}\right)(1-\Delta\Lambda)d\Lambda \\
&= \sigma_*^2.
\end{aligned}$$

Lemma 7.2.1 implies that $\hat{\sigma}_{k,\ell}$ in Eq. (5.2) is a consistent estimator of $\sigma_{k,\ell}$ in (5.1). □

Theorem 7.5.1 can be used in test procedures based on the value of the vector $\{W_{K_1}, \ldots, W_{K_L}\}$. Suppose one computes the standardized statistics $\{W_{K_\ell}/(\hat{\sigma}_{\ell,\ell})^{1/2} : \ell = 1, \ldots, L\}$ and corresponding p-values $\{p_\ell : \ell = 1, \ldots, L\}$, then rejects $H_0 : F_1 = F_2$ if $\min\{p_\ell : \ell = 1, \ldots, L\} \leq p(\alpha)$, where $p(\alpha)$ satisfies $P\{\min\{p_\ell : \ell = 1, \ldots, L\} \leq p(\alpha)\} = \alpha$ under H_0. Theorem 7.5.1 implies that

$$\max\{(\hat{\sigma}_{11})^{-1/2} W_{K_1}, \ldots, (\hat{\sigma}_{LL})^{-1/2} W_{K_L}\}$$

is distributed approximately as $\max\{Z_1, \ldots, Z_L\}$, where $(Z_1, \ldots, Z_L)$ is a mean zero L-variate Gaussian vector, with $\operatorname{corr}(Z_k, Z_\ell)$ consistently estimated by $\hat{\sigma}_{k,\ell}(\hat{\sigma}_{k,k}\hat{\sigma}_{\ell,\ell})^{-1/2}$.

For any c, $P\{\max\{Z_1, \ldots, Z_L\} \leq c\}$ can be computed using a program for calculating multivariate normal probabilities. One can then find $c(\alpha)$ such that $P\{\max\{Z_1, \ldots, Z_L\} \geq c(\alpha)\} = \alpha$, and, in turn, solve for $p(\alpha)$, since $P\{Z \geq c(\alpha)\} = p(\alpha)$ where $Z \sim N(0, 1)$.

Insight can be obtained into the power properties of procedures based upon clusters of statistics in $\mathcal{K}$ by studying their large-sample correlations under varying degrees of censorship. The correlation between W_{K_ℓ} and W_{K_k},

$$\operatorname{corr}(W_{K_\ell}, W_{K_k}) = \frac{\sigma_{k,\ell}}{\sqrt{\sigma_{k,k} * \sigma_{\ell,\ell}}},$$

follows from Eq. (5.3). Simple expressions for this correlation are obtained by assuming $1 - L_{ij}^n = 1 - L_i = C_i = (S)^\delta$ where the constant $\delta \geq 0$. While C is rarely a power of S, values of δ help infer qualitative information about the effect of censorship on the correlation of statistics in $\mathcal{K}$.

For example, consider two statistics G^{ρ_1} and G^{ρ_2}, so $w_1(s) = \{S(s-)\}^{\rho_1}$ and $w_2(s) = \{S(s-)\}^{\rho_2}$. By straightforward calculation, then

$$\operatorname{corr}(G^{\rho_1}, G^{\rho_2}) = \frac{\{(2\rho_1 + 1 + \delta)(2\rho_2 + 1 + \delta)\}^{1/2}}{(\rho_1 + \rho_2 + 1 + \delta)},$$

so G^ρ statistics are highly correlated. Thus, a procedure based upon a cluster of G^ρ statistics (say, for $\rho \in [0, 2]$) would not be much more versatile than the use of the Wilcoxon (G^1) alone.

As expected, the correlation between the two G^ρ statistics converges to unity as censorship increases. Another interesting observation is that $G^{1/2}$ is not a balanced compromise between the Wilcoxon and logrank statistics. Indeed, $G^{1/2}$ is closer to the Wilcoxon than to the logrank since $\text{corr}(G^{1/2}, G^1) > \text{corr}(G^{1/2}, G^0)$.

If W_K is in $\mathcal{K}$, then yet another test of H_0 can be based on

$$\widetilde{W}_K = \sup_{0 \le t \le T} \frac{W_K(t)}{\hat{\sigma}(T)},$$

where, as before, $T = \sup\{t : \overline{Y}_1(t) \wedge \overline{Y}_2(t) > 0\}$.

A test based on $\widetilde{W}_K$ should be more sensitive than W_K to large early survival differences, especially if such differences disappear. Furthermore, $\widetilde{W}_K$ should still be sensitive to alternatives to H_0 against which W_K is efficient. We show below $\widetilde{W}_K \xrightarrow{\mathcal{D}} \sup_{0 \le t \le 1} W_0(t)$, where W_0 is Brownian motion. Since, for $x > 0$,

$$P\left\{ \sup_{0 \le t \le 1} W_0(t) > x \right\} = 2P\{W_0(1) > x\},$$

a size α test based on $\widetilde{W}_K$ will be at least as powerful as a size $\alpha/2$ test based on W_K. This lower bound is conservative in most applications.

The next result is used for one- and two-sided procedures based on $\widetilde{W}_K$ and on the cluster $(W_K, \widetilde{W}_K)$, and follows directly from the Continuous Mapping Theorem (Appendix B, Theorem B.1.1) and from Corollary 7.2.1.

Theorem 7.5.2. Assume the conditions of Corollary 7.2.1 are satisfied for a statistic W_K in $\mathcal{K}$. Let W_0 represent Brownian Motion, and $\hat{\sigma}^2$ be as defined in Eq. (2.6). Then

$$\widetilde{W}_K(\infty) \equiv \sup_{0 \le t \le \infty} \frac{W_K(t)}{\{\hat{\sigma}^2(\infty)\}^{1/2}} \xrightarrow{\mathcal{D}} \sup_{t \in A} W_0(t), \tag{5.8}$$

$$\widetilde{\widetilde{W}}_K(\infty) \equiv \sup_{0 \le t \le \infty} \left| \frac{W_K(t)}{\{\hat{\sigma}^2(\infty)\}^{1/2}} \right| \xrightarrow{\mathcal{D}} \sup_{t \in A} |W_0(t)|, \tag{5.9}$$

$$\left(\frac{W_K(\infty)}{\{\hat{\sigma}^2(\infty)\}^{1/2}}, \widetilde{W}_K(\infty) \right) \xrightarrow{\mathcal{D}} \left(W_0(1), \sup_{t \in A} W_0(t) \right), \tag{5.10}$$

and

$$\left(\frac{|W_K(\infty)|}{\{\hat{\sigma}^2(\infty)\}^{1/2}}, \widetilde{\widetilde{W}}_K(\infty) \right) \xrightarrow{\mathcal{D}} \left(|W_0(1)|, \sup_{t \in A} |W_0(t)| \right), \tag{5.11}$$

where

$$A = \{\sigma^2(t)/\sigma^2(\infty) : 0 \le t \le \infty\}.$$

□

If $1 - F = S$ is continuous, then so is $\sigma^2 = \int (h_1 + h_2)(1 - \Delta\Lambda)d\Lambda$. Then $A = [0, 1]$ and the following result yields expressions for the asymptotic distributions in Eqs. (5.8) through (5.11). Otherwise $A \subseteq [0, 1]$ so that the result yields a conservative approximation to the asymptotic distributions of tests based on (5.8) through (5.11).

Lemma 7.5.1. Let W_0 represent Brownian motion. Then

$$P\left\{\sup_{0 \le t \le 1} W_0(t) > y\right\} = 2\{1 - \Phi(y)\}, \tag{5.12}$$

$$P\left\{\sup_{0 \le t \le 1} |W_0(t)| > y\right\} = 1 - \frac{4}{\pi}\sum_{k=0}^{\infty} \frac{(-1)^k}{2k+1} \exp\{-\pi^2(2k+1)^2/8y^2\}, \tag{5.13}$$

$$P\left\{W_0(1) \le x,\ \sup_{0 \le t \le 1} W_0(t) \le y\right\} = -1 + \Phi(x) + \Phi(2y - x) \tag{5.14}$$

for $x \le y$ and $y \ge 0$, and

$$\begin{aligned} P\left\{|W_0(1)| \le x,\ \sup_{0 \le t \le 1} |W_0(t)| \le y\right\} &= \sum_{k=-\infty}^{\infty} \{\Phi(4ky + x) - \Phi(4ky - x)\} \\ &- \sum_{k=-\infty}^{\infty} \{\Phi(2y + x + 4ky) - \Phi(2y - x + 4ky)\} \end{aligned} \tag{5.15}$$

for $0 \le x \le y$, where Φ is a standard normal distribution function.

Proof. These relationships are presented in Billingsley (1968, Section 11).

□

Suppose now that $W(t) = f\{\hat{S}(t-)\}$ or $W(t) = f\{\hat{\pi}(t)\}$, and that $S = 1 - F$ is continuous. Then Eqs. (5.8) and (5.12) and the proof of Theorem 7.2.1 provide the large sample distribution for the one-sided supremum statistic $\widetilde{W_K}(\infty)$, while Eqs. (5.9) and (5.13) give the large sample distribution for the two-sided statistic. Fleming, Harrington, and O'Sullivan (1987) have found that the supremum versions of the logrank ($W(t) = 1$) and Wilcoxon ($W(t) = \hat{S}(t-)$) statistics are more versatile than their linear rank analogues in uncensored or moderately censored data.

Results from Theorem 7.5.2 and Lemma 7.5.1 make possible α-level test procedures based on the simultaneous use of a statistic and its supremum version. Let $X \equiv W_K(\infty)/\{\hat{\sigma}^2(\infty)\}^{1/2}$ denote a one-sided statistic in $\mathcal{K}$ satisfying the conditions of Corollary 7.2.1, and let $Y \equiv \widetilde{W}_K$ be its supremum version. When S is continuous, Lemma 7.5.1 implies that, for any x, y such that $x \leq y$ and $y \geq 0$,

$$P\{X \leq x, Y \leq y\} = -1 + \Phi(x) + \Phi(2y - x). \tag{5.16}$$

This joint distribution of X and Y provides an easy method for constructing a one-sided α-level procedure based on the value of (X, Y). If (x_0, y_0) satisfies

$$-1 + \Phi(x_0) + \Phi(2y_0 - x_0) = 1 - \alpha,$$

then a test which rejects H_0 whenever either $X > x_0$ or $Y > y_0$ will have size α. The smaller the value of x_0, the greater the weight given to the X statistic of the (X, Y) pair. Full weight is given to the X statistic at the minimum value of x_0, where $x_0 = \Phi^{-1}(1 - \alpha)$ and $y_0 = \infty$. At the maximum value for x_0 (i.e., at $x_0 = y_0 = \Phi^{-1}(1 - \alpha/2)$), total weight is given to the Y statistic.

Suppose for example that an investigator computes one-sided p values for both X and Y, denoted by p_x and p_y, then rejects H_0 when the minimum p value is less than $p(\alpha)$, where $p(\alpha)$ satisfies $P[\min(p_x, p_y) < p(\alpha)] = \alpha$. Using (5.16),

$$\alpha = p(\alpha) + \Phi[2\Phi^{-1}\{p(\alpha)/2\} - \Phi^{-1}\{p(\alpha)\}]. \tag{5.17}$$

Replacing $p(\alpha)$ on the right-hand side of Eq.(5.17) by $\min(p_x, p_y)$ provides an "adjusted" one-sided p value for the test based on (X, Y). Conversely, Eq. (5.17) can be used to solve for $p(\alpha)$ for a specified α. For α = .05 and .01, the corresponding $p(\alpha)$ are .0406 and .0081, respectively.

Consequently, a .05-level one-sided test procedure is obtained if H_0 is rejected whenever the one-sided p value from either a W_K statistic or its supremum analogue is less than .04, regardless of censoring patterns or degree of censorship, relative sample sizes, or the weight function, W.

Two-sided test procedures can also be based on a two-sided statistic and its supremum version. Again denote these by X and Y and suppose that S is continuous. Then, by Lemma 7.5.1, for any x, y such that $0 \leq x \leq y$,

$$P\{X \leq x, Y \leq y\} = \sum_{k=-\infty}^{\infty} \{\Phi(4ky + x) - \Phi(4ky - x)\} - \sum_{k=-\infty}^{\infty} \{\Phi(2y + x + 4ky) - \Phi(2y - x + 4ky)\}. \tag{5.18}$$

Suppose that a two-sided procedure is based on the minimum two-sided p value; or

$$\min\{2\{1 - \Phi(x)\}, 1 - H(y)\} \equiv p,$$

where $H(y) \equiv P\{\sup_{0\le t\le 1} |W_0(t)| \le y\}$, (for tabled values of H see, e.g., Schumacher 1984). One minus the adjusted p value is obtained by replacing x and y in Eq. (5.18) by $\Phi^{-1}(1-p/2)$ and $H^{-1}(1-p)$, respectively. For values of $p \le .10$, the infinite sum in Eq. (5.18) is accurately approximated by $1-2\{\Phi(-x)+\Phi(-2y+x)\}$. Working in the opposite direction, for two-sided values of α = .05 and .01, $p(\alpha)$ = .0405 and .0080, respectively.

7.6 BIBLIOGRAPHIC NOTES

Section 7.1

Many textbooks provide detailed presentations of statistical methods based on ranks. These include Hollander and Wolfe (1973) and Lehmann (1975), and mathematically advanced works by Lehmann (1959), Hájek and Šidák (1967), Randles and Wolfe (1979), and Hettmansperger (1984). Kalbfleisch and Prentice (1980, Chapter 6) and Gill (1980) provide thorough presentations of rank based methods for censored survival data.

Section 2

Gill (1980, Section 4.3) established uniform consistency of two types of variance estimators for tests of the class $\mathcal{K}$, thereby proving a more general result than that given by Lemma 7.2.1. He also established very weak conditions under which asymptotic normality holds for the logrank statistic and the generalized Wilcoxon statistics of Gehan (1965) and Efron (1967).

A k-sample logrank statistic was proposed and studied by Peto and Peto (1972) and by Crowley and Thomas (1975). Censored data generalizations of the multi-sample Wilcoxon (i.e., Kruskal-Wallis) statistic were proposed by Prentice (1978) and by Breslow (1970), with the latter author providing a multi-sample version of Gehan's (1965) statistic. One-sample logrank statistics were proposed by Breslow (1975), and by Hyde (1977), while Harrington and Fleming (1982) provided the one-sample G^ρ statistics. Andersen, Borgan, Gill and Keiding (1982) surveyed linear nonparametric one- and multi-sample tests for counting processes, and presented asymptotic results for censored data linear rank tests under very general censoring patterns.

Several authors have considered the survival analysis of censored matched pair data. Holt and Prentice (1974), Breslow (1975), and Kalbfleisch and Prentice (1980, Section 8.1) fit a proportional hazards model

$$\lambda_\ell(t|\mathbf{Z}) = \lambda_{0\ell}(t)\exp(\boldsymbol{\beta}'\mathbf{Z}),$$

where the ℓth pair has its own baseline hazard function. When the covariate vector $\mathbf{Z}$ simply denotes group membership for each component of the matched pair in a two group problem, the partial likelihood score statistic reduces to the censored data Sign test of Armitage (1959). These authors also consider log-linear failure time models for paired data. O'Brien and Fleming (1987) propose a class of censored paired data statistics by indicating how classical linear rank statistics, generalized to censored data by Prentice (1978), in turn can be generalized to paired censored data. These paired data statistics make use of interblock information. One member of this class is the paired Prentice-Wilcoxon (PPW) statistic, which can be viewed as a censored data generalization of the Conover-Iman (1981) "paired t-test on the ranks" or of the Lam-Longnecker (1983) modified Wilcoxon rank-sum statistic. The PPW provides an alternative to Wei's (1980) paired data version of Gehan's (1965) statistic, while the O'Brien-Fleming paired data version of the logrank statistic is similar to a statistic proposed by Mantel and Ciminera (1979). O'Brien and Fleming evaluate the relative properties of the PPW and Sign statistics, and of Woolson and Lachenbruch's (1980) censored data generalization of the Wilcoxon signed rank statistic.

Sections 7.3 and 7.4

The theorems in Section 7.3 relating to consistency against ordered hazards and stochastic ordering alternatives follow directly from results in Gill (1980, Section 4.1). Chapter 5 of Gill (1980) provides a mathematically detailed development of the efficiency properties of statistics of the class $\mathcal{K}$.

Section 7.5

Classes of Renyi-type statistics for censored survival data have been proposed by Gill (1980), Fleming and Harrington (1981), and Fleming, Harrington and O'Sullivan (1987), with the latter authors providing a comparison of relative properties. Other censored data Kolmogorov-Smirnov-type statistics have been developed by Fleming et al. (1980) and Schumacher (1984). Cramér-Von Mises-type statistics have been proposed by Koziol (1978) and by Schumacher (1984), who also provided a detailed comparison of many of these tests.

Gastwirth (1985) has studied tests based on linear combinations of G^{ρ} statistics that maximize minimum ARE against a range of alternatives.

Recently, a class of procedures intended to be sensitive against the alternative of stochastic ordering have been developed by Pepe and Fleming (1989, 1990). These statistics are based on the integrated weighted difference in Kaplan-Meier estimators, hence are called weighted Kaplan-Meier statistics.

The small sample properties of test procedures based on the use of clusters of linear rank statistics have been investigated by Tarone (1981) and by Fleming and Harrington (1984b).

CHAPTER 8

Distribution Theory for Proportional Hazards Regression

8.1 INTRODUCTION

Apart from technical regularity conditions, large-sample properties of likelihood-based statistics follow from straightforward arguments when data consist of independent observations. If $\mathbf{Y} = (Y_1, \ldots, Y_n)'$ is a vector of observations with likelihood $L(\boldsymbol{\theta})$ depending on a p-dimensional parameter $\boldsymbol{\theta}$, then the score vector at $\boldsymbol{\theta} = \boldsymbol{\theta}_0$, $\mathbf{U}(\boldsymbol{\theta}_0) = \frac{\partial}{\partial \boldsymbol{\theta}} \log L(\boldsymbol{\theta})|_{\boldsymbol{\theta}=\boldsymbol{\theta}_0}$, can be written $\sum_{i=1}^n \mathbf{U}_i(\boldsymbol{\theta}_0)$, where $\mathbf{U}_i(\boldsymbol{\theta}_0) = \frac{\partial}{\partial \boldsymbol{\theta}} \log L_i(\boldsymbol{\theta})|_{\boldsymbol{\theta}=\boldsymbol{\theta}_0}$ is the score from the contribution of Y_i to the likelihood. (To simplify the notation, the dependence of $\mathbf{U}$ and L on $\mathbf{Y}$ has been suppressed, as has the dependence of $\mathbf{U}_i$ and L_i on Y_i.) The $\mathbf{U}_i$ are independent, have mean zero when the true parameter is $\boldsymbol{\theta}_0$ and, under sufficient conditions for the Lindeberg-Feller Multivariate Central Limit Theorem, $n^{-1/2}\mathbf{U}(\boldsymbol{\theta}_0)$ has approximately a mean zero p-variate normal distribution with covariance matrix $\boldsymbol{\Sigma}_{\boldsymbol{\theta}_0}$, whose j, kth term is

$$\begin{aligned}
\sigma_{jk}(\boldsymbol{\theta}_0) &= E\left[\left\{n^{-1/2}\frac{\partial}{\partial \theta_j} \log L(\boldsymbol{\theta})\bigg|_{\boldsymbol{\theta}=\boldsymbol{\theta}_0}\right\}\left\{n^{-1/2}\frac{\partial}{\partial \theta_k} \log L(\boldsymbol{\theta})\bigg|_{\boldsymbol{\theta}=\boldsymbol{\theta}_0}\right\}\right] \\
&= -\frac{1}{n}E\left\{\frac{\partial^2}{\partial \theta_j \partial \theta_k} \log L(\boldsymbol{\theta})\bigg|_{\boldsymbol{\theta}=\boldsymbol{\theta}_0}\right\} \\
&= -\frac{1}{n}\sum_{i=1}^n E\left\{\frac{\partial^2}{\partial \theta_i \partial \theta_j} \log L_i(\boldsymbol{\theta})\bigg|_{\boldsymbol{\theta}=\boldsymbol{\theta}_0}\right\}.
\end{aligned}$$

When the Y_i are also identically distributed, σ_{jk} will not depend on n and, by the Strong Law of Large Numbers, this constant will be the limiting value of

$$\hat{\sigma}_{jk}(\boldsymbol{\theta}_0) = -\frac{1}{n}\sum_{i=1}^n \frac{\partial^2}{\partial \theta_j \partial \theta_k} \log L_i(\boldsymbol{\theta})\bigg|_{\boldsymbol{\theta}=\boldsymbol{\theta}_0}.$$

For concave likelihoods L, the maximum likelihood estimator $\hat{\theta}$ for θ will be the solution to the p score equations $\mathbf{U}(\theta) = \mathbf{0}$. The asymptotic distribution properties of $\hat{\theta}$ are suggested by examining a first order Taylor series expansion for $\mathbf{U}$ centered at the true value θ_0 of θ. The score $\mathbf{U}$ may be written

$$\mathbf{U}(\hat{\theta}) = \mathbf{U}(\theta_0) - \mathcal{I}(\theta^*)(\hat{\theta} - \theta_0)$$

where $\mathcal{I}(\theta^*)$ is the observed information matrix at θ^* with j, kth element $n\hat{\sigma}_{jk}(\theta^*)$, and θ^* is on a line segment between θ and θ_0. Since $\mathbf{U}(\hat{\theta}) = \mathbf{0}$, we have

$$n^{-1/2}\mathbf{U}(\theta_0) = n^{-1}\mathcal{I}(\theta^*)n^{1/2}(\hat{\theta} - \theta_0),$$

and $n^{1/2}(\hat{\theta} - \theta_0)$ will be asymptotically multivariate normal with mean zero whenever $n^{-1/2}\mathbf{U}(\theta_0)$ converges in distribution to a mean-zero multivariate normal and $n^{-1}\mathcal{I}(\theta^*)$ converges in probability or almost surely to a nonsingular (constant) matrix (more precisely, whenever the elements of $n^{-1}\mathcal{I}(\theta^*)$ converge to the elements of a nonsingular matrix). If $\hat{\theta}$ is consistent, i.e., if $\hat{\theta}$ converges in probability to θ_0 as $n \to \infty$, and if $n^{-1}\mathcal{I}(\theta_0)$ converges in probability to Σ_{θ_0}, the approximate covariance matrix of $n^{1/2}(\hat{\theta} - \theta_0)$ will be

$$\begin{aligned}\text{cov}\left\{n\mathcal{I}^{-1}(\hat{\theta})n^{-1/2}\mathbf{U}(\hat{\theta})\right\} &\approx n\mathcal{I}^{-1}(\hat{\theta})\textstyle\sum_{\hat{\theta}} n\mathcal{I}^{-1}(\hat{\theta})' \\ &\approx n\mathcal{I}^{-1}(\hat{\theta}),\end{aligned}$$

when n is large.

Loosely speaking, $n^{1/2}(\hat{\theta}_j - \theta_{0j})$ should be approximately normal with variance $\hat{\sigma}^{jj}(\hat{\theta})$, where $n^{-1}\hat{\sigma}^{jj}$ is the jth diagonal element in the inverse of the observed information matrix.

Although we have used these results in the examples in Section 4.4, an important step in the above argument breaks down for partial likelihoods and must be examined more closely. As pointed out in Section 4.3, the partial likelihood score statistic $\mathbf{U}(\beta, t)$ (Equation 3.24 in Chapter 4) is not a sum of independent terms. The martingale representation (3.25 in Chapter 4) of the score does, however, imply that $\mathbf{U}(\beta, t)$ is a sum of uncorrelated terms, and will allow the general argument sketched above to be used.

Throughout this chapter, we will assume the multiplicative intensity model of Definition 4.2.1. The next section establishes the asymptotic normality for the partial likelihood score vector $\mathbf{U}(\beta, t)$, for appropriate values of t, using Theorem 5.3.5. In Section 8.3, these results are then used to derive the asymptotic distribution of the maximum partial likelihood estimator $\hat{\beta}$ and for the estimator $\hat{\Lambda}_0$ of the baseline cumulative hazard function. Our approach here closely follows that of Andersen and Gill (1982), who were among the first to apply martingale methods to censored data regression models. The proofs in Sections 8.2 and 8.3 apply to the general multiplicative intensity model specified in Definition 4.2.1. Section 8.4 explores a simpler form of sufficient conditions for convergence in less general situations, such as the proportional hazards regression model for independent, identically distributed failure time observations. Section 8.5 examines the asymptotic efficiency of partial likelihood estimates of regression coefficients in the proportional hazards model. The asymptotic distribution theory derived here and in Chapters 6 and 7 is, of course, for a sequence of statistics and probability models indexed by increasing sample size n, and this dependence on n was reflected explicitly in the notation used in Theorem 5.3.5. For simplicity, we will usually suppress the dependence on n in the notation that follows, except in situations where n plays an explicit role as a normalizing factor.

8.2 THE PARTIAL LIKELIHOOD SCORE STATISTIC

The notation used here will be the same as in Section 4.3.

Let β_0 denote the true value of a p-dimensional regression parameter in the multiplicative intensity model. For any scalar y, let $||y|| = |y|$, and for a vector $\mathbf{y}$ or matrix $\mathbf{Y}$ let $||\mathbf{y}|| = \max_i |(\mathbf{y})_i|$ and $||\mathbf{Y}|| = \max_{i,j} |(\mathbf{Y})_{ij}|$. For any vector $\mathbf{y}$, $|\mathbf{y}|$ will denote the Euclidean norm

$$|\mathbf{y}| = (\Sigma y_i^2)^{1/2}.$$

A closed interval $[\mathbf{a}, \mathbf{b}]$ of p-dimensional Euclidean space R^p will be the set of points $\{\mathbf{x} \in R^p : a_i \leq x_i \leq b_i, i = 1, \ldots, p\}$.

Definition 8.2.1. Let $\{f_\alpha : \alpha \in \mathcal{A}\}$ be a family of real-valued functions from an interval $[\mathbf{a}, \mathbf{b}]$ of R^p to the real line R. The family is said to be *equicontinuous* at $\mathbf{x} \in [\mathbf{a}, \mathbf{b}]$ if, for any $\epsilon > 0$, there is a $\delta > 0$ such that $|\mathbf{x} - \mathbf{y}| < \delta$ implies $|f_\alpha(\mathbf{x}) - f_\alpha(\mathbf{y})| < \epsilon$, for all $\alpha \in \mathcal{A}$. The family is called equicontinuous on $[\mathbf{a}, \mathbf{b}]$ if it is equicontinuous at each $\mathbf{x} \in [\mathbf{a}, \mathbf{b}]$. □

We begin with some conditions that will be sufficient to imply Eqs. (3.17) and (3.18) in Theorem 5.3.5. For convenience, these conditions are numbered on the left-hand margin.

(2.1) The time τ is such that $\int_0^\tau \lambda_0(x)dx < \infty$.

(2.2) For $S^{(j)}, j = 0, 1$, and 2, defined in (3.18)–(3.20) of Chapter 4, there exists a neighborhood $\mathcal{B}$ of β_0 and, respectively, scalar, vector, and matrix functions $s^{(0)}, \mathbf{s}^{(1)}$ and $\mathbf{s}^{(2)}$ defined on $\mathcal{B} \times [0, \tau]$ such that, for $j = 0, 1, 2$,

$$\sup_{x \in [0,\tau], \beta \in \mathcal{B},} ||\mathbf{S}^{(j)}(\beta, x) - \mathbf{s}^{(j)}(\beta, x)|| \to 0$$

in probability as $n \to \infty$.

(2.3) There exists a $\delta > 0$ such that

$$n^{-1/2} \sup_{\substack{1 \leq i \leq n \\ 0 \leq x \leq \tau}} |\mathbf{Z}_i(x)| Y_i(x) I_{\{\beta_0' \mathbf{Z}_i(x) > -\delta|\mathbf{Z}_i(x)|\}} \to 0$$

in probability as $n \to \infty$.

(2.4) Let $\mathcal{B}$ and $\mathbf{s}^{(j)}, j = 0, 1, 2$, be as defined in (2.2), and let

$$\mathbf{e} = \mathbf{s}^{(1)}/s^{(0)}$$

and

$$\mathbf{v} = \frac{\mathbf{s}^{(2)}}{s^{(0)}} - \mathbf{e}^{\otimes 2}.$$

Then, for all $\beta \in \mathcal{B}$ and $0 \le x \le \tau$,

$$\frac{\partial}{\partial \beta} s^{(0)}(\beta, x) = \mathbf{s}^{(1)}(\beta, x)$$

and

$$\frac{\partial^2}{\partial \beta^2} s^{(0)}(\beta, x) = \mathbf{s}^{(2)}(\beta, x).$$

(2.5) The functions $\mathbf{s}^{(j)}$ are bounded and $s^{(0)}$ is bounded away from 0 on $\mathcal{B} \times [0, \tau]$; for $j = 0, 1, 2$, the family of functions $\mathbf{s}^{(j)}(\cdot, x), 0 \le x \le \tau$, is an equicontinuous family at β_0.

(2.6) The matrix

$$\Sigma(\beta_0, \tau) = \int_0^\tau \mathbf{v}(\beta_0, x) s^{(0)}(\beta_0, x) \lambda_0(x) dx$$

is positive definite.

We will show later how these conditions can be checked in some specific situations, and relaxed in some other cases. Briefly, Condition (2.5) – that $s^{(0)}$ be bounded away from 0 in $\mathcal{B} \times [0, \tau]$ – and (2.1) provide the most significant constraints. Analogous to results under Condition (1) of Theorem 6.2.1, weak convergence results obtained under these conditions are restricted in most settings to sub-intervals of the support of the distribution for the data, often excluding the right-hand tail. These restrictions are removed in some special cases in Section 8.4. Condition (2.3) is important in establishing the Lindeberg condition, Eq. (3.18) of Chapter 5, and trivially holds when covariates are bounded. Condition (2.2) is an asymptotic stability condition for the functions $\mathbf{S}^{(j)}$, while Conditions (2.4)–(2.6) are regularity conditions similar to those found in standard asymptotic likelihood theory. Since Condition (2.4) holds for the functions $\mathbf{S}^{(j)}$, it simply insures that differentiation with respect to β and limits with respect to n may be interchanged.

The main result about the score vector can now be stated. The vector $\mathbf{E}$ and matrix $\mathbf{V}$ in Theorem 8.2.1 are defined in Eqs. (3.22) and (3.23) of Chapter 4, respectively.

Theorem 8.2.1. When Conditions (2.1)–(2.6) are added to the multiplicative intensity model of Definition 4.2.1:

1. The normalized vector score process $\{n^{-1/2}\mathbf{U}(\beta_0, t) : 0 \le t \le \tau\}$, whose value at time t is

$$n^{-1/2}\mathbf{U}(\beta_0, t) = n^{-1/2} \sum_{i=1}^n \int_0^t \{\mathbf{Z}_i(x) - \mathbf{E}(\beta_0, x)\} dN_i(x),$$

converges weakly in $D[0,\tau]^p$ to a mean zero p-variate Gaussian process such that

a) each component process has independent increments, and
b) the covariance function at t for components ℓ and ℓ' is

$$\{\Sigma(\beta_0, t)\}_{\ell\ell'} = \int_0^t \{\mathbf{v}(\beta_0, x)\}_{\ell\ell'} s^{(0)}(\beta_0, x)\lambda_0(x)dx.$$

2. If $\hat{\beta}$ is any consistent estimator of β_0, then

$$\sup_{0\le t\le \tau}\left\|\frac{1}{n}\int_0^t \sum_{i=1}^n \mathbf{V}(\hat{\beta}, x)dN_i(x) - \Sigma(\beta_0, t)\right\| \to 0$$

in probability as $n \to \infty$. □

Theorem 8.3.1 establishes that the maximum partial likelihood estimator $\hat{\beta}$ is consistent. Since

$$\int_0^t \sum_{i=1}^n \mathbf{V}(\beta, x)dN_i(x) = \mathcal{I}(\beta, t),$$

Theorem 8.2.1, part (2) states that both $n^{-1}\mathcal{I}(\beta_0, t)$ and $n^{-1}\mathcal{I}(\hat{\beta}, t)$ are uniformly (on $[0,\tau]$) consistent estimators of the variance function $\Sigma(\beta_0, t)$ of the score process at $\beta = \beta_0$.

The proof uses the following results which are special cases of Lenglart's Inequality stated in Theorem 3.4.1 and Corollary 3.4.1.

Lemma 8.2.1. Let N be a univariate counting process with continuous compensator A, let $M = N - A$, and let H be a locally bounded, predictable process. Then, for all $\delta, \rho > 0$ and any $t \ge 0$,

1.

$$P\{N(t) \ge \rho\} \le \frac{\delta}{\rho} + P\{A(t) \ge \delta\},$$

2.

$$P\left\{\sup_{0\le y\le t}\left|\int_0^y H(x)dM(x)\right| \ge \rho\right\} \le \frac{\delta}{\rho^2} + P\left\{\int_0^t H^2(x)dA(x) \ge \delta\right\}.$$

Proof. Part (2) follows immediately from Corollary 3.4.1 by taking the stopping time T in the proof of that corollary to be t, and by taking $\delta = \eta$ and $\rho = \sqrt{\epsilon}$.

Part (1) is established from the original Theorem 3.4.1 in much the same way as the proof of Corollary 3.4.1, and we only provide an outline. As in Corollary 3.4.1, we can find a localizing sequence $\{\tau_k\}$ so that $M(\cdot \wedge \tau_k)$ is a square integrable martingale for each k. For any finite stopping time T, the Optional Stopping Theorem implies that

$$E\{N(T \wedge \tau_k)\} = E\{A(T \wedge \tau_k)\},$$

and, by taking $N(\cdot \wedge \tau_k) = X(\cdot)$ and $A(\cdot \wedge \tau_k) = Y(\cdot)$ in Theorem 3.4.1, then for any k,

$$P\left\{\sup_{s \leq T} N(s \wedge \tau_k) > \rho\right\} \leq \frac{\delta}{\rho} + P\{A(T) > \delta\}.$$

Since $\tau_k \uparrow \infty$ a.s., the term on the left side of the inequality is increasing in k, and hence $\lim_{k\to\infty} P\{\sup_{s\leq T} N(s \wedge \tau_k) > \rho\}$ exists and equals $P\{\sup_{s\leq T} N(s) > \rho\}$. Taking $T = t$ completes the proof. □

Proof of Theorem 8.2.1. Part (1) follows from an application of Theorem 5.3.5, while (2) uses Lenglart's inequality. Let $t \in [0, \tau]$ in all that follows.

1. By expression (3.25), Chapter 4, **U** may be written

$$\mathbf{U}(\beta_0, t) = \sum_{i=1}^{n} \int_0^t \{\mathbf{Z}_i(x) - \mathbf{E}(\beta_0, x)\} dM_i(x),$$

where

$$M_i(t) = N_i(t) - \int_0^t Y_i(x) e^{\beta_0' \mathbf{Z}_i(x)} \lambda_0(x) dx.$$

The ℓth component, $U_\ell(\beta_0, t)$, of the score process can then be written as

$$U_\ell(\beta_0, t) = \sum_{i=1}^{n} \int_0^t \{Z_{i\ell}(x) - E_\ell(\beta_0, x)\} dM_i(x),$$

where

$$E_\ell(\beta_0, x) = \frac{\sum_{j=1}^{n} Y_j(x) Z_{j\ell}(x) \exp\{\beta_0' \mathbf{Z}_j(x)\}}{\sum_{j=1}^{n} Y_j(x) \exp\{\beta_0' \mathbf{Z}_j(x)\}}.$$

Let

$$H_{i\ell}(x) = Z_{i\ell}(x) - E_\ell(\beta_0, x),$$

so that

$$U_\ell(\beta_0, t) = \sum_{i=1}^{n} \int_0^t H_{i\ell}(x) dM_i(x) \tag{2.7}$$

and

$$\mathbf{U}(\beta_0, t) = \sum_{i=1}^{n} \int_0^t \mathbf{H}_i(x) dM_i(x). \tag{2.8}$$

In this representation for the components of **U**, Conditions (1.1)–(1.3) in Chapter 5 all hold by the assumptions imposed in Chapter 4 and need not be checked. The remaining conditions, (3.17) and (3.18) in Chapter 5, involve predictable covariation processes.

Let $\mathbf{U}^{(n)}(\cdot,\cdot) = n^{-1/2}\mathbf{U}(\cdot,\cdot)$. We first note that since no two counting processes N_i jump at the same time, we have that

$$\begin{aligned}\langle U_\ell^{(n)}(\beta_0,\cdot), U_\ell^{(n)}(\beta_0,\cdot)\rangle(t) &\\ &= n^{-1}\sum_{i=1}^{n}\int_0^t H_{i\ell}^2(x)Y_i(x)e^{\beta_0'\mathbf{Z}_i(x)}\lambda_0(x)dx\\ &= \int_0^t n^{-1}\sum_{i=1}^{n}\{Z_{i\ell}(x) - E_\ell(\beta_0,x)\}^2 Y_i(x)e^{\beta_0'\mathbf{Z}_i(x)}\lambda_0(x)dx.\end{aligned}$$

By Eq. (3.23) in Chapter 4, the integrand in the last integral above is the ℓth diagonal element of the matrix $\mathbf{V}(\beta_0,x)S^{(0)}(\beta_0,x)\lambda_0(x)$, and by Conditions (2.1), (2.2), and (2.5), this integral converges in probability to the corresponding diagonal element in

$$\int_0^t \mathbf{v}(\beta_0,x)s^{(0)}(\beta_0,x)\lambda_0(x)dx,$$

thus establishing Condition (3.17) in Chapter 5 for $\ell = \ell'$. When $\ell \neq \ell'$, Condition (3.17) in Chapter 5, follows similarly, after one notes that

$$\langle U_\ell^{(n)}(\beta_0,\cdot), U_{\ell'}^{(n)}(\beta_0,\cdot)\rangle(t)$$

is the element in row ℓ, column ℓ' in the matrix

$$\int_0^t \mathbf{V}(\beta_0,x)S^{(0)}(\beta_0,x)\lambda_0(x)dx,$$

and converges to the corresponding element in

$$\int_0^t \mathbf{v}(\beta_0,x)s^{(0)}(\beta_0,x)\lambda_0(x)dx.$$

It only remains to check the Lindeberg condition, Eq. (3.18) in Chapter 5. Let $\mathbf{H}_i^{(n)} = n^{-1/2}\mathbf{H}_i$. Then

$$U_\ell^{(n)}(\beta_0,t) = \sum_{i=1}^{n}\int_0^t H_{i\ell}^{(n)}(x)dM_i(x)$$

and

$$U_{\ell,\epsilon}^{(n)}(\beta_0,t) = \sum_{i=1}^{n}\int_0^t H_{i\ell}^{(n)}(x)I_{\{|H_{i\ell}^{(n)}(x)|\geq\epsilon\}}dM_i(x),$$

and hence

$$\langle U_{\ell,\epsilon}^{(n)}, U_{\ell,\epsilon}^{(n)}\rangle(t)$$

$$= \sum_{i=1}^{n} \int_0^t \{H_{i\ell}^{(n)}(x)\}^2 I_{\{|H_{i\ell}^{(n)}(x)|\geq\epsilon\}} Y_i(x) e^{\beta_0' \mathbf{Z}_i(x)} \lambda_0(x) dx$$

$$= n^{-1} \sum_{i=1}^{n} \int_0^t \{H_{i\ell}(x)\}^2 I_{\{n^{-1/2}|H_{i\ell}(x)|\geq\epsilon\}} Y_i(x) e^{\beta_0' \mathbf{Z}_i(x)} \lambda_0(x) dx.$$

For any real numbers a and b, one can show

$$|a-b|^2 I_{\{|a-b|>\epsilon\}} \leq 4|a|^2 I_{\{|a|>\epsilon/2\}} + 4|b|^2 I_{\{|b|>\epsilon/2\}},$$

so the last line above is bounded by

$$4n^{-1} \sum_{i=1}^{n} \int_0^{\tau} |Z_{i\ell}(x)|^2 I_{\{n^{-1/2}|Z_{i\ell}(x)|>\epsilon/2\}} Y_i(x) e^{\beta_0' \mathbf{Z}_i(x)} \lambda_0(x) dx$$

$$+4n^{-1} \sum_{i=1}^{n} \int_0^{\tau} |E_\ell(\beta_0, x)|^2 I_{\{n^{-1/2}|E_\ell(\beta_0,x)|>\epsilon/2\}} Y_i(x) e^{\beta_0' \mathbf{Z}_i(x)} \lambda_0(x) dx. \quad (2.9)$$

The second term in expression (2.9) may be written

$$4 \int_0^{\tau} |E_\ell(\beta_0, x)|^2 I_{\{n^{-1/2}|E_\ell(\beta_0,x)|>\epsilon/2\}} S^{(0)}(\beta_0, x) \lambda_0(x) dx.$$

Fix $\epsilon' > 0$. Conditions (2.2) and (2.5) imply that, and for n sufficiently large, there exists a set A_1 with $P\{A_1\} \geq 1 - \epsilon'$ on which

$$I_{\{n^{-1/2}|E_\ell(\beta_0,x)|>\epsilon/2\}} = 0,$$

for all $0 \leq x \leq \tau$. Thus, the second term in expression (2.9) converges to zero in probability.

A more delicate argument is needed for convergence to zero in probability of the first term in (2.9).

We first write

$$\{n^{-1/2}|Z_{i\ell}(x)| > \epsilon/2\} = B_{1i}^{\ell}(x) \bigcup B_{2i}^{\ell}(x),$$

where $B_{1i}^{\ell}(x)$ and $B_{2i}^{\ell}(x)$ are the disjoint events

$$B_{1i}^{\ell}(x) = \{n^{-1/2}|Z_{i\ell}(x)| > \epsilon/2, \beta_0'\mathbf{Z}_i(x) > -\delta|\mathbf{Z}_i(x)|\}$$

and

$$B_{2i}^{\ell}(x) = \{n^{-1/2}|Z_{i\ell}(x)| > \epsilon/2, \beta_0'\mathbf{Z}_i(x) \leq -\delta|\mathbf{Z}_i(x)|\}.$$

Condition (2.3) implies that there is at least one $\delta > 0$ so that, for a fixed $\epsilon' > 0$ and for n sufficiently large, there again exists a set A_2 with $P\{A_2\} \geq 1-\epsilon'$ and on which $I_{\{B_{1_i}^{\ell}(x)\}} Y_i(x) = 0$ for $0 \leq x \leq \tau, 1 \leq i \leq n$. Consequently

$$\int_0^{\tau} n^{-1} \sum_{i=1}^{n} |Z_{i\ell}(x)|^2 I_{\{B_{1_i}^{\ell}(x)\}} Y_i(x) e^{\beta_0' \mathbf{Z}_i(x)} \lambda_0(x) dx \to 0$$

in probability.

To show that

$$\int_0^{\tau} n^{-1} \sum_{i=1}^{n} |Z_{i\ell}(x)|^2 I_{\{B_{2_i}^{\ell}(x)\}} Y_i(x) e^{\beta_0' \mathbf{Z}_i(x)} \lambda_0(x) dx \to 0$$

in probability, we note that this term is bounded by

$$\int_0^{\tau} n^{-1} \sum_{i=1}^{n} |Z_{i\ell}(x)|^2 e^{-\delta |\mathbf{Z}_i(x)|} I_{\{|Z_{i\ell}(x)| > n^{1/2}\epsilon/2\}} \lambda_0(x) dx. \tag{2.10}$$

Since $\lim_{x\to\infty} x^2 e^{-\delta x} = 0$ when $\delta > 0$, for any fixed $\eta > 0$ and all n sufficiently large

$$I_{\{|Z_{i\ell}(x)| > n^{1/2}\epsilon/2\}} |Z_{i\ell}(x)|^2 e^{-\delta |\mathbf{Z}_i(x)|} < \eta \quad \text{for all } i.$$

Thus (2.10) will be bounded by $\eta \int_0^{\tau} \lambda_0(x) dx$, and hence (2.10) must converge to zero in probability as $n \to \infty$.

2. The counting process $\overline{N} = \sum_{i=1}^{n} N_i$ has compensator $\sum_{i=1}^{n} A_i$. Set $M = \overline{N} - \sum_{i=1}^{n} A_i$. By Lemma 8.2.1 (1),

$$\begin{aligned} P\{n^{-1}\overline{N}(\tau) > c\} &\leq \frac{\delta}{c} + P\left\{\int_0^{\tau} n^{-1} \sum_{i=1}^{n} Y_i(x) e^{\beta_0' \mathbf{Z}_i(x)} \lambda_0(x) dx > \delta\right\} \\ &= \frac{\delta}{c} + P\left\{\int_0^{\tau} S^{(0)}(\beta_0, x) \lambda_0(x) dx > \delta\right\}. \end{aligned}$$

By the Strong Law of Large Numbers, $n^{-1}\overline{N}(\tau)$ converges almost surely, and hence $\lim_{n\to\infty} P\{n^{-1}\overline{N}(\tau) > c\}$ exists for any $c < \infty$. By Condition (2.2), the right side of the above inequality converges to

$$\frac{\delta}{c} + P\left\{\int_0^{\tau} s^{(0)}(\beta_0, x) \lambda_0(x) dx > \delta\right\}.$$

Since δ is arbitrary, taking $\delta > \int_0^{\tau} s^{(0)}(\beta_0, x) \lambda_0(x) dx$ shows

$$\lim_{c\uparrow\infty} \lim_{n\to\infty} P\{n^{-1}\overline{N}(\tau) > c\} = 0. \tag{2.11}$$

Note now that for any estimator $\hat{\beta}$

$$\begin{aligned}
&\|n^{-1}\mathcal{I}(\hat{\beta},\tau) - \Sigma\| \\
&\quad \le \left\| \int_0^\tau \{\mathbf{V}(\hat{\beta},x) - \mathbf{v}(\hat{\beta},x)\} n^{-1} d\overline{N}(x) \right\| \\
&\quad + \left\| \int_0^\tau \{\mathbf{v}(\hat{\beta},x) - \mathbf{v}(\beta_0,x)\} n^{-1} d\overline{N}(x) \right\| \qquad (2.12) \\
&\quad + \left\| \int_0^\tau \mathbf{v}(\beta_0,x) n^{-1} \{d\overline{N}(x) - \sum_{i=1}^n Y_i(x) \exp\{\beta_0' \mathbf{Z}_i(x)\} \lambda_0(x) dx\} \right\| \\
&\quad + \left\| \int_0^\tau \mathbf{v}(\beta_0,x) \{S^{(0)}(\beta_0,x) - s^{(0)}(\beta_0,x)\} \lambda_0(x) dx\} \right\| .
\end{aligned}$$

Condition (2.2) and the boundedness of Condition (2.5) imply that, for a consistent estimator $\hat{\beta}$,

$$\sup_{0 \le x \le \tau} \|\mathbf{V}(\hat{\beta},x) - \mathbf{v}(\hat{\beta},x)\| \to 0$$

in probability as $n \to \infty$. This together with Eq. (2.11) establishes that the first term on the right-hand side of Inequality (2.12) converges to zero in probability as $n \to \infty$.

Condition (2.11) and the equicontinuity of the families $\mathbf{s}^{(j)}$ imply that the second term on the right-hand side of (2.12) also converges to zero, and the asymptotic negligibility of the fourth term follows from a direct application of Conditions (2.1), (2.2), and (2.5).

Part (2) of Lemma 8.2.1 implies that, for each pair j, k,

$$\begin{aligned}
&P\left\{ \left| \int_0^\tau \{\mathbf{v}(\beta_0,x)\}_{jk} n^{-1} M(x) \right| > \rho \right\} \\
&\quad \le \frac{\delta}{\rho^2} + P\left\{ n^{-1} \int_0^\tau [\{\mathbf{v}(\beta_0,x)\}_{jk}]^2 S^{(0)}(\beta_0,x) \lambda_0(x) dx > \delta \right\}.
\end{aligned}$$

Conditions (2.1), (2.2) and (2.5) also imply that the left-hand side of this inequality must converge to zero as $n \to \infty$. □

8.3 ESTIMATORS OF THE REGRESSION PARAMETERS AND THE CUMULATIVE HAZARD FUNCTION

The asymptotic properties of the score statistics lead directly to results for maximum partial likelihood estimators (MPLE's) of regression coefficients, and less directly to properties of estimators of the baseline cumulative hazard function. This section presents the details of those results.

We begin with the consistency of the MPLE of β, stating first a result on finding extrema in sequences of concave functions. This extension of a result from standard convexity theory is drawn from the Appendix in Andersen and Gill (1982), and its proof can be found there.

Lemma 8.3.1. Let E be an open convex subset of R^p, and let $F_1, F_2, \ldots,$ be a sequence of random concave functions on E and f a real-valued function on E such that, for all $x \in E$,

$$\lim_{n \to \infty} F_n(x) = f(x)$$

in probability. Then:

1. The function f is concave.
2. For all compact subsets A of E,

$$\sup_{x \in A} |F_n(x) - f(x)| \to 0$$

   in probability, as $n \to \infty$.
3. If F_n has a unique maximum at X_n and f has one at x, then $X_n \to x$ in probability as $n \to \infty$. □

Theorem 8.3.1. Let $\hat{\beta}$ denote the MPLE of β and β_0 the true value of β in the multiplicative intensity model. Then

$$\lim_{n \to \infty} \hat{\beta} = \beta_0$$

in probability, i.e., $\hat{\beta}$ is consistent.

Proof. The proof is a simple argument based on the convergence of the log partial likelihood to a function maximized at $\beta = \beta_0$ and on Lemma 8.3.1. We first establish convergence of the log partial likelihood by expressing a closely related term as a martingale.

Let $nX_n(\beta, \cdot)$ denote the process which, at time t, is the difference in log partial likelihoods over $[0, t]$, evaluated at an arbitrary β and the true value β_0. By Eq. (3.13) in Chapter 4,

$$\begin{aligned} X_n(\beta, t) &= n^{-1}\{\mathcal{L}(\beta, t) - \mathcal{L}(\beta_0, t)\} \\ &= n^{-1}\left[\sum_{i=1}^{n} \int_0^t (\beta - \beta_0)'\mathbf{Z}_i(x) dN_i(x) \right. \\ &\quad \left. - \int_0^t \log\left\{\frac{\sum_{i=1}^n Y_i(x) e^{\beta'\mathbf{Z}_i(x)}}{\sum_{i=1}^n Y_i(x) e^{\beta_0'\mathbf{Z}_i(x)}}\right\} d\overline{N}(x)\right] \end{aligned}$$

If

$$A_n(\beta, t) = n^{-1}\left[\sum_{i=1}^{n}\int_0^t (\beta - \beta_0)'\mathbf{Z}_i(x)Y_i(x)\exp\{\beta_0'\mathbf{Z}_i(x)\}\lambda_0(x)dx \right.$$

$$\left. - \int_0^t \sum_{i=1}^{n} \log\left\{\frac{S^{(0)}(\beta, x)}{S^{(0)}(\beta_0, x)}\right\} Y_i(x)\exp\{\beta_0'\mathbf{Z}_i(x)\}\lambda_0(x)dx\right],$$

then

$$X_n(\beta, t) - A_n(\beta, t)$$

$$= n^{-1}\left[\sum_{i=1}^{n}\int_0^t \left\{(\beta - \beta_0)'\mathbf{Z}_i(x) - \log\frac{S^{(0)}(\beta, x)}{S^{(0)}(\beta_0, x)}\right\} dM_i(x)\right]. \quad (3.1)$$

The process $X_n(\beta, \cdot) - A_n(\beta, \cdot)$ is then a locally square integrable martingale—it is trivial that the integrand in Eq. (3.1) is locally bounded and predictable—with predictable variation process at t

$$\langle X_n(\beta, \cdot) - A_n(\beta, \cdot), X_n(\beta, \cdot) - A_n(\beta, \cdot)\rangle(t)$$

$$= n^{-2}\sum_{i=1}^{n}\int_0^t \left[(\beta - \beta_0)'\mathbf{Z}_i(x) - \log\left\{\frac{S^{(0)}(\beta, x)}{S^{(0)}(\beta_0, x)}\right\}\right]^2 Y_i(x)\exp\{\beta_0'\mathbf{Z}_i(x)\}\lambda_0(x)dx$$

$$= n^{-1}\int_0^t \left[(\beta - \beta_0)'\mathbf{S}^{(2)}(\beta_0, x)(\beta - \beta_0) - 2(\beta - \beta_0)'\mathbf{S}^{(1)}(\beta_0, x)\log\left\{\frac{S^{(0)}(\beta, x)}{S^{(0)}(\beta_0, x)}\right\}\right.$$

$$\left. + \left[\log\left\{\frac{S^{(0)}(\beta, x)}{S^{(0)}(\beta_0, x)}\right\}\right]^2 S^{(0)}(\beta_0, x)\right]\lambda_0(x)dx.$$

By Conditions (2.1), (2.2), and (2.5),

$$n\langle X_n(\beta, \cdot) - A_n(\beta, \cdot), X_n(\beta, \cdot) - A_n(\beta, \cdot)\rangle(\tau)$$

converges to a finite limit, and hence Lemma 8.2.1 (2) implies

$$\lim_{n\to\infty} X_n(\beta, \tau) - A_n(\beta, \tau) = 0$$

in probability. Since $A_n(\beta, \tau)$ converges to

$$A(\beta, \tau) = \int_0^\tau \left[(\beta - \beta_0)'\mathbf{s}^{(1)}(\beta_0, x) - \log\left\{\frac{s^{(0)}(\beta, x)}{s^{(0)}(\beta_0, x)}\right\} s^{(0)}(\beta_0, x)\right]\lambda_0(x)dx,$$

$X_n(\beta, \tau)$ must converge in probability to the same limit, as long as $\beta \in \mathcal{B}$. It is not difficult to show (Exercise 8.1) that $X_n(\beta, \tau)$ is a concave function of β with a unique maximum, and (using Conditions (2.4) and (2.5)) that $A(\beta, \tau)$ has a unique maximum at $\beta = \beta_0$. The Theorem now follows from Lemma 8.3.1 (3). □

The asymptotic normality of $\hat{\beta}$ is now easy to establish.

Theorem 8.3.2. Let $\Sigma(\beta_0, t)$ be as defined in Condition (2.6). Then $n^{1/2}(\hat{\beta} - \beta_0)$ converges in distribution as $n \to \infty$ to a mean zero p-variate Gaussian variable with covariance matrix $\{\Sigma(\beta_0, \tau)\}^{-1}$.

Proof. Recall that for the score statistic $\mathbf{U}$,

$$\mathbf{U}(\hat{\beta}, \tau) = \mathbf{U}(\beta_0, \tau) - \mathcal{I}(\beta^*, \tau)(\hat{\beta} - \beta_0),$$

where β^* is on a line segment between $\hat{\beta}$ and β_0. Since $\mathbf{U}(\hat{\beta}, \tau) = 0$,

$$n^{-1}\mathcal{I}(\beta^*, \tau)n^{1/2}(\hat{\beta} - \beta_0) = n^{-1/2}\mathbf{U}(\beta_0, \tau).$$

Theorem (8.2.1) (1) established that $n^{-1/2}\mathbf{U}(\beta_0, \tau)$ is asymptotically normal with covariance matrix $\Sigma(\beta_0, \tau)$. Since $\hat{\beta}$ is consistent, we have that $\beta^* \to \beta$ in probability as $n \to \infty$, and the proof of Theorem 8.2.1 (2) shows that $\mathcal{I}(\beta^*, \tau)$ converges in probability to the nonsingular matrix $\Sigma(\beta_0, \tau)$. The result now follows from Slutsky's Theorem. □

Complete estimation in the multiplicative intensity model requires an estimate of λ_0 or of the closely related term $\Lambda_0 = \int \lambda_0$. An estimator proposed by Breslow (1972, 1974) for Λ_0 is the one most commonly used, and it resembles the Nelson cumulative hazard rate estimator. A heuristic derivation of Breslow's estimator,

$$\hat{\Lambda}_0(t) = \int_0^t \left\{\sum_{i=1}^n Y_i(x)e^{\hat{\beta}_0'\mathbf{Z}_i(x)}\right\}^{-1} d\overline{N}(x),$$

can be found in Section 4.3.

Johansen (1983) has shown that $\hat{\Lambda}_0$ can be derived as a generalized maximum likelihood estimator, but those arguments are not explored here.

The asymptotic distribution theory for $\hat{\Lambda}_0$ is established using an approach similar to that for the score and the MPLE $\hat{\beta}$. Martingale decompositions are the main tool, and ϵ, δ arguments are required to check the details. The main results are stated here, with only an outline of the proof. Interested readers can supply intermediate steps.

Theorem 8.3.3. Assume as before that Conditions (2.1)–(2.6) hold in the multiplicative intensity model given by Definition 4.2.1. Let $\hat{\Lambda}_0$ be the estimator of the baseline cumulative hazard function Λ_0 given at time t by

$$\hat{\Lambda}_0(t) = \int_0^t \left\{\sum_{i=1}^n Y_i(x)e^{\hat{\beta}'\mathbf{Z}_i(x)}\right\}^{-1} d\overline{N}(x).$$

Then $n^{1/2}\{\hat{\Lambda}_0 - \Lambda_0\}$ converges weakly on $D[0, \tau]$ to a Gaussian process with mean zero, independent increments, and variance function

$$\int_0^t \frac{\lambda_0(x)}{s^{(0)}(\beta_0, x)}dx + \mathbf{Q}'(\beta_0, t)\Sigma^{-1}(\beta_0, \tau)\mathbf{Q}(\beta_0, t),$$

where the vector function $\mathbf{Q}$ is given by

$$\mathbf{Q}(\beta_0, t) = \int_0^t \mathbf{e}(\beta_0, x)\lambda_0(x)dx.$$

Proof (Outline). The decomposition of $\hat{\Lambda}_0 - \Lambda_0$ used in the proof is a natural one:

$$\begin{aligned} & n^{1/2}\{\hat{\Lambda}_0(t) - \Lambda_0(t)\} \\ &= n^{1/2} \int_0^t \left[\left\{ \sum_{i=1}^n Y_i(x) e^{\hat{\beta}'\mathbf{Z}_i(x)} \right\}^{-1} d\overline{N}(x) - \lambda_0(x)dx \right] \\ &= n^{1/2} \int_0^t \left[\left\{ \sum_{i=1}^n Y_i(x) e^{\hat{\beta}'\mathbf{Z}_i(x)} \right\}^{-1} - \left\{ \sum_{i=1}^n Y_i(x) e^{\beta_0'\mathbf{Z}_i(x)} \right\}^{-1} \right] d\overline{N}(x) \\ &\quad + n^{1/2} \left[\int_0^t \left\{ \sum_{i=1}^n Y_i(x) e^{\beta_0'\mathbf{Z}_i(x)} \right\}^{-1} d\overline{N}(x) - \Lambda_0^*(t) \right] \\ &\quad + n^{1/2}\{\Lambda_0^*(t) - \Lambda_0(t)\}, \end{aligned}$$

where

$$\Lambda_0^*(t) = \int_0^t I_{\{\overline{Y}(x)>0\}} \lambda_0(x)dx.$$

The central idea of the proof (again, due to Andersen and Gill, 1982) is to show that the first two terms converge in distribution to asymptotically independent Gaussian processes, while the third term is asymptotically negligible. The asymptotic negligibility of the third term is easy (cf. the proof of Theorem 3.2.1 for a similar result), and the second term is of the form $\int_0^t H(x)dM(x)$. Martingale methods applied earlier are easily used to show that the second term converges to an independent increment Gaussian process with variance function $\int_0^t \lambda_0(x)\{s^{(0)}(x)\}^{-1}dx$.

Convergence of first term and the asymptotic independence of the first and second terms are more difficult. A Taylor expansion of

$$\left\{ \sum_{i=1}^n Y_i(x) e^{\beta_0'\mathbf{Z}_i(x)} \right\}^{-1}$$

about $\beta = \beta_0$ shows that the first term can be written as

$$\mathbf{H}'(\beta^*, t) n^{1/2}(\hat{\beta} - \beta_0),$$

where β^* is on the line segment between $\hat{\beta}$ and β_0 and $\mathbf{H}$ is the column vector

$$\mathbf{H}(\beta, t) = -\int_0^t n^{-1} \frac{\mathbf{S}^{(1)}(\beta, x)}{\{S^{(0)}(\beta, x)\}^2} d\overline{N}(x).$$

A key step involves establishing the convergence of the vector $\mathbf{H}(\beta^*, \cdot)$ to

$$\int \mathbf{s}^{(1)}(\beta_0, x)\{s^{(0)}(\beta_0, x)\}^{-1}\lambda_0(x)dx.$$

First, since β^* is on the line segment between $\hat{\beta}$ and β_0 and $\hat{\beta}$ is consistent, it is reasonable to investigate the limit of $\mathbf{H}(\beta_0, \cdot)$. Now

$$\begin{aligned}\mathbf{H}(\beta_0, t) &= -\int_0^t n^{-1}\frac{\mathbf{S}^{(1)}(\beta_0, x)}{\{S^{(0)}(\beta_0, x)\}^2}d\overline{N}(x) \\ &= -\int_0^t n^{-1}\frac{\mathbf{S}^{(1)}(\beta_0, x)}{\{S^{(0)}(\beta_0, x)\}^2}dM(x) \\ &\quad -\int_0^t n^{-1}\frac{\mathbf{S}^{(1)}(\beta_0, x)}{\{S^{(0)}(\beta_0, x)\}^2}\sum_{i=1}^n Y_i(x)e^{\beta_0'\mathbf{Z}_i(x)}\lambda_0(x)dx.\end{aligned}$$

The first term in the expression for $\mathbf{H}(\beta_0, t)$ is a locally square integrable martingale with predictable quadratic variation process

$$\begin{aligned}&\int_0^t n^{-2}\left[\frac{\mathbf{S}^{(1)}(\beta_0, x)}{\{S^{(0)}(\beta_0, x)\}^2}\right]^2\sum_{i=1}^n Y_i(x)e^{\beta_0'\mathbf{Z}_i(x)}\lambda_0(x)dx \\ &= \int_0^t n^{-1}\left[\frac{\mathbf{S}^{(1)}(\beta_0, x)}{\{S^{(0)}(\beta_0, x)\}^2}\right]^2 S^{(0)}(\beta_0, x)\lambda_0(x)dx,\end{aligned}$$

which converges to zero in probability for each t. This first term is therefore a mean zero process converging to a normal process with zero variance, and thus is converging in probability to zero. The second term in the expression for $\mathbf{H}(\beta_0, \cdot)$ is

$$-\int_0^{\cdot}\frac{\mathbf{S}^{(1)}(\beta_0, x)}{S^{(0)}(\beta_0, x)}\lambda_0(x)dx,$$

and, subject to the regularity Conditions (2.1)–(2.6), will converge to

$$-\int_0^{\cdot}\frac{\mathbf{s}^{(1)}(\beta_0, x)}{s^{(0)}(\beta_0, x)}\lambda_0(x)dx.$$

Since $\mathbf{H}$ converges to the deterministic function

$$-\int_0^{\cdot}\frac{\mathbf{s}^{(1)}(\beta_0, x)}{s^{(0)}(\beta_0, x)}\lambda_0(x)dx,$$

and since $n^{1/2}(\hat{\beta}-\beta_0)$ is asymptotically normal with covariance matrix $\Sigma^{-1}(\beta_0, \tau)$, the asymptotic variance matrix of the first term in the decomposition of $n^{1/2}\{\hat{\Lambda}_0(t)-\Lambda_0(t)\}$ is $\mathbf{Q}'(\beta_0, t)\Sigma^{-1}(\beta_0, \tau)\mathbf{Q}(\beta_0, t)$. Except for the missing details then, a complete proof need only establish the asymptotic independence of the first two terms. Since $\hat{\beta}$ is a function of the score $\mathbf{U}(\beta, \tau)$ at time τ, it would be sufficient to show that the second term is independent of the score vector. The score process $\mathbf{U}(\beta_0, \cdot)$ is a p-dimensional local martingale, with integral representation for its ℓth component

$$U_\ell(\beta_0, \cdot) = \sum_{i=1}^{n} \int_0^{\cdot} \{Z_{i\ell}(x) - E_\ell(\beta_0, x)\} dM_i(x),$$

and consequently

$$\left\langle U_\ell(\beta_0, \cdot), \int_0^{\cdot} \left\{ \sum_{i=1}^{n} Y_i(x) e^{\beta_0' \mathbf{Z}_i(x)} \right\}^{-1} dM(x) \right\rangle (t)$$

$$= \sum_{i=1}^{n} \int_0^t \left[\frac{\{Z_{i\ell}(x) - E_\ell(\beta_0, x)\}}{\sum_{j=1}^{n} Y_j(x) e^{\beta_0' \mathbf{Z}_j(x)}} \right] Y_i(x) e^{\beta_0' \mathbf{Z}_i(x)} \lambda_0(x) dx$$

$$= \sum_{i=1}^{n} \int_0^t \left\{ \frac{S_\ell^{(1)}(\beta_0, x)}{S^{(0)}(\beta_0, x)} - E_\ell(\beta_0, x) \right\} \lambda_0(x) dx$$

$$= 0. \qquad \square$$

There are several important points to note here in passing. First, we have obscured one detail in the proof of Theorem 8.3.3. Strictly speaking, the variance formula derived above relies on the joint asymptotic normality of the score process and $\int \{\sum_{i=1}^{n} Y_i e^{\beta_0' \mathbf{Z}_i}\}^{-1} dM$, and the outline of the proof alludes only to the marginal asymptotic distributions of these terms. The joint weak convergence can be shown, however, using the multivariate martingale central limit theorem (Theorem 5.3.5).

Second, and perhaps more important, this is a good time to re-emphasize the heuristic value of the martingale methods with censored and counting process data. Those methods can be used to provide quick derivations of several important steps in the above proof, and of a natural estimator for the variance function of $\hat{\Lambda}_0$.

The variance function of $\hat{\Lambda}_0$ itself contains unknown parameters and must be estimated, as has been the case with all censored data statistics. The stochastic integral representation of the variance suggests how to compute a natural estimator: replace each $\mathbf{s}^{(j)}$ by its empirical version $\mathbf{S}^{(j)}$, and $\lambda_0(u)du$ by $d\hat{\Lambda}_0(u)$. The variance estimator then becomes

$$\widehat{\text{var}} \left[n^{1/2} \{\hat{\Lambda}_0(t) - \Lambda_0(t)\} \right] = \int_0^t \frac{d\hat{\Lambda}(x)}{S^{(0)}(x)} + \mathbf{H}'(\hat{\beta}, t) \hat{\Sigma}^{-1}(\hat{\beta}, \tau) \mathbf{H}(\hat{\beta}, t),$$

where

$$\hat{\Sigma}(\hat{\beta}, \tau) = \frac{1}{n} \int_0^{\tau} \sum_{i=1}^{n} \mathbf{V}(\hat{\beta}, x) dN_i(x).$$

Finally, the same difficulty is encountered in estimating Λ_0 that was discussed in the single sample problem in Chapter 3. Namely, only Λ_0^* can be estimated, so $\hat{\Lambda}_0$ contains no information about Λ_0 for values t larger than $\sup\{s : \overline{Y}(s) > 0\}$.

The asymptotic distribution of the log likelihood ratio statistic follows directly from the distribution of the score with the same arguments used in standard likelihood theory (see, e.g. Rao, 1973, Chapter 6), and we do not reproduce the details here. The main results are summarized in Theorem 8.3.4.

Theorem 8.3.4. Assume that conditions (2.1)–(2.6) hold in the multiplicative intensity model.

1. Under the hypothesis $\beta = \beta_0$, twice the negative log likelihood ratio test statistic,
$$-2\{\log L(\beta_0, \tau) - \log L(\hat{\beta}, \tau)\},$$
converges in distribution, as $n \to \infty$, to a χ_p^2 random variable.
2. Let $(\beta_{10}', \hat{\beta}_2')$ denote the MPLE of β' under the constraint $\beta_1 = \beta_{10}$, where β_1 is a vector consisting of the first r components of β. Then
$$-2\left[\log L\{(\beta_{10}', \hat{\beta}_2')', \tau\} - \log L(\hat{\beta}, \tau)\right]$$
is asymptotically distributed as a χ_r^2 variate. □

The asymptotic results in this section and the previous are the basis for the approximate tests and confidence intervals used in the example in Section 4.4. There are, however, a few remaining unanswered questions. The Conditions (2.1)–(2.6) are reasonably general, but can be difficult or tedious to check in specific cases. In the next section we show these conditions hold or can be weakened in several commonly used models.

8.4 THE ASYMPTOTIC THEORY FOR SIMPLE MODELS

Conditions (2.1)–(2.6) are easier to verify in some simple cases than might be expected, and we illustrate in this section how that is done when observations are independent and identically distributed. This section also contains results showing when the asymptotic distribution theory for estimators from the proportional hazards and multiplicative intensity models can be applied to data not restricted to intervals on which the cumulative hazard (or cumulative intensity) function is bounded or on which $s^{(0)}$ is bounded away from zero.

The theorems in this section are of two types. Complete proofs are given for independent, identically distributed observations from the proportional hazards model when all covariates are constant (in time) and bounded. This restriction leads to direct proofs with simple structure, and these conditions arise often enough in practice to make the theorems useful. Theorem 8.4.1 shows that Conditions (2.1)–(2.6) are easily verified under this restriction, and Theorem 8.4.3 extends Theorem 8.4.1 by showing that these conditions can be relaxed in this setting. Complete proofs of Theorems 8.4.1 and 8.4.3 in the multiplicative intensity model are more difficult; Theorems 8.4.2 and 8.4.4 generalize 8.4.1 and 8.4.3, respectively, and are stated without proof. These more general theorems are proven in Andersen and Gill (1982).

The functions that appear in Condition (2.2) take simple forms with useful interpretations when observations are independent and identically distributed. Each of the $\mathbf{S}^{(j)}$ is a sum of independent terms and, subject to boundedness conditions on the time-independent covariates, converges almost surely at each point t to its expected value. For example, $\mathbf{S}^{(1)}(\beta, t)$ converges to

$$\begin{aligned} E\{\mathbf{S}^{(1)}(\beta, t)\} &= E\{\mathbf{Z}_i Y_i(t) e^{\beta' \mathbf{Z}_i}\} \\ &= E[\mathbf{Z}_i e^{\beta' \mathbf{Z}_i} P\{Y_i(t) = 1 | Z_i\}]. \end{aligned} \tag{4.1}$$

If $Y_i(t) = I_{\{X_i \geq t\}}$, where T_i and U_i are failure and censoring time variables and $X_i = \min(T_i, U_i)$, (4.1) becomes

$$E\{\mathbf{Z}_i e^{\beta' \mathbf{Z}_i} \pi(t | \mathbf{Z}_i)\}, \tag{4.2}$$

where $\pi(t|\mathbf{Z}_i) = P\{X_i \geq t | \mathbf{Z}_i\}$. Similarly, $S^{(0)}(\beta, t)$ and $\mathbf{S}^{(2)}(\beta, t)$ converge to

$$E\{S^{(0)}(\beta, t)\} = E\{e^{\beta' \mathbf{Z}_i} \pi(t | \mathbf{Z}_i)\}, \tag{4.3}$$

and

$$E\mathbf{S}^{(2)}(\beta, t) = E\{\mathbf{Z}_i^{\otimes 2} e^{\beta' \mathbf{Z}_i} \pi(t | \mathbf{Z}_i)\}. \tag{4.4}$$

When $\beta = \mathbf{0}$, $\mathbf{s}^{(1)}(t)/s^{(0)}(t)$ becomes

$$\frac{E[\mathbf{Z}_i P\{X_i \geq t | \mathbf{Z}_i\}]}{E[P\{X_i \geq t | \mathbf{Z}_i\}]}.$$

This is a weighted average over the covariate space with weights given by the conditional probability that a case with covariate vector $\mathbf{Z}_i$ will be at risk at t.

Theorem 8.4.1 uses several conditions to simplify (2.1)–(2.6). In addition to assuming independent, identically distributed observations in a proportional hazards model, we take the covariate vector $\mathbf{Z}$ to be bounded, and assume that failure and censoring variables are conditionally independent, given $\mathbf{Z}$, a slightly more restrictive form of Condition (3.1) in Chapter 1. We also will need to assume Condition (2.6) to insure invertibility of the asymptotic covariance of the score vector. While both $\Sigma(\beta_0, t)$ and its estimator $\mathcal{I}(\hat{\beta}, t)$ are symmetric, neither has the direct interpretation as a covariance matrix of some random vector with linearly independent components and thus neither is necessarily positive definite. However the matrix $\mathcal{I}(\hat{\beta}, t)$ is a sum, over event times in $[0, t]$, of empirical covariance matrices of covariate vectors for subjects at risk at observed failure times (or at jumps of the processes N_i in the general multiplicative intensity model). Each of the empirical covariance matrices will be positive semi-definite (positive definite if the $\mathbf{Z}_i$ of items in a risk set are linearly independent), and the linear combination must be at least positive semi-definite. The convergence of $\mathcal{I}(\hat{\beta}, t)$ to $\Sigma(\beta_0, t)$ implies that Σ will always be positive semi-definite.

Theorem 8.4.1. Suppose in the proportional hazards model that $(T_i, U_i, \mathbf{Z}_i : i = 1, \ldots, n)$ are independent, identically distributed replicates of $(T, U, \mathbf{Z})$, where the failure and censoring time variables T_i and U_i are conditionally independent, given $\mathbf{Z}_i$. Let $X_i = \min(T_i, U_i)$, $N_i(t) = I_{\{X_i \le t, \delta_i = 1\}}$, and $Y_i(t) = I_{\{X_i \ge t\}}$. Assume the covariate vector $\mathbf{Z}_i$ is constant (in time) and bounded. Conditions (2.1)–(2.6) will be satisfied if $P\{Y_i(\tau) > 0\} > 0$ and $\boldsymbol{\Sigma}(\beta_0, \tau)$ is positive definite.

Proof. We have

$$\begin{aligned} 0 &< P\{Y_i(\tau) > 0\} \\ &= P\{T_i \ge \tau, U_i \ge \tau\} \\ &= EP\{T_i \ge \tau, U_i \ge \tau | \mathbf{Z}_i\} \\ &= E[P\{T_i \ge \tau | \mathbf{Z}_i\} P\{U_i \ge \tau | \mathbf{Z}_i\}] \\ &= E[S_0(\tau)^{\exp\{\beta_0' \mathbf{Z}_i\}} P\{U_i \ge \tau | \mathbf{Z}_i\}]. \end{aligned}$$

Consequently,

$$P\{S_0(\tau)^{\exp\{\beta_0' \mathbf{Z}_i\}} > 0\} > 0,$$

so $S_0(\tau) > 0$ and $\int_0^\tau \lambda_0(u)du < \infty$. Condition (2.1) is established.

To establish Condition (2.2), we note that, for any $t \in [0, \tau]$ and $\beta \in R^p$, $\mathbf{S}^{(j)}(\beta, t)$ converges to $\mathbf{s}^{(j)}(\beta, t)$ by the Strong Law of Large Numbers. We also must show there exists a neighborhood $\mathcal{B}$ of β_0 such that, with probability one, the convergence is uniform on the set $\mathcal{B} \times [0, \tau]$.

When $\pi(t|\mathbf{Z}_i)$ is continuous in t, proving uniformity for $t \in [0, \tau]$ is particularly simple, and we give that proof in detail for $s^{(0)}$. If $\pi(t|\mathbf{Z}_i)$ can have discontinuities, the argument is nearly identical to that used in the proof of the Glivenko-Cantelli Theorem (cf. Chung 1974, Theorem 5.5.1). The details are long, however, so we only outline that argument for $s^{(0)}$. Establishing the uniform convergence of $\mathbf{S}^{(1)}$ and $\mathbf{S}^{(2)}$ is done similarly.

Suppose first that $P\{U_i \ge t|\mathbf{Z}_i\}$ is continuous in t for any value of $\mathbf{Z}_i$. Thus,

$$\pi(t|\mathbf{Z}_i) = S_0(\tau)^{\exp(\beta_0' \mathbf{Z}_i)} P\{U_i \ge t | \mathbf{Z}_i\}$$

also will be continuous in t for each value of $\mathbf{Z}_i$. The Bounded Convergence Theorem then implies that

$$s^{(0)}(\beta, t) = E\{e^{\beta' \mathbf{Z}_i} \pi(t|\mathbf{Z}_i)\}$$

is continuous in t. For any $t \in [0, \tau]$, the Strong Law of Large Numbers implies that, with probability one,

$$S^{(0)}(\beta, t) \to s^{(0)}(\beta, t), \qquad \text{as } n \to \infty.$$

Since $S^{(0)}(\beta, \cdot), n = 1, 2, \ldots$, is a sequence of bounded monotonic functions converging pointwise to the bounded monotonic continuous function $s^{(0)}(\beta, \cdot)$, the convergence must be uniform over $t \in [0, \tau]$.

If $P\{U_i \geq t|\mathbf{Z}_i\}$ has discontinuities, then both $\pi(t|\mathbf{Z}_i)$ and $s^{(0)}(\beta, t) = E\{e^{\beta'\mathbf{Z}_i}\pi(t|\mathbf{Z}_i)\}$ may not be continuous. Since $\mathbf{Z}_i$ is bounded, $s^{(0)}(\beta, t)$ is a bounded left-continuous nonincreasing function on $[0, \infty)$, and hence will have at most a countable number of jumps on $[0, \infty)$, denoted by $J = \{t_1, t_2, \ldots\}$. Again, let Q denote the rational numbers. As in the proof of the Glivenko-Cantelli Theorem, one can show that there is a set of probability one on which both

$$\frac{1}{n}\sum_{i=1}^{n} e^{\beta'\mathbf{Z}_i} Y_i(t) \to s^{(0)}(\beta, t)$$

for all $t \in Q \cap [0, \tau]$, and

$$\frac{1}{n}\sum_{i=1}^{n} e^{\beta'\mathbf{Z}_i}\{Y_i(t^+) - Y_i(t)\} \to E\left\{e^{\beta'\mathbf{Z}_i}\left(P\{X_i > t|\mathbf{Z}_i\} - P\{X_i \geq t|\mathbf{Z}_i\}\right)\right\}$$

for any $t \in J \cap [0, \tau]$. Convergence on a dense set in $[0, \tau]$ and at all the discontinuities of $s^{(0)}$ is then used to show that the convergence on $[0, \tau]$ is uniform on this same set of probability one, again as in the proof of the Glivenko-Cantelli Theorem.

If $\mathcal{B}$ is any compact neighborhood of β_0, the boundedness of $\mathbf{Z}_i$ and the above results imply that

$$\lim_{n\to\infty} \sup_{0\leq t\leq\tau, \beta\in\mathcal{B}} ||S^{(0)}(\beta, t) - s^{(0)}(\beta, t)|| = 0 \quad \text{a.s.}$$

As stated above, the bound for $\mathbf{Z}_i$ eliminates the need for checking Condition (2.3). The boundedness of the $\mathbf{Z}_i$ and the Dominated Convergence Theorem are sufficient to justify the interchange of differentiation and expectation required in Condition (2.4). The $s^{(j)}$ are clearly bounded on $\mathcal{B} \times [0, \tau]$. To see that $s^{(0)}(\beta, t)$ is bounded away from zero on $\mathcal{B} \times [0, \tau]$, observe that the compactness of $\mathcal{B}$ and boundedness of $\mathbf{Z}_i$ imply that a constant k exists such that $|\beta'\mathbf{Z}_i| \leq k$ for every $\beta \in \mathcal{B}$. Then, for $0 \leq t \leq \tau$,

$$\begin{aligned} s^{(0)}(\beta, t) &= E\left[e^{\beta'\mathbf{Z}_i} P\{X_i \geq t|\mathbf{Z}_i\}\right] \\ &\geq e^{-k} E\left[S_0(\tau)^{\exp(\beta_0'\mathbf{Z}_i)} P\{U_i \geq \tau|\mathbf{Z}_i\}\right] \\ &\geq e^{-k} S_0(\tau)^{\exp(k)} P\{U_i \geq \tau\} \\ &> 0. \end{aligned}$$

Finally, we show that the continuity of $s^{(0)}(\beta, t)$ in β at the true value β_0 is uniform in t for $0 \leq t \leq \tau$. (Similar arguments work for $s^{(1)}$ and $s^{(2)}$.) Consider a sequence $\{\beta_m\}$ such that $\{\beta_m\} \to \beta_0$ componentwise as $m \to \infty$. It follows that

$$\begin{aligned} \sup_{0\leq t\leq\tau} |s^{(0)}(\beta_m, t) - s^{(0)}(\beta_0, t)| &\leq \sup_{0\leq t\leq\tau} E\{|e^{\beta_m'\mathbf{Z}_i} - e^{\beta_0'\mathbf{Z}_i}|\pi(t|\mathbf{Z}_i)\} \\ &\leq E|e^{\beta_m'\mathbf{Z}_i} - e^{\beta_0'\mathbf{Z}_i}| \to 0 \end{aligned}$$

as $m \to \infty$ by the boundedness of $\mathbf{Z}_i$. □

Andersen and Gill (1982) give a proof for a version of Theorem 8.4.1 for the multiplicative intensity model which relaxes the boundedness of the covariates, and we state their result without proof.

Theorem 8.4.2. Let $(N_i, Y_i, \mathbf{Z}_i : i = 1, \ldots, n)$ be independent, identically distributed processes satisfying the conditions of Definition 4.2.1. Suppose the covariate processes $\mathbf{Z}_i$ satisfy the conditions of Theorem 4.2.3, and that the Y_i are left-continuous processes. Then conditions (2.1)–(2.6) are satisfied if

$$\int_0^\tau \lambda_0(s)ds < \infty,$$

there exists a neighborhood $\mathcal{B}$ of β_0 such that

$$E\left\{\sup_{0\leq s\leq\tau,\beta\in\mathcal{B}} Y(s)|\mathbf{Z}(s)|^2 e^{\beta'\mathbf{Z}(s)}\right\} < \infty, \tag{4.5}$$

$$P\{Y(s) = 1, 0 \leq s \leq \tau\} > 0, \tag{4.6}$$

and $\Sigma(\beta_0, \tau)$ is positive definite. □

From the data analyst's perspective, the most serious restriction in the hypotheses of Theorems 8.4.1 and 8.4.2 is that data must be restricted to an interval $[0, \tau]$, with $P\{Y_i(\tau) > 0\} > 0$. In a randomized clinical trial, for instance, with uniform accrual over the interval $[0, a]$ and with a period of additional follow-up of length $b-a$, τ would have to be chosen strictly smaller than b and, strictly speaking, should be chosen in advance. It is more reasonable to expect to use any observation time (censored or failed) in the interval $[0, b)$, without having to specify τ in advance. Conceptually, then, it is natural to seek a convergence theorem for the score statistic and maximum partial likelihood estimators when data are used on an interval $[0, u]$ where $u = \sup\{t : P\{Y(t) > 0\} > 0\}$, much like the weak convergence result in Theorem 6.2.1. Theorem 8.4.3 provides such a result. Loosely speaking, Theorem 8.4.3 allows a data analyst to use all of the failure and follow-up times in a data set.

A complete proof of the asymptotic sampling distribution of $\hat{\beta}$ when all the failure and follow-up times on $[0, \infty)$ are used would provide conditions for the asymptotic normality of the score $\mathbf{U}_n(\beta, u)$, for consistency of the estimator $\mathcal{I}(\hat{\beta}, u)$ of its covariance, and for consistency and asymptotic normality of $\hat{\beta}$ when using data on $[0, u]$. We illustrate here only how the asymptotic normality of the score is established. Similar arguments are used for the rest of the pieces.

Theorem 8.4.3. Suppose in the proportional hazards model that $(T_i, U_i, \mathbf{Z}_i : i = 1, \ldots, n)$ are independent, identically distributed replicates where the failure and censoring time variables T_i and U_i are conditionally independent, given $\mathbf{Z}_i$. Let $X_i = \min(T_i, U_i)$, $N_i(t) = I_{\{X_i \leq t, \delta_i = 1\}}$, and $Y_i(t) = I_{\{X_i \geq t\}}$. Assume that the covariate vector $\mathbf{Z}_i$ is constant (in time) and bounded. Suppose u is such that, for all $\tau < u$, $P\{Y_i(\tau) > 0\} > 0$ and $\Sigma(\beta_0, \tau)$ is positive definite. Then the normalized score vector $\mathbf{U}^{(n)}(\beta_0, u) = n^{-1/2}\mathbf{U}(\beta_0, u)$ converges in distribution to a p-variate normal random vector with covariance matrix

$$\Sigma(\beta_0, u) = \int_0^u \mathbf{v}(\beta_0, t)s^{(0)}(\beta_0, t)\lambda_0(t)dt.$$

Proof. The asymptotic normality of $\mathbf{U}^{(n)}(\beta_0, \tau)$ for $\tau < u$ follows from Theorem 8.4.1. Suppose $\mathbf{U}^{\infty}(\beta_0, \tau)$ denotes the limit (in distribution) of $\mathbf{U}^n(\beta_0, \tau)$, and let $\mathbf{U}^{\infty}(\beta_0, u)$ denote a p–variate normal vector with covariance matrix $\Sigma(\beta_0, u)$. To complete the proof, we must show that

$$\lim_{\tau \uparrow u} \mathbf{U}^{\infty}(\beta_0, \tau) = \mathbf{U}^{\infty}(\beta_0, u) \quad \text{in distribution}$$

and

$$\lim_{\tau \uparrow u} \limsup_{n \to \infty} P\left\{ \sup_{\tau \le s \le u} ||\mathbf{U}^n(\beta_0, s) - \mathbf{U}^n(\beta_0, \tau)|| > \epsilon \right\} = 0.$$

The convergence of $\mathbf{U}^{\infty}(\beta_0, \tau)$ to $\mathbf{U}^{\infty}(\beta_0, u)$ as $\tau \uparrow u$ is easy to establish since both vectors are normally distributed with zero mean, and since $\Sigma(\beta_0, \tau) \to \Sigma(\beta_0, u)$ as $\tau \uparrow u$. Establishing the second condition, i.e., establishing tightness, is more difficult. Since it will be sufficient to show tightness for each component of the score vector, we assume without loss of generality in the rest of the proof that $p = 1$ and drop the vector notation.

We have that

$$U^{(n)}(\beta_0, s) - U^{(n)}(\beta_0, \tau) = n^{-1/2} \sum_{i=1}^{n} \int_{\tau}^{s} \{Z_i - E(\beta_0, x)\} dM_i(x).$$

By a straightforward extension of Corollary 3.4.1, for any $\epsilon, \delta > 0$,

$$P\left\{ \sup_{\tau \le s \le u} n^{-1/2} \left| \sum_{i=1}^{n} \int_{\tau}^{s} \{Z_i - E(\beta_0, x)\} dM_i(x) \right| > \epsilon \right\}$$

$$\le \frac{\delta}{\epsilon^2} + P\left\{ \frac{1}{n} \sum_{i=1}^{n} \int_{\tau}^{u} \{Z_i - E(\beta_0, s)\}^2 d\langle M_i, M_i\rangle(s) > \delta \right\}$$

$$\le \frac{\delta}{\epsilon^2} + \frac{1}{\delta} E\left[\frac{1}{n} \sum_{i=1}^{n} \int_{\tau}^{u} \{Z_i - E(\beta_0, s)\}^2 Y_i(s) e^{\beta_0 Z_i} \lambda_0(s) ds \right]$$

$$= \frac{\delta}{\epsilon^2} + \frac{1}{\delta} \int_{\tau}^{u} E\left[\{Z_i - E(\beta_0, s)\}^2 Y_i(s) e^{\beta_0, Z_i} \right] \lambda_0(s) ds.$$

The proof will be complete if

$$\lim_{\tau \uparrow u} \int_{\tau}^{u} E\left[\{Z_i - E(\beta_0, s)\}^2 Y_i(s) e^{\beta_0 Z_i} \right] \lambda_0(s) ds = 0.$$

For any $s \geq 0$,

$$E(\beta_0, s) = \frac{\sum_{i=1}^n Z_i Y_i(s) e^{\beta_0 Z_i}}{\sum_{i=1}^n Y_i(s) e^{\beta_0 Z_i}}$$

is an expectation of the covariates for cases at risk at s, computed with respect to the discrete distribution

$$h(Z_i) = \frac{Y_i(s) e^{\beta_0 Z_i}}{\sum_{i=1}^n Y_i(s) e^{\beta_0 Z_i}}.$$

For any distribution with mean μ, $E(Z - \mu)^2 \leq EZ^2$, and consequently,

$$\frac{\sum_{i=1}^n \{Z_i - E(\beta_0, s)\}^2 Y_i(s) e^{\beta_0 Z_i}}{\sum_{i=1}^n Y_i(s) e^{\beta_0 Z_i}} \leq \frac{\sum_{i=1}^n Z_i^2 Y_i(s) e^{\beta_0 Z_i}}{\sum_{i=1}^n Y_i(s) e^{\beta_0 Z_i}}.$$

Since the numerators on both sides of this inequality will be zero whenever the common denominator is zero,

$$\sum_{i=1}^n \{Z_i - E(\beta_0, s)\}^2 Y_i(s) e^{\beta_0 Z_i} \leq \sum_{i=1}^n Z_i^2 Y_i(s) e^{\beta_0 Z_i}.$$

The summands on the left–hand side of the inequality are identically distributed, as are those on the right, and, consequently,

$$E\{Z_i - E(\beta_0, s)\}^2 Y_i(s) e^{\beta_0 Z_i} \leq E\{Z_i^2 Y_i(s) e^{\beta_0 Z_i}\}.$$

It is sufficient to prove, then, that

$$\lim_{\tau \uparrow u} \int_\tau^u E\{Z_i^2 Y_i(s) e^{\beta_0 Z_i}\} \lambda_0(s) ds = 0,$$

so we need only show

$$\int_0^u E\{Z_i^2 Y_i(s) e^{\beta_0 Z_i}\} \lambda_0(s) ds < \infty.$$

But, if $|Z_i| \leq c$,

$$\begin{aligned}
\int_0^u E\{Z_i^2 Y_i(s) e^{\beta_0 Z_i}\} \lambda_0(s) ds &\leq c^2 E \int_0^u Y_i(s) e^{\beta_0 Z_i} \lambda_0(s) ds \\
&= c^2 EN(u) \\
&\leq c^2.
\end{aligned}$$

□

Andersen and Gill proved a version of Theorem 8.4.3 for the multiplicative intensity model which we give in slightly modified form here, stating it in terms of convergence of the score vector on $[0, u]$.

Theorem 8.4.4. Let $(N_i, Y_i, \mathbf{Z}_i : i = 1, \ldots, n)$ be independent, identically distributed processes satisfying the conditions of Definition 4.2.1. Suppose the covariate processes $\mathbf{Z}_i$ are bounded, satisfying the conditions of Theorem 4.2.3, and that the Y_i are left-continuous processes. Suppose u is such that, for all $\tau < u$, $P\{Y(t) = 1 \text{ for all } t \leq \tau\} > 0$ and $\Sigma(\beta_0, \tau)$ is positive definite, and suppose $EN(u) < \infty$. Then the normalized score vector $n^{-1/2}\mathbf{U}(\beta_0, u)$ converges in distribution to a p-variate normal random vector with covariance matrix $\Sigma(\beta_0, u)$. □

Theorems 8.4.1 to 8.4.4 all give convergence in distribution of the score vector evaluated at a fixed time to a multivariate normal vector. This is the most frequently used distribution result for the proportional hazards and multiplicative intensity models, since it is the basis for the asymptotic normality of maximum partial likelihood estimators. The martingale convergence theorems in Chapter 5, however, can be used to obtain convergence in distribution, or weak convergence, of the score process. In fact, since Theorems 8.4.1 to 8.4.4 provide conditions under which Conditions (2.1)–(2.6) hold, these theorems establish weak convergence of the score process when observations are independent and identically distributed, covariates are bounded, Σ is positive definite, and cases have positive probability of being at risk at each $t < u$. Theorem 8.4.4 establishes weak convergence on $[0, u]$, and since the score will be constant on $[u, \infty]$, in effect establishes convergence on $[0, \infty]$. Theorem 8.4.4 can be used for the long overdue proof of Lemma 4.5.1, promised in Chapter 4.

Proof of Lemma 4.5.1.

1. A Taylor series expansion implies that

$$n^{-1/2}\mathbf{U}(\hat{\beta}, t) = n^{-1/2}\mathbf{U}(\beta_0, t) - n^{-1}\mathcal{I}(\beta_t^*, t)\sqrt{n}(\hat{\beta} - \beta_0) \tag{4.7}$$

for some β_t^* on a line segment connecting β_0 and $\hat{\beta}$. In turn,

$$\sqrt{n}(\hat{\beta} - \beta_0) = \{n^{-1}\mathcal{I}(\beta_\infty^*, \infty)\}^{-1} n^{-1/2}\mathbf{U}(\beta_0, \infty).$$

Inserting this into Eq. (4.7) yields

$$n^{-1/2}\mathbf{U}(\hat{\beta}, t)$$
$$= n^{-1/2}\mathbf{U}(\beta_0, t) - \{n^{-1}\mathcal{I}(\beta_t^*, t)\}\{n^{-1}\mathcal{I}(\beta_\infty^*, \infty)\}^{-1} n^{-1/2}\mathbf{U}(\beta_0, \infty).$$

The weak convergence of $\{n^{-1/2}\mathbf{U}(\hat{\beta},t) : 0 \leq t \leq \infty\}$ in Eq. (5.21) of Chapter 4, now follows from Theorems 8.4.4 and 8.2.1, and from the fact that $n^{-1}\mathcal{I}(\beta_t^*, t)$ converges in probability to $\Sigma(\beta_0, t)$ for any $t \in [0, \infty]$.

2. Suppose as in Lemma 4.5.1 that $\mathbf{B}$ is a mean zero vector of Gaussian processes with covariance matrix Σ. To obtain the large sample covariance structure of $n^{-1/2}\mathbf{U}(\hat{\beta}, \cdot)$, observe that, for any $s \leq t$,

$$E\{[\mathbf{B}(s) - \Sigma(\beta_0, s)\{\Sigma(\beta_0, \infty)\}^{-1}\mathbf{B}(\infty)]$$
$$[\mathbf{B}(t) - \Sigma(\beta_0, t)\{\Sigma(\beta_0, \infty)\}^{-1}\mathbf{B}(\infty)]'\}$$
$$= \Sigma(\beta_0, s) - \Sigma(\beta_0, s)\{\Sigma(\beta_0, \infty)\}^{-1}\Sigma(\beta_0, t). \tag{4.8}$$

Recall that $\Sigma(\beta_0, \cdot) = \int_0^{\cdot} \mathbf{v}(\beta_0, x)\, s^{(0)}(\beta_0, x)\, \lambda_0(x)dx$. For some $j = 1, \ldots, p$, if $\{\mathbf{v}(\beta_0, \cdot)\}_{jk}$ is proportional to $\{\mathbf{v}(\beta_0, \cdot)\}_{jj}$ for every k (over the region where $s^{(0)}(\beta_0, \cdot)\ \lambda_0(\cdot) > 0$), then Exercise 8.2 establishes that Eq. (5.22) in Chapter 4 follows from Eq. (4.8) and the consistency of $n^{-1}\mathcal{I}(\hat{\beta}, \infty)$. □

8.5 ASYMPTOTIC RELATIVE EFFICIENCY OF PARTIAL LIKELIHOOD INFERENCE IN THE PROPORTIONAL HAZARDS MODEL

The derivation of the partial likelihood shows that terms are discarded which normally would be present in a full likelihood for censored observations from the Cox proportional hazards model defined in Example 1.5.2. This causes some possible loss of information about the regression coefficients β. While there are appealing arguments that this loss may be negligible in many situations, it is still instructive to study the loss more precisely.

In this section we examine the asymptotic relative efficiency of the MPLE $\hat{\beta}$ compared to estimation using the full likelihood under parametric assumptions on the baseline hazard λ_0. The main ideas here were first explored by Efron (1977) and Oakes (1977), and discussed further by Cox and Oakes (1984), but we have recast the derivations to use the martingale methods. The relative efficiency of the two likelihood based methods for inference about the parameter β can be studied, in large samples, by examining the asymptotic standardized information matrices for β from the partial likelihood and from a full likelihood, under parametric assumptions on λ_0.

When λ_0 is characterized by a finite number of unknown parameters, an informal derivation of the full likelihood for β is straightforward. We assume for simplicity that the conditional distribution of failure, given the covariate vector, is absolutely continuous. If in the model $\lambda(t|\mathbf{Z}) = \lambda_0(t)\exp\{\beta'\mathbf{Z}(t)\}$ we parameterize $\lambda_0(t) = \lambda_0(t, \alpha)$, where $\alpha' = (\alpha_1, \ldots, \alpha_q)$, then the conditional hazard and survivor functions for T given $\mathbf{Z}$ will depend on the $p + q$ dimensional parameter $\phi' = (\beta', \alpha')$. We assume the distribution of $\mathbf{Z}$ does not depend on α or β.

When the observations $(X_1, \delta_1, \mathbf{Z}_1), \ldots, (X_n, \delta_n, \mathbf{Z}_n)$ are independent, and censoring is independent and uninformative, the full likelihood for the data is

$$L_f(\phi) = \prod_{i=1}^{n} \{\lambda(X_i, \phi|\mathbf{Z}_i)\}^{\delta_i} S(X_i, \phi|\mathbf{Z}_i).$$

The log likelihood $\mathcal{L}_f(\phi)$ and the score $\mathbf{U}_f(\phi)$ will be

$$\mathcal{L}_f(\phi) = \sum_{i=1}^{n} \{\delta_i \log \lambda(X_i, \phi|\mathbf{Z}_i) - \Lambda(X_i, \phi|\mathbf{Z}_i)\},$$

and

$$\mathbf{U}_f(\phi) = \sum_{i=1}^{n} \left\{\delta_i \frac{\partial}{\partial \phi} \log \lambda(X_i, \phi|\mathbf{Z}_i) - \frac{\partial}{\partial \phi} \int_0^\infty Y_i(s)\lambda(s, \phi|\mathbf{Z}_i) ds\right\}.$$

If $\lambda(s, \phi|\mathbf{Z}_i) > 0$, then the score can be written

$$\begin{aligned}
\mathbf{U}_f(\phi) &= \sum_{i=1}^{n} \int_0^\infty \frac{\partial}{\partial \phi} \log \lambda(s, \phi|\mathbf{Z}_i) dN_i(s) \\
&\quad - \sum_{i=1}^{n} \int_0^\infty \frac{\partial}{\partial \phi} \log \lambda(s, \phi|\mathbf{Z}_i) Y_i(s) \lambda(s, \phi|\mathbf{Z}_i) ds \\
&= \sum_{i=1}^{n} \int_0^\infty \frac{\partial}{\partial \phi} \log \lambda(s, \phi|\mathbf{Z}_i) dM_i(s).
\end{aligned}$$

Thus, even if $\mathbf{Z}$ is a time-dependent covariate, the score $\mathbf{U}$ will be a mean zero martingale as long as $\mathbf{Z}$ is a predictable process. Using formulas for quadratic variation, the covariance matrix for $\mathbf{U}_f$ is

$$\begin{aligned}
E\left\{\mathbf{U}_f(\phi)\mathbf{U}_f'(\phi)\right\} &= E\sum_{i=1}^{n} \int_0^\infty \left\{\frac{\partial}{\partial \phi} \log \lambda(s, \phi|\mathbf{Z}_i)\right\}^{\otimes 2} Y_i(s)\lambda(s, \phi|\mathbf{Z}_i) ds \\
&\equiv E\mathcal{I}_f(\phi). \qquad (5.1)
\end{aligned}$$

It is an easy exercise to show (as expected) that

$$\begin{aligned}
En^{-1}\mathcal{I}_f(\phi) &= En^{-1}\{\mathbf{U}_f(\phi)\mathbf{U}_f'(\phi)\} \\
&= -En^{-1} \frac{\partial^2}{\partial \phi^2} \mathcal{L}_f(\phi),
\end{aligned}$$

so the standardized covariance matrix for $\mathbf{U}_f$ is equal to the standardized information matrix for ϕ based on the full likelihood.

Recall now that the normalized score vector $n^{-1/2}\mathbf{U}(\beta)$ from the partial likelihood for β has standardized information matrix $\Sigma(\beta, n)$ given by

$$E\left[\left\{n^{-1/2}\mathbf{U}(\beta)\right\}^{\otimes 2}\right] = E\int_0^\infty n^{-1}\sum_{i=1}^n \mathbf{V}(\beta, s)dN_i(s)$$

$$= En^{-1}\sum_{i=1}^n\int_0^\infty \mathbf{V}(\beta, s)Y_i(s)\lambda(s, \phi|\mathbf{Z}_i)ds,$$

where $\mathbf{V}(\beta, s)$ is given by Eq. (3.23) in Chapter 4. The standardized information matrix for $\mathbf{U}_f$ is a $(p+q)\times(p+q)$ matrix and that for $\mathbf{U}$ is $p \times p$, but the two expressions are obviously similar.

Suppose now that

$$\lambda(t, \phi|\mathbf{Z}_i) = \exp\{\alpha'\mathbf{y}(t)\}\exp\{\beta'\mathbf{Z}_i(t)\},$$

where $\mathbf{y}$ is a specified set of functions. The first factor represents the parametrization for $\lambda_0(t)$, which we assume is independent of β. In this case, $\left\{\frac{\partial}{\partial\phi}\log\lambda(s, \phi|\mathbf{Z}_i)\right\}^{\otimes 2}$ can be written as the partitioned matrix

$$\mathbf{A}(\mathbf{Z}, s) = \begin{pmatrix} \mathbf{y}(s)\mathbf{y}'(s) & \mathbf{y}(s)\mathbf{Z}'(s) \\ \mathbf{Z}(s)\mathbf{y}'(s) & \mathbf{Z}(s)\mathbf{Z}'(s) \end{pmatrix},$$

and

$$E\{n^{-1}\mathbf{U}_f(\phi)^{\otimes 2}\} = En^{-1}\sum_{i=1}^n\int_0^\infty \mathbf{A}(\mathbf{Z}_i, s)Y_i(s)\lambda(s, \phi|\mathbf{Z}_i)ds.$$

To obtain the information matrix for β when the full likelihood is maximized with respect to α, partition $\mathcal{I}_f(\phi)$ into

$$\mathcal{I}_f(\phi) = \begin{pmatrix} \mathcal{I}_{\alpha\alpha}(\phi) & \mathcal{I}_{\alpha\beta}(\phi) \\ \mathcal{I}'_{\alpha\beta}(\phi) & \mathcal{I}_{\beta\beta}(\phi) \end{pmatrix},$$

so

$$\mathcal{I}_{\alpha\alpha}(\phi) = \sum_{i=1}^n\int_0^\infty \mathbf{y}(s)\mathbf{y}'(s)Y_i(s)\lambda(s, \phi|\mathbf{Z}_i)ds,$$

$$\mathcal{I}_{\beta\beta}(\phi) = \sum_{i=1}^n\int_0^\infty \mathbf{Z}_i(s)\mathbf{Z}_i'(s)Y_i(s)\lambda(s, \phi|\mathbf{Z}_i)ds,$$

and

$$\mathcal{I}_{\alpha\beta}(\phi) = \sum_{i=1}^n\int_0^\infty \mathbf{y}(s)\mathbf{Z}_i'(s)Y_i(s)\lambda(s, \phi|\mathbf{Z}_i)ds.$$

If we define

$$\{\mathcal{I}_f(\phi)\}^{-1} = \begin{pmatrix} \mathcal{I}^{\alpha\alpha}(\phi) & \mathcal{I}^{\alpha\beta}(\phi) \\ \mathcal{I}'^{\alpha\beta}(\phi) & \mathcal{I}^{\beta\beta}(\phi) \end{pmatrix},$$

then

$$n^{-1}\{\mathcal{I}^{\beta\beta}(\phi)\}^{-1} = n^{-1}\left\{\mathcal{I}_{\beta\beta}(\phi) - \mathcal{I}'_{\alpha\beta}(\phi)\mathcal{I}^{-1}_{\alpha\alpha}(\phi)\mathcal{I}_{\alpha\beta}(\phi)\right\}$$

and, as $n \to \infty$,

$$En^{-1}\left\{\mathcal{I}_{\beta\beta}(\phi) - \mathcal{I}'_{\alpha\beta}(\phi)\mathcal{I}^{-1}_{\alpha\alpha}(\phi)\mathcal{I}_{\alpha\beta}(\phi)\right\} \to \Sigma_f(\beta),$$

which is the asymptotic standardized information matrix for β in the full likelihood when α is replaced by $\hat{\alpha}$.

For further insight into the relationship between $\Sigma(\beta) \equiv \lim_{n\to\infty} \Sigma(\beta, n)$ and $\Sigma_f(\beta)$, consider the special case in which $\{(X_i, \delta_i, \mathbf{Z}_i) : i = 1, \ldots, n\}$ are independent and identically distributed, $q = 1$ and $y \equiv 1$ (so $\lambda_0(t) = e^{\alpha}$), and $p = 1$ (so β also is a scalar). Then

$$n^{-1}\left\{\mathcal{I}_{\beta\beta}(\phi) - \mathcal{I}'_{\alpha\beta}(\phi)\mathcal{I}^{-1}_{\alpha\alpha}(\phi)\mathcal{I}_{\alpha\beta}(\phi)\right\}$$

$$= n^{-1}\left[\sum_{i=1}^{n}\int_0^\infty Z_i^2(s)Y_i(s)\lambda(s,\phi|Z_i)ds - \frac{\left\{\sum_{i=1}^n\int_0^\infty Z_i(s)Y_i(s)\lambda(s,\phi|Z_i)ds\right\}^2}{\sum_{i=1}^n\int_0^\infty Y_i(s)\lambda(s,\phi|Z_i)ds}\right]$$

$$= \left[\frac{\sum_{i=1}^n\int_0^\infty Z_i^2(s)Y_i(s)\lambda(s,\phi|Z_i)ds}{\sum_{i=1}^n\int_0^\infty Y_i(s)\lambda(s,\phi|Z_i)ds} - \left\{\frac{\sum_{i=1}^n\int_0^\infty Z_i(s)Y_i(s)\lambda(s,\phi|Z_i)ds}{\sum_{i=1}^n\int_0^\infty Y_i(s)\lambda(s,\phi|Z_i)ds}\right\}^2\right]$$

$$\times \frac{\sum_{i=1}^n\int_0^\infty Y_i(s)\lambda(s,\phi|Z_i)ds}{n}.$$

As $n \to \infty$,

$$\begin{aligned} &En^{-1}\left[\mathcal{I}_{\beta\beta}(\phi) - \mathcal{I}'_{\alpha\beta}(\phi)\mathcal{I}^{-1}_{\alpha\alpha}(\phi)\mathcal{I}_{\alpha\beta}(\phi)\right] \\ &\to \left[E\{Z^2(T)|\delta = 1\} - \{E\,Z(T)|\delta = 1\}^2\right] E\,N(\infty) \\ &= \text{var}\{Z(T)|\delta = 1\}E\,N(\infty). \end{aligned}$$

Since

$$\Sigma_f(\beta) = \text{var}\{Z(T)|\delta = 1\}E\,N(\infty),$$

we can write

$$\begin{aligned} \{E\,N(\infty)\}^{-1}\Sigma_f(\beta) &= \text{var}\{Z(T)|\delta = 1\} \\ &= E[\text{var}\{Z(T)|T, \delta = 1\}|\delta = 1] \\ &\quad +\text{var}[E\{Z(T)|T, \delta = 1\}|\delta = 1]. \end{aligned} \tag{5.2}$$

The asymptotic standardized information matrix from the partial likelihood for β, $\Sigma(\beta) = \lim_{n\to\infty} E\,\{n^{-1/2}U(\beta)\}^2$, can also be written using terms in Eq. (5.2). We have, as $n \to \infty$,

$$n^{-1}\mathcal{I}(\beta) \to \Sigma(\beta) \quad \text{a.s.},$$

where

$$n^{-1}\mathcal{I}(\beta)$$

$$= n^{-1}\sum_{i=1}^{n}\int_0^\infty V(\beta, s)dN_i(s)$$

$$= n^{-1}\sum_{i=1}^{n}\int_0^t \left[\frac{\sum_{\ell=1}^n Y_\ell(s)Z_\ell^2(s)e^{\beta Z_\ell(s)}}{\sum_{\ell=1}^n Y_\ell(s)e^{\beta Z_\ell(s)}} - \left\{\frac{\sum_{\ell=1}^n Y_\ell(s)Z_\ell(s)e^{\beta Z_\ell(s)}}{\sum_{\ell=1}^n Y_\ell(s)e^{\beta Z_\ell(s)}}\right\}^2\right] dN_i(s).$$

Thus,

$$\begin{aligned}\Sigma(\beta) &= E[\{E(Z^2(T)|T, \delta = 1) - \{E(Z(T)|T, \delta = 1)\}^2\}|\delta = 1]EN(\infty)\\ &= E[\text{var}\{Z(T)|T, \delta = 1\}|\delta = 1\}EN(\infty). \qquad (5.3)\end{aligned}$$

The asymptotic relative efficiency of the partial likelihood based estimator for β with respect to an estimator based on the full likelihood is the ratio $\Sigma(\beta)/\Sigma_f(\beta)$. By Eqs. (5.2) and (5.3),

$$\Sigma(\beta)/\Sigma_f(\beta) = \frac{E[\text{var}\{Z(T)|T, \delta = 1\}|\delta = 1]}{E[\text{var}\{Z(T)|T, \delta = 1\}|\delta = 1] + \text{var}[E\{Z(T)|T, \delta = 1\}|\delta = 1]}.$$

This asymptotic relative efficiency is bounded above by unity, and equals one when

$$\text{var}[E\{Z(T)|T, \delta = 1\}|\delta = 1] = 0. \qquad (5.4)$$

In general, the partial likelihood based estimator of β will have high efficiency relative to the full likelihood approach if the ratio of the "between time" component of $\text{var}\{Z(T)|\delta = 1\}$ to the "within time" component is small. The within time component represents the average (over observed failure times) of the variability of Z over a weighting of the risk set at each observed failure time, while the between time component represents the variability (over observed failure times) of the weighted average of Z over the risk set at each observed failure time. Intuitively, the between time variability (appearing in the left hand side of Eq. (5.4)) should be small when Z is not predictive of failure time (i.e., $\beta = 0$), Z is independent of the censoring mechanism, and Z does not vary with time.

The left-hand side of Eq. (5.4) can be rewritten as

$$E[[E\{Z(T)|T, \delta = 1\}]^2|\delta = 1] - [E[E\{Z(T)|T, \delta = 1\}|\delta = 1]]^2$$

$$= \frac{\int_0^\infty \left[\frac{E\{Z(s)Y(s)e^{\beta Z(s)}\}}{E\{Y(s)e^{\beta Z(s)}\}}\right]^2 E\{Y(s)e^{\beta Z(s)}\}\lambda_0(s)ds}{\int_0^\infty E\{Y(s)e^{\beta Z(s)}\}\lambda_0(s)ds}$$

$$- \left\{\frac{\int_0^\infty \left[\frac{E\{Z(s)Y(s)e^{\beta Z(s)}\}}{E\{Y(s)e^{\beta Z(s)}\}}\right] E\{Y(s)e^{\beta Z(s)}\}\lambda_0(s)ds}{\int_0^\infty E\{Y(s)e^{\beta Z(s)}\}\lambda_0(s)ds}\right\}^2. \qquad (5.5)$$

The right–hand side of Eq. (5.5) is zero when (1) $\beta = 0$, *and* (2) censoring is independent of Z, *and* (3) Z is non–time varying. Note that conditions (1) and (2) are analogous to the condition $\pi_1 = \pi_2$ required in Theorem 7.4.2 for full efficiency of the G^ρ statistics.

8.6 BIBLIOGRAPHIC NOTES

Section 8.2

A variety of approaches for obtaining asymptotic distribution theory in the Cox model were tried prior to Andersen and Gill (1982). Some of the earlier important contributions were by Cox (1975), Liu and Crowley (1978), Tsiatis (1981), Sen (1981), Næs (1982), and Bailey (1983). While not explicitly indicated, martingale-like ideas appeared as early as Cox's paper in 1975 in establishing that the partial likelihood score vector can be written as a sum of uncorrelated terms. That work preceded the martingale central limit theorems discussed in Chapter 5, however, and consequently could not be used for rigorous proofs of the asymptotic normality of the score. Both Sen and Næs used martingale methods explicitly, but they made use of discrete time martingales instead of the continuous time versions used in the treatment described here. Tsiatis and Bailey both used more traditional approaches. Tsiatis showed that the score statistic is asymptotically equivalent to a sum of independent, identically distributed terms, and established asymptotic normality under the assumption that $E(\mathbf{Z}^{\otimes 2} e^{\beta' \mathbf{Z}}) < \infty$. Bailey, on the other hand, used methods based on the notion of generalized likelihood. All of these papers imposed conditions that can be relaxed in the Andersen-Gill approach, usually either requiring that all covariates be bounded or by not allowing covariates to vary with time.

Self and Prentice (1982) gave an interesting discussion of the subtleties associated with time-dependent covariates in the martingale approach to the proportional hazards model. In a 1983 paper, Prentice and Self used the martingale approach to the Cox model to derive asymptotic distributions of regression coefficients when the hazard ratio is more general than $\exp(\beta' \mathbf{Z})$.

Section 8.3

There is little to add to this section, which is primarily an application of the results in Section 2. Surprisingly little is known about the sampling distributions of estimates from the Cox model in small samples. A simulation study shedding some light on this subject can be found in Johnson et al., (1982).

Section 8.5

Several authors have used asymptotic methods to study the efficiency of regression parameter estimates in the Cox model. Efron (1977) and Oakes (1977) both showed that the partial likelihood produces efficient estimates when $|\beta|$ is small and the baseline hazard function is constant. The approaches in these two papers are quite different and provide an interesting insight into the variety of arguments used with the Cox model before the application of martingale methods. Greenwood and Wefelmeyer (1989a,b) have used more abstract methods to prove efficiency results for parameter estimates in classes of stochastic processes that include the Cox model and general relative risk models.

APPENDIX A

Some Results from Stieltjes Integration and Probability Theory

A.1 LEBESGUE-STIELTJES INTEGRATION

The martingale calculus for counting processes relies heavily on integrals of the form $\int f dG$. Such integrals are called Stieltjes integrals, and the functions f and G are, respectively, the integrand and integrator. Although a dummy variable of integration is not necessary, we will sometimes write $\int f dG$ as $\int f(x)dG(x)$ for increased clarity. In the applications in this book, G usually is a smooth function with derivative $G' \equiv g$, or a right-continuous step function, or some simple combination of a differentiable function and a step function. Stieltjes integrals are particularly easy to calculate in those cases.

This section summarizes the important aspects of Stieltjes integrals, including definitions and some useful computational formulas. These integrals can be defined using either the Riemann or Lebesgue theories of integration. We outline the Lebesgue approach here, since it leads to less ambiguity about the class of functions that are integrable for a given integrator G. We do not provide proofs for most of the important results; the details can be found in Royden (1968) or Ash (1972). Apostol (1974) gives a detailed account of Riemann-Stieltjes integration.

The treatment here is far from complete, and presumes that the reader is familiar with at least the basic concepts of Lebesgue integration. The definitions here are enough to show the clear and direct link between Stieltjes and general Lebesgue integrals, and to outline the derivation of the two computational results used frequently in the book: the formula for $\int f dG$ when G consists of absolutely continuous and purely discontinuous components, and the integration by parts formula in Theorem A.1.2.

The second section of this Appendix summarizes some results from probability theory and necessarily recalls a few definitions from integration and measure theory, so it might be logically more consistent for that section to precede this one. Stieltjes integrals appear as early as Chapter 0, however, so this review is given first for readers who need it for that introductory material.

Lebesgue-Stieltjes integration is based on a one-to-one correspondence between right-continuous increasing functions and a class of measures on the real line.

Definition A.1.1. A function $G : R \to R$ is *increasing* if $a < b$ implies $G(a) \leq G(b)$. □

Definition A.1.2. A *Borel measure* on $R = (-\infty, \infty)$ is a nonnegative set function, μ, defined for all Borel sets of R such that

1. $\mu(\emptyset) = 0$ where $\emptyset$ denotes the empty set,
2. $\mu(I) < \infty$ for each bounded interval I, and
3. $\mu\left(\bigcup_{i=1}^{\infty} B_i\right) = \sum_{i=1}^{\infty} \mu(B_i)$ for any sequence of disjoint Borel sets.

Property (3) is called *countable additivity.* □

Theorem A.1.1.

1. Let μ be a Borel measure on R and let G be defined, up to an additive constant, by $G(b) - G(a) = \mu(a, b]$. Then G is right-continuous and increasing.
2. Let $G : R \to R$ be a right-continuous, increasing function and, for any interval $(a, b], -\infty < a < b < \infty$, let $\mu(a, b] = G(b) - G(a)$. Then there is a unique extension of μ to a Borel measure on R.

Proof. To prove (1), it suffices to show right-continuity of G, since it clearly is increasing. If $\{x_n\}$ is a sequence of points decreasing to a limit x, then $\bigcap_{n=1}^{\infty}(x, x_n] = \emptyset$ and

$$\begin{aligned}\lim_{n\to\infty} G(x_n) - G(x) &= \lim_{n\to\infty} \mu(x, x_n] \\ &= \mu\left\{\bigcap_{n=1}^{\infty}(x, x_n]\right\} \\ &= 0,\end{aligned}$$

since countably additive measures are continuous at $\emptyset$, (see Royden 1968, Proposition 11.2).

The proof of (2) uses the Carathéodory Extension Theorem, and can be found in Ash (1972) or Royden (1968). □

The following relationships between a right-continuous increasing function G and its associated Borel measure μ are easy to establish. In each of these equations, $G(t-) = \lim_{x\uparrow t} G(x)$ for $t = a$ or $t = b$.

1. $\mu(a,b) = G(b-) - G(a)$.
2. $\mu[a,b] = G(b) - G(a-)$.
3. $\mu[a,b) = G(b-) - G(a-)$.
4. $\mu\{a\} = G(a) - G(a-)$.
5. G is continuous at a if and only if $\mu\{a\} = 0$.

Equation (1) is established by observing that

$$\begin{aligned}\mu(a,b) &= \lim_{n\to\infty} \mu(a, b - \frac{1}{n}] \\ &= \lim_{n\to\infty} \left\{ G(b - \frac{1}{n}) - G(a) \right\} \\ &= G(b-) - G(a).\end{aligned}$$

The other equations are proved similarly. Equations (4) and (5) taken together imply that when G is continuous at the point a, then a has μ-measure 0, and when G is discontinuous at a, the magnitude of the discontinuity is the measure of a.

The definition of a Lebesgue-Stieltjes integral now follows from the general definition of a Lebesgue integral.

Definition A.1.3. Let $f : R \to R$ be a Borel measurable function, $G : R \to R$ a right-continuous increasing function, and μ the Borel measure corresponding to G. For any Borel set $B \subset R$, define $\int_B f dG$ to be $\int_B f d\mu$. □

This abstract definition quickly translates into straightforward formulas when G has certain properties. When G is a step function, it will have at most countably many jumps $\{x_1, x_2, \ldots\}$ where $\Delta G(x_n) = G(x_n) - G(x_n-) > 0$. The measure μ will be discrete with positive measure at each of the points $x_1, x_2, \ldots$, so

$$\int_B f dG = \sum_{n : x_n \in B} f(x_n) \Delta G(x_n).$$

When integrating over an interval, the notation $\int_s^t f dG$ may lead to ambiguities if either s or t is a discontinuity of G, so by convention we take

$$\begin{aligned}\int_s^t f dG &= \int_{(s,t]} f dG \\ &= \sum_{n : s < x_n \le t} f(x_n) \Delta G(x_n).\end{aligned}$$

In this case, the Stieltjes integral is a convenient notation for a sum with a finite or countably infinite number of terms.

When G has derivative g at each point in an interval $(s,t]$, then $\mu(s,t] = \int_s^t g(x)dx$ and μ is absolutely continuous with respect to Lebesgue measure. Thus

$$\begin{aligned}\int_s^t f(x)dG(x) &= \int_s^t f(x)d\mu(x)\\ &= \int_s^t f(x)g(x)dx.\end{aligned}$$

Stieltjes integrals in this book often use integrators G that map $[0,\infty)$ into R, are right-continuous on $[0,\infty)$, increasing, and are differentiable on R except at points in a countably infinite set $\{x_1, x_2, \ldots\}$, where each $x_i > 0$. Then G can be written

$$G(t) = G(0) + \int_0^t g(x)dx + \sum_{n:0<x_n\le t} \Delta G(x_n).$$

In these cases

$$\int_0^t f(x)dG(x) = \int_0^t f(x)g(x)dx + \sum_{n:0<x_n\le t} f(x_n)\Delta G(x_n).$$

Obvious modifications of the above formula are needed for integrals over $(s,t]$.

When G is a right-continuous function of bounded variation on each finite interval I, then G can be written $G_1 - G_2$, where both G_1 and G_2 are increasing right-continuous functions. In this case $\int fdG = \int fdG_1 - \int fdG_2$. Using this decomposition, it is easy to show that all of the standard results from Lebesgue integration, such as the convergence theorems for integrals and Fubini's Theorem for iterated integrals over product spaces, hold for $\int fdG$ when f is Borel measurable and G is of bounded variation on finite intervals.

If F and G are differentiable functions with respective derivatives f and g, then the "integration by parts" formula from calculus states that, for $s < t$,

$$\begin{aligned}F(t)G(t) - F(s)G(s) &= \int_s^t F(x)g(x)dx + \int_s^t G(x)f(x)dx\\ &= \int_s^t F(x)dG(x) + \int_s^t G(x)dF(x).\end{aligned}$$

If either F or G have discontinuities, then a bit more care must be taken with the Stieltjes integrals. Theorem A.1.2 gives a more general formula for integration by parts over $(0,t]$ when integrands and integrators have discontinuities. Only a small change is needed to adapt the formula to integrals over $(s,t]$.

Theorem A.1.2. Let $F : [0,\infty) \to R$ and $G : [0,\infty) \to R$ be right-continuous functions of bounded variation on any finite interval. Then

$$F(t)G(t) - F(0)G(0) = \int_0^t F(x-)dG(x) + \int_0^t G(x)dF(x), \tag{1.1}$$

and

$$F(t)G(t)-F(0)G(0) = \int_0^t F(x-)dG(x)+\int_0^t G(x-)dF(x)+\sum_{0<x\leq t} \Delta F(x)\Delta G(x). \tag{1.2}$$

Proof. Note that, whenever $s \leq t$,

$$F(s) - F(0) = \int_0^s dF(x)$$
$$= \int_0^t I_{\{0<x\leq s\}} dF(x),$$

and

$$F(s-) - F(0) = \int_0^t I_{\{0<x<s\}} dF(x),$$

where $I_A = 1$ if the inequality in A is true, and $I_A = 0$ otherwise. By Fubini's Theorem,

$$\{F(t) - F(0)\}\{G(t) - G(0)\}$$
$$= \int_0^t \int_0^t dF(x)dG(y)$$
$$= \int_0^t \int_0^t I_{\{x<y\}} dF(x)dG(y) + \int_0^t \int_0^t I_{\{y\leq x\}} dF(x)dG(y)$$
$$= \int_0^t \{F(y-) - F(0)\}dG(y) + \int_0^t \{G(x) - G(0)\}dF(x)$$
$$= \int_0^t F(y-)dG(y) - F(0)\{G(t) - G(0)\} + \int_0^t G(x)dF(x) - G(0)[F(t) - F(0)].$$

Algebraic simplification leads to Eq. (1.1).

To prove Eq. (1.2) we must show

$$\int_0^t G(x)dF(x) = \int_0^t G(x-)dF(x) + \sum_{0<x\leq t} \Delta F(x)\Delta G(x),$$

or

$$\int_0^t \Delta G(x)dF(x) = \sum_{0<x\leq t} \Delta F(x)\Delta G(x).$$

Let $F^c(x) = F(x) - \sum_{s \leq x} \Delta F(s)$ be the continuous part of F. Then

$$\int_0^t \Delta G(x) dF(x) = \int_0^t \Delta G(x) dF^c(x) + \sum_{0<x\leq t} \Delta G(x)\Delta F(x)$$

$$= \sum_{0<x\leq t} \Delta G(x)\Delta F(x),$$

since F^c is continuous (and hence assigns measure 0 to any single point) and $\Delta G = 0$ at all but at most a countable number of points in $(0, t]$. □

A.2 A REVIEW OF PROBABILITY THEORY

This section provides a short review of some concepts from probability theory. We do little more than recall a few definitions and major theorems, most of which are stated without proof. A more extended review can be found in Appendix A1 of Brémaud (1981) or in Chapter 1 of Liptser and Shiryayev (1977) and, of course, the real thing is the subject of many excellent texts, including that by Chung (1974). The review here establishes some common notation and terminology and provides enough background to prove a few technical results needed in Chapters 1 and 2.

We begin with some notions from set theory.

Definition A.2.1. Let Ω be a nonempty set of elements, each denoted generically by ω, and $\mathcal{A}$ a nonempty collection of subsets of Ω.

1. $\mathcal{A}$ is a *monotone class* if $E_j \in \mathcal{A}, E_j \subset E_{j+1}, 1 \leq j < \infty$, implies $\bigcup_{j=1}^{\infty} E_j \in \mathcal{A}$; and if $E_j \in \mathcal{A}, E_j \supset E_{j+1}, 1 \leq j < \infty$, implies $\bigcap_{j=1}^{\infty} E_j \in \mathcal{A}$.
2. $\mathcal{A}$ is an *algebra* if the complement $\bar{E}$ of E is a member of $\mathcal{A}$ whenever $E \in \mathcal{A}$, and $E_1 \cup E_2 \in \mathcal{A}$ if $E_1 \in \mathcal{A}$ and $E_2 \in \mathcal{A}$.
3. $\mathcal{A}$ is a *σ-algebra* if $\bar{E} \in \mathcal{A}$ when $E \in \mathcal{A}$, and $\bigcup_{j=1}^{\infty} E_j \in \mathcal{A}$ whenever $E_j \in \mathcal{A}, 1 \leq j < \infty$. □

For a σ-algebra $\mathcal{A}$, it is not hard to show that if $E_j \in \mathcal{A}, 1 \leq j < \infty$, then $\bigcap_{j=1}^{\infty} E_j \in \mathcal{A}$. A σ-algebra is thus closed under countable unions and intersections. It is also clear that both the empty set $\emptyset$ and Ω are members of $\mathcal{A}$. It is not difficult to show as well that an algebra is closed under finite unions and intersections.

If $\mathcal{A}$ is an arbitrary collection of subsets of Ω, then there will always be a smallest σ-algebra of subsets of Ω that contains $\mathcal{A}$, denoted by $\sigma(\mathcal{A})$. The Borel σ-algebra $\mathcal{B}$ of the real line R is the smallest σ-algebra containing all the open intervals or, equivalently, all subsets of the form $(a, b], -\infty < a < b < \infty$. One version of the Monotone Class Theorem states that if $\mathcal{A}_0$ is an algebra of sets then the smallest σ-algebra and the smallest monotone class containing $\mathcal{A}_0$ coincide. The proof of Lemma 1.5.2 uses a slightly different version, so we state both versions here. Proofs may be found in Chung (1974) and Appendix A1 of Bremaud (1981). The proof of Lemma 1.5.2 is given later in this Appendix, after the definition of random variables.

Theorem A.2.1. (Monotone Class Theorem)

1. Let $\mathcal{A}_0$ be an algebra of subsets of a space Ω, and $\mathcal{A}$ and $\mathcal{M}$ the smallest σ-algebra and monotone class, respectively, containing $\mathcal{A}_0$. Then $\mathcal{A} = \mathcal{M}$.
2. Let $\mathcal{A}_0$ be a class of subsets of a space Ω that is closed under finite intersection. Let $\mathcal{A}$ be the smallest σ-algebra containing $\mathcal{A}_0$, and let $\mathcal{N}$ be the smallest collection of subsets of Ω satisfying:
   a) $\mathcal{A}_0 \subset \mathcal{N}$
   b) $\mathcal{N}$ contains both Ω and $\emptyset$.
   c) $E_1 \cap \bar{E}_2 \in \mathcal{N}$ whenever $E_1, E_2 \in \mathcal{N}$ and $E_2 \subset E_1$.
   d) $\bigcup_{n=1}^{\infty} E_j \in \mathcal{N}$ when $E_j \in \mathcal{N}$ and $E_j \subset E_{j+1}, j = 1, 2, \ldots$.
   Then $\mathcal{N} = \mathcal{A}$. □

Definition A.2.2. A *probability space* $(\Omega, \mathcal{F}, P)$ is an abstract space Ω equipped with a σ-algebra $\mathcal{F}$ and a set function P defined on $\mathcal{F}$ satisfying

1. $P\{E\} \geq 0$ when $E \in \mathcal{F}$.
2. $P\{\Omega\} = 1$.
3. $P\{\bigcup_{j=1}^{\infty} E_j\} = \sum_{j=1}^{\infty} P\{E_j\}$ whenever $E_j \in \mathcal{F}, 1 \leq j < \infty$, and $E_i \cap E_j = \emptyset, i \neq j$. □

The most common probability measures are those on $(R, \mathcal{B})$, the real line equipped with the Borel σ-algebra, and are specified either through a density function f, so that $P\{(a, b]\} = \int_a^b f(x)dx$, or through a probability mass function p, so that $P\{a_1, \ldots, a_k\} = \sum_{j=1}^{k} p(a_j)$. Much of applied probability and statistical inference is a study of probability and statistical models built from particular densities and mass functions, such as the normal density, the Poisson probability function, etc. The reader should be well versed in these models.

Definition A.2.3. Let $(\Omega, \mathcal{F}, P)$ be a probability space. A *random variable* $X : \Omega \to R$ is a real-valued function from Ω to R satisfying $\{\omega : -\infty < X(\omega) \leq b\} \equiv X^{-1}\{(-\infty, b]\} \in \mathcal{F}$. □

In the language of measure theory, a random variable is a measurable mapping from $(\Omega, \mathcal{F})$ to $(R, \mathcal{B})$, and the probability measure P_X on $(R, \mathcal{B})$ given by $P_X\{(a, b]\} = P\{\omega : a < X(\omega) \leq b\}$ is the probability induced on R by X. Convention dictates that the argument of X be suppressed whenever this does not lead to ambiguity, so it is more common to write $P_X\{(a, b]\} = P\{a < X \leq b\}$. The function $F_X(b) = P\{X \le b\}, -\infty < b \le \infty$, is called the cumulative distribution function of X.

Measurable functions are the core of the Lebesgue theory of integration and, because of Kolmogorov's axioms, they are the centerpiece of the abstract approach to probability theory. There are many reasons for their prominence, but the most important may be that measurability is preserved under many limiting operations. That is, if $X_n, n = 1, 2, \ldots$, is a sequence of measurable functions, then $\sup_n X_n$, $\inf_n X_n$, $\limsup_n X_n$ and $\liminf_n X_n$ are all measurable, as is $\lim_n X_n$ when it exists. This result is used in the proof of Lemma 1.5.2, which we give here. The notation used in the proof is from Section 5 of Chapter 1. The proof can be skipped without any loss of continuity by readers interested in only the main results of the Appendix.

Proof of Lemma 1.5.2. We show first that $\mathcal{H}$ contains all the mappings $I_A, A \in \sigma(\mathcal{S})$, then extend to all the measurable mappings from $(D_0, \sigma(\mathcal{S}))$. The first step is established by showing that $\mathcal{D} = \{A : I_A \in \mathcal{H}\}$ contains $\sigma(\mathcal{S})$.

Note first that $\mathcal{S} \subset \mathcal{D}$ by Assumption (1) in the statement of the lemma. Since $\mathcal{H}$ contains the constant functions, Ω and $\emptyset$ are both in $\mathcal{D}$. If $E_1, E_2 \in \mathcal{D}, E_2 \subset E_1$, then $I_{E_1 \cap \bar{E}_2} = I_{E_1} - I_{E_2}$ and, since $\mathcal{H}$ is a vector space, $E_1 \cap \bar{E}_2 \in \mathcal{D}$. Suppose $\{E_j, j = 1, 2, \ldots\}$ is a collection of subsets of $\mathcal{D}$ with $E_j \subset E_{j+1}, j = 1, 2, \ldots$. Then $I_{\{\cup_j E_j\}} = \lim_{j\to\infty} I_{E_j} = \sup_j I_{E_j}$, and $\bigcup_{j=1}^{\infty} E_j \in \mathcal{D}$ by Assumption (2) of Lemma 1.5.2. By Theorem A.2.1(2), $\sigma(\mathcal{S}) \subset \mathcal{D}$.

Now let $X : (D_0, \sigma(\mathcal{S})) \to (R, \mathcal{B})$ be measurable, and let $X^{\oplus} = \max(X, 0)$, $X^{\ominus} = \max(-X, 0)$. Then $X = X^{\oplus} - X^{\ominus}$ and $X \in \mathcal{H}$ whenever $X^{\oplus}$ and $X^{\ominus}$ are in $\mathcal{H}$. Thus we must only show the nonnegative measurable mappings of $(D_0, \sigma(\mathcal{S}))$ are contained in $\mathcal{H}$. But for any nonnegative mapping X it is possible to find a sequence $\{X_n, n = 1, 2, \ldots\}$ such that $X_{n+1} \ge X_n, X_n = \sum_{j=1}^{k_n} a_{j,n} I_{E_{j,n}}, E_{j,n} \in \mathcal{S}$, and $\lim_{n\to\infty} X_n = X$. (cf Ash, 1972, Theorem 1.5.5a). Since $\mathcal{H}$ is a vector space, $X_n \in \mathcal{H}$, and $X = \sup_n X_n \in \mathcal{H}$ by Assumption (2). If X is bounded, then the X_n can be taken to be bounded and Lemma 1.5.2 is proved. □

In the abstract approach to probability theory, the expectation or expected value $E(X)$ of a random variable X on $(\Omega, \mathcal{F}, P)$ is $\int_\Omega X dP$, as long as $\int_\Omega |X| dP < \infty$. The integral here is a Lebesgue integral, but of course when X has a density function f or probability mass function p this reduces to $\int_{-\infty}^{\infty} x f(x) dx$ or $\sum_j x_j p(x_j)$. Using the Lebesgue integral often allows easy applications of the convergence theorems for sequences of integrals of measurable functions, i.e. of random variables, and those convergence theorems are used throughout the book. We summarize the most important of those theorems here, giving the versions for expectations of random variables. The versions for integrals of arbitrary measurable functions differ only in the language used, and can be found in Royden (1968). Each of the theorems depends on the notion of almost sure convergence and convergence in probability, which are the probabilistic analogues of convergence almost everywhere and convergence in measure.

Definition A.2.4. Let X and $X_n, n = 1, 2, \ldots$, be random variables defined on a probability space $(\Omega, \mathcal{F}, P)$.

1. The sequence $\{X_n : n = 1, 2, \ldots\}$ converges to X *almost surely* (a.s.) if $P\{\omega : \lim_{n\to\infty} X_n(\omega) = X(\omega)\} = 1$.
2. The sequence $\{X_n : n = 1, 2, \ldots\}$ converges to X *in probability* if

$$\lim_{n\to\infty} P\{\omega : |X_n(\omega) - X(\omega)| \geq \epsilon\} = 0$$

for all $\epsilon > 0$. In this case we write $X_n \xrightarrow{P} X$. □

As in the definition of almost sure convergence, any condition is said to hold almost surely if it holds on a set of probability one.

A third type of convergence, convergence in distribution, is discussed in Appendix B.

Theorem A.2.2. Let X and $X_n, n = 1, 2, \ldots$, be a sequence of random variables defined on $(\Omega, \mathcal{F}, P)$.

1. *Fatou's Lemma:* If $X_n \geq 0$ a.s., then

$$E(\liminf_{n\to\infty} X_n) \leq \liminf_{n\to\infty} E(X_n).$$

If $X_n \to X$ either almost surely or in probability, then $E(X_n) \to E(X)$ whenever at least one of the following holds:

2. *Dominated Convergence Theorem:* $|X_n| \leq Y$ a.s. and $E(Y) < \infty$.
3. *Bounded Convergence Theorem:* There is a constant k such that $|X_n| \leq k$ a.s. for all n.
4. *Monotone Convergence Theorem:* $X_n \geq 0$ a.s. and $X_n \leq X_{n+1}$ for all n.

□

The abstract definition of a stochastic process uses the general definition of a random variable.

Definition A.2.5. A *stochastic process* is a family of random variables $X = \{X(t) : t \in \Gamma\}$ indexed by a set Γ, all defined on the same probability space $(\Omega, \mathcal{F}, P)$. □

Both the theory and the applications of stochastic processes are extensive. The index set Γ usually denotes time, and is either $[0, \infty)$ or $\{0, 1, 2, \ldots\}$. Most of the applied literature is devoted to methods for calculating or approximating the distribution of the random variable $X(t)$ or the joint distribution of $\{X(t_1), X(t_2), \ldots, X(t_k)\}$. Karlin and Taylor (1975) provide an excellent introduction to this approach. Rather than study the properties of the random variable $X(t)$ for fixed t, the modern approach to the general theory of processes relies on properties of the sample path $X(t, \omega), t \in \Gamma$, for fixed ω. The modern perspective is explored in Dellacherie and Meyer (1975). The seminal text in the study of stochastic processes is by Doob (1953). The approach in this book relies much more on the general theory than on the more traditional approaches, and we summarize here a few of the commonly used results. While this material duplicates some of the discussion in Chapter 1, the pace here is slightly more relaxed.

The processes we study almost always have index set $\Gamma = [0, \infty)$, and much of the development in this book depends on the sample path behavior $\{X(t) : 0 \leq t < \infty\}$ for fixed ω, or on the measurability properties that a process X has when considered as a mapping from $R^+ \times \Omega$ to R, where $R^+ = [0, \infty)$. The sample path properties of X are easiest to define, and are introduced in the beginning of Chapter 1. A process is called continuous, or left- or right-continuous, or of bounded variation, if the subset of Ω on which a particular path property holds has probability one.

While sample path properties are easy to define, there is a subtlety that should not be overlooked, as the following example from Elliott (1982) shows.

Example A.2.1. Let $\Omega = [0, 1], \mathcal{F}$ the Borel subsets of Ω, and P Lebesgue measure. Define the process $Y = \{Y(t) : 0 \leq t < \infty\}$ by

$$\begin{aligned} Y(t, \omega) &= 1 \qquad \text{if } t - [t] = \omega \\ &= 0 \qquad \text{otherwise,} \end{aligned}$$

where $[t]$ is the greatest integer less than or equal to t. For any $\omega \in \Omega$, the path $Y(t), t \geq 0$, will have a countable number of discontinuities. For any fixed t, however, the set of paths on which Y is continuous at t will have probability one, since Y will be continuous at t for all $\omega \neq \omega_t = t - [t]$. Moreover, if X is the process on the same probability space defined by $X(t) = 0$ for all t and all ω then, for any fixed $t, P\{X(t) = Y(t)\} = 1$ but the probability of the set on which the paths of X and Y coincide is zero. □

Example A.2.1 illustrates the difficulty involved in saying that two processes X and Y are the "same" process. Two random variables X and Y are equivalent if $P\{X = Y\} = 1$. A process Y is called a modification of a process X if $P\{X(t) = Y(t)\} = 1$ for all t in the index set. Two processes X and Y are called indistinguishable if $P\{\omega : X(t) = Y(t) \text{ for all } t \in \Gamma\} = 1$, that is, the paths of X and Y coincide with probability one. Fortunately, this subtle distinction is unnecessary for the processes with well-behaved paths that are used in this book, as the next theorem shows.

Theorem A.2.3. Suppose the process $\{Y(t) : t \geq 0\}$ is a modification of $\{X(t) : t \geq 0\}$ and that both processes are right-continuous or that both are left-continuous. Then the two processes are indistinguishable.

Proof. Let Q denote the set of rational numbers. For each $r \in Q, P\{X(r) \neq Y(r)\} = 0$. If $D \subset \Omega$ is defined by

$$D = \bigcup_{r \in Q} \{X(r) \neq Y(r)\},$$

then $P(D) = 0$. Suppose both X and Y are right-continuous, and consider a sample path $\omega \notin D$ for the two processes. At any $t \geq 0$, let $r_n, n = 1, 2, \ldots$, be a sequence of rational numbers decreasing to t. Then by right-continuity

$$X(t) = \lim_{n \to \infty} X(r_n) = \lim_{n \to \infty} Y(r_n) = Y(t).$$

The same argument works if both X and Y are left-continuous. □

In the counting process approach to survival analysis, the paths of stochastic processes are often used in integrals $\int X dY$ as either integrands or integrators. Both X and Y must satisfy certain measurability and sample path properties to insure that $\int_s^t X(\cdot, \omega) dY(\cdot, \omega)$ is well-defined as a Lebesgue-Stieltjes integral for each ω. The sample path of X for each ω must be a measurable function on $(s, t]$, and each path of Y must have bounded variation on $(s, t]$. Nearly all the processes encountered in the book are of bounded variation on finite intervals, and that condition will always be easy to verify when needed. A result from the theory of stochastic processes can be used to check the measurability of X. We lead up to that result with definitions.

Definition A.2.6. A family of σ-algebras $\{\mathcal{F}_t : t \geq 0\}$ is called a *filtration* of a probability space $(\Omega, \mathcal{F}, P)$ if $\mathcal{F}_t \subset \mathcal{F}$ for all t and $\mathcal{F}_s \subset \mathcal{F}_t$ for $s \leq t$. □

A filtration $\{\mathcal{F}_t : t \geq 0\}$ is often used as an abstraction for information that is increasing with time; that notion is explored in detail throughout the book.

Definition A.2.7. A process $\{X(t) : t \geq 0\}$ is said to be *adapted* to $\{\mathcal{F}_t : t \geq 0\}$ if $X(t)$ is $\mathcal{F}_t$-measurable for each t. □

Theorem A.2.4. Suppose X is a stochastic process adapted to a filtration on a probability space $(\Omega, \mathcal{F}, P)$. If X is either a right-continuous process or a left-continuous process, the restricted mapping $X : [0, t] \times \Omega \to R$ is a measurable mapping when $[0, t] \times \Omega$ is given the product σ-algebra $\mathcal{B}_t \times \mathcal{F}_t$, where $\mathcal{B}_t$ consists of the Borel sets of $[0, t]$.

Proof. For any $t > 0$, partition the interval $[0, t]$ with $0 = a_0^n < a_1^n < \ldots < a_{2^n}^n = t$, where $a_j^n = t2^{-n}j, j = 0, 1, \ldots, 2^n$. Suppose X is right-continuous. Define the process X_n at $s \in [0, t)$ by

$$X_n(s) = \sum_{j=1}^{2^n} I_{\{a_{j-1}^n \leq s < a_j^n\}} X(a_j^n),$$

and take $X_n(t) = X(t)$. If B is any Borel subset of R, then $X_n(s,\omega) \in B$ if $s < t$ and $a^n_{j-1} \leq s < a^n_j$ and $X(a^n_j) \in B$, or if $s = t$ and $X(t) \in B$. Thus

$$X_n^{-1}(B) = \bigcup_{j=1}^{2^n} [[a^n_{j-1}, a^n_j) \times \{X(a^n_j, \cdot)\}^{-1}(B)]$$
$$\bigcup \ [\{t\} \times \{X(t,\cdot)\}^{-1}(B)]$$
$$\in \mathcal{B}_t \times \mathcal{F}_t.$$

Since $\lim_{n\to\infty} X_n(s,\omega) = X(s,\omega)$ for $(s,\omega) \in [0,t]\times\Omega$, X is the limit of $\mathcal{B}_t \times \mathcal{F}_t$-measurable mappings and so must itself be $\mathcal{B}_t \times \mathcal{F}_t$-measurable. A similar argument works when X is left-continuous. □

Theorem A.2.4 and Fubini's Theorem imply that when X is an adapted process that is either left-continuous or right-continuous, for any fixed ω, $X(\cdot,\omega)$ will be a measurable function on $[0,t]$ for any $t < \infty$ and can be used as an integrand in any integral $\int_s^t X dY$ for $t < \infty$. Since $\int_s^\infty X dY$ is a limit of integrals over finite intervals, integrals over infinite intervals are also well defined.

The next theorem summarizes a measurability property that Stieltjes stochastic integrals have. The theorem is similar to Proposition 3.2.4 in Jacobsen (1982) and can be proved using the arguments in that text. The definition of a right-continuous filtration appears in Definition A.2.8 following the theorem.

Theorem A.2.5. Let X and Y be stochastic processes on $(\Omega, \mathcal{F}, P)$, both adapted to a right-continuous filtration $\{\mathcal{F}_t : t \geq 0\}$. Suppose the paths of X are right- or left-continuous, and the paths of Y are right-continuous and of bounded variation on finite intervals. Then the process whose value at time t is $\int_0^t X dY$ is also adapted to $\{\mathcal{F}_t : t \geq 0\}$. □

Note that Theorem A.2.5 implies that for any t, $\int_0^t X dY$ will be a random variable, and consequently for any two time points s and t, $s \leq t$, $\int_s^t X dY$ is a random variable.

Many results from the general theory of stochastic processes and consequently in this book rely on two conditions that filtrations may satisfy.

Definition A.2.8.

1. A filtration $\{\mathcal{F}_t : t \geq 0\}$ on a probability space $(\Omega, \mathcal{F}, P)$ is called *complete* if $\mathcal{F}$ is complete (i.e., all subsets of P-null sets are in $\mathcal{F}$) and if each $\mathcal{F}_t$ contains all the P-null sets of $\mathcal{F}$.
2. A filtration $\{\mathcal{F}_t : t \geq 0\}$ is called *right-continuous* if $\mathcal{F}_t = \bigcap_{s>0} \mathcal{F}_{t+s}$. □

It is not difficult to see why completeness can be important. If X is a process adapted to $\{\mathcal{F}_t : t \geq 0\}$, then a modification Y of X will be adapted if each $\mathcal{F}_t$ contains the P-null subsets of $\mathcal{F}$. In general, an incomplete filtration can always be completed by replacing $\mathcal{F}$ with its completion $\mathcal{F}^P$, and replacing each $\mathcal{F}_t$ with the σ-algebra generated by $\mathcal{F}_t$ and the P-null subsets of $\mathcal{F}^P$. Since completing a filtration changes no important features of P, we can assume that all filtrations encountered are complete and do so throughout the book.

More care must be taken with right-continuity. It is a difficult condition to check, and surprisingly elusive (cf. Exercise 1.5). We will rely throughout on the following definition and theorem for verifying right-continuity of a filtration.

Definition A.2.9. Let $X = \{X(t) : t \geq 0\}$ be a stochastic process, and let $\mathcal{F}_t = \sigma\{X(s) : 0 \leq s \leq t\}$, the smallest σ-algebra making all of the random variables $X(s), 0 \leq s \leq t$, measurable. The filtration $\{\mathcal{F}_t : t \geq 0\}$ is called the *history* of X. □

The proof of the next theorem may be found in Appendix A2 of Bremaud (1981).

Theorem A.2.6. Let $X = \{X(t) : t \geq 0\}$ be a stochastic process on $(\Omega, \mathcal{F}, P)$, and suppose that for all $t \geq 0$ and all $\omega \in \Omega$ there exists a number $\epsilon(t, \omega) > 0$ such that

$$X(t + s, \omega) = X(t, \omega), \qquad 0 \leq s < \epsilon(t, \omega).$$

Then the history of X is a right-continuous filtration. □

APPENDIX B

An Introduction to Weak Convergence

In Chapter 5, the martingale structure for stochastic processes of the form $\sum_{i=1}^{n} \int H_i dM_i$ yields central limit theorems and weak convergence results characterizing the large-sample distributions of corresponding statistics. These asymptotic results often provide an adequate approximation to small sample distributions.

For readers unfamiliar with convergence in distribution of stochastic processes, this appendix provides background for Chapter 5 by briefly outlining that concept, its importance, and methods for establishing convergence of stochastic processes. Billingsley (1968) and Pollard (1984) provide detailed accounts of this topic.

Weak convergence (i.e., convergence in distribution) of stochastic processes generalizes convergence in distribution of real-valued random variables. We begin with a convergence criterion for a sequence of distribution functions of ordinary random variables.

Definition B.1.1. For arbitrary distribution functions on the real line, F and F_n, $n = 1, 2, \ldots$, F_n *converges weakly* to F as $n \to \infty$ (written $F_n \Longrightarrow F$) if and only if $F_n(x) \to F(x)$ at all continuity points of F. □

A simple example illustrates a weakly convergent sequence F_n which does not converge at the discontinuity point of its limit.

Example B.1.1. Suppose F_n and F are the distribution functions displayed in Figure B.1.1 (a) and (b), respectively. Then $F_n \Longrightarrow F$ as $n \to \infty$, even though $F_n(0) = 0 \neq 1 = F(0)$. □

Definition B.1.2. Consider a sequence of random variables $\{X, X_n, n = 1, 2, \ldots\}$, with each member taking on values on the real line $R = (-\infty, \infty)$, and let $F_n(t) = P\{X_n \leq t\}$, $F(t) = P\{X \leq t\}$. We say X_n *converges in distribution* to X (written $X_n \xrightarrow{\mathcal{D}} X$) if and only if $F_n \Longrightarrow F$. □

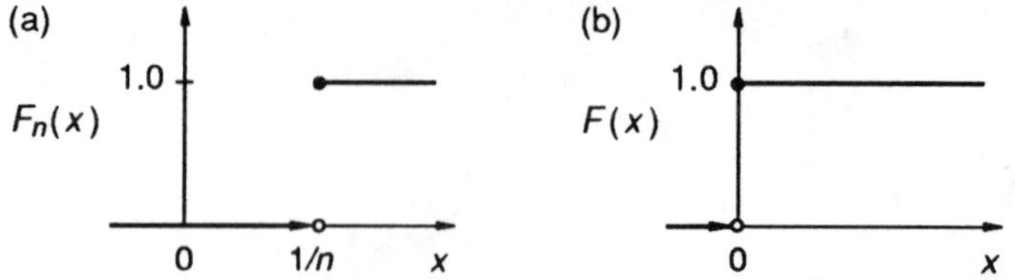

Figure B.1.1 A weakly convergent sequence, pictured in (a), which does not converge at the discontinuity point ($x = 0$) of the limit F, shown in (b).

The central limit theorem is the most common instance of convergence in distribution: Let $X_1, X_2, \ldots$ be independent and identically distributed random variables with $E(X_i) = \mu$ and $\text{var}(X_i) = \sigma^2 < \infty$. Let

$$Z_n = \left\{ \sum_{i=1}^{n} X_i - n\mu \right\} / (\sigma\sqrt{n})$$

and let $Z \sim N(0, 1)$. Then

$$Z_n \xrightarrow{\mathcal{D}} Z;$$

equivalently, for any $t \in R$,

$$P\{Z_n \leq t\} \to \int_{-\infty}^{t} (2\pi)^{-1/2} e^{-s^2/2} ds = P\{Z \leq t\} \quad \text{as } n \to \infty.$$

Our interest in large-sample distributions extends beyond results for real-valued random variables, however, to large-sample properties of functions of processes $\{X_n(t) : 0 \leq t \leq u\}$ of the form $X_n(t) = \sum_{i=1}^{n} \int_0^t H_i(s) dM_i(s)$. With the right generalization of weak convergence from convergence in distribution of a sequence of random variables, $X_n \xrightarrow{\mathcal{D}} X$, to convergence of a sequence of stochastic processes $\{X_n(t) : 0 \leq t \leq u\}$ to $\{X(t) : 0 \leq t \leq u\}$, the Continuous Mapping Theorem, (given later) yields convergence of $f\{X_n(t) : 0 \leq t \leq u\}$ to $f\{X(t) : 0 \leq t \leq u\}$ for "continuous functionals" f. Examples of the single- and multi-dimensional convergence results which would follow include

$$\sup_{0 \leq t \leq u} X_n(t) \xrightarrow{\mathcal{D}} \sup_{0 \leq t \leq u} X(t),$$

$$\{X_n(t_1), X_n(t_2), \ldots, X_n(t_k)\} \xrightarrow{\mathcal{D}} \{X(t_1), X(t_2), \ldots, X(t_k)\}$$

for

$$0 \leq t_1 \leq t_2 \leq \ldots \leq t_k \leq u,$$

and

$$\{\sup_{0 \leq t \leq u} X_n(t), X_n(u)\} \xrightarrow{\mathcal{D}} \{\sup_{0 \leq t \leq u} X(t), X(u)\}.$$

This more general concept of weak convergence thus yields a range of convergence in distribution results, many being very difficult to establish using other methods.

Let's return to Definition B.1.2. Since X_n is a random variable with respect to $(\Omega, \mathcal{F}, P)$, we have $\{X_n^{-1}(B)\} \equiv \{\omega : X_n(\omega) \in B\} \in \mathcal{F}$ for any $B \in \mathcal{B}$, where $\mathcal{B}$ is the Borel σ-algebra of R. Thus,

$$(\Omega, \mathcal{F}, P) \stackrel{X_n}{\to} (R, \mathcal{B}, \mathcal{P}_n)$$

where $\mathcal{P}_n$ is defined by $\mathcal{P}_n\{B\} = P\{X_n^{-1}(B)\}$ for any $B \in \mathcal{B}$, and $\mathcal{P}\{B\} = P\{X^{-1}(B)\}$.

When $X_n \stackrel{\mathcal{D}}{\to} X$ and the set $B = (a, b]$, then

$$\begin{aligned} \mathcal{P}_n\{B\} &= P\{a < X_n \leq b\} \\ &= F_n(b) - F_n(a) \\ &\to F(b) - F(a) \\ &= P\{a < X \leq b\} \\ &= \mathcal{P}\{B\}, \end{aligned}$$

whenever $\mathcal{P}\{\{a, b\}\} = P\{X = a \text{ or } X = b\} = 0$. Thus $X_n \stackrel{\mathcal{D}}{\to} X$ implies $\mathcal{P}_n\{B\} \to \mathcal{P}\{B\}$ for all Borel sets $B = (a, b]$ whose boundaries $\{a, b\}$ have probability zero with respect to the measure $\mathcal{P}$. We have motivated a definition of weak convergence in terms of convergence of probability measures.

Definition B.1.3. Suppose $\mathcal{B}$ is the Borel σ-algebra of R and let $\mathcal{P}$ and $\mathcal{P}_n$ be probability measures on $(R, \mathcal{B})$. Let ∂B denote the boundary of any set $B \epsilon \mathcal{B}$. We say $\mathcal{P}_n$ *converges weakly* to $\mathcal{P}$ (written $\mathcal{P}_n \Longrightarrow \mathcal{P}$) as $n \to \infty$ if and only if $\mathcal{P}_n\{B\} \to \mathcal{P}\{B\}$ for each $B \epsilon \mathcal{B}$ such that $\mathcal{P}\{\partial B\} = 0$. □

The next lemma has a straightforward proof.

Lemma B.1.1. Let $\mathcal{P}$ and $\mathcal{P}_n$ be the probability measures on $(R, \mathcal{B})$ corresponding to distribution functions F and $F_n, n = 1, 2, \ldots$. Then $F_n \Longrightarrow F$ if and only if $\mathcal{P}_n \Longrightarrow \mathcal{P}$. □

Lemma B.1.1 suggests that Definition B.1.3 can be used to extend weak convergence to more general spaces of random objects (e.g., stochastic processes) in a way that reduces to Definition B.1.2 for real-valued random variables. We now give a more precise statement of weak convergence, $\mathcal{P}_n \Longrightarrow \mathcal{P}$, in an arbitrary metric space.

Definition B.1.4. Suppose Γ is a metric space and let $\mathcal{S}_0$ denote the smallest σ-algebra containing all open sets. Let $\mathcal{P}_n$ and $\mathcal{P}$ be probability measures on $(\Gamma, \mathcal{S}_0)$. Then we say $\mathcal{P}_n \Longrightarrow \mathcal{P}$ if and only if either

1. $\mathcal{P}_n\{A\} \to \mathcal{P}\{A\}$ for each set $A \in \mathcal{S}_0$ such that $\mathcal{P}\{\partial A\} = 0$, or
2. $\liminf_n \mathcal{P}_n\{\mathcal{O}\} \geq \mathcal{P}\{\mathcal{O}\}$ for each open set $\mathcal{O} \in \mathcal{S}_0$. □

Conditions (1) and (2) are equivalent (cf. Billingsley, 1968, Theorem 2.1). Generally, Γ is a metric space with metric $\rho(x, y)$ denoting the distance from x to y, for each $x, y \in \Gamma$. One then uses ρ to define the open sets and hence the σ-algebra $\mathcal{S}_0$ generated by the open sets. However, it is important to realize that it is the definition of the open sets (i.e., the topology), rather than the definition of the metric, that is intrinsically important to the definition of weak convergence.

The Continuous Mapping Theorem extends weak convergence from a sequence of measures on one metric space $(\Gamma, \mathcal{S}_0)$ to an induced sequence of measures on another $(\Gamma', \mathcal{S}'_0)$.

Theorem B.1.1 (The Continuous Mapping Theorem). Suppose f is a continuous mapping from one metric space $(\Gamma, \mathcal{S}_0)$ into another $(\Gamma', \mathcal{S}'_0)$. If $\mathcal{P}_n \Longrightarrow \mathcal{P}$ on Γ, then $\mathcal{P}_n f^{-1} \Longrightarrow \mathcal{P} f^{-1}$ on Γ'.

Continuous functions on an interval $[0, u]$ are said to belong to the space $C[0, u]$. Right-continuous functions on $[0, u]$ which have finite left-hand limits at each point of $[0, u]$ are said to belong to $D[0, u]$. We will use the general definition of weak convergence for stochastic processes whose paths are in $C[0, u]$ with probability one, or are in $D[0, u]$. We begin with $C[0, u]$.

Consider a stochastic process $X = \{X(t) : 0 \leq t \leq u\}$ for a positive constant u, and suppose X has continuous sample paths. Then

$$\Omega \xrightarrow{X} C[0, u].$$

The uniform metric, ρ, on $C[0, u]$ is defined by

$$\rho(x, y) = \sup_{0 \leq t \leq u} | x(t) - y(t) |$$

for continuous functions x and y on $[0, u]$. A set $\mathcal{O}$ will be open in the topology generated by ρ if, for each $x \in \mathcal{O}$, there is a $\delta > 0$ such that each $y \in C[0, u]$ with $\sup_{0 \leq t \leq u} |x(t) - y(t)| < \delta$ belongs to $\mathcal{O}$. Let $\mathcal{S}$ denote the smallest σ-algebra containing the open sets generated by the uniform metric.

An example illustrates one important use of weak convergence on $C[0, u]$.

Example B.1.2. Consider a stochastic process $X = \{X(t) : 0 \leq t \leq u\}$ and a sequence of processes $X_n = \{X_n(t) : 0 \leq t \leq u\}$, $n = 1, 2, \ldots$, each having continuous sample paths. For any $A \in \mathcal{S}$, let

$$\mathcal{P}_n\{A\} = P\{X_n^{-1}(A)\} = P\{\omega : X_n(\omega) \in A\}$$

and

$$\mathcal{P}\{A\} = P\{X^{-1}(A)\} = P\{\omega : X(\omega) \in A\},$$

where Billingsley (1968, p. 57) establishes $X_n^{-1}(A) \in \mathcal{F}$ and $X^{-1}(A) \in \mathcal{F}$. The function $f : C[0,u] \to R$ defined by $f(x) = \sup_{0\leq t\leq u} |x(t)|$ is continuous, since $f(x) = \rho(x, 0)$ and any metric is a continuous function of its arguments. Consequently the Continuous Mapping Theorem can be used to determine when $f(X_n) = \sup_{0\leq t\leq u} X_n(t) \xrightarrow{\mathcal{D}} \sup_{0\leq t\leq u} X(t) = f(X)$.

A diagram helps explain what is happening:

$$(\Omega, \mathcal{F}, P) \xrightarrow{X_n} (C[0,u], \mathcal{S}, \mathcal{P}_n) \xrightarrow{f} (R, \mathcal{B}, \mathcal{P}_n^*).$$

In this diagram $\mathcal{P}_n\{A\} \equiv P\{X_n^{-1}(A)\}$ for any $A \epsilon \mathcal{S}$. Since f is continuous, for any $B \epsilon \mathcal{B}$ there exists $A \epsilon \mathcal{S}$ such that $f^{-1}(B) = A$, so it is appropriate to write

$$\mathcal{P}_n^*\{B\} \equiv P\{X_n^{-1}[f^{-1}(B)]\} = \mathcal{P}_n f^{-1}(B). \tag{1.1}$$

Now suppose $\mathcal{P}_n \Longrightarrow \mathcal{P}$ (these are measures on $(C[0,u], \mathcal{S})$). By the Continuous Mapping Theorem (B.1.1),

$$\mathcal{P}_n f^{-1} \Longrightarrow \mathcal{P} f^{-1} \tag{1.2}$$

(these are measures on $(R, \mathcal{B})$). Taking $B \equiv (-\infty, v]$ for any real number v such that $P\{f(X) = v\} = 0$, we have by Eqs. (1.1) and (1.2) that

$$P\{\omega : f(X_n)(\omega) \leq v\} \to P\{\omega : f(X)(\omega) \leq v\}$$

so

$$\sup_{0\leq t\leq u} X_n(t) \xrightarrow{\mathcal{D}} \sup_{0\leq t\leq u} X(t). \tag{1.3}$$

In conclusion, weak convergence, $\mathcal{P}_n \Longrightarrow \mathcal{P}$, is sufficient to prove Eq. (1.3). □

Weak convergence of a sequence of processes with paths in $C[0,u]$, then, is specified through the behavior of the associated sequence of probability measures on the topological space $(C[0,u], \mathcal{S})$, where $\mathcal{S}$ is the smallest σ-algebra containing the open sets generated by the uniform metric. Although it is not obvious, weak convergence is stronger than convergence of the finite-dimensional distribution functions for the sequence of processes, and it is useful to examine the relationship between the two criteria for convergence.

Let R^k denote k-dimensional Euclidean space with elements $\mathbf{x}' = (x_1, x_2, \ldots, x_k)$. For any k time points $t_1, t_2, \ldots, t_k \in [0,u]$, the finite-dimensional distribution of a process $X \in C[0,u]$ at $t_1, \ldots, t_k$ is

$$F(\mathbf{x}) = P\{X(t_1) \leq x_1, X(t_2) \leq x_2, \ldots, X(t_k) \leq x_k\}.$$

Knowledge of $\{X(t) : 0 \leq t \leq u\}$ completely specifies $\{X(t_1), \ldots, X(t_k)\}$ for any k-tuple $(t_1, \ldots, t_k)$, and consequently all finite-dimensional distributions of a process $X \in C[0,u]$ are characterized by its probability measure on the space $(C[0,u], \mathcal{S})$.

A sequence of probability measures $\mathcal{P}_n$ on the topological space $(C[0,u],\mathcal{S})$ is admittedly abstract and difficult to work with. The finite-dimensional distribution functions are more familiar items, and can sometimes be computed explicitly. Unfortunately, the next theorem and example establish the previous claim that convergence of all sequences of finite-dimensional distributions does not imply weak convergence.

Let $\mathcal{B}^k$ be the usual Borel sets of R^k.

Theorem B.1.2. Let $\pi_{t_1,\ldots,t_k}$ be the projection mapping

$$\pi_{t_1,\ldots,t_k} : (C[0,u],\mathcal{S},\mathcal{P}) \to (R^k,\mathcal{B}^k,\mathcal{P}\pi^{-1}_{t_1,\ldots,t_k})$$

that carries the element $X(\omega) \equiv x \in C[0,u]$ to the element $\{x(t_1),\ldots,x(t_k)\}$ of R^k. Let $H \in \mathcal{B}^k$. Since $\pi_{t_1,\ldots,t_k}$ is continuous, $\pi^{-1}_{t_1,\ldots,t_k}(H) \equiv A \in \mathcal{S}$, where $\mathcal{S}$ is the σ-algebra generated by open sets on $C[0,u]$. Thus, we can define $\mathcal{P}\pi^{-1}_{t_1,\ldots,t_k}(H) \equiv \mathcal{P}(A)$. (The set A is called a "finite-dimensional set" in $C[0,u]$.)

For every n, let $\mathcal{P}_n, \mathcal{P}$ and $\mathcal{Q}$ denote probability measures on $(C[0,u],\mathcal{S})$. Suppose for any k and for any $0 \le t_1 \le t_2 \le \ldots \le t_k \le u$,

1. $\mathcal{P}\pi^{-1}_{t_1,\ldots,t_k} = \mathcal{Q}\pi^{-1}_{t_1,\ldots,t_k}$, or
2. $\mathcal{P}_n\pi^{-1}_{t_1,\ldots,t_k} \Longrightarrow \mathcal{P}\pi^{-1}_{t_1,\ldots,t_k}$.

Then, under (1), $\mathcal{P} = \mathcal{Q}$. However, under (2), $\mathcal{P}_n \Longrightarrow \mathcal{P}$ need not hold. □

Note that all measures in (1) and (2) of Theorem B.1.2 are probability measures on $(R^k,\mathcal{B}^k)$. Consider probability measures $\mathcal{P}$ and $\mathcal{Q}$ on $C[0,u]$. Theorem B.1.2 states that if the corresponding finite-dimensional distribution functions (on R^k) are always equal, then $\mathcal{P} = \mathcal{Q}$. However, if $\{\mathcal{P}_n : n \ge 0\}$ and $\mathcal{P}$ are probability measures on $C[0,u]$, convergence of all corresponding finite-dimensional distribution functions (on R^k, for all k), need not imply $\mathcal{P}_n \Longrightarrow \mathcal{P}$.

In the following example, finite-dimensional distribution functions of processes on $C[0,u]$ converge, but the processes do not converge weakly.

Example B.1.3. Suppose X and X_n are continuous deterministic processes x and x_n, with $x(t) = 0$ for all $t \in [0,u]$ and x_n defined in Figure B.1.2. The corresponding probability measures $\mathcal{P}\{\cdot\} \equiv P\{X^{-1}(\cdot)\}$ and $\mathcal{P}_n\{\cdot\} \equiv P\{X_n^{-1}(\cdot)\}$ on $(C[0,u],\mathcal{S})$ place unit mass on the functions x and x_n, respectively. Choose k and $0 \le t_1 \le t_2 \le \ldots \le t_k \le u$. For any n such that u/n is smaller than the least non-zero t_i,

$$\mathcal{P}_n\pi^{-1}_{t_1,\ldots,t_k}\{H\} = \mathcal{P}\pi^{-1}_{t_1,\ldots,t_k}\{H\}$$

for any $H \in \mathcal{B}^k$, so

$$\mathcal{P}_n\pi^{-1}_{t_1,\ldots,t_k} \Longrightarrow \mathcal{P}\pi^{-1}_{t_1,\ldots,t_k}.$$

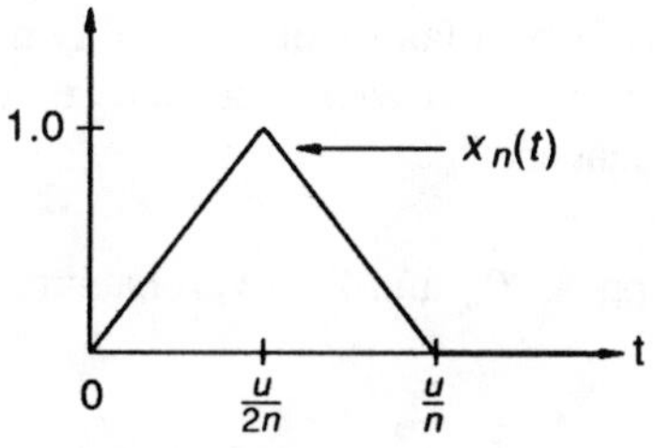

Figure B.1.2 The finite dimensional distribution functions of the sequence $X_n \equiv x_n$ converge to those of $X = 0$, but $X_n \not\Rightarrow X$. (See Example B.1.3)

Equivalently, the finite dimensional distribution functions do converge.

However, let k_0 be a constant $\in (0, 1)$ and consider the open set $\mathcal{O} \in \mathcal{S}$ defined by

$$\mathcal{O} = \{y : \rho(0, y) < k_0\} = \{y \in C[0, u] : \sup_{0 \leq t \leq u} |y(t)| \leq k_0\}.$$

Then, for any $n \geq 1$,

$$\begin{aligned} \mathcal{P}_n\{\mathcal{O}\} &= P\{\omega : X_n(\omega) \in \mathcal{O}\} \\ &= P\{\omega : \sup_{0 \leq t \leq u} X_n(t, \omega) \leq k_0\} \\ &= 0, \end{aligned}$$

while

$$\begin{aligned} \mathcal{P}\{\mathcal{O}\} &= P\{\omega : X(\omega) \in \mathcal{O}\} \\ &= P\{\omega : \sup_{0 \leq t \leq u} X(t, \omega) \leq k_0\} \\ &= 1. \end{aligned}$$

Thus

$$\begin{aligned} \liminf_n \mathcal{P}_n\{\mathcal{O}\} &= 0 \\ &< \mathcal{P}\{\mathcal{O}\} \end{aligned}$$

and Condition (2) in Definition B.1.4 is violated. Interestingly, we see that

$$\sup_{0 \leq t \leq u} X_n(t) \xrightarrow{\mathcal{D}} \sup_{0 \leq t \leq u} X(t)$$

fails as well. □

Convergence of finite-dimensional distributions does imply weak convergence when the probability space of paths of a stochastic process has some additional structure.

Definition B.1.5. Let Π be a family of probability measures on a metric space $(\Gamma, \mathcal{S}_0)$. Π is *relatively compact* if every sequence of elements of Π contains a weakly convergent subsequence. □

Theorem B.1.3. Suppose $\mathcal{P}_n$ and $\mathcal{P}$ are probability measures on $(C[0,u], \mathcal{S})$, for $n = 1, 2 \ldots$.

Suppose

1. $$\mathcal{P}_n \pi^{-1}_{t_1,\ldots,t_k} \Longrightarrow \mathcal{P}\pi^{-1}_{t_1,\ldots,t_k} \tag{1.4}$$
   (i.e., the finite-dimensional distributions converge), and
2. $\{\mathcal{P}_n\}$ is relatively compact.

Then $\mathcal{P}_n \Longrightarrow \mathcal{P}$.

Proof. Each subsequence $\{\mathcal{P}'_n\}$ contains a further subsequence $\{\mathcal{P}''_n\}$ converging weakly to some $\mathcal{Q}$. Thus, for each $(t_1, \ldots, t_k)$, the Continuous Mapping Theorem implies

$$\mathcal{P}''_n \pi^{-1}_{t_1,\ldots,t_k} \Longrightarrow \mathcal{Q}\pi^{-1}_{t_1,\ldots,t_k}.$$

By (1),

$$\mathcal{Q}\pi^{-1}_{t_1,\ldots,t_k} = \mathcal{P}\pi^{-1}_{t_1,\ldots,t_k},$$

so Theorem B.1.2 implies,

$$\mathcal{Q} = \mathcal{P}.$$

Thus, each subsequence $\{\mathcal{P}'_n\}$ contains a further subsequence $\{\mathcal{P}''_n\}$ converging weakly to $\mathcal{P}$. It follows directly from Theorem 2.3 of Billingsley (1968) that $\mathcal{P}_n \Longrightarrow \mathcal{P}$. □

Again let u be a fixed constant, and consider a process $X = \{X(t) : 0 \le t \le u\}$ with sample paths in $D[0,u]$, the space of right-continuous sample paths which have finite left-hand limits. Then we can write

$$\Omega \xrightarrow{X} D[0,u].$$

Our processes of principal interest, $X_n(t) = \sum_{i=1}^n \int_0^t H_i(s)dM_i(s)$, have sample paths in $D[0,u]$. We need a version of Theorem B.1.3 applying to measures on $D[0,u]$.

The open sets in $D[0,u]$ are defined using a metric d in which two sample paths x and y are close if they can be made equal by a uniformly small perturbation of the abscissa and/or the ordinate. More specifically, let Λ be the class of strictly increasing, continuous mappings of $[0,u]$ onto itself where, for any $\lambda \in \Lambda$, $\lambda(0) = 0$ and $\lambda(u) = u$. Now define $d(x,y)$ to be the infimum of all $\epsilon > 0$ such that there exists $\lambda \in \Lambda$ with

$$\sup_{0 \le t \le u} \mid \lambda(t) - t \mid \le \epsilon$$

and

$$\sup_{0 \le t \le u} | x(t) - y\{\lambda(t)\} | \le \epsilon.$$

The metric d is called the Skorohod metric and the resulting open sets the Skorohod topology. The smallest σ-algebra generated by the open sets of the Skorohod topology will be denoted by $\mathcal{S}^*$. An example helps clarify the definition of d.

Example B.1.4. In all three cases in Figure B.1.3, $d(x, y) = \epsilon$. In (2) and (3), take λ to be piecewise linear with $\lambda(0) = 0$, $\lambda(u/2) = u/2 - \epsilon$, and $\lambda(u) = u$. □

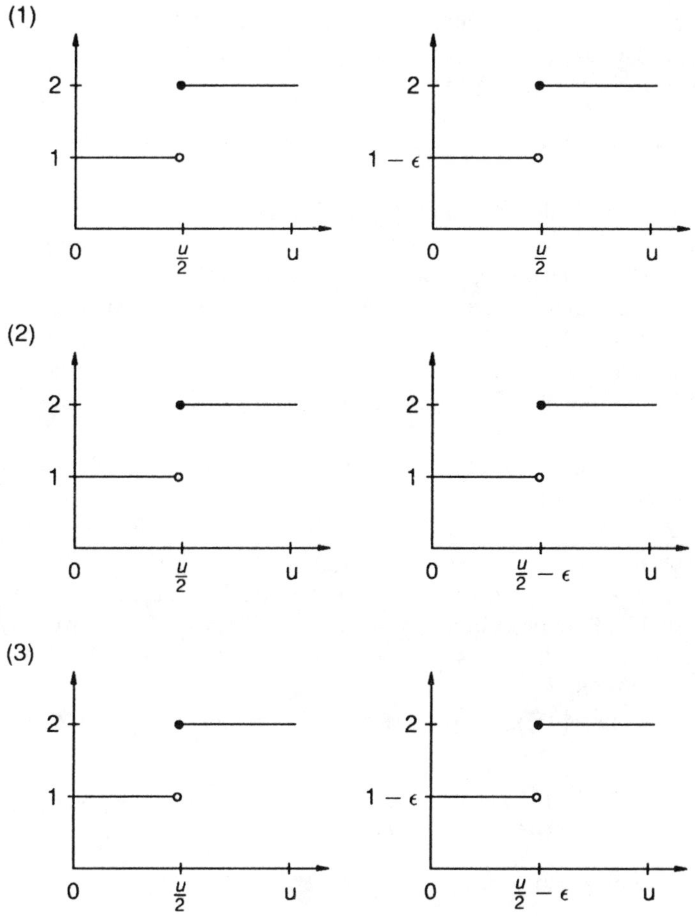

Figure B.1.3 In each of the three situations (1), (2), and (3), the Skorohod distance between the function x on the left and y on the right satisfies $d(x, y) = \epsilon$. (See Example B.1.4)

The next result follows by establishing that the projection mapping

$$\pi_{t_1,\ldots,t_k} : (D[0,u], \mathcal{S}^*, \mathcal{P}) \to (R^k, \mathcal{B}^k, \mathcal{P}\pi^{-1}_{t_1,\ldots,t_k})$$

is continuous except on a set of $\mathcal{P}$-measure 0, for a sufficiently large set of $\{t_1, \ldots, t_k\}$. Billingsley (1968, Sections 14–15) provides the details.

Theorem B.1.4. The results of Theorem B.1.3 hold when $(C[0,u], \mathcal{S})$ is replaced by $(D[0,u], \mathcal{S}^*)$, and when Conditions (1) and (2) are met. □

It is clear that Conditions (1) and (2) in Theorems B.1.3 and B.1.4 are both necessary as well as sufficient for weak convergence $\mathcal{P}_n \Longrightarrow \mathcal{P}$. There is a sufficient condition for relative compactness which is easier to verify in practice.

Definition B.1.6. A family Π of probability measures on a metric space, $(\Gamma, \mathcal{S}_0)$, is said to be *tight* if, for any $\epsilon > 0$, there exists a compact set K such that $\mathcal{P}\{K\} > 1 - \epsilon$ for all $\mathcal{P} \in \Pi$. □

Theorem B.1.5. If Π is tight, it is relatively compact. □

Tightness and relative compactness are nearly equivalent since the converse to Theorem B.1.5 holds on any complete and separable metric space. The space $(D[0,u], \mathcal{S}^*)$ is separable, is not complete under d, but is complete under another metric, d_0, which generates the same Skorohod topology as d does. As a result, tightness and relative compactness are equivalent for $(D[0,u], \mathcal{S}^*)$. By the remark following Theorem B.1.4, it follows that tightness and convergence of finite-dimensional distributions are necessary and sufficient conditions for weak convergence $\mathcal{P}_n \Longrightarrow \mathcal{P}$ on $(D[0,u], \mathcal{S}^*)$.

The next result, proved by Stone (1963), provides a sufficient condition for tightness in $(D[0,u], \mathcal{S}^*)$.

Theorem B.1.6. Let $W_n = \{W_n(t) : 0 \le t \le u\}, n = 1, 2\ldots,$ be a sequence of stochastic processes, each having sample paths in $(D[0,u])$, and let the sequence $\{\mathcal{P}_n\}$ of probability measures on $(D[0,u], \mathcal{S}^*)$ be given by $\mathcal{P}_n(A) \equiv P\{\omega : W_n(\omega) \in A\}$ for any n and any $A \in \mathcal{S}^*$. Note $W_n^{-1}(A) \in \mathcal{F}$, (see Billingsley, 1968, p. 128).

The sequence $\{\mathcal{P}_n\}$ is tight if, for any $\epsilon > 0$ and for $0 \le s, t \le u$,

$$\lim_{\delta \to 0} \limsup_{n \to \infty} P\left\{ \sup_{|s-t|<\delta} | W_n(s) - W_n(t) |> \epsilon \right\} = 0; \tag{1.5}$$

i.e., for any $\epsilon > 0$, $\eta > 0$, there exists δ^* such that

$$\limsup_{n \to \infty} P\left\{ \sup_{|s-t|<\delta^*} | W_n(s) - W_n(t) |> \epsilon \right\} < \eta.$$ □

Condition (1.5) indicates that, for large n, $W_n(s)$ and $W_n(t)$ are probabilistically close if s and t are close; intuitively, it is the condition one might expect should be added to finite-dimensional convergence to produce weak convergence.

The next result follows immediately from Theorems B.1.4, B.1.5, and B.1.6.

Corollary B.1.1. Let $W_n = \{W_n(t) : 0 \leq t \leq u\}, n = 1, 2, \ldots,$ and $W = \{W(t) : 0 \leq t \leq u\}$ be stochastic processes having sample paths in $D[0, u]$, and having probability measures on $(D[0, u], \mathcal{S}^*)$ defined by $\mathcal{P}_n\{A\} \equiv P\{\omega : W_n(\omega) \in A\}$ and $\mathcal{P}\{A\} = P\{\omega : W(\omega) \in A\}$, for $A \in \mathcal{S}^*$ and for any n. Suppose (1.4) and (1.5) hold. Then $\mathcal{P}_n \Longrightarrow \mathcal{P}$. We also write $W_n \Longrightarrow W$. □

To show $W_n \Longrightarrow W$, then, we need only establish convergence of finite-dimensional distributions, and also Condition (1.5) (Stone's sufficient condition for tightness). This will be our approach whenever we establish weak convergence on $D[0, u]$.

APPENDIX C

The Martingale Central Limit Theorem: Some Preliminaries

This appendix establishes results required to complete the development in Chapter 5 of the martingale central limit theorem.

C.1 ESTABLISHING TIGHTNESS IN THE PROOF OF THEOREM 5.2.4

Lemma C.1.1. In the proof of Theorem 5.2.4, the tightness Condition (2.8) in Chapter 5 holds for the process $\widetilde{W}_n$. □

We first establish a lemma to be used in the proof of Lemma C.1.1.

Lemma C.1.2. Let $\{U_i, i = 0, 1, \ldots, n\}$ be a martingale with $U_0 \equiv 0$. Then for any positive constant c,

$$P\left\{\max_{0 \leq j \leq n} |U_j| > 2c\right\} \leq E\left(\frac{2}{c}|U_n| I_{\{|U_n| \geq c\}}\right).$$

Proof. The number of upcrossings of the interval $[a,b]$ by the sequence $\{X_i, i = 0, 1, \ldots, n\}$, denoted $\beta_x^{[a,b]}$, is the number of times the sequence crosses from $\leq a$ to $\geq b$. By Theorem 9.4.2 of Chung (1974), when X is a sub-martingale,

$$E\beta_x^{[a,b]} \leq \frac{E(X_n - a)^{\oplus} - E(X_0 - a)^{\oplus}}{b - a} \tag{1.1}$$

where $Y^{\oplus} \equiv \max(Y, 0)$ for any random variable, Y.

Now,

$$\begin{aligned} P\left\{\min_{k\leq n} U_k < -2c\right\} &= P\left\{\min_{k\leq n} U_k < -2c \text{ and } U_n \geq -c\right\} \\ &\quad + P\left\{\min_{k\leq n} U_k < -2c \text{ and } U_n < -c\right\} \\ &\leq P\left\{\beta_u^{[-2c,-c]} > 0\right\} + P\{U_n < -c\} \\ &\leq E(\beta_u^{[-2c,-c]}) + P\{U_n < -c\}. \end{aligned}$$

It follows that

$$\begin{aligned} P\{\max U_k > 2c\} &= P\{\min_{k\leq n} -U_k < -2c\} \\ &\leq E(\beta_{-u}^{[-2c,-c]}) + P\{U_n > c\}. \end{aligned}$$

Hence,

$$P\{\max_{k\leq n} |U_k| > 2c\} \leq P\{|U_n| > c\} + E(\beta_u^{[-2c,-c]}) + E(\beta_{-u}^{[-2c,-c]}).$$

Both $\{U_i\}$ and $\{-U_i\}$ are martingales with first terms equal to zero. Thus, by Inequality (1.1),

$$\begin{aligned} E(\beta_u^{[-2c,-c]}) + E(\beta_{-u}^{[-2c,-c]}) &\leq \frac{E(U_n + 2c)^{\oplus} - 2c}{c} + \frac{E(-U_n + 2c)^{\oplus} - 2c}{c} \\ &= \frac{1}{c} E(U_n I_{\{U_n \geq -2c\}} - U_n I_{\{U_n \leq 2c\}}) \\ &\quad - 2(P\{U_n < -2c\} + P\{U_n > 2c\}) \\ &\leq \frac{1}{c} E(U_n I_{\{U_n \geq 2c\}} - U_n I_{\{U_n \leq -2c\}}) \\ &= \frac{1}{c} E(|U_n| I_{\{|U_n| \geq 2c\}}) \\ &\leq \frac{1}{c} E(|U_n| I_{\{|U_n| \geq c\}}). \end{aligned}$$

Also,

$$P\{|U_n| \geq c\} \leq \frac{1}{c} E(|U_n| I_{\{|U_n| \geq c\}}).$$

Hence,

$$P\{\max_{k\leq n} |U_k| > 2c\} \leq \frac{2}{c} E(|U_n| I_{\{|U_n| \geq c\}}). \qquad \square$$

We will now use Lemma C.1.2 to prove Lemma C.1.1.

Proof of Lemma C.1.1. Let $[x]$ denote the largest integer less than or equal to x. Observe that

$$
\begin{aligned}
&P\left\{\sup_{\substack{|s-t|<\delta\\ 0\le s,t\le T}} |\widetilde{W}_n(s) - \widetilde{W}_n(t)| > \epsilon\right\}\\
&\quad = P\left\{\sup_{\substack{|s-t|<\delta\\ 0\le s,t\le T}} \left|\sum_{i=r_n(s)+1}^{r_n(t)} \widetilde{X}_{n,i}\right| > \epsilon\right\}\\
&\quad \le \sum_{j=0}^{[T/\delta]} P\left\{\sup_{j\delta\le t\le (j+1)\delta} \left|\sum_{i=r_n(j\delta)}^{r_n(t)} \widetilde{X}_{n,i}\right| > \frac{\epsilon}{3}\right\}\\
&\quad \le \sum_{j=0}^{[T/\delta]} E\left[\frac{12}{\epsilon}\left|\sum_{i=r_n(j\delta)}^{r_n((j+1)\delta)} \widetilde{X}_{n,i}\right| I_{\left\{\left|\sum_{i=r_n(j\delta)}^{r_n((j+1)\delta)} \widetilde{X}_{n,i}\right|\ge \frac{\epsilon}{6}\right\}}\right]\\
&\quad \le \frac{12}{\epsilon}\sum_{j=0}^{[T/\delta]} \left[E\left\{\sum_{i=r_n(j\delta)}^{r_n((j+1)\delta)} \widetilde{X}_{n,i}\right\}^2\right]^{1/2} \left[P\left\{\left|\sum_{i=r_n(j\delta)}^{r_n((j+1)\delta)} \widetilde{X}_{n,i}\right| \ge \frac{\epsilon}{6}\right\}\right]^{1/2},
\end{aligned}
$$

the last inequality following from the Cauchy-Schwartz inequality. The second to last inequality follows from Lemma C.1.2 by identifying the martingale in Lemma C.1.2, $\{U_i, i = 0, 1, \ldots, n\}$, with the martingale

$$
0, \widetilde{X}_{n,r_n(j\delta)}, \widetilde{X}_{n,r_n(j\delta)} + \widetilde{X}_{n,r_n(j\delta)+1}, \ldots, \sum_{r_n(j\delta)}^{r_n((j+1)\delta)} \widetilde{X}_{n,i}.
$$

Note also that, since $U_0 \equiv 0$,

$$
P\left\{\max_{0\le j\le n} |U_j| > 2c\right\} = P\left\{\max_{1\le j\le n} |U_j| > 2c\right\}.
$$

Now

$$
\begin{aligned}
E\left(\sum_{i=r_n(j\delta)}^{r_n((j+1)\delta)} \widetilde{X}_{n,i}\right)^2 &= E\sum_{i=r_n(j\delta)}^{r_n((j+1)\delta)} \widetilde{X}_{n,i}^2 + 2E\sum_{i=r_n(j\delta)}^{r_n((j+1)\delta)} \sum_{i'<i} E_{i-1}\left(\widetilde{X}_{n,i}, \widetilde{X}_{n,i'}\right)\\
&= E\sum_{i=r_n(j\delta)}^{r_n((j+1)\delta)} \widetilde{X}_{n,i}^2 - (j+1)\delta - j\delta\\
&= \delta
\end{aligned}
$$

by Condition (2.10) in Chapter 5. We have

$$P\left\{\left|\sum_{i=r_n(j\delta)}^{r_n((j+1)\delta)} \widetilde{X}_{n,i}\right| \geq \frac{\epsilon}{6}\right\} \to 2\left\{1-\Phi\left(\frac{\epsilon}{6\sqrt{\delta}}\right)\right\}$$

since, by the arguments proving convergence of the finite-dimensional distributions,

$$\sum_{i=r_n(j\delta)}^{r_n((j+1)\delta)} \widetilde{X}_{n,i} \xrightarrow{\mathcal{D}} N(0,(j+1)\delta - j\delta) = N(0,\delta).$$

Hence, it remains to prove that

$$\lim_{\delta\to 0}\sum_{j=0}^{[T/\delta]} \delta^{1/2}\left\{1-\Phi\left(\frac{\epsilon}{6}\frac{1}{\sqrt{\delta}}\right)\right\}^{1/2} = 0.$$

For $\delta \leq 1$, we have

$$\left|\sum_{j=0}^{[T/\delta]} \delta^{1/2}\left\{1-\Phi\left(\frac{\epsilon}{6}\frac{1}{\sqrt{\delta}}\right)\right\}^{1/2}\right| \leq (T+1)\frac{1}{\sqrt{\delta}}\left\{1-\Phi\left(\frac{\epsilon}{6}\frac{1}{\sqrt{\delta}}\right)\right\}^{1/2},$$

and, by L'Hôpital's rule,

$$\begin{aligned}
\lim_{\delta\to 0}\frac{1-\Phi(\frac{\epsilon}{6}\frac{1}{\sqrt{\delta}})}{\delta} &= \lim_{\delta\to 0}\Phi'\left(\frac{\epsilon}{6\sqrt{\delta}}\right)\delta^{-3/2}\left(\frac{\epsilon}{12}\right) \\
&= \lim_{\delta\to 0}\frac{\epsilon}{12}\delta^{-3/2}\frac{e^{-\frac{\epsilon^2}{72\delta}}}{\sqrt{2\pi}} \\
&= \frac{\epsilon}{12\sqrt{2\pi}}\lim_{x\to\infty} x^{3/2}e^{-\frac{\epsilon^2}{72}x} \\
&= 0.
\end{aligned}$$

□

C.2 THE PROOF OF COROLLARY 5.2.1

Proof of Corollary 5.2.1. As indicated in Chapter 5, the proof has two major steps. We first verify that the array $X_{n,j}$ satisfies the following three conditions for all $t \in [0,\tau]$:

$$\max_{j\leq r_n(t)} |X_{n,j}| \xrightarrow{P} 0, \tag{2.1}$$

$$\sum_{j=1}^{r_n(t)} (X_{n,j})^2 \xrightarrow{P} t, \tag{2.2}$$

and

$$\sum_{j=1}^{r_n(t)} |E_{j-1}X_{n,j}I_{\{|X_{n,j}|\le c\}}| \xrightarrow{P} 0 \text{ for all } c \in (0,\infty). \tag{2.3}$$

By Lemma 3.5 of Dvoretsky (1972), for all $\epsilon, \eta > 0$,

$$P\left\{\max_{j\le r_n(t)} |X_{n,j}| > \epsilon\right\} \le \eta + P\left\{\sum_{j=1}^{r_n(t)} P\{|X_{n,j}| > \epsilon|\mathcal{F}_{n,j-1}\} > \eta\right\}. \tag{2.4}$$

Since

$$\begin{aligned} E\left(X_{n,j}^2 I_{\{|X_{n,j}|>\epsilon\}}|\mathcal{F}_{n,j-1}\right) &\ge E\left(\epsilon^2 I_{\{|X_{n,j}|>\epsilon\}}|\mathcal{F}_{n,j-1}\right) \\ &= \epsilon^2 P\{|X_{n,j}| > \epsilon|\mathcal{F}_{n,j-1}\}, \end{aligned}$$

Eq. (2.1) follows directly from Condition (2.13) of Chapter 5.

To verify Eq. (2.2), begin by selecting arbitrary $\epsilon > 0$, and $t \in [0, \tau]$, and define

$$W_{n,j} = X_{n,j} I_{\{|X_{n,j}|\le\epsilon, \sum_{k=1}^{j} E_{k-1}X_{n,k}^2 \le t+1\}}.$$

Note that as $n \to \infty$,

$$P\{X_{n,j} \ne W_{n,j} \text{ for some } j \le r_n(t)\} \to 0, \tag{2.5}$$

since $P\{X_{n,j} \ne W_{n,j} \text{ for some } j\} \le P\{\max_j |X_{n,j}| > \epsilon\} + P\{\sum_{j=1}^{r_n(t)} E_{j-1}X_{n,j}^2 \ge t+1\}$, where the first term on the right-hand side converges to zero by Eq. (2.1) and the second term converges to zero by Condition (2.14) of Chapter 5.

A second preliminary result,

$$\sum_{j=1}^{r_n(t)} E_{j-1}\left(X_{n,j}^2 - W_{n,j}^2\right) \xrightarrow{P} 0 \text{ as } n \to \infty, \tag{2.6}$$

can be verified by observing that

$$\begin{aligned} \sum_{j=1}^{r_n(t)} E_{j-1}\left(X_{n,j}^2 - W_{n,j}^2\right) &\le \sum_{j=1}^{r_n(t)} E_{j-1}\left(X_{n,j}^2 I_{\{|X_{n,j}|>\epsilon\}}\right) \\ &\quad + \sum_{j=1}^{r_n(t)} E_{j-1}\left(X_{n,j}^2 I_{\{\sum_{k=1}^{j} E_{k-1}X_{n,k}^2 > t+1\}}\right). \end{aligned}$$

Both terms on the right-hand side converge in probability to zero, the first by Condition (2.13) of Chapter 5, while convergence of the second term follows from Condition (2.14) of Chapter 5 since

$$\sum_{j=1}^{r_n(t)} E_{j-1}\left(X_{n,j}^2 I_{\{\sum_{k=1}^{j} E_{k-1}X_{n,k}^2>t+1\}}\right)$$

$$= \sum_{j=1}^{r_n(t)} (E_{j-1}X_{n,j}^2)\, I_{\{\sum_{k=1}^{j} E_{k-1}X_{n,k}^2>t+1\}}$$

$$\leq \left(\sum_{j=1}^{r_n(t)} E_{j-1}X_{n,j}^2\right) I_{\{\sum_{j=1}^{r_n(t)} E_{j-1}X_{n,j}^2>t+1\}}.$$

Now, to verify Eq. (2.2), it suffices by Condition (2.14) of Chapter 5 to show

$$\sum_{j=1}^{r_n(t)} X_{n,j}^2 - \sum_{j=1}^{r_n(t)} E_{j-1}X_{n,j}^2 \xrightarrow{P} 0.$$

In turn, by Eqs. (2.5) and (2.6), this will follow if we can show, for any fixed η, that

$$P\left\{\left|\sum_{j=1}^{r_n(t)} W_{n,j}^2 - \sum_{j=1}^{r_n(t)} E_{j-1}W_{n,j}^2\right| > \eta\right\}$$

can be made arbitrarily small though proper choice of ϵ.

Now

$$E\left\{\sum_{j=1}^{r_n(t)} (W_{n,j}^2 - E_{j-1}W_{n,j}^2)\right\}^2$$

$$= E\sum_{j=1}^{r_n(t)} E_{j-1}\left(W_{n,j}^2 - E_{j-1}W_{n,j}^2\right)^2$$

$$+2E\sum_{j<k} E_{k-1}\left\{\left(W_{n,j}^2 - E_{j-1}W_{n,j}^2\right)\left(W_{n,k}^2 - E_{k-1}W_{n,k}^2\right)\right\}$$

$$= E\sum_{j=1}^{r_n(t)} \left\{E_{j-1}(W_{n,j}^2)^2 - (E_{j-1}W_{n,j}^2)^2\right\}$$

$$+2E\sum_{j<k} \left\{\left(W_{n,j}^2 - E_{j-1}W_{n,j}^2\right) E_{k-1}\left(W_{n,k}^2 - E_{k-1}W_{n,k}^2\right)\right\}$$

$$= \sum_{j=1}^{r_n(t)} EW_{n,j}^4 - \sum_{j=1}^{r_n(t)} E(E_{j-1}W_{n,j}^2)^2$$

$$\leq \sum_{j=1}^{r_n(t)} EW_{n,j}^4$$

$$\leq \epsilon^2 E \sum_{j=1}^{r_n(t)} E_{j-1} W_{n,j}^2$$
$$\leq \epsilon^2 (t+1).$$

By Chebyshev's Inequality and the above inequality,

$$P\left\{\left|\sum_{j=1}^{r_n(t)} W_{n,j}^2 - \sum_{j=1}^{r_n(t)} E_{j-1} W_{n,j}^2\right| > \eta\right\} \leq \frac{\epsilon^2(t+1)}{\eta^2},$$

which establishes Eq. (2.2).

To verify Eq. (2.3), pick any $0 < c < \infty$. By Condition (2.15) of Chapter 5, it is sufficient to show

$$\sum_j |E_{j-1}(X_{n,j} I_{\{|X_{n,j}|>c\}})| \xrightarrow{P} 0.$$

This follows from Condition (2.13) of Chapter 5 since

$$|E_{j-1}(X_{n,j} I_{\{|X_{n,j}|>c\}})| \leq \frac{1}{c} E_{j-1}(X_{n,j}^2 I_{\{|X_{n,j}|>c\}}).$$

Proceeding to the second major step of the proof, we will show the newly verified conditions, (2.1), (2.2), and (2.3), are sufficient to establish the desired weak convergence result, $W_n \Longrightarrow W$. Begin by defining $Y_{n,j} = X_{n,j} I_{(|X_{n,j}| \leq c)}$. Condition (2.1) gives Condition (2.4) of Chapter 5, which in turn implies

$$\sum_{j=1}^{r_n(\tau)} |X_{n,j}| I_{(|X_{n,j}|>c)} \xrightarrow{P} 0,$$

so

$$P\{Y_{n,j} \neq X_{n,j} \text{ for some } j \leq r_n(\tau)\} \to 0. \tag{2.7}$$

Thus, to show $W_n \Longrightarrow W$, it suffices to show

$$\sum_{j=1}^{r_n(\cdot)} Y_{n,j} \Longrightarrow W. \tag{2.8}$$

Further, since

$$\sup_{0 \leq s \leq \tau} \left|\sum_{j=1}^{r_n(s)} E_{j-1} Y_{n,j}\right| \leq \sum_{j=1}^{r_n(\tau)} |E_{j-1} Y_{n,j}| \xrightarrow{P} 0$$

by Eq. (2.3), Theorem 4.1 of Billingsley (1968) implies Eq. (2.8) will hold if

$$\sum_{j=1}^{r_n(\cdot)} (Y_{n,j} - E_{j-1} Y_{n,j}) \Longrightarrow W. \tag{2.9}$$

By establishing that the martingale difference array $(Y_{n,j} - E_{j-1}Y_{n,j})$ satisfies conditions (1) and (2) of Theorem 5.2.4, the convergence in Eq. (2.9) will follow.

For condition (2), $\sum_{j=1}^{r_n(t)}(Y_{n,j} - E_{j-1}Y_{n,j})^2 \xrightarrow{P} t$, observe that

$$\sum_{j=1}^{r_n(t)}(Y_{n,j} - E_{j-1}Y_{n,j})^2 = \sum_{j=1}^{r_n(t)} Y_{n,j}^2 - 2\sum_{j=1}^{r_n(t)}(Y_{n,j}E_{j-1}Y_{n,j}) + \sum_{j=1}^{r_n(t)}(E_{j-1}Y_{n,j})^2.$$

Clearly

$$\sum_{j=1}^{r_n(t)} Y_{n,j}^2 \xrightarrow{P} t$$

by Eqs. (2.2) and (2.7). Now,

$$\sum_{j=1}^{r_n(t)}(E_{j-1}Y_{n,j})^2 \le c\sum_{j=1}^{r_n(t)} |E_{j-1}Y_{n,j}| \xrightarrow{P} 0$$

by Eq. (2.3). The proof of Condition (2) is completed by noting that

$$\begin{aligned}\left|\sum_{j=1}^{r_n(t)}(Y_{n,j}E_{j-1}Y_{n,j})\right| &\le \sum_{j=1}^{r_n(t)} |Y_{n,j}||E_{j-1}Y_{n,j}| \\ &\le \left(\sum_{j=1}^{r_n(t)} Y_{n,j}^2\right)^{1/2} \sum_{j=1}^{r_n(t)}(E_{j-1}Y_{n,j})^2.\end{aligned}$$

To satisfy Condition (1), we need to verify

$$E\left(\max_{j\le r_n(t)} |Y_{n,j} - E_{j-1}Y_{n,j}|\right)^2 \to 0.$$

Noting that $(A - B)^2 \le 2A^2 + 2B^2$ for any real A, B,

$$E\left(\max_{j\le r_n(t)} |Y_{n,j} - E_{j-1}Y_{n,j}|\right)^2 \le 2E\left\{\max_{j\le r_n(t)} (Y_{n,j})^2\right\} + 2E\left\{\max_{j\le r_n(t)} (E_{j-1}Y_{n,j})^2\right\}. \tag{2.10}$$

For the first term on the right-hand side of Inequality (2.10),

$$\begin{aligned}\max_{j\le r_n(t)} (Y_{n,j})^2 &\le c \max_{j\le r_n(t)} |Y_{n,j}| \\ &\le c \max_{j\le r_n(t)} |X_{n,j}| \xrightarrow{P} 0\end{aligned}$$

by Eq. (2.1). Since $\max_{j\le r_n(t)}(Y_{n,j})^2 \le c^2$, Chung's Theorem 4.1.4 implies

$$E \max_{j\le r_n(t)} (Y_{n,j})^2 \to 0.$$

For the second term on the right-hand side of Inequality (2.10),

$$\max_{j \le r_n(t)} (E_{j-1}Y_{n,j})^2 \le c \max_{j \le r_n(t)} |E_{j-1}Y_{n,j}| \xrightarrow{P} 0$$

by Eq. (2.3). Since $\max_{j \le r_n(t)}(E_{j-1}Y_{n,j})^2 \le c^2$,

$$E \max_{j \le r_n(t)} (E_{j-1}Y_{n,j})^2 \to 0. \qquad \square$$

C.3 THE PROOF OF THEOREM 5.2.5

We now provide a proof of Theorem 5.2.5, which is the multivariate generalization of the weak convergence result in Corollary 5.2.1, which in turn is a generalization of the central limit theorem of Dvoretzky (1972). We will begin with two preliminary lemmas.

Lemma C.3.1. Consider a sequence of processes $\{(Z_{1,n}, Z_{2,n}, \ldots, Z_{r,n}), n = 1, 2, \ldots\}$ and a vector of limiting processes $(Z_1, Z_2, \ldots, Z_r)$. Then

$$(Z_{1,n}, Z_{2,n}, \ldots, Z_{r,n}) \Longrightarrow (Z_1, Z_2, \ldots, Z_r) \text{ in } (D[0, \tau])^r \tag{3.1}$$

if and only if

$$\sum_{\ell=1}^{r} \int_0^{\cdot} c_\ell(s) dZ_{\ell,n}(s) \Longrightarrow \sum_{\ell=1}^{r} \int_0^{\cdot} c_\ell(s) dZ_\ell(s) \text{ in } D[0, \tau] \tag{3.2}$$

for any bounded left-continuous step functions c_ℓ on $[0, \tau]; \ell = 1, \ldots, r$.

Proof. That Eq. (3.1) implies Eq. (3.2) is clear. If Eq. (3.2) holds then, by the Cramer-Wold Lemma 5.2.1, we have that the finite-dimensional distributions of $(Z_{1,n}, Z_{2,n}, \ldots, Z_{r,n})$ converge to those of $(Z_1, Z_2, \ldots, Z_r)$ since Eq. (3.2) implies

$$\sum_{\ell=1}^{r} \sum_{i=1}^{k} c_{i\ell}\{Z_{\ell,n}(t_i) - Z_{\ell,n}(t_{i-1})\} \xrightarrow{\mathcal{D}} \sum_{\ell=1}^{r} \sum_{i=1}^{k} c_{i\ell}\{Z_\ell(t_i) - Z_\ell(t_{i-1})\}$$

for any $\{t_0, t_1, \ldots, t_k\} \subset [0, \tau]$.

As a special case of Eq. (3.2), we have that $Z_{\ell,n} \Longrightarrow Z_\ell$ in $D[0, \tau]$, $\ell = 1, \ldots, r$. In particular, for each $\ell = 1, \ldots, r$, $Z_{\ell,n}$ is tight in $(D[0, u], \mathcal{S}^*, d_0)$, where $\mathcal{S}^*$ and d_0 are the σ-algebra generated by the Skorohod topology and the metric yielding completeness which are discussed after the statement of Theorem B.1.5 in Appendix B. Since the joint distributions are tight if and only if the marginals are tight, $\{Z_{1,n}, \ldots, Z_{r,n}\}$ is relatively compact in $(D[0, \tau])^r$. $\square$

Lemma C.3.2. Let f be a measurable nonnegative function such that $\int_0^t f^2(s)ds < \infty$ for all $t > 0$. If c is a bounded left-continuous step function on $[0, \tau]$ and we have some random variables $\{Z_{n,j}, j = 1, \ldots, r_n(t), n = 1, 2, \ldots\}$ that satisfy

$$\sum_{j=1}^{r_n(t)} Z_{n,j} \xrightarrow{P} \int_0^t f^2(s)ds \tag{3.3}$$

for any $t \in [0, \tau]$, then

$$\sum_{j=1}^{r_n(t)} c^2\{r_n^{-1}(j)\} Z_{n,j} \xrightarrow{P} \int_0^t c^2(s) f^2(s) ds \tag{3.4}$$

for any $t \in [0, \tau]$, where $r_n^{-1}(j) \equiv \inf\{t : r_n(t) \geq j\}$.

Proof. Since c is a step function, there are a finite number of points $\{t_1, \ldots, t_k\} \subset [0, \tau]$ where c has a jump. Let $t_0 = 0$ and $t_{k+1} = \tau$ without loss of generality. Define $c_i = c(t_{i+1}) = c(t_i + 1), i = 0, \ldots, k$. We have

$$\sum_{j=r_n(s)+1}^{r_n(t)} Z_{n,j} \xrightarrow{P} \int_s^t f^2(u) du$$

for any $s, t \in [0, \tau]$. Thus,

$$\begin{aligned}
\sum_{j=1}^{r_n(t)} c^2\{r_n^{-1}(j)\} Z_{n,j} &= \sum_{i=0}^{k} c_i^2 \sum_{j=r_n(t_i \wedge t)+1}^{r_n(t_{i+1} \wedge t)} Z_{n,j} \\
&\xrightarrow{P} \sum_{i=0}^{k} c_i^2 \int_{t_i \wedge t}^{t_{i+1} \wedge t} f^2(u) du \\
&= \int_0^t c^2(u) f^2(u) du
\end{aligned}$$

□

Proof of Theorem 5.2.5. Let $c_\ell, \ell = 1, \ldots, r$, be left-continuous step functions on $[0, \tau]$. We will show that, in $D[0, \tau]$,

$$V_n \equiv \sum_{\ell=1}^{r} \int_0^{\cdot} c_\ell(s) dW_{\ell,n}(s) \Longrightarrow \sum_{\ell=1}^{r} \int_0^{\cdot} c_\ell(s) dW^{(\ell)}(s),$$

where we have used the notation $W^{(\ell)}(s) \equiv \int_0^s f_\ell(s) dW_\ell(s)$. Without loss of generality, assume $\sum_{\ell=1}^r |c_\ell(s)| \leq 1$ on $[0, \tau]$. We can rewrite $V_n(t)$ as

$$\begin{aligned}
V_n(t) &= \sum_{\ell=1}^{r} \sum_{j=1}^{r_n(t)} c_\ell\{r_n^{-1}(j)\} X_{\ell,n,j} \\
&= \sum_{j=1}^{r_n(t)} \sum_{\ell=1}^{r} c_\ell\{r_n^{-1}(j)\} X_{\ell,n,j}.
\end{aligned}$$

Thus, $V_n(t) = \sum_{j=1}^{r_n(t)} Y_{n,j}$, where $Y_{n,j} \equiv \sum_{\ell=1}^r c_\ell\{r_n^{-1}(j)\} X_{\ell,n,j}$. Now, $\{Y_{n,j}\}$ is a martingale difference array with respect to $\{\mathcal{F}_{n,j}\}$.

Suppose Conditions (2.21), (2.22), and (2.23) of Chapter 5 hold. Observe that

$$\sum_{j=1}^{r_n(t)} E_{j-1}(Y_{n,j}^2 I_{\{|Y_{n,j}| > \epsilon\}}) \leq \sum_{j=1}^{r_n(t)} \sum_{\ell=1}^{r} E_{j-1}(X_{\ell,n,j} I_{\{|X_{\ell,n,j}| > \epsilon/r\}})$$

since $\sum |c_\ell(s)| \le 1$ on $[0, \tau]$ implies $|Y_{n,j}| \le \sum_{\ell=1}^r |X_{\ell,n,j}|$. Thus, Condition (2.22) of Chapter 5 holds for $\{Y_{n,j}\}$ when it holds for each $\{X_{\ell,n,j}\}$. When Conditions (2.21) and (2.23) of Chapter 5 hold, then Lemma C.3.2 implies

$$\begin{aligned}\sum_{j=1}^{r_n(t)} E_{j-1}Y_{n,j}^2 &= \sum_{j=1}^{r_n(t)}\sum_{\ell=1}^{r} c_\ell^2\{r_n^{-1}(j)\}E_{j-1}(X_{\ell,n,j})^2 \\ &\quad + 2\sum_{j=1}^{r_n(t)}\sum_{\ell=2}^{r}\sum_{\ell'=1}^{\ell-1} c_\ell\{r_n^{-1}(j)\}c_{\ell'}\{r_n^{-1}(j)\}E_{j-1}(X_{\ell,n,j}X_{\ell',n,j}) \\ &\xrightarrow{P} \sum_{\ell=1}^{r}\int_0^t c_\ell^2(s)f_\ell^2(s)ds + 0,\end{aligned}$$

and so Condition (2.21) of Chapter 5 also holds for $\{Y_{n,j}\}$. Hence, Corollary 5.2.1 implies

$$V_n \Longrightarrow \int_0 \left\{\sum_{\ell=1}^{r} c_\ell^2(s)f_\ell^2(s)\right\}^{1/2} dW(s)$$

on $D[0, \tau]$. It is easily seen that the finite-dimensional distributions of $\int \left\{\sum_{\ell=1}^r c_\ell^2(s)f_\ell^2(s)\right\}^{1/2} dW(s)$ and $\sum_{\ell=1}^r \int c_\ell(s)f_\ell(s)dW_\ell(s)$ are the same. Hence, in $D[0, \tau]$,

$$\begin{aligned}V_n &\Longrightarrow \sum_{\ell=1}^{r}\int_0 c_\ell(s)f_\ell(s)dW_\ell(s) \\ &= \sum_{\ell=1}^{r}\int_0 c_\ell(s)dW^{(\ell)}(s)\end{aligned}$$

for all bounded left-continuous step functions $\{c_\ell, \ell = 1, \ldots, r\}$ on $[0, \tau]$. Thus Lemma C.3.1 implies the desired result,

$$(W_{1,n}, W_{2,n}, \ldots, W_{r,n}) \Longrightarrow (\int f_1 dW_1, \ldots, \int f_r dW_r)$$

in $(D[0, \tau])^r$. □

C.4 THE PROOF OF LEMMA 5.3.1

Proof of Lemma 5.3.1. Beginning with Part (1) of this Lemma, observe that the assumption $Y_{n,k} \xrightarrow{P} Y_n$ as $k \to \infty$ for each fixed n implies one can choose k_n sufficiently large such that

$$P\{|Y_{n,k} - Y_n| > 1/n\} < 1/n \text{ for all } k \ge k_n, \tag{4.1}$$

and such that $k_n \geq k_{n-1}$. Choose an arbitrary $\epsilon > 0$ and note that

$$P\{|Y_{n,k} - Y_n| > \epsilon\} \leq P\{|Y_{n,k} - Y_n| > \epsilon/2\} + P\{|Y_n - Y| > \epsilon/2\}.$$

It follows from Inequality (4.1) and the assumption $\lim_{n\to\infty} Y_n = Y$ in probability that one can choose n_0 sufficiently large that both probabilities on the right-hand side are bounded by $\epsilon/2$ for all $n \geq n_0$ and $k \geq k_n$. Thus $Y_{n,k_n} \xrightarrow{P} Y$ as $n \to \infty$ for any increasing sequence $\{k_n\}$ such that $k_n \to \infty$ fast enough.

To prove Part (2), we only have to show that a single increasing sequence $\{k_n\}$ can cover every t in $[0, c]$. By Part (1), one can select $\{k_n\}$ such that, if $A = \{0, \epsilon, 2\epsilon, \ldots, ([c/\epsilon] + 1)\epsilon\}$, $\max_{t\in A} |Y_{n,k}(t) - Y(t)|$ is close to zero, in probability, for any $k \geq k_n$ when n is sufficiently large. Then, using the inequalities

$$Y_{n,k}([t/\epsilon]\epsilon) \leq Y_{n,k}(t) \leq Y_{n,k}([t/\epsilon]\epsilon + \epsilon)$$

and the continuity of $Y(\cdot)$, it follows that $\max_{0\leq t\leq c} |Y_{n,k}(t) - Y(t)|$ is close to zero, in probability, for any $k \geq k_n$ when n is sufficiently large. □

C.5 THE PROOF OF LEMMA 5.3.2

Proof of Lemma 5.3.2. We need to show that, as $n \to \infty$,

$$P\left\{\bigcup_{k=1}^{K_n} A_{n,k}\right\} \to 0 \text{ if and only if } \sum_{k=1}^{K_n} E_{k-1} I_{\{A_{n,k}\}} \xrightarrow{P} 0.$$

Replacing $A_{n,k}$ with $\widetilde{A}_{n,k} \equiv A_{n,k} \cap I_{\{K_n \geq k\}}$, observe that it is enough to prove this result for $K_n = \infty$ since $\widetilde{A}_{n,k} \in \mathcal{F}_{n,k}$. Let $M_n(\omega) \equiv \min\{k : \omega \in A_{n,k}\}$ for $\omega \in \bigcup_{k=1}^{\infty} A_{n,k}$ and $M_n(\omega) = \infty$ for $\omega \in \left(\bigcup_{k=1}^{\infty} A_{n,k}\right)^c = \bigcap_{k=1}^{\infty} A_{n,k}^c$, where A^c denotes the complement of the set A. Observe that

$$\begin{aligned}
P\left\{\bigcup_{k=1}^{\infty} A_{n,k}\right\} &= \sum_{k=1}^{\infty} P\left\{\left(\bigcap_{j=1}^{k-1} A_{n,j}^c\right) \cap A_{n,k}\right\} \\
&= E\sum_{k=1}^{\infty}\left(I_{\{\bigcap_{j=1}^{k-1} A_{n,j}^c\}} E_{k-1} I_{\{A_{n,k}\}}\right) \qquad (5.1) \\
&= E\sum_{k=1}^{M_n} E_{k-1} I_{\{A_{n,k}\}}.
\end{aligned}$$

Thus, for any $\epsilon > 0$,

$$P\left\{\sum_{k=1}^{\infty} E_{k-1} I_{\{A_{n,k}\}} > \epsilon\right\} \leq P\{M_n < \infty\} + P\left\{\sum_{k=1}^{M_n} E_{k-1} I_{\{A_{n,k}\}} > \epsilon\right\}$$
$$\leq P\left\{\bigcup_{k=1}^{\infty} A_{n,k}\right\} + \frac{1}{\epsilon} E\left\{\sum_{k=1}^{M_n} E_{k-1} I_{\{A_{n,k}\}}\right\}$$
$$= \left(1 + \frac{1}{\epsilon}\right) P(\bigcup_{k=1}^{\infty} A_{n,k})$$

by Eq. (5.1). Thus, the ($\Rightarrow$) direction is established.

To prove the converse, observe that

$$\sum_{k=1}^{\infty} E_{k-1} I_{\{A_{n,k}\}} \xrightarrow{P} 0 \text{ implies } \sum_{k=1}^{M_n} E_{k-1} I_{\{A_{n,k}\}} \xrightarrow{P} 0.$$

Since $\sum_{k=1}^{M_n} E_{k-1} I_{\{A_{n,k}\}} \in L^1$ by Eq. (5.1), it follows from Theorem 4.1.4 of Chung (1974) that, as $n \to \infty$,

$$E \sum_{k=1}^{M_n} E_{k-1} I_{\{A_{n,k}\}} \to 0. \tag{5.2}$$

By Eqs. (5.1) and (5.2), the proof is complete. □

C.6 THE PROOF OF LEMMA 5.3.3

Proof of Lemma 5.3.3. If we define

$$a_k \equiv \max_{j \leq k(t)} \sup_{u,v \in [t_{j,k}, t_{j+1,k}]} |x(u) - x(v)|,$$

then we wish to prove

$$\lim_{k \to \infty} a_k = \max_{s \leq t} |\Delta x(s)|. \tag{6.1}$$

Clearly, a_k is nonincreasing and, for any k,

$$a_k \geq \max_{s \leq t_{k(t)+1,k}} |\Delta x(s)| \geq \max_{s \leq t} |\Delta x(s)|.$$

Thus, it remains to prove that, given $\epsilon > 0$, there exists $k_0(\epsilon)$ such that

$$a_{k_0} \leq \max_{s \leq t} |\Delta x(s)| + \epsilon.$$

Since $x \in D[0,t]$, Lemma 14.1 of Billingsley (1968) implies there exists a set of real numbers $\{p_i, i = 0, \ldots, r\}$ such that $0 = p_0 < p_1 \ldots < p_r = t$ with $\sup_{[p_{i-1}, p_i)} |x(u) - x(v)| < \epsilon/2, i = 1, \ldots r$. Choose δ and k_0 so that $0 < \delta < \min_{i=1,\ldots,r} |p_i - p_{i-1}|$ and

$$\max_{j \leq k_0(t)-1} (t_{j+1,k_0} - t_{j,k_0}) < \delta.$$

Then, for each $j \leq k_0(t) - 1$, there exists $i \in \{0, \ldots, r-2\}$ such that $[t_{j,k_0}, t_{j+1,k_0}]$ is contained in $[p_i, p_{i+2})$. This implies

$$\begin{aligned}\sup_{u,v \in [t_{j,k_0}, t_{j+1,k_0}]} |x(u) - x(v)| &\leq \sup_{[p_i, p_{i+1})} |x(u) - x(v)| + \sup_{[p_{i+1}, p_{i+2})} |x(u) - x(v)| \\ &\quad + |\Delta x(p_{i+1})| \\ &\leq \frac{2\epsilon}{2} + \max_{s \leq t} |\Delta x(s)|\end{aligned}$$

for any $j \leq k_0(t) - 1$. For $j = k_0(t)$,

$$\begin{aligned}\sup_{[t_{j,k_0}, t_{j+1,k_0}]} |x(u) - x(v)| &\leq \sup_{[p_{r-1}, p_r)} |x(u) - x(v)| + |\Delta x(t)| + \sup_{[t, t_{j+1,k_0}]} |x(t) - x(u)| \\ &\leq |\Delta x(t)| + \frac{2\epsilon}{2}\end{aligned}$$

if k_0 is large enough, since x is right-continuous at t and $t_{k_0(t)+1,k_0}$ decreases monotonically to t as k_0 increases. Thus, if k_0 is large enough,

$$a_k \leq \max_{s \leq t} |\Delta x(s)| + \epsilon,$$

so Eq. (6.1) follows. □

C.7 THE PROOF OF THEOREM 5.3.4

We now provide a proof of Theorem 5.3.4, which is the multivariate generalization of the weak convergence result in Theorem 5.3.3. We begin with a preliminary lemma which is a simple corollary to Doléans-Dade's Corollary 5.3.1.

Lemma C.7.1. Let M_1 and M_2 be square-integrable martingales with continuous predictable variation processes. Then

$$\sum_{j=0}^{m-1} E\left[\{M_1(t_{j+1}) - M_1(t_j)\}\{M_2(t_{j+1}) - M_2(t_j)\} | \mathcal{F}_{t_j}\right] \xrightarrow{L^1} \langle M_1, M_2 \rangle(t) \tag{7.1}$$

as the partition $(t_0, t_1, \ldots, t_m)$ of $[0, t]$ becomes sufficiently fine.

Proof. For any constants, $a, b, ab = \frac{1}{4}\{(a+b)^2 - (a-b)^2\}$. Thus the left-hand side of Eq. (7.1) can be rewritten

$$\begin{aligned}
&\frac{1}{4}\left[\sum_{j=0}^{m-1} E\left[\{(M_1+M_2)(t_{j+1}) - (M_1+M_2)(t_j)\}^2 | \mathcal{F}_{t_j}\right]\right. \\
&\qquad \left. - \sum_{j=0}^{m-1} E\left[\{(M_1-M_2)(t_{j+1}) - (M_1-M_2)(t_j)\}^2 | \mathcal{F}_{t_j}\right]\right] \\
&\xrightarrow{L^1} \frac{1}{4}\{\langle M_1+M_2\rangle(t) - \langle M_1 - M_2\rangle(t)\} \\
&= \langle M_1, M_2\rangle(t).
\end{aligned}$$

where the convergence follows from Corollary 5.3.1, and where the last equality follows by definition. □

Proof of Theorem 5.3.4. We will extend the proof of Theorem 5.1.1 to the multivariate case. The proof of Theorem 5.3.4 can then be completed in the same manner that Theorem 5.3.3 was established from Theorem 5.1.1. Theorem 5.1.1 was proved by finding a sequence k_n such that $W^{(n)}$ satisfied Conditions (2.13) and (2.14) of Chapter 5 and

$$\sup_{0\le t\le \tau} |U^{(n)}(t) - W^{(n)}(t)| \xrightarrow{P} 0.$$

We will find a sequence k_n such that

$$\sup_{0\le t\le \tau} |U_\ell^{(n)}(t) - W_\ell^{(n)}(t)| \xrightarrow{P} 0, \ell = 1, \ldots, r, \tag{7.2}$$

and such that $\{W_1^{(n)}, \ldots, W_r^{(n)}\}$ satisfies Conditions (2.21), (2.22), and (2.23) of Chapter 5. Then, by Theorem 5.2.5, it follows that

$$(W_1^{(n)}, \ldots, W_r^{(n)}) \Longrightarrow \{\int f_1 dW_1, \ldots, \int f_r dW_r\}$$

in $(D[0, \tau])^r$ and, by (7.2),

$$(U_1^{(n)}, \ldots, U_r^{(n)}) \Longrightarrow \{\int f_1 dW_1, \ldots, \int f_r dW_r\}$$

in $(D[0, \tau])^r$.

By proceeding as in the proof of Theorem 5.1.1, one can use Conditions (3.14) and (3.15) of Chapter 5 to establish that a sequence $k_n^{(\ell)}(t)$ can be found for each $\ell = 1, \ldots, r$, such that Conditions (2.21) and (2.22) of Chapter 5 hold for

$$\begin{aligned} W_\ell^{(n)}(t) &= \sum_{j=0}^{k_n^{(\ell)}(t)} X_{\ell,j}^{(n)} \\ &= \sum_{j=0}^{k_n^{(\ell)}(t)} \{U_\ell^{(n)}(t_{j+1,k_n^{(\ell)}}) - U_\ell^{(n)}(t_{j,k_n^{(\ell)}})\}, \end{aligned}$$

and such that

$$\sup_{0\le t\le \tau} |U_\ell^{(n)}(t) - W_\ell^{(n)}(t)| \xrightarrow{P} 0.$$

For any sequence $k_n^* \geq \max_{1\le\ell\le r}(k_n^{(\ell)})$, if we define

$$W_\ell^{(n)}(t) \equiv \sum_{j=0}^{k_n^*(t)} X_{\ell,j}^{(n)},$$

we will have Conditions (2.21) and (2.22) of Chapter 5 holding for each $W_\ell^{(n)}$ and Eq. (7.2) will also be valid.

It remains to show one can find $k_n^* \geq \max_{1\le\ell\le r}(k_n^{(\ell)})$ such that Condition (2.23) of Chapter 5 also holds for each $W_\ell^{(n)}$. In this setting, Condition (2.23) of Chapter 5 can be rewritten as

$$\begin{aligned} &\sum_{j=1}^{k_n(t)+1} E\left(X_{\ell,j-1}^{(n)} X_{\ell',j-1}^{(n)} | \mathcal{F}_{t_{j-1,k_n}}^{(n)}\right) \\ &= \sum_{j=0}^{k_n(t)} E\left[\{U_\ell^{(n)}(t_{j+1,k_n}) - U_\ell^{(n)}(t_{j,k_n})\}\{U_{\ell'}^{(n)}(t_{j+1,k_n}) - U_{\ell'}^{(n)}(t_{j,k_n})\} | \mathcal{F}_{t_{j,k_n}}^{(n)}\right] \\ &\xrightarrow{P} 0 \end{aligned} \tag{7.3}$$

for each $\ell \neq \ell'$ and for all $t \in [0, \tau]$. For each pair $\ell \neq \ell'$, by Lemma C.7.1 and Condition (3.16) of Chapter 5, we can use Lemma 5.3.1(2) to show one sequence $k_n^{\ell,\ell'}$ can be found such that Eq. (7.3) holds uniformly for $t \in [0, \tau]$. If we choose

$$k_n^* = \max\left\{ \max_{1\le\ell\le r} (k_n^{(\ell)}),\ \max_{\substack{1\le\ell,\ell'\le r\\ \ell\neq\ell'}} (k_n^{\ell\ell'}) \right\},$$

then $W_\ell^{(n)}(t) \equiv \sum_{j=0}^{k_n^*(t)} X_{\ell,j}^{(n)}$ satisfies Conditions (2.21), (2.22), (2.23) of Chapter 5 for all $\ell = 1, \dots, r$, and Eq. (7.2) also is valid. □

APPENDIX D

Data

D.1 PBC DATA

The following pages contain the data from the Mayo Clinic trial in primary biliary cirrhosis (PBC) of the liver conducted between 1974 and 1984. A description of the clinical background for the trial and the covariates recorded here is in Chapter 0, especially Section 0.2. A more extended discussion can be found in Dickson, et al. (1989) and Markus, et al. (1989).

A total of 424 PBC patients, referred to Mayo Clinic during that ten-year interval, met eligibility criteria for the randomized placebo controlled trial of the drug D-penicillamine. The first 312 cases in the data set participated in the randomized trial, and contain largely complete data. The additional 112 cases did not participate in the clinical trial, but consented to have basic measurements recorded and to be followed for survival. Six of those cases were lost to follow-up shortly after diagnosis, so there are data here on an additional 106 cases as well as the 312 randomized participants. Missing data items are denoted by ".".

The variables contained here are:

- N : Case number.
- X : The number of days between registration and the earlier of death, liver transplantation, or study analysis time in July, 1986.
- δ : 1 if X is time to death, 0 if time to censoring, with an asterisk denoting that censoring was due to liver transplantation.
- Z_1 : Treatment Code, 1 = D-penicillamine, 2 = placebo.
- Z_2 : Age in years. For the first 312 cases, age was calculated by dividing the number of days between birth and study registration by 365.
- Z_3 : Sex, 0 = male, 1 = female.
- Z_4 : Presence of ascites, 0 = no, 1 = yes.
- Z_5 : Presence of hepatomegaly, 0 = no, 1 = yes.
- Z_6 : Presence of spiders, 0 = no, 1 = yes.

Z_7 : Presence of edema, 0 = no edema and no diuretic therapy for edema; 0.5 = edema present for which no diuretic therapy was given, or edema resolved with diuretic therapy; 1 = edema despite diuretic therapy.
Z_8 : Serum bilirubin, in mg/dl.
Z_9 : Serum cholesterol, in mg/dl.
Z_{10} : Albumin, in gm/dl.
Z_{11} : Urine copper, in μg/day.
Z_{12} : Alkaline phosphatase, in U/liter.
Z_{13} : SGOT, in U/ml.
Z_{14} : Triglycerides, in mg/dl.
Z_{15} : Platelet count; coded value is number of platelets per-cubic-milliliter of blood divided by 1000.
Z_{16} : Prothrombin time, in seconds.
Z_{17} : Histologic stage of disease, graded 1, 2, 3, or 4.

N	X	δ	Z_1	Z_2	Z_3	Z_4	Z_5	Z_6	Z_7	Z_8	Z_9	Z_{10}	Z_{11}	Z_{12}	Z_{13}	Z_{14}	Z_{15}	Z_{16}	Z_{17}
1	400	1	1	58.7652	1	1	1	1	1.0	14.5	261	2.60	156	1718.0	137.95	172	190	12.2	4
2	4500	0	1	56.4463	1	0	1	1	0.0	1.1	302	4.14	54	7394.8	113.52	88	221	10.6	3
3	1012	1	1	70.0726	0	0	0	0	0.5	1.4	176	3.48	210	516.0	96.10	55	151	12.0	4
4	1925	1	1	54.7406	1	0	1	1	0.5	1.8	244	2.54	64	6121.8	60.63	92	183	10.3	4
5	1504	0*	2	38.1054	1	0	1	1	0.0	3.4	279	3.53	143	671.0	113.15	72	136	10.9	3
6	2503	1	2	66.2587	1	0	1	0	0.0	0.8	248	3.98	50	944.0	93.00	63	.	11.0	3
7	1832	0	2	55.5346	1	0	1	0	0.0	1.0	322	4.09	52	824.0	60.45	213	204	9.7	3
8	2466	1	2	53.0568	1	0	0	0	0.0	0.3	280	4.00	52	4651.2	28.38	189	373	11.0	3
9	2400	1	1	42.5079	1	0	0	1	0.0	3.2	562	3.08	79	2276.0	144.15	88	251	11.0	2
10	51	1	2	70.5599	1	1	0	1	1.0	12.6	200	2.74	140	918.0	147.25	143	302	11.5	4
11	3762	1	2	53.7139	1	0	1	1	0.0	1.4	259	4.16	46	1104.0	79.05	79	258	12.0	4
12	304	1	2	59.1376	1	0	0	1	0.0	3.6	236	3.52	94	591.0	82.15	95	71	13.6	4
13	3577	0	2	45.6893	1	0	0	0	0.0	0.7	281	3.85	40	1181.0	88.35	130	244	10.6	3
14	1217	1	2	56.2218	0	1	1	0	1.0	0.8	.	2.27	43	728.0	71.00	.	156	11.0	4
15	3584	1	1	64.6461	1	0	0	0	0.0	0.8	231	3.87	173	9009.8	127.71	96	295	11.0	3
16	3672	0	2	40.4435	1	0	0	0	0.0	0.7	204	3.66	28	685.0	72.85	58	198	10.8	3
17	769	1	2	52.1834	1	0	1	0	0.0	2.7	274	3.15	159	1533.0	117.80	128	224	10.5	4
18	131	1	1	53.9302	1	0	1	1	1.0	11.4	178	2.80	588	961.0	280.55	200	283	12.4	4
19	4232	0	1	49.5606	1	0	1	0	0.5	0.7	235	3.56	39	1881.0	93.00	123	209	11.0	3
20	1356	1	2	59.9535	1	0	1	0	0.0	5.1	374	3.51	140	1919.0	122.45	135	322	13.0	4
21	3445	0	2	64.1889	0	0	1	1	0.0	0.6	252	3.83	41	843.0	65.10	83	336	11.4	4
22	673	1	1	56.2765	1	0	0	1	0.0	3.4	271	3.63	464	1376.0	120.90	55	173	11.6	4
23	264	1	2	55.9671	1	1	1	1	1.0	17.4	395	2.94	558	6064.8	227.04	191	214	11.7	4
24	4079	1	1	44.5202	0	0	1	0	0.0	2.1	456	4.00	124	5719.0	221.88	230	70	9.9	2
25	4127	0	2	45.0732	1	0	0	0	0.0	0.7	298	4.10	40	661.0	106.95	66	324	11.3	2
26	1444	1	2	52.0246	1	0	1	1	0.0	5.2	1128	3.68	53	3228.0	165.85	166	421	9.9	3
27	77	1	2	54.4394	1	1	1	1	0.5	21.6	175	3.31	221	3697.4	101.91	168	80	12.0	4
28	549	1	2	44.9473	1	1	1	1	1.0	17.2	222	3.23	209	1975.0	189.10	195	144	13.0	4
29	4509	0	2	63.8768	1	0	0	0	0.0	0.7	370	3.78	24	5833.0	73.53	86	390	10.6	2

N	X	δ	Z_1	Z_2	Z_3	Z_4	Z_5	Z_6	Z_7	Z_8	Z_9	Z_{10}	Z_{11}	Z_{12}	Z_{13}	Z_{14}	Z_{15}	Z_{16}	Z_{17}
30	321	1	2	41.3854	1	0	1	1	0.0	3.6	260	2.54	172	7277.0	121.26	158	124	11.0	4
31	3839	1	2	41.5524	1	0	1	0	0.0	4.7	296	3.44	114	9933.2	206.40	101	195	10.3	2
32	4523	0	2	53.9959	1	0	1	0	0.0	1.8	262	3.34	101	7277.0	82.56	158	286	10.6	4
33	3170	1	2	51.2827	1	0	0	0	0.0	0.8	210	3.19	82	1592.0	218.55	113	180	12.0	3
34	3933	0	1	52.0602	1	0	0	0	0.0	0.8	364	3.70	37	1840.0	170.50	64	273	10.5	2
35	2847	1	2	48.6188	1	0	0	0	0.0	1.2	314	3.20	201	12258.8	72.24	151	431	10.6	3
36	3611	0	2	56.4107	1	0	0	0	0.0	0.3	172	3.39	18	558.0	71.30	96	311	10.6	2
37	223	1	1	61.7276	1	1	1	0	1.0	7.1	334	3.01	150	6931.2	180.60	118	102	12.0	4
38	3244	1	2	36.6270	1	0	1	1	0.0	3.3	383	3.53	102	1234.0	137.95	87	234	11.0	4
39	2297	1	1	55.3922	1	0	1	0	0.0	0.7	282	3.00	52	9066.8	72.24	111	563	10.6	4
40	4467	0	1	46.6694	1	0	0	0	0.0	1.3	.	3.34	105	11046.6	104.49	.	358	11.0	4
41	1350	1	1	33.6345	1	0	1	0	0.0	6.8	.	3.26	96	1215.0	151.90	.	226	11.7	4
42	4453	0	2	33.6947	1	0	1	1	0.0	2.1	.	3.54	122	8778.0	56.76	.	344	11.0	4
43	4556	0	1	48.8706	1	0	0	0	0.0	1.1	361	3.64	36	5430.2	67.08	89	203	10.6	2
44	3428	1	2	37.5825	1	0	1	1	1.0	3.3	299	3.55	131	1029.0	119.35	50	199	11.7	3
45	4025	0	2	41.7933	1	0	0	0	0.0	0.6	.	3.93	19	1826.0	71.30	.	474	10.9	2
46	2256	1	1	45.7988	1	0	1	0	0.0	5.7	482	2.84	161	11552.0	136.74	165	518	12.7	3
47	2576	0	2	47.4278	1	0	0	0	0.0	0.5	316	3.65	68	1716.0	187.55	71	356	9.8	3
48	4427	0	2	49.1362	0	0	0	0	0.0	1.9	259	3.70	281	10396.8	188.34	178	214	11.0	3
49	708	1	2	61.1526	1	0	1	0	0.0	0.8	.	3.82	58	678.0	97.65	.	233	11.0	4
50	2598	1	1	53.5086	1	0	1	0	0.0	1.1	257	3.36	43	1080.0	106.95	73	128	10.6	4
51	3853	1	2	52.0876	1	0	0	0	0.0	0.8	276	3.60	54	4332.0	99.33	143	273	10.6	2
52	2386	1	1	50.5407	0	0	0	0	0.0	6.0	614	3.70	158	5084.4	206.40	93	362	10.6	1
53	1000	1	1	67.4086	1	0	1	0	0.0	2.6	.	3.10	94	6456.2	56.76	.	214	11.0	4
54	1434	1	1	39.1978	1	1	1	1	1.0	1.3	288	3.40	262	5487.2	73.53	125	254	11.0	4
55	1360	1	1	65.7632	0	0	0	0	0.0	1.8	416	3.94	121	10165.0	79.98	219	213	11.0	3
56	1847	1	2	33.6181	1	0	1	1	0.0	1.1	498	3.80	88	13862.4	95.46	319	365	10.6	2
57	3282	1	1	53.5715	1	0	1	0	0.5	2.3	260	3.18	231	11320.2	105.78	94	216	12.4	3
58	4459	0	1	44.5695	0	0	0	0	0.0	0.7	242	4.08	73	5890.0	56.76	118	.	10.6	1

N	X	δ	Z_1	Z_2	Z_3	Z_4	Z_5	Z_6	Z_7	Z_8	Z_9	Z_{10}	Z_{11}	Z_{12}	Z_{13}	Z_{14}	Z_{15}	Z_{16}	Z_{17}
59	2224	1	1	40.3943	1	0	1	1	0.0	0.8	329	3.50	49	7622.8	126.42	124	321	10.6	3
60	4365	0	1	58.3819	1	0	0	0	0.0	0.9	604	3.40	82	876.0	71.30	58	228	10.3	3
61	4256	0	2	43.8987	0	0	0	0	0.0	0.6	216	3.94	28	601.0	60.45	188	211	13.0	1
62	3090	1	2	60.7064	1	1	0	0	0.0	1.3	302	2.75	58	1523.0	43.40	112	329	13.2	4
63	859	1	2	46.6283	1	0	0	1	1.0	22.5	932	3.12	95	5396.0	244.90	133	165	11.6	3
64	1487	1	2	62.9076	1	0	1	0	0.0	2.1	373	3.50	52	1009.0	150.35	188	178	11.0	3
65	3992	0	1	40.2026	1	0	0	0	0.0	1.2	256	3.60	74	724.0	141.05	108	430	10.0	1
66	4191	1	1	46.4531	0	0	1	0	0.0	1.4	427	3.70	105	1909.0	182.90	171	123	11.0	3
67	2769	1	2	51.2882	1	0	0	0	0.0	1.1	466	3.91	84	1787.0	328.60	185	261	10.0	3
68	4039	0	1	32.6133	1	0	0	0	0.0	0.7	174	4.09	58	642.0	71.30	46	203	10.6	3
69	1170	1	1	49.3388	1	0	1	1	0.5	20.0	652	3.46	159	3292.0	215.45	184	227	12.4	3
70	3458	0	1	56.3997	1	0	0	0	0.0	0.6	.	4.64	20	666.0	54.25	.	265	10.6	2
71	4196	0	2	48.8460	1	0	1	0	0.0	1.2	258	3.57	79	2201.0	120.90	76	410	11.5	4
72	4184	0	2	32.4928	1	0	0	0	0.0	0.5	320	3.54	51	1243.0	122.45	80	225	10.0	3
73	4190	0	2	38.4942	1	0	0	0	0.0	0.7	132	3.60	17	423.0	49.60	56	265	11.0	1
74	1827	1	1	51.9206	1	0	1	1	0.0	8.4	558	3.99	280	967.0	89.90	309	278	11.0	4
75	1191	1	1	43.5181	1	1	1	1	0.5	17.1	674	2.53	207	2078.0	182.90	598	268	11.5	4
76	71	1	1	51.9425	1	0	1	1	0.5	12.2	394	3.08	111	2132.0	155.00	243	165	11.6	4
77	326	1	2	49.8261	1	0	1	1	0.5	6.6	244	3.41	199	1819.0	170.50	91	132	12.1	3
78	1690	1	1	47.9452	1	0	1	0	0.0	6.3	436	3.02	75	2176.0	170.50	104	236	10.6	4
79	3707	0	1	46.5161	1	0	1	0	0.0	0.8	315	4.24	13	1637.0	170.50	70	426	10.9	3
80	890	1	2	67.4114	0	0	1	0	0.0	7.2	247	3.72	269	1303.0	176.70	91	360	11.2	4
81	2540	1	1	63.2635	1	0	1	1	0.0	14.4	448	3.65	34	1218.0	60.45	318	385	11.7	4
82	3574	1	1	67.3101	1	0	0	0	0.0	4.5	472	4.09	154	1580.0	117.80	272	412	11.1	3
83	4050	0	1	56.0137	1	0	1	0	0.5	1.3	250	3.50	48	1138.0	71.30	100	81	12.9	4
84	4032	0	2	55.8303	1	0	0	0	0.0	0.4	263	3.76	29	1345.0	137.95	74	181	11.2	3
85	3358	1	2	47.2170	1	0	1	0	0.0	2.1	262	3.48	58	2045.0	89.90	84	225	11.5	4
86	1657	1	1	52.7584	1	0	1	1	0.0	5.0	1600	3.21	75	2656.0	82.15	174	181	10.9	3
87	198	1	1	37.2786	1	0	0	0	0.0	1.1	345	4.40	75	1860.0	218.55	72	447	10.7	3

N	X	δ	Z_1	Z_2	Z_3	Z_4	Z_5	Z_6	Z_7	Z_8	Z_9	Z_{10}	Z_{11}	Z_{12}	Z_{13}	Z_{14}	Z_{15}	Z_{16}	Z_{17}
88	2452	0	2	41.3936	1	0	0	0	0.5	0.6	296	4.06	37	1032.0	80.60	83	442	12.0	3
89	1741	1	1	52.4435	1	0	1	0	0.0	2.0	408	3.65	50	1083.0	110.05	98	200	11.4	2
90	2689	1	1	33.4757	0	0	0	0	0.0	1.6	660	4.22	94	1857.0	151.90	155	337	11.0	2
91	460	1	2	45.6071	1	0	1	1	0.5	5.0	325	3.47	110	2460.0	246.45	56	430	11.9	4
92	388	1	1	76.7091	1	1	0	0	1.0	1.4	206	3.13	36	1626.0	86.80	70	145	12.2	4
93	3913	0	1	36.5339	1	0	0	0	0.0	1.3	353	3.67	73	2039.0	232.50	68	380	11.1	2
94	750	1	1	53.9165	1	0	1	1	0.0	3.2	201	3.11	178	1212.0	159.65	69	188	11.8	4
95	130	1	2	46.3901	1	1	1	1	1.0	17.4	.	2.64	182	559.0	119.35	.	401	11.7	2
96	3850	0	1	48.8460	1	0	0	0	0.0	1.0	.	3.70	33	1258.0	99.20	.	338	10.4	3
97	611	1	2	71.8932	0	0	1	0	0.5	2.0	420	3.26	62	3196.0	77.50	91	344	11.4	3
98	3823	0	1	28.8843	1	0	0	0	0.0	1.0	239	3.77	77	1877.0	97.65	101	312	10.2	1
99	3820	0	2	48.4682	0	0	0	0	0.0	1.8	460	3.35	148	1472.0	108.50	118	172	10.2	2
100	552	1	2	51.4689	0	0	1	0	0.0	2.3	178	3.00	145	746.0	178.25	122	119	12.0	4
101	3581	0	2	44.9500	1	0	0	0	0.0	0.9	400	3.60	31	1689.0	164.30	166	327	10.4	3
102	3099	0	1	56.5695	1	0	0	0	0.0	0.9	248	3.97	172	646.0	62.00	84	128	10.1	1
103	110	1	2	48.9637	1	1	1	1	1.0	2.5	188	3.67	57	1273.0	119.35	102	110	11.1	4
104	3086	1	1	43.0171	1	0	0	0	0.0	1.1	303	3.64	20	2108.0	128.65	53	349	11.1	2
105	3092	0*	2	34.0397	1	0	1	0	0.0	1.1	464	4.20	38	1644.0	151.90	102	348	10.3	3
106	3222	1	1	68.5092	1	1	1	0	0.0	2.1	.	3.90	50	1087.0	103.85	.	137	10.6	2
107	3388	0	2	62.5216	1	0	0	0	0.0	0.6	212	4.03	10	648.0	71.30	77	316	17.1	1
108	2583	1	1	50.3573	1	0	0	0	0.0	0.4	127	3.50	14	1062.0	49.60	84	334	10.3	2
109	2504	0	2	44.0630	1	0	0	0	0.0	0.5	120	3.61	53	804.0	110.05	52	271	10.6	3
110	2105	1	1	38.9103	1	0	1	1	0.0	1.9	486	3.54	74	1052.0	108.50	109	141	10.9	3
111	2350	0*	1	41.1526	1	0	0	0	0.0	5.5	528	4.18	77	2404.0	172.05	78	467	10.7	3
112	3445	1	2	55.4579	1	0	1	1	0.0	2.0	267	3.67	89	754.0	196.85	90	136	11.8	4
113	980	1	1	51.2334	1	0	1	1	0.0	6.7	374	3.74	103	979.0	128.65	100	266	11.1	4
114	3395	1	2	52.8268	0	0	0	0	0.0	3.2	259	4.30	208	1040.0	110.05	78	268	11.7	3
115	3422	0	2	42.6393	1	0	0	1	0.0	0.7	303	4.19	81	1584.0	111.60	156	307	10.3	3
116	3336	0	1	61.0705	1	0	0	1	0.5	3.0	458	3.63	74	1588.0	106.95	382	438	9.9	3
117	1083	1	1	49.6564	1	0	1	1	0.0	6.5	950	3.11	111	2374.0	170.50	149	354	11.0	4

N	X	δ	Z_1	Z_2	Z_3	Z_4	Z_5	Z_6	Z_7	Z_8	Z_9	Z_{10}	Z_{11}	Z_{12}	Z_{13}	Z_{14}	Z_{15}	Z_{16}	Z_{17}
118	2288	1	1	48.8542	1	0	1	0	0.0	3.5	390	3.30	67	878.0	137.95	93	207	10.2	3
119	515	1	1	54.2560	1	0	0	1	0.0	0.6	636	3.83	129	944.0	97.65	114	306	9.5	3
120	2033	0*	1	35.1513	0	0	0	0	0.0	3.5	325	3.98	444	766.0	130.20	210	344	10.6	3
121	191	1	2	67.9069	0	1	1	0	1.0	1.3	151	3.08	73	1112.0	46.50	49	213	13.2	4
122	3297	0	1	55.4360	1	0	0	0	0.0	0.6	298	4.13	29	758.0	65.10	85	256	10.7	3
123	971	1	1	45.8207	1	0	1	1	1.0	5.1	.	3.23	18	790.0	179.80	.	104	13.0	4
124	3069	0	1	52.8898	0	0	1	0	0.0	0.6	251	3.90	25	681.0	57.35	107	182	10.8	4
125	2468	0*	2	47.1814	1	0	1	0	0.0	1.3	316	3.51	75	1162.0	147.25	137	238	10.0	4
126	824	1	1	53.5989	1	1	1	1	0.0	1.2	269	3.12	.	1441.0	165.85	68	166	11.1	4
127	3255	0	2	44.1040	1	0	0	0	0.0	0.5	268	4.08	9	1174.0	86.80	95	453	10.0	2
128	1037	1	1	41.9493	1	0	1	1	0.0	16.2	.	2.89	42	1828.0	299.15	.	123	12.6	4
129	3239	0	1	63.6140	1	0	1	0	0.0	0.9	420	3.87	30	1009.0	57.35	232	.	9.7	3
130	1413	1	2	44.2272	1	0	1	1	0.0	17.4	1775	3.43	205	2065.0	165.85	97	418	11.5	3
131	850	1	2	62.0014	1	0	1	1	0.0	2.8	242	3.80	74	614.0	136.40	104	121	13.2	4
132	2944	0	1	40.5530	1	0	0	0	0.0	1.9	448	3.83	60	1052.0	127.10	175	181	9.8	3
133	2796	1	2	62.6448	0	0	0	0	0.0	1.5	331	3.95	13	577.0	128.65	99	165	10.1	4
134	3149	0	2	42.3354	1	0	0	0	0.0	0.7	578	3.67	35	1353.0	127.10	105	427	10.7	2
135	3150	0	1	42.9678	1	0	0	0	0.0	0.4	263	3.57	123	836.0	74.40	121	445	11.0	2
136	3098	0	1	55.9617	1	0	0	0	0.0	0.8	263	3.35	27	1636.0	116.25	69	206	9.8	2
137	2990	0	1	62.8611	1	0	0	0	0.0	1.1	399	3.60	79	3472.0	155.00	152	344	10.1	2
138	1297	1	1	51.2498	0	0	1	0	0.0	7.3	426	3.93	262	2424.0	145.70	218	252	10.5	3
139	2106	0	2	46.7625	1	0	1	0	0.0	1.1	328	3.31	159	1260.0	94.55	134	142	11.6	4
140	3059	0	1	54.0753	1	0	1	0	0.0	1.1	290	4.09	38	2120.0	186.00	146	318	10.0	3
141	3050	0	1	47.0363	1	0	0	0	0.0	0.9	346	3.77	59	794.0	125.55	56	336	10.6	2
142	2419	1	2	55.7262	1	0	1	0	0.0	1.0	364	3.48	20	720.0	134.85	88	283	9.9	2
143	786	1	2	46.1027	1	0	1	0	0.0	2.9	332	3.60	86	1492.0	134.85	103	277	11.0	4
144	943	1	2	52.2875	1	0	1	0	0.5	28.0	556	3.26	152	3896.0	198.40	171	335	10.0	3
145	2976	0	2	51.2005	1	0	0	1	0.0	0.7	309	3.84	96	858.0	41.85	106	253	11.4	3
146	2615	0	2	33.8645	1	0	0	0	0.5	1.2	.	3.89	58	1284.0	173.60	.	239	9.4	3

N	X	δ	Z_1	Z_2	Z_3	Z_4	Z_5	Z_6	Z_7	Z_8	Z_9	Z_{10}	Z_{11}	Z_{12}	Z_{13}	Z_{14}	Z_{15}	Z_{16}	Z_{17}
147	2995	0	1	75.0116	1	0	0	0	0.5	1.2	288	3.37	32	791.0	57.35	114	213	10.7	2
148	1427	1	2	30.8638	1	0	1	0	0.0	7.2	1015	3.26	247	3836.0	198.40	280	330	9.8	3
149	762	1	1	61.8042	0	0	1	1	0.5	3.0	257	3.79	290	1664.0	102.30	112	140	9.9	4
150	2891	0	2	34.9870	1	0	0	1	0.0	1.0	.	3.63	57	1536.0	134.85	.	233	10.0	1
151	2870	0	1	55.0418	1	0	0	0	0.0	0.9	460	3.03	57	721.0	85.25	174	301	9.4	2
152	1152	1	1	69.9411	0	0	1	0	0.0	2.3	586	3.01	243	2276.0	114.70	126	339	10.9	3
153	2863	0	1	49.6044	1	0	0	0	0.0	0.5	217	3.85	68	453.0	54.25	68	270	11.1	1
154	140	1	1	69.3771	0	0	0	1	1.0	2.4	168	2.56	225	1056.0	120.90	75	108	14.1	3
155	2666	0	2	43.5565	1	0	1	1	0.5	0.6	220	3.35	57	1620.0	153.45	80	311	11.2	4
156	853	1	2	59.4086	1	0	1	0	0.0	25.5	358	3.52	219	2468.0	201.50	205	151	11.5	2
157	2835	0	2	48.7584	1	0	0	0	0.0	0.6	286	3.42	34	1868.0	77.50	206	487	10.0	2
158	2475	0*	1	36.4928	1	0	0	0	0.0	3.4	450	3.37	32	1408.0	116.25	118	313	11.2	2
159	1536	1	2	45.7604	0	0	0	0	0.0	2.5	317	3.46	217	714.0	130.20	140	207	10.1	3
160	2772	0	2	57.3717	1	0	0	0	0.0	0.6	217	3.62	13	414.0	75.95	119	224	10.5	3
161	2797	0	2	42.7433	1	0	0	0	0.0	2.3	502	3.56	4	964.0	120.90	180	269	9.6	2
162	186	1	2	58.8172	1	0	1	1	0.0	3.2	260	3.19	91	815.0	127.10	101	160	12.0	4
163	2055	1	1	53.4976	1	0	0	0	0.0	0.3	233	4.08	20	622.0	66.65	68	358	9.9	3
164	264	1	2	43.4141	1	0	1	1	0.5	8.5	.	3.34	161	1428.0	181.35	.	88	13.3	4
165	1077	1	1	53.3060	0	0	1	0	0.0	4.0	196	3.45	80	2496.0	133.30	142	212	11.3	4
166	2721	0	2	41.3552	1	0	1	0	0.0	5.7	1480	3.26	84	1960.0	457.25	108	213	9.5	2
167	1682	1	1	60.9582	0	0	1	0	0.0	0.9	376	3.86	200	1015.0	83.70	154	238	10.3	4
168	2713	0	2	47.7536	1	0	1	0	0.0	0.4	257	3.80	44	842.0	97.65	110	.	9.2	2
169	1212	1	2	35.4908	1	0	0	0	0.0	1.3	408	4.22	67	1387.0	142.60	137	295	10.1	3
170	2692	0	1	48.6626	1	0	0	0	0.0	1.2	390	3.61	32	1509.0	88.35	52	263	9.0	3
171	2574	0	1	52.6680	1	0	0	0	0.0	0.5	.	4.52	31	784.0	74.40	.	361	10.1	3
172	2301	0	2	49.8700	1	0	0	1	0.0	1.3	205	3.34	65	1031.0	91.45	126	217	9.8	3
173	2657	0	1	30.2752	1	0	1	1	0.0	3.0	236	3.42	76	1403.0	89.90	86	493	9.8	2
174	2644	0	1	55.5674	1	0	0	0	0.0	0.5	.	3.85	63	663.0	79.05	.	311	9.7	1
175	2624	0	2	52.1533	1	0	0	0	0.0	0.8	283	3.80	152	718.0	108.50	168	340	10.1	3

N	X	δ	Z_1	Z_2	Z_3	Z_4	Z_5	Z_6	Z_7	Z_8	Z_9	Z_{10}	Z_{11}	Z_{12}	Z_{13}	Z_{14}	Z_{15}	Z_{16}	Z_{17}
176	1492	1	1	41.6099	1	0	1	1	0.0	3.2	.	3.56	77	1790.0	139.50	.	149	10.1	4
177	2609	0	2	55.4524	1	0	0	0	0.0	0.9	258	4.01	49	559.0	43.40	133	277	10.4	2
178	2580	0	1	70.0041	1	0	0	0	0.0	0.6	.	4.08	51	665.0	74.40	.	325	10.2	4
179	2573	0	2	43.9425	1	0	1	0	0.0	1.8	396	3.83	39	2148.0	102.30	133	278	9.9	4
180	2563	0	2	42.5681	1	0	0	0	0.0	4.7	478	4.38	44	1629.0	237.15	76	175	10.4	3
181	2556	0	1	44.5695	1	0	1	1	0.0	1.4	248	3.58	63	554.0	75.95	106	79	10.3	4
182	2555	0	1	56.9446	1	0	1	0	0.0	0.6	.	3.69	161	674.0	26.35	.	539	9.9	2
183	2241	0*	2	40.2601	1	0	0	0	0.0	0.5	201	3.73	44	1345.0	54.25	145	445	10.1	2
184	974	1	2	37.6071	1	0	1	0	0.0	11.0	674	3.55	358	2412.0	167.40	140	471	9.8	3
185	2527	0	1	48.3614	1	0	0	0	0.0	0.8	256	3.54	42	1132.0	74.40	94	192	10.5	3
186	1576	1	1	70.8364	1	0	0	1	0.5	2.0	225	3.53	51	933.0	69.75	62	200	12.7	3
187	733	1	2	35.7919	1	0	1	0	0.0	14.0	808	3.43	251	2870.0	153.45	137	268	11.5	3
188	2332	0	1	62.6229	1	0	1	0	0.0	0.7	187	3.48	41	654.0	120.90	98	164	11.0	4
189	2456	0	2	50.6475	1	0	1	0	0.0	1.3	360	3.63	52	1812.0	97.65	164	256	9.9	3
190	2504	0	1	54.5270	1	0	0	1	0.0	2.3	.	3.93	24	1828.0	133.30	.	327	10.2	2
191	216	1	2	52.6927	1	1	1	1	0.0	24.5	1092	3.35	233	3740.0	147.25	432	399	15.2	4
192	2443	0	1	52.7201	1	0	1	0	0.0	0.9	308	3.69	67	696.0	51.15	101	344	9.8	4
193	797	1	2	56.7721	1	0	0	0	0.0	10.8	932	3.19	267	2184.0	161.20	157	382	10.4	4
194	2449	0	1	44.3970	1	0	0	0	0.0	1.5	293	4.30	50	975.0	125.55	56	336	9.1	2
195	2330	0	1	29.5551	1	0	1	0	0.0	3.7	347	3.90	76	2544.0	221.65	90	129	11.5	4
196	2363	0	1	57.0404	1	0	1	1	0.0	1.4	226	3.36	13	810.0	72.85	62	117	11.6	4
197	2365	0	1	44.6270	1	0	0	0	0.0	0.6	266	3.97	25	1164.0	102.30	102	201	10.1	2
198	2357	0	2	35.7974	1	0	0	1	0.0	0.7	286	2.90	38	1692.0	141.05	90	381	9.6	2
199	1592	0	1	40.7173	1	0	0	0	0.0	2.1	392	3.43	52	1395.0	184.45	194	328	10.2	3
200	2318	0	2	32.2327	1	0	0	1	0.0	4.7	236	3.55	112	1391.0	137.95	114	332	9.9	3
201	2294	0	2	41.0924	1	0	1	0	0.0	0.6	235	3.20	26	1758.0	106.95	67	228	10.8	4
202	2272	0	1	61.6400	1	0	0	0	0.0	0.5	223	3.80	15	1044.0	80.60	89	514	10.0	2
203	2221	0	2	37.0568	1	0	1	0	0.0	0.5	149	4.04	227	598.0	52.70	57	166	9.9	2
204	2090	1	2	62.5791	1	0	0	0	0.0	0.7	255	3.74	23	1024.0	77.50	58	281	10.2	3

N	X	δ	Z_1	Z_2	Z_3	Z_4	Z_5	Z_6	Z_7	Z_8	Z_9	Z_{10}	Z_{11}	Z_{12}	Z_{13}	Z_{14}	Z_{15}	Z_{16}	Z_{17}
205	2081	1	1	48.9774	1	1	0	0	0.0	2.5	382	3.55	108	1516.0	238.70	.	126	10.3	3
206	2255	0	1	61.9904	1	0	0	0	0.0	0.6	213	4.07	12	5300.0	57.35	68	240	11.0	1
207	2171	0	1	72.7721	1	0	0	0	0.5	0.6	.	3.33	14	733.0	85.25	.	259	10.1	4
208	904	1	1	61.2950	1	0	1	0	0.0	3.9	396	3.20	58	1440.0	153.45	131	156	10.0	4
209	2216	0	2	52.6242	1	0	1	1	0.0	0.7	252	4.01	11	1210.0	72.85	58	309	9.5	2
210	2224	0	2	49.7632	0	0	1	0	0.0	0.9	346	3.37	81	1098.0	122.45	90	298	10.0	2
211	2195	0	2	52.9144	1	0	0	0	0.0	1.3	.	3.76	27	1282.0	100.75	.	114	10.3	3
212	2176	0	2	47.2635	1	0	0	0	0.0	1.2	232	3.98	11	1074.0	100.75	99	223	9.9	3
213	2178	0	1	50.2040	1	0	0	1	0.0	0.5	400	3.40	9	1134.0	96.10	55	356	10.2	3
214	1786	1	2	69.3470	1	0	1	0	0.0	0.9	404	3.43	34	1866.0	79.05	224	236	9.9	3
215	1080	1	2	41.1691	1	0	0	0	0.0	5.9	1276	3.85	141	1204.0	203.05	157	216	10.7	3
216	2168	0	1	59.1650	1	0	0	0	0.0	0.5	.	3.68	20	856.0	55.80	.	146	10.4	3
217	790	1	2	36.0794	1	0	1	0	0.0	11.4	608	3.31	65	1790.0	151.90	210	298	10.8	4
218	2170	0	1	34.5955	1	0	0	0	0.0	0.5	.	3.89	29	897.0	66.65	.	423	10.1	1
219	2157	0	2	42.7132	1	0	0	0	0.0	1.6	215	4.17	67	936.0	134.85	85	176	9.6	3
220	1235	1	1	63.6304	1	0	0	1	0.0	3.8	426	3.22	96	2716.0	210.80	113	228	10.6	2
221	2050	0	2	56.6297	1	0	1	0	0.0	0.9	360	3.65	72	3186.0	94.55	154	269	9.7	4
222	597	1	2	46.2642	1	0	1	0	0.0	4.5	372	3.38	227	2310.0	167.40	135	240	12.4	3
223	334	1	1	61.2430	1	1	1	0	1.0	14.1	448	2.43	123	1833.0	134.00	155	210	11.0	4
224	1945	0	1	38.6201	1	0	0	0	0.0	1.0	309	3.66	67	1214.0	158.10	101	309	9.7	3
225	2022	0	1	38.7707	1	0	0	0	0.0	0.7	274	3.66	108	1065.0	88.35	135	251	10.1	2
226	1978	0	2	56.6954	1	0	1	0	0.0	0.5	223	3.70	39	884.0	75.95	104	231	9.6	3
227	999	1	1	58.9514	0	0	0	0	0.0	2.3	316	3.35	172	1601.0	179.80	63	394	9.7	2
228	1967	0	2	36.9227	1	0	0	0	0.0	0.7	215	3.35	41	645.0	93.00	74	165	9.6	3
229	348	1	1	62.4148	1	1	1	0	0.5	4.5	191	3.05	200	1020.0	175.15	118	139	11.4	4
230	1979	0	2	34.6092	1	0	1	1	0.0	3.3	302	3.41	51	310.0	83.70	44	95	11.5	4
231	1165	1	2	58.3354	1	0	1	1	0.0	3.4	518	1.96	115	2250.0	203.05	90	190	10.7	4
232	1951	0	1	50.1821	1	0	1	0	0.0	0.4	267	3.02	47	1001.0	133.30	87	265	10.6	3
233	1932	0	1	42.6858	1	0	1	1	0.0	0.9	514	3.06	412	2622.0	105.40	87	284	9.8	4

N	X	δ	Z_1	Z_2	Z_3	Z_4	Z_5	Z_6	Z_7	Z_8	Z_9	Z_{10}	Z_{11}	Z_{12}	Z_{13}	Z_{14}	Z_{15}	Z_{16}	Z_{17}
234	1776	0	2	34.3792	1	0	0	0	0.0	0.9	578	3.35	78	976.0	116.25	177	322	11.2	2
235	1882	0	2	33.1828	1	0	1	0	0.0	13.0	1336	4.16	71	3510.0	209.25	111	338	11.9	3
236	1908	0	1	38.3819	1	0	1	1	0.0	1.5	253	3.79	67	1006.0	139.50	106	341	9.7	3
237	1882	0	1	59.7618	1	0	1	0	0.0	1.6	442	2.95	105	820.0	85.25	108	181	10.1	3
238	1874	0	2	66.4120	1	0	0	0	0.5	0.6	280	3.35	.	1093.0	128.65	81	295	9.8	2
239	694	1	1	46.7899	1	0	1	1	0.0	0.8	300	2.94	231	1794.0	130.20	99	319	11.2	4
240	1831	0	1	56.0794	1	0	0	0	0.0	0.4	232	3.72	24	369.0	51.15	139	326	10.1	3
241	837	0*	2	41.3744	1	0	1	1	0.0	4.4	316	3.62	308	1119.0	114.70	322	282	9.8	4
242	1810	0	1	64.5722	1	0	1	0	0.0	1.9	354	2.97	86	1553.0	196.85	152	277	9.9	3
243	930	1	2	67.4880	1	0	1	0	0.0	8.0	468	2.81	139	2009.0	198.40	139	233	10.0	4
244	1690	1	1	44.8296	1	0	0	1	0.0	3.9	350	3.22	121	1268.0	272.80	231	270	9.6	3
245	1790	0	2	45.7714	1	0	1	0	0.0	0.6	273	3.65	48	794.0	52.70	214	305	9.6	3
246	1435	0*	1	32.9500	1	0	1	0	0.0	2.1	387	3.77	63	1613.0	150.35	33	185	10.1	4
247	732	0*	1	41.2211	1	0	1	0	0.0	6.1	1712	2.83	89	3681.0	158.10	139	297	10.0	3
248	1785	0	2	55.4168	1	0	1	0	0.0	0.8	324	3.51	39	1237.0	66.65	146	371	10.0	3
249	1783	0	1	47.9808	1	0	0	1	0.0	1.3	242	3.20	35	1556.0	175.15	71	195	10.6	4
250	1769	0	2	40.7912	1	0	1	0	0.0	0.6	299	3.36	23	2769.0	220.10	85	303	10.9	4
251	1457	0	1	56.9747	1	0	0	0	0.0	0.5	227	3.61	40	676.0	83.00	120	249	9.9	2
252	1770	0	1	68.4627	1	0	1	1	0.0	1.1	246	3.35	116	924.0	113.15	90	317	10.0	4
253	1765	0	1	78.4394	0	1	1	1	0.0	7.1	243	3.03	380	983.0	158.10	154	97	11.2	4
254	737	0*	1	39.8576	1	0	1	1	0.0	3.1	227	3.75	121	1136.0	110.00	91	264	10.0	3
255	1735	0	2	35.3101	1	0	1	1	0.0	0.7	193	3.85	35	466.0	53.00	118	156	10.3	3
256	1701	0	1	31.4442	1	0	0	0	0.0	1.1	336	3.74	48	823.0	84.00	108	242	9.7	3
257	1614	0	1	58.2642	1	0	0	0	0.0	0.5	280	4.23	36	377.0	56.00	146	227	10.6	2
258	1702	0	1	51.4880	1	0	0	0	0.0	1.1	414	3.44	80	1003.0	99.00	55	271	9.6	1
259	1615	0	2	59.9699	1	0	1	0	0.0	3.1	277	2.97	42	1110.0	125.00	126	221	9.8	3
260	1656	0	2	74.5243	0	0	1	0	0.0	5.6	232	3.59	188	1120.0	98.00	128	248	10.9	4
261	1677	0	2	52.3641	1	0	1	1	0.0	3.2	375	3.14	129	857.0	89.00	.	375	9.5	3
262	1666	0	2	42.7871	1	0	1	0	0.0	2.8	322	3.06	65	2562.0	91.00	209	231	9.5	3

N	X	δ	Z_1	Z_2	Z_3	Z_4	Z_5	Z_6	Z_7	Z_8	Z_9	Z_{10}	Z_{11}	Z_{12}	Z_{13}	Z_{14}	Z_{15}	Z_{16}	Z_{17}
263	1301	0*	2	34.8747	1	0	1	1	0.5	1.1	432	3.57	45	1406.0	190.00	77	248	11.4	4
264	1542	0*	2	44.1396	1	0	1	1	0.0	3.4	356	3.12	188	1911.0	92.00	130	318	11.2	3
265	1084	0*	2	46.3819	1	0	1	0	0.0	3.5	348	3.20	121	938.0	120.00	146	296	10.0	4
266	1614	0	1	56.3094	1	0	0	0	0.0	0.5	318	3.32	52	613.0	70.00	260	279	10.2	3
267	179	1	1	70.9076	1	1	1	1	1.0	6.6	222	2.33	138	620.0	106.00	91	195	12.1	4
268	1191	1	1	55.3949	1	1	1	0	0.5	6.4	344	2.75	16	834.0	82.00	179	149	11.0	4
269	1363	0	2	45.0842	1	0	0	0	0.0	3.6	374	3.50	143	1428.0	188.00	44	151	10.1	2
270	1568	0	1	26.2779	1	0	1	1	0.0	1.0	448	3.74	102	1128.0	71.00	117	228	10.2	3
271	1569	0	2	50.4723	1	0	1	0	0.0	1.0	321	3.50	94	955.0	111.00	177	289	9.7	3
272	1525	0	1	38.3984	1	0	0	0	0.0	0.5	226	2.93	22	674.0	58.00	85	153	9.8	1
273	1558	0	2	47.4196	1	0	0	1	0.0	2.2	328	3.46	75	1677.0	87.00	116	202	9.6	3
274	1447	0*	1	47.9808	1	0	0	0	0.0	1.6	.	3.07	136	1995.0	128.00	.	372	9.6	4
275	1349	0	1	38.3162	1	0	0	0	0.0	2.2	572	3.77	77	2520.0	92.00	114	309	9.5	4
276	1481	0	1	50.1081	1	0	0	0	0.0	1.0	219	3.85	67	640.0	145.00	108	95	10.7	2
277	1434	0	2	35.0883	1	0	0	0	0.5	1.0	317	3.56	44	1636.0	84.00	111	394	9.8	3
278	1420	0	2	32.5038	1	0	0	0	0.0	5.6	338	3.70	130	2139.0	185.00	193	215	9.9	4
279	1433	0	2	56.1533	1	0	0	0	0.0	0.5	198	3.77	38	911.0	57.00	56	280	9.8	2
280	1412	0	1	46.1547	1	0	0	0	0.0	1.6	325	3.69	69	2583.0	142.00	140	284	9.6	3
281	41	1	1	65.8836	1	1	0	0	1.0	17.9	175	2.10	220	705.0	338.00	229	62	12.9	4
282	1455	0	2	33.9439	1	0	1	0	0.0	1.3	304	3.52	97	1622.0	71.00	169	255	9.5	4
283	1030	0	2	62.8611	1	0	0	0	0.0	1.1	412	3.99	103	1293.0	91.00	113	422	9.6	4
284	1418	0	2	48.5640	1	0	0	0	0.0	1.3	291	3.44	75	1082.0	85.00	195	251	9.5	3
285	1401	0	1	46.3491	1	0	0	0	0.0	0.8	253	3.48	65	688.0	57.00	80	252	10.0	1
286	1408	0	1	38.8528	1	0	1	1	0.0	2.0	310	3.36	70	1257.0	122.00	118	143	9.8	3
287	1234	0	1	58.6475	1	0	0	1	0.0	6.4	373	3.46	155	1768.0	120.00	151	258	10.1	4
288	1067	0*	2	48.9363	1	0	1	0	0.5	8.7	310	3.89	107	637.0	117.00	242	298	9.6	2
289	799	1	1	67.5729	0	0	1	0	0.5	4.0	416	3.99	177	960.0	86.00	242	269	9.8	2
290	1363	0	1	65.9849	1	0	0	0	0.0	1.4	294	3.57	33	722.0	93.00	69	283	9.8	3
291	901	0*	1	40.9008	1	0	0	0	0.0	3.2	339	3.18	123	3336.0	205.00	84	304	9.9	4

N	X	δ	Z_1	Z_2	Z_3	Z_4	Z_5	Z_6	Z_7	Z_8	Z_9	Z_{10}	Z_{11}	Z_{12}	Z_{13}	Z_{14}	Z_{15}	Z_{16}	Z_{17}
292	1329	0	2	50.2450	0	0	1	0	0.0	8.6	546	3.73	84	1070.0	127.00	153	291	11.2	3
293	1320	0	2	57.1964	1	0	1	1	1.0	8.5	194	2.98	196	815.0	163.00	78	122	12.3	4
294	1302	0	1	60.5366	0	0	1	0	0.0	6.6	1000	3.07	88	3150.0	193.00	133	299	10.9	4
295	877	0*	1	35.3511	0	0	0	0	0.0	2.4	646	3.83	102	855.0	127.00	194	306	10.3	3
296	1321	0	2	31.3812	1	0	0	0	0.0	0.8	328	3.31	62	1105.0	137.00	95	293	10.9	4
297	533	0*	1	55.9863	0	0	1	0	0.0	1.2	275	3.43	100	1142.0	75.00	91	217	11.3	4
298	1300	0	2	52.7255	1	0	1	0	0.0	1.1	340	3.37	73	289.0	97.00	93	243	10.2	3
299	1293	0	1	38.0917	1	0	0	0	0.0	2.4	342	3.76	90	1653.0	150.00	127	213	10.8	3
300	207	1	2	58.1711	1	0	1	0	0.0	5.2	.	2.23	234	601.0	135.00	.	206	12.3	4
301	1295	0	2	45.2101	1	0	0	0	0.0	1.0	393	3.57	50	1307.0	74.00	103	295	10.5	4
302	1271	0	1	37.7988	1	0	0	0	0.0	0.7	335	3.95	43	657.0	52.00	104	268	10.6	2
303	1250	0	2	60.6598	1	0	1	1	0.0	1.0	372	3.25	108	1190.0	140.00	55	248	10.6	4
304	1230	0	1	35.5346	1	0	0	0	0.0	0.5	219	3.93	22	663.0	45.00	75	246	10.8	3
305	1216	0	2	43.0664	1	0	1	1	0.0	2.9	426	3.61	73	5184.0	288.00	144	275	10.6	3
306	1216	0	2	56.3915	1	0	1	0	0.0	0.6	239	3.45	31	1072.0	55.00	64	227	10.7	2
307	1149	0	2	30.5736	1	0	0	0	0.0	0.8	273	3.56	52	1282.0	130.00	59	344	10.5	2
308	1153	0	1	61.1828	1	0	1	0	0.0	0.4	246	3.58	24	797.0	91.00	113	288	10.4	2
309	994	0	2	58.2998	1	0	0	0	0.0	0.4	260	2.75	41	1166.0	70.00	82	231	10.8	2
310	939	0	1	62.3326	1	0	0	0	0.0	1.7	434	3.35	39	1713.0	171.00	100	234	10.2	2
311	839	0	1	37.9986	1	0	0	0	0.0	2.0	247	3.16	69	1050.0	117.00	88	335	10.5	2
312	788	0	2	33.1526	1	0	0	1	0.0	6.4	576	3.79	186	2115.0	136.00	149	200	10.8	2
313	4062	0	.	60.0000	.	.	.	.	0.0	0.7	.	3.65	.	.	.	.	378	11.0	.
314	3561	1	.	65.0000	.	.	.	.	0.5	1.4	.	3.04	.	.	.	.	331	12.1	.
315	2844	0	.	54.0000	.	.	.	.	0.0	0.7	.	4.03	.	.	.	.	226	9.8	.
316	2071	1	.	75.0000	.	.	.	.	0.5	0.7	.	3.96	.	.	.	.	.	11.3	.
317	3030	0	.	62.0000	.	.	.	.	0.0	0.8	.	2.48	.	.	.	.	273	10.0	.
318	1680	0	.	43.0000	.	.	.	.	0.0	0.7	.	3.68	.	.	.	.	306	9.5	.
319	41	1	.	46.0000	.	.	.	.	0.0	5.0	.	2.93	.	.	.	.	260	10.4	.
320	2403	0	.	44.0000	.	.	.	.	0.5	0.4	.	3.81	.	.	.	.	226	10.5	.

N	X	δ	Z_1	Z_2	Z_3	Z_4	Z_5	Z_6	Z_7	Z_8	Z_9	Z_{10}	Z_{11}	Z_{12}	Z_{13}	Z_{14}	Z_{15}	Z_{16}	Z_{17}
321	1170	0	.	61.0000	.	.	.	.	0.5	1.3	.	3.41	.	.	.	.	259	10.9	.
322	2011	1	.	64.0000	.	.	.	.	0.0	1.1	.	3.69	.	.	.	.	139	10.5	.
323	3523	0	.	40.0000	.	.	.	.	0.0	0.6	.	4.04	.	.	.	.	130	11.2	.
324	3468	0	.	63.0000	.	.	.	.	0.0	0.6	.	3.94	.	.	.	.	234	11.5	.
325	4795	0	.	34.0000	.	.	.	.	0.0	1.8	.	3.24	.	.	.	.	.	18.0	.
326	1236	0	.	52.0000	.	.	.	.	0.0	1.5	.	3.42	.	.	.	.	246	10.3	.
327	4214	0	.	49.0000	.	.	.	.	0.0	1.2	.	3.99	.	.	.	.	.	11.2	.
328	2111	1	.	54.0000	.	.	.	.	0.0	1.0	.	3.60	.	.	.	.	.	12.1	.
329	1462	1	.	63.0000	.	.	.	.	0.0	0.7	.	3.40	.	.	.	.	371	10.1	.
330	1746	1	.	54.0000	.	.	.	.	0.0	3.5	.	3.63	.	.	.	.	325	10.3	.
331	94	1	.	46.0000	.	.	.	.	0.5	3.1	.	3.56	.	.	.	.	142	13.6	.
332	785	1	.	53.0000	.	.	.	.	0.0	12.6	.	2.87	.	.	.	.	114	11.8	.
333	1518	1	.	56.0000	.	.	.	.	0.0	2.8	.	3.92	.	.	.	.	.	10.6	.
334	466	1	.	56.0000	.	.	.	.	0.0	7.1	.	3.51	.	.	.	.	721	11.8	.
335	3527	0	.	55.0000	.	.	.	.	0.0	0.6	.	4.15	.	.	.	.	280	10.1	.
336	2635	0	.	65.0000	.	.	.	.	0.0	2.1	.	3.34	.	.	.	.	155	10.1	.
337	2286	1	.	56.0000	.	.	.	.	0.0	1.8	.	3.64	.	.	.	.	141	10.0	.
338	791	1	.	47.0000	.	.	.	.	0.0	16.0	.	3.42	.	.	.	.	475	13.8	.
339	3492	0	.	60.0000	.	.	.	.	0.0	0.6	.	4.38	.	.	.	.	269	10.6	.
340	3495	0	.	53.0000	.	.	.	.	0.0	5.4	.	4.19	.	.	.	.	141	11.2	.
341	111	1	.	54.0000	.	.	.	.	0.0	9.0	.	3.29	.	.	.	.	286	13.1	.
342	3231	0	.	50.0000	.	.	.	.	0.0	0.9	.	4.01	.	.	.	.	244	10.5	.
343	625	1	.	48.0000	.	.	.	.	0.0	11.1	.	2.84	.	.	.	.	.	12.2	.
344	3157	0	.	36.0000	.	.	.	.	0.0	8.9	.	3.76	.	.	.	.	209	10.6	.
345	3021	0*	.	48.0000	.	.	.	.	0.0	0.5	.	3.76	.	.	.	.	388	10.1	.
346	559	1	.	70.0000	.	.	.	.	0.5	0.6	.	3.81	.	.	.	.	160	11.0	.
347	2812	1	.	51.0000	.	.	.	.	0.0	3.4	.	3.92	.	.	.	.	.	9.3	.
348	2834	0	.	52.0000	.	.	.	.	0.0	0.9	.	3.14	.	.	.	.	191	12.3	.
349	2855	0	.	54.0000	.	.	.	.	0.0	1.4	.	3.82	.	.	.	.	249	10.3	.

N	X	δ	Z_1	Z_2	Z_3	Z_4	Z_5	Z_6	Z_7	Z_8	Z_9	Z_{10}	Z_{11}	Z_{12}	Z_{13}	Z_{14}	Z_{15}	Z_{16}	Z_{17}
350	662	1	.	48.0000	.	.	.	.	0.0	2.1	.	4.10	.	.	.	.	200	9.0	.
351	727	1	.	66.0000	.	.	.	.	0.0	15.0	.	3.40	.	.	.	.	150	11.1	.
352	2716	0	.	53.0000	.	.	.	.	0.0	0.6	.	4.19	.	.	.	.	330	9.9	.
353	2698	0	.	62.0000	.	.	.	.	0.0	1.3	.	3.40	.	.	.	.	167	10.6	.
354	990	1	.	59.0000	.	.	.	.	0.0	1.3	.	3.12	.	.	.	.	125	9.6	.
355	2338	0	.	39.0000	.	.	.	.	0.0	1.6	.	3.75	.	.	.	.	145	10.4	.
356	1616	1	.	67.0000	.	.	.	.	0.5	2.2	.	3.26	.	.	.	.	171	11.1	.
357	2563	0	.	58.0000	.	.	.	.	0.0	3.0	.	3.46	.	.	.	.	109	10.4	.
358	2537	0	.	64.0000	.	.	.	.	0.0	0.8	.	3.49	.	.	.	.	314	10.3	.
359	2534	0	.	46.0000	.	.	.	.	0.0	0.8	.	2.89	.	.	.	.	419	.	.
360	778	1	.	64.0000	.	.	.	.	0.0	1.8	.	3.15	.	.	.	.	183	10.4	.
361	617	0*	.	41.0000	.	.	.	.	0.0	5.5	.	2.31	.	.	.	.	517	10.4	.
362	2267	0*	.	49.0000	.	.	.	.	0.0	18.0	.	3.04	.	.	.	.	432	9.7	.
363	2249	0	.	44.0000	.	.	.	.	0.0	0.6	.	3.50	.	.	.	.	150	9.9	.
364	359	1	.	59.0000	.	.	.	.	0.0	2.7	.	3.35	.	.	.	.	142	11.5	.
365	1925	0	.	63.0000	.	.	.	.	0.0	0.9	.	3.58	.	.	.	.	224	10.0	.
366	249	1	.	61.0000	.	.	.	.	0.0	1.3	.	3.01	.	.	.	.	223	10.7	.
367	2202	0	.	64.0000	.	.	.	.	0.0	1.1	.	3.49	.	.	.	.	166	9.8	.
368	43	1	.	49.0000	.	.	.	.	0.0	13.8	.	2.77	.	.	.	.	388	.	.
369	1197	1	.	42.0000	.	.	.	.	0.0	4.4	.	4.52	.	.	.	.	102	10.8	.
370	1095	1	.	50.0000	.	.	.	.	0.0	16.0	.	3.36	.	.	.	.	384	10.0	.
371	489	1	.	51.0000	.	.	.	.	0.5	7.3	.	3.52	.	.	.	.	265	11.1	.
372	2149	0	.	37.0000	.	.	.	.	0.0	0.6	.	3.55	.	.	.	.	248	10.3	.
373	2103	0	.	62.0000	.	.	.	.	0.0	0.7	.	3.29	.	.	.	.	190	9.8	.
374	1980	0	.	51.0000	.	.	.	.	0.0	0.7	.	3.10	.	.	.	.	274	10.6	.
375	1347	0*	.	52.0000	.	.	.	.	0.0	1.7	.	3.24	.	.	.	.	231	10.5	.
376	1478	1	.	44.0000	.	.	.	.	0.0	9.5	.	3.63	.	.	.	.	292	10.2	.
377	1987	0	.	33.0000	.	.	.	.	0.0	2.2	.	3.76	.	.	.	.	253	9.9	.
378	1168	1	.	60.0000	.	.	.	.	0.5	1.8	.	3.62	.	.	.	.	225	9.9	.

N	X	δ	Z_1	Z_2	Z_3	Z_4	Z_5	Z_6	Z_7	Z_8	Z_9	Z_{10}	Z_{11}	Z_{12}	Z_{13}	Z_{14}	Z_{15}	Z_{16}	Z_{17}
379	597	1	.	63.0000	.	.	.	.	0.5	3.3	.	2.73	.	.	.	.	224	11.1	.
380	1725	0*	.	33.0000	.	.	.	.	0.0	2.9	.	4.08	.	.	.	.	418	10.5	.
381	1899	0	.	41.0000	.	.	.	.	0.0	1.7	.	3.66	.	.	.	.	92	11.0	.
382	221	1	.	51.0000	.	.	.	.	0.0	14.0	.	2.58	.	.	.	.	190	11.6	.
383	1022	0*	.	37.0000	.	.	.	.	0.5	0.8	.	3.00	.	.	.	.	76	10.8	.
384	1639	0	.	59.0000	.	.	.	.	0.0	1.3	.	3.40	.	.	.	.	243	9.7	.
385	1635	0	.	55.0000	.	.	.	.	0.0	0.7	.	2.93	.	.	.	.	209	10.6	.
386	1654	0	.	54.0000	.	.	.	.	0.0	1.7	.	2.38	.	.	.	.	166	9.8	.
387	1653	0	.	49.0000	.	.	.	.	0.5	13.6	.	3.00	.	.	.	.	233	9.9	.
388	1560	0	.	40.0000	.	.	.	.	0.0	0.9	.	3.50	.	.	.	.	117	10.9	.
389	1581	0	.	67.0000	.	.	.	.	0.0	0.7	.	3.06	.	.	.	.	165	10.0	.
390	1419	0	.	68.0000	.	.	.	.	0.0	3.0	.	3.15	.	.	.	.	139	10.0	.
391	1443	0	.	41.0000	.	.	.	.	0.0	1.2	.	2.80	.	.	.	.	120	11.0	.
392	1368	0	.	69.0000	.	.	.	.	0.0	0.4	.	3.03	.	.	.	.	173	10.9	.
393	193	1	.	52.0000	.	.	.	.	0.5	0.7	.	2.96	.	.	.	.	319	9.9	.
394	1367	0	.	57.0000	.	.	.	.	0.5	2.0	.	3.07	.	.	.	.	80	12.1	.
395	1329	0	.	36.0000	.	.	.	.	0.0	1.4	.	3.98	.	.	.	.	402	11.0	.
396	1343	0	.	50.0000	.	.	.	.	0.0	1.6	.	3.48	.	.	.	.	277	10.2	.
397	1328	0	.	64.0000	.	.	.	.	0.0	0.5	.	3.65	.	.	.	.	425	10.2	.
398	1375	0	.	62.0000	.	.	.	.	0.0	7.3	.	3.49	.	.	.	.	189	10.9	.
399	1260	0	.	42.0000	.	.	.	.	0.0	8.1	.	2.82	.	.	.	.	193	10.4	.
400	1223	0	.	44.0000	.	.	.	.	0.0	0.5	.	3.34	.	.	.	.	258	10.6	.
401	935	1	.	69.0000	.	.	.	.	0.0	4.2	.	3.19	.	.	.	.	120	11.1	.
402	943	0	.	52.0000	.	.	.	.	0.0	0.8	.	3.01	.	.	.	.	256	10.6	.
403	1141	0	.	66.0000	.	.	.	.	0.0	2.5	.	3.33	.	.	.	.	256	10.8	.
404	1092	0	.	40.0000	.	.	.	.	0.0	4.6	.	3.60	.	.	.	.	337	10.4	.
405	1150	0	.	52.0000	.	.	.	.	0.0	1.0	.	3.64	.	.	.	.	340	10.6	.
406	703	1	.	46.0000	.	.	.	.	0.0	4.5	.	2.68	.	.	.	.	219	11.5	.
407	1129	0	.	54.0000	.	.	.	.	0.0	1.1	.	3.69	.	.	.	.	220	10.8	.

N	X	δ	Z_1	Z_2	Z_3	Z_4	Z_5	Z_6	Z_7	Z_8	Z_9	Z_{10}	Z_{11}	Z_{12}	Z_{13}	Z_{14}	Z_{15}	Z_{16}	Z_{17}
408	1086	0	.	51.0000	.	.	.	.	0.5	1.9	.	3.17	.	.	.	.	162	10.7	.
409	1067	0	.	43.0000	.	.	.	.	0.0	0.7	.	3.73	.	.	.	.	214	10.8	.
410	1072	0	.	39.0000	.	.	.	.	0.0	1.5	.	3.81	.	.	.	.	255	10.8	.
411	1119	0	.	51.0000	.	.	.	.	0.0	0.6	.	3.57	.	.	.	.	286	10.6	.
412	1097	0	.	67.0000	.	.	.	.	0.0	1.0	.	3.58	.	.	.	.	244	10.8	.
413	989	0	.	35.0000	.	.	.	.	0.0	0.7	.	3.23	.	.	.	.	312	10.8	.
414	681	1	.	67.0000	.	.	.	.	0.0	1.2	.	2.96	.	.	.	.	174	10.9	.
415	1103	0	.	39.0000	.	.	.	.	0.0	0.9	.	3.83	.	.	.	.	180	11.2	.
416	1055	0	.	57.0000	.	.	.	.	0.0	1.6	.	3.42	.	.	.	.	143	9.9	.
417	691	0	.	58.0000	.	.	.	.	0.0	0.8	.	3.75	.	.	.	.	269	10.4	.
418	976	0	.	53.0000	.	.	.	.	0.0	0.7	.	3.29	.	.	.	.	350	10.6	.

D.2 CGD DATA

The following data are from a placebo controlled randomized trial of gamma interferon in chronic granulotomous disease (CGD). The CGD study, described in Section 4.4 and in a report by the International CGD Cooperative Study Group (1991), was designed to have a single interim analysis when the follow-up data as of July 15, 1989 were complete. The monitoring committee for the trial terminated the trial at a meeting on September 22, 1989, based on the analysis described in Section 4.4. The treatment given each patient was unblinded at the first scheduled visit for that patient following the decision of the monitoring committee.

The data for each case give the time to initial and any recurrent serious infections, from study entry until the first scheduled visit of the patient after the decision of the monitoring committee. These infections are those observed through the interim analysis date of record (7/15/89) as well as the additional data on occurrence of serious infections between the interim analysis cutoff and the final blinded study visit for each patient. There is a minimum of 1 record per patient, with an additional record for each serious infection occurring up to the study completion date.

The variables contained here are:

ID : Case identification number. The first 3 digits denote a hospital code.
RDT : Date randomized onto study (for $S = 1$ record), in the format mmddyy.
IDT : Date of onset of a serious infection, or date the patient was taken off the study.
Z_1 : Treatment Code, 1 = rIFN-γ, 2 = placebo.
Z_2 : Pattern of inheritance, 1 = X-linked, 2 = autosomal recessive.
Z_3 : Age, in years.
Z_4 : Height, in cm.
Z_5 : Weight, in kg.
Z_6 : Using corticosteroids at time of study entry, 1 = yes, 2 = no.
Z_7 : Using prophylactic antibiotics at time of study entry, 1 = yes, 2 = no.
Z_8 : 1 = male, 2 = female.
Z_9 : Hospital category, 1 = US - NIH, 2 = US - other, 3 = Europe–Amsterdam, 4 = Europe–other.
T_1 : Elapsed time (in days) from randomization (from $S = 1$ record) to diagnosis of a serious infection, or if a censored observation, elapsed time from randomization to censoring date. Computed as $IDT - RDT$ (from $S = 1$ record).
T_2 : 0, for $S = 1$. If $S > 1$, $T_2 = T_1$ (from previous record) + 1.
δ : Censoring indicator, 1 = Non-censored observation (ie, a serious infection occurred at the date specified by the IDT field), 2 = censored observation (i.e., the patient was taken off study at the date specified by the IDT field.) Note that only patient ID = 238087 had a serious infection diagnosed on the date he was taken off study.
S : Sequence number. For each patient, the infection records are in sequence number order.

ID	RDT	IDT	Z_1	Z_2	Z_3	Z_4	Z_5	Z_6	Z_7	Z_8	Z_9	T_1	T_2	δ	S
174054	120688	092589	1	2	38	152.20	66.70	2	1	2	2	293	0	2	1
174077	011389	092589	2	1	14	144.00	32.80	2	1	1	2	255	0	2	1
174109	022489	092589	2	1	26	81.25	55.00	2	1	1	2	213	0	2	1
174111	030689	092589	2	1	26	178.50	69.30	2	1	1	2	203	0	2	1
204001	082888	040489	1	2	12	147.00	62.00	2	2	2	2	219	0	1	1
204001	040589	090589	1	2	12	147.00	62.00	2	2	2	2	373	220	1	2
204001	090689	101689	1	2	12	147.00	62.00	2	2	2	2	414	374	2	3
204002	082888	090588	2	2	15	159.00	47.50	2	1	1	2	8	0	1	1
204002	090688	092388	2	2	15	159.00	47.50	2	1	1	2	26	9	1	2
204002	092488	012789	2	2	15	159.00	47.50	2	1	1	2	152	27	1	3
204002	012889	042689	2	2	15	159.00	47.50	2	1	1	2	241	153	1	4
204002	042789	050489	2	2	15	159.00	47.50	2	1	1	2	249	242	1	5
204002	050589	071689	2	2	15	159.00	47.50	2	1	1	2	322	250	1	6
204002	071789	081389	2	2	15	159.00	47.50	2	1	1	2	350	323	1	7
204002	081489	111089	2	2	15	159.00	47.50	2	1	1	2	439	351	2	8
204003	082988	091589	1	1	19	171.00	72.70	2	1	1	2	382	0	2	1
204004	091388	100689	1	1	12	142.00	34.00	2	1	1	2	388	0	2	1
204014	101188	051089	2	1	1	79.00	10.50	2	1	1	2	211	0	1	1
204014	051189	062889	2	1	1	79.00	10.50	2	1	1	2	260	212	1	2
204014	062989	070389	2	1	1	79.00	10.50	2	1	1	2	265	261	1	3
204014	070489	070789	2	1	1	79.00	10.50	2	1	1	2	269	266	1	4
204014	070889	081489	2	1	1	79.00	10.50	2	1	1	2	307	270	1	5
204014	081589	100989	2	1	1	79.00	10.50	2	1	1	2	363	308	2	6
204015	101788	010789	1	2	9	134.50	32.70	2	1	1	2	82	0	1	1
204015	010889	020889	1	2	9	134.50	32.70	2	1	1	2	114	83	1	2
204015	020989	091989	1	2	9	134.50	32.70	2	1	1	2	337	115	1	3
204015	092089	101989	1	2	9	134.50	32.70	2	1	1	2	367	338	2	4
204018	101988	110688	2	1	1	79.00	11.47	2	1	1	2	18	0	1	1
204018	110788	101689	2	1	1	79.00	11.47	2	1	1	2	362	19	2	2

ID	RDT	IDT	Z_1	Z_2	Z_3	Z_4	Z_5	Z_6	Z_7	Z_8	Z_9	T_1	T_2	δ	S
204029	111588	080989	1	1	5	102.00	18.00	2	1	1	2	267	0	1	1
204029	081089	111089	1	1	5	102.00	18.00	2	1	1	2	360	268	2	2
204056	120888	111089	1	2	22	169.00	52.20	2	1	1	2	337	0	2	1
204085	012389	102089	2	2	19	159.00	46.00	2	2	2	2	270	0	2	1
204088	012489	102589	2	1	7	115.50	19.50	2	1	1	2	274	0	2	1
204102	021789	111589	1	1	25	185.00	58.40	2	1	1	2	271	0	2	1
204103	021789	102789	2	1	31	170.00	80.50	2	1	1	2	252	0	2	1
204131	031789	111589	1	1	37	155.00	67.50	2	1	2	2	243	0	2	1
204134	032189	070389	2	2	6	130.00	21.60	2	1	2	2	104	0	1	1
204134	070489	110389	2	2	6	130.00	21.60	2	1	2	2	227	105	2	2
204135	032189	110389	2	2	3	96.00	13.10	2	1	2	2	227	0	2	1
222121	031789	100189	2	2	22	163.00	49.10	2	1	2	4	198	0	2	1
222122	031789	101089	2	1	17	169.00	63.50	2	1	1	4	207	0	2	1
222123	031789	090189	2	1	19	182.00	63.90	2	1	1	4	168	0	1	1
222123	090289	100389	2	1	19	182.00	63.90	2	1	1	4	200	169	2	2
222125	031789	093089	1	2	36	167.00	60.80	2	1	2	4	197	0	2	1
238005	092888	060189	2	1	17	162.50	52.70	2	1	1	1	246	0	1	1
238005	060289	060889	2	1	17	162.50	52.70	2	1	1	1	253	247	1	2
238005	060989	101689	2	1	17	162.50	52.70	2	1	1	1	383	254	2	3
238009	100488	072589	2	2	27	176.00	82.80	2	1	1	1	294	0	1	1
238009	072689	091889	2	2	27	176.00	82.80	2	1	1	1	349	295	2	2
238010	100488	101089	1	1	5	113.00	19.50	2	1	1	1	371	0	2	1
238011	100488	102388	2	1	2	93.00	13.20	2	1	1	1	19	0	1	1
238011	102488	011489	2	1	2	93.00	13.20	2	1	1	1	102	20	2	2
238012	101088	101889	1	1	8	124.00	25.40	2	1	1	1	373	0	1	1
238012	101989	110289	1	1	8	124.00	25.40	2	1	1	1	388	374	2	2
238013	101088	110289	1	1	12	144.00	36.90	2	1	1	1	388	0	2	1
238016	101788	101789	2	1	27	174.00	67.80	2	1	1	1	365	0	2	1
238017	101888	091789	2	1	14	143.50	33.40	2	1	1	1	334	0	1	1
238017	091889	102389	2	1	14	143.50	33.40	2	1	1	1	370	335	1	2

ID	RDT	IDT	Z_1	Z_2	Z_3	Z_4	Z_5	Z_6	Z_7	Z_8	Z_9	T_1	T_2	δ	S
238017	102489	110489	2	1	14	143.50	33.40	2	1	1	1	382	371	2	3
238019	102788	110489	1	2	11	149.00	50.90	2	1	1	1	373	0	2	1
238020	110288	080989	2	2	29	175.00	73.10	2	1	1	1	280	0	1	1
238020	081089	112289	2	2	29	175.00	73.10	2	1	1	1	385	281	2	2
238021	110288	111389	1	2	31	167.00	51.80	2	1	2	1	376	0	2	1
238022	110988	110489	1	1	7	121.00	19.90	2	1	1	1	360	0	2	1
238023	110988	091189	2	2	26	153.00	46.90	2	1	2	1	306	0	2	1
238036	112388	032189	1	1	13	145.20	36.20	2	1	1	1	118	0	1	1
238036	032289	072189	1	1	13	145.20	36.20	2	1	1	1	240	119	1	2
238036	072289	080189	1	1	13	145.20	36.20	2	1	1	1	251	241	2	3
238044	113088	060589	1	2	25	168.00	68.90	2	1	2	1	187	0	1	1
238044	060689	112189	1	2	25	168.00	68.90	2	1	2	1	356	188	2	2
238045	113088	110489	1	2	9	140.00	36.00	2	1	2	1	339	0	2	1
238046	113088	120688	2	1	28	174.00	63.70	2	1	1	1	6	0	1	1
238046	120788	092789	2	1	28	174.00	63.70	2	1	1	1	301	7	2	2
238051	120588	110489	1	2	13	139.00	34.80	2	1	2	1	334	0	2	1
238065	121688	091189	1	2	24	177.00	78.40	2	1	1	1	269	0	2	1
238076	011089	101089	2	1	11	123.00	24.30	2	1	1	1	273	0	2	1
238086	012489	051789	1	1	4	103.80	16.80	2	1	1	1	113	0	1	1
238086	051889	102489	1	1	4	103.80	16.80	2	1	1	1	273	114	2	2
238087	012489	050389	2	1	19	170.00	71.20	2	1	1	1	99	0	1	1
238087	050489	112689	2	1	19	170.00	71.20	2	1	1	1	306	100	1	2
238099	021489	110489	1	1	18	166.00	58.10	2	1	1	1	263	0	2	1
238100	021589	080189	1	2	7	135.00	42.90	2	1	1	1	167	0	1	1
238100	080289	101389	1	2	7	135.00	42.90	2	1	1	1	240	168	2	2
238106	022389	112189	2	2	12	166.00	51.90	2	1	2	1	271	0	2	1
238107	022389	110489	1	1	10	129.00	27.40	2	1	1	1	254	0	2	1
242091	012789	110189	2	1	9	129.40	28.70	2	1	1	2	278	0	2	1
242094	020389	102689	1	1	5	112.30	20.70	2	1	1	2	265	0	2	1

ID	RDT	IDT	Z_1	Z_2	Z_3	Z_4	Z_5	Z_6	Z_7	Z_8	Z_9	T_1	T_2	δ	S
242101	021789	092289	1	1	1	76.30	11.30	2	1	1	2	217	0	2	1
242110	030389	081589	1	1	7	119.00	20.60	2	1	1	2	165	0	1	1
242110	081689	092989	1	1	7	119.00	20.60	2	1	1	2	210	166	2	2
242117	031489	092289	2	2	11	137.50	40.30	2	1	1	2	192	0	2	1
242119	031689	032789	2	2	4	98.30	14.40	2	1	1	2	11	0	1	1
242119	032889	040789	2	2	4	98.30	14.40	2	1	1	2	22	12	1	2
242119	040889	090189	2	2	4	98.30	14.40	2	1	1	2	169	23	1	3
242119	090289	092789	2	2	4	98.30	14.40	2	1	1	2	195	170	2	4
242132	031789	071589	2	2	7	113.00	20.40	2	1	2	2	120	0	1	1
242132	071689	100689	2	2	7	113.00	20.40	2	1	2	2	203	121	2	2
242133	032089	100389	1	1	15	178.70	60.50	2	1	1	2	197	0	2	1
243030	111588	060989	2	1	1	79.00	12.20	2	1	1	2	206	0	1	1
243030	061089	102889	2	1	1	79.00	12.20	2	1	1	2	347	207	2	2
243040	112888	102989	1	1	7	116.60	23.30	2	1	1	2	335	0	2	1
243041	112888	082989	1	1	17	170.20	47.90	2	1	1	2	274	0	1	1
243041	083089	112489	1	1	17	170.20	47.90	2	1	1	2	361	275	1	2
243041	112589	112889	1	1	17	170.20	47.90	2	1	1	2	365	362	2	3
243042	112888	103089	1	1	8	125.40	27.70	2	1	1	2	336	0	2	1
243052	120588	012689	2	1	20	173.00	68.40	2	1	1	2	52	0	1	1
243052	012789	020889	2	1	20	173.00	68.40	2	1	1	2	65	53	1	2
243052	020989	081789	2	1	20	173.00	68.40	2	1	1	2	255	66	1	3
243052	081889	090189	2	1	20	173.00	68.40	2	1	1	2	270	256	2	4
243053	120588	021089	2	1	5	114.50	23.00	2	2	1	2	67	0	1	1
243053	021189	081089	2	1	5	114.50	23.00	2	2	1	2	248	68	1	2
243053	081189	081289	2	1	5	114.50	23.00	2	2	1	2	250	249	1	3
243053	081389	091589	2	1	5	114.50	23.00	2	2	1	2	284	251	1	4
243053	091689	111789	2	1	5	114.50	23.00	2	2	1	2	347	285	2	5
243060	121288	102689	1	1	6	105.50	19.50	1	1	1	2	318	0	2	1
243061	121288	102689	2	1	9	129.60	29.30	2	2	1	2	318	0	1	1

ID	RDT	IDT	Z_1	Z_2	Z_3	Z_4	Z_5	Z_6	Z_7	Z_8	Z_9	T_1	T_2	δ	S
243061	102789	120689	2	1	9	129.60	29.30	2	2	1	2	359	319	2	2
243097	020789	102489	1	1	5	110.40	19.10	2	1	1	2	259	0	2	1
245006	093088	092989	1	2	44	153.30	45.00	2	2	2	2	364	0	2	1
245007	093088	071989	2	1	22	175.00	59.70	2	1	1	2	292	0	1	1
245007	072089	092989	2	1	22	175.00	59.70	2	1	1	2	364	293	2	2
245008	093088	092889	1	1	7	111.00	17.40	2	1	1	2	363	0	2	1
245043	112988	111489	1	1	19	173.60	61.40	2	2	1	2	350	0	2	1
248078	011389	100989	1	1	34	182.60	94.80	2	2	1	2	269	0	2	1
248108	022489	082889	1	1	32	177.90	63.35	2	2	1	2	185	0	2	1
248115	031089	060989	2	1	25	185.00	74.55	2	2	1	2	91	0	2	1
248116	031089	060989	2	1	21	189.00	101.50	2	2	1	2	91	0	2	1
249028	111188	110389	2	1	7	109.00	14.65	2	1	1	2	357	0	2	1
249031	111588	050989	2	1	5	101.00	15.40	2	1	1	2	175	0	1	1
249031	051089	082289	2	1	5	101.00	15.40	2	1	1	2	280	176	1	2
249031	082389	110389	2	1	5	101.00	15.40	2	1	1	2	353	281	2	3
249035	112288	103189	2	1	24	169.70	58.95	2	1	1	2	343	0	2	1
249055	120888	083089	1	1	12	138.00	28.10	2	1	1	2	265	0	1	1
249055	083189	100789	1	1	12	138.00	28.10	2	1	1	2	303	266	2	2
249063	121388	072789	2	1	5	97.10	15.30	2	1	1	2	226	0	1	1
249063	072889	103189	2	1	5	97.10	15.30	2	1	1	2	322	227	2	2
249105	022189	110389	1	1	24	171.00	55.10	2	1	1	2	255	0	2	1
328064	121488	021789	1	1	26	176.80	66.00	2	1	1	4	65	0	1	1
328064	021889	112289	1	1	26	176.80	66.00	2	1	1	4	343	66	2	2
328073	010989	103089	2	1	6	104.40	13.80	2	1	1	4	294	0	2	1
328074	010989	110889	1	2	9	122.80	20.10	2	1	2	4	303	0	2	1
328079	011689	020889	2	1	25	176.90	73.50	2	2	1	4	23	0	1	1
328079	020989	101389	2	1	25	176.90	73.50	2	2	1	4	270	24	2	2
328080	011689	101389	1	1	2	93.50	14.20	2	1	1	4	270	0	2	1
328083	012389	092589	2	1	8	121.60	22.40	2	1	1	4	245	0	2	1

ID	RDT	IDT	Z_1	Z_2	Z_3	Z_4	Z_5	Z_6	Z_7	Z_8	Z_9	T_1	T_2	δ	S
328084	012389	101189	2	1	10	125.90	29.70	2	1	1	4	261	0	2	1
328092	013089	111089	1	1	20	169.80	50.10	2	1	1	4	284	0	2	1
328093	013089	110289	2	2	34	166.50	58.20	2	1	1	4	276	0	2	1
328095	020689	112789	1	1	6	119.90	26.20	2	1	1	4	294	0	2	1
328096	020689	111089	1	1	11	139.20	34.90	2	1	1	4	277	0	2	1
328098	021389	021789	2	2	3	91.70	14.40	2	1	1	4	4	0	1	1
328098	021889	072289	2	2	3	91.70	14.40	2	1	1	4	159	5	1	2
328098	072389	091489	2	2	3	91.70	14.40	2	1	1	4	213	160	1	3
328098	091589	112789	2	2	3	91.70	14.40	2	1	1	4	287	214	2	4
328104	022089	011790	1	1	9	131.30	24.00	2	1	1	4	331	0	2	1
328112	030789	122089	2	2	11	138.60	36.10	2	1	2	4	288	0	2	1
328113	030789	120189	2	1	17	156.70	36.80	2	1	1	4	269	0	2	1
328114	030789	120189	2	1	10	143.00	31.30	2	1	1	4	269	0	2	1
331047	120188	102789	2	2	17	171.00	46.10	2	1	2	2	330	0	2	1
331062	121388	020889	2	2	8	115.00	19.00	2	1	2	2	57	0	1	1
331062	020989	041389	2	2	8	115.00	19.00	2	1	2	2	121	58	1	2
331062	041489	112989	2	2	8	115.00	19.00	2	1	2	2	351	122	2	3
331069	123088	102389	1	1	6	111.90	17.90	2	1	1	2	297	0	2	1
331075	011089	101889	2	2	7	116.20	22.20	2	1	1	2	281	0	2	1
331081	011789	102089	1	1	7	119.70	21.50	2	1	1	2	276	0	2	1
331089	012489	103089	1	1	8	116.80	20.00	2	1	1	2	279	0	2	1
331090	012589	020889	2	1	11	141.50	36.00	2	1	1	2	14	0	1	1
331090	020989	103089	2	1	11	141.50	36.00	2	1	1	2	278	15	2	2
331118	031489	092989	1	1	4	108.10	17.30	2	1	1	2	199	0	2	1
332032	111788	092189	2	2	18	179.00	67.00	2	1	1	3	308	0	2	1
332033	111788	101089	1	1	13	151.00	49.00	2	1	1	3	327	0	2	1
332034	111788	101289	2	1	11	136.50	31.20	2	2	1	3	329	0	2	1
332037	112888	101289	1	1	2	86.00	13.50	2	1	1	3	318	0	2	1
332038	112888	092889	1	1	17	180.00	68.00	2	1	1	3	304	0	2	1

ID	RDT	IDT	Z_1	Z_2	Z_3	Z_4	Z_5	Z_6	Z_7	Z_8	Z_9	T_1	T_2	δ	S
332039	112888	101089	2	2	35	181.50	80.00	2	2	1	3	316	0	2	1
332048	120288	092889	2	1	25	172.50	50.00	2	1	1	3	300	0	2	1
332049	120288	042789	1	1	14	145.00	29.00	2	2	1	3	146	0	1	1
332049	042889	060889	1	1	14	145.00	29.00	2	2	1	3	188	147	1	2
332049	060989	092889	1	1	14	145.00	29.00	2	2	1	3	300	189	2	3
332050	120288	100289	2	2	25	187.50	74.00	2	1	1	3	304	0	1	1
332050	100389	101089	2	2	25	187.50	74.00	2	1	1	3	312	305	2	2
332057	120988	031089	2	1	27	169.50	65.00	1	2	2	3	91	0	1	1
332057	031189	040989	2	1	27	169.50	65.00	1	2	2	3	121	92	1	2
332057	041089	063089	2	1	27	169.50	65.00	1	2	2	3	203	122	1	3
332057	070189	092289	2	1	27	169.50	65.00	1	2	2	3	287	204	2	4
332058	120988	092889	2	2	32	185.00	95.00	2	1	1	3	293	0	2	1
332059	120988	092889	1	1	6	120.00	23.00	2	1	1	3	293	0	2	1
332066	121688	090689	2	2	8	138.00	31.00	2	2	1	3	264	0	1	1
332066	090789	092889	2	2	8	138.00	31.00	2	2	1	3	286	265	2	2
332067	121688	092889	1	2	9	144.00	34.40	2	1	1	3	286	0	2	1
332068	122988	092889	1	2	23	170.00	49.00	2	2	2	3	273	0	2	1
332070	010589	082989	2	2	23	122.00	31.20	1	1	2	3	236	0	1	1
332070	083089	100589	2	2	23	122.00	31.20	1	1	2	3	273	237	2	2
332071	010589	100589	1	2	17	127.00	27.00	2	1	2	3	273	0	2	1
332072	010589	073189	1	2	21	158.00	49.00	2	1	1	3	207	0	1	1
332072	080189	100589	1	2	21	158.00	49.00	2	1	1	3	273	208	2	2
332082	012089	101189	2	1	1	81.00	10.40	2	1	1	3	264	0	2	1
336024	110988	041889	1	1	12	136.50	30.00	2	1	1	2	160	0	2	1
336025	110988	040489	2	1	7	120.00	23.00	2	1	1	2	146	0	1	1
336025	040589	092189	2	1	7	120.00	23.00	2	1	1	2	316	147	2	2
336026	110988	092189	2	1	1	79.00	10.60	2	1	1	2	316	0	2	1
336027	111088	092189	1	1	3	97.80	13.00	2	1	1	2	315	0	2	1

APPENDIX E

Exercises

0.1 In Example 0.2.2, for any $k \neq k' \in \{1, \ldots, L\}$, prove $(D_{1k} - E_{1k})$ and $(D_{1k'} - E_{1k'})$ are uncorrelated mean-zero random variables, but show they are not independent.

1.1 **a.** Show that any martingale must have constant expected value.

b. A Poisson process, N^*, with rate λ is defined by

i. $N^*(0) = 0$ a.s.

ii. $N^*(t) - N^*(s)$ has a Poisson distribution with parameter $\lambda(t - s)$ for any $0 \leq s \leq t$,

iii. $N^*(t_2) - N^*(t_1), N^*(t_3) - N^*(t_2), \ldots, N^*(t_n) - N^*(t_{n-1})$ are independent for any $0 \leq t_1 \leq t_2 \leq \ldots \leq t_n$.

Show that $\{N^*(t) - \lambda t : t \geq 0\}$ is a martingale.

c. Use (a) and (b) to obtain $EN^*(t)$.

1.2 **a.** Suppose X is an adapted increasing process with $E|X(t)| < \infty$ for all t. Prove X is a submartingale.

b. Let M be a martingale with $EM^2(t) < \infty$ for all $t \geq 0$. Prove M^2 is a submartingale.

1.3 Let I be an index set, and $\{\mathcal{F}_i, i \in I\}$ a family of sub-σ-algebras of $\mathcal{F}$ on a probability space Ω.

a. Show that $\bigcap_{i \in I} \mathcal{F}_i$ is a σ-algebra. ($A \in \bigcap_{i \in I} \mathcal{F}_i$ if and only if $A \in \mathcal{F}_i$ for each i.)

b. Show by counterexample that $\bigcup_{i \in I} \mathcal{F}_i$ need not be a σ-algebra.

c. The smallest σ-algebra containing each of the $\mathcal{F}_i$ is denoted by $\sigma(\bigcup_{i \in I} \mathcal{F}_i)$ or by $\bigvee_{i \in I} \mathcal{F}_i$. Show that $\bigvee_{i \in I} \mathcal{F}_i$ always exists and is unique.

1.4 One implication of the Martingale Convergence Theorem is Levy's Theorem: Let $\ldots, \mathcal{F}_{-2}, \mathcal{F}_{-1}, \mathcal{F}_0, \mathcal{F}_1, \mathcal{F}_2, \ldots$ be a nondecreasing sequence of sub-σ-algebras of a probability space $(\Omega, \mathcal{F}, P)$. Let

$$\mathcal{F}_\infty = \sigma\left(\bigcup_{n=-\infty}^{\infty} \mathcal{F}_n\right),$$

$$\mathcal{F}_{-\infty} = \bigcap_{n=-\infty}^{\infty} \mathcal{F}_n,$$

and let X be a random variable with $E|X| < \infty$. Then, with probability one,

$$E(X|\mathcal{F}_n) \rightarrow E(X|\mathcal{F}_\infty), \qquad \text{as } n \rightarrow \infty,$$

and

$$E(X|\mathcal{F}_n) \rightarrow E(X|\mathcal{F}_{-\infty}), \qquad \text{as } n \rightarrow -\infty.$$

Use Levy's Theorem to prove Proposition 1.2.1.

1.5 Consider a probability space $(\Omega, \mathcal{F}, P)$ and a set $A \in \mathcal{F}$ such that $0 < P(A) < 1$. Suppose a filtration $\{\mathcal{F}_t : t \geq 0\}$ is the history of a stochastic process X. Prove that this filtration is not right-continuous at t_0 when

a. X is the right-continuous process $X(t) = t + \{(t \wedge t_0) - t\}I_A$;

b. X is the step function process $X(t) = I_{\{t > t_0\}} I_A$.

1.6 Let $\mathcal{F}_s = \sigma\{N(u), I_{\{X \leq u, \delta=0\}}, 0 \leq u \leq s\}$ and $\mathcal{F}_{s-} = \bigvee_{u<s} \mathcal{F}_u$. Show that

$$\mathcal{F}_{s-} = \sigma\{N(u), I_{\{X \leq u, \delta=0\}}, 0 \leq u < s\}.$$

1.7 It is tempting to define independent censoring with the condition

$$P\{s \leq T < s + ds | T \geq s\} = P\{s \leq T < s + ds, U \geq T | T \geq s, U \geq s\},$$

i.e., that the conditional probability of failure when censoring is not present is the same as the conditional probability of an observable failure in the presence of censoring.

a. Show that for T discrete and U arbitrary,

$$P\{T = s | T \geq s, U \geq s\} = P\{T = s, U \geq T | T \geq s, U \geq s\}.$$

b. Show that when T and U both have density functions,

$$\lim_{h \downarrow 0} \frac{1}{h} P\{s \leq T < s + h | T \geq s, U \geq s\}$$
$$= \lim_{h \downarrow 0} \frac{1}{h} P\{s \leq T < s + h, U \geq T | T \geq s, U \geq s\}.$$

c. Suppose that T has a density function, but that U has positive probability at a point s, i.e., $P\{U = s\} > 0$. Show with an example that

$$\lim_{h \downarrow 0} \frac{1}{h} P\{s \leq T < s + h | T \geq s, U \geq s\}$$

need not equal

$$\lim_{h \downarrow 0} \frac{1}{h} P\{s \leq T < s + h, U \geq T | T \geq s, U \geq s\}.$$

d. In spite of (c) above, why is it not necessary to re-formulate Condition (3.1) of Chapter 1 as

$$\lambda(t) = \lim_{h \downarrow 0} \frac{1}{h} P\{t \le T < t + h, U \ge T | T \ge t, U \ge t\}$$

whenever $P\{X > t\} > 0$?

1.8 **a.** Suppose $\{T_i, i = 1, \ldots, n\}$ is a sequence of independent and identically distributed failure random variables having an absolutely continuous distribution. Suppose we have type II censoring, i.e. for every i, $U_i \equiv T_d^o$, the time of the dth failure for fixed $d \le n$. Certainly, T_i and U_i are dependent. Does $\lambda(t) = \lambda^\#(t)$ in Eq. (3.1) from Chapter 1 hold? (Hint: Use Eq. (3.2) from Chapter 1.)

b. Suppose Y_1, Y_2 and Y_{12} are independent exponentially distributed variables with hazards $\lambda_1(t) = \lambda_1$, $\lambda_2(t) = \lambda_2$, and $\lambda_{12}(t) = \lambda_{12}$ respectively. Let $T \equiv \min(Y_1, Y_{12})$ and $U \equiv \min(Y_2, Y_{12})$. Then (T, U) have the bivariate exponential distribution discussed by Barlow and Proschan (1975, p. 128). Do the dependent variables T and U satisfy Condition (3.1) from Chapter 1?

1.9 Suppose (T, U) have the joint distribution $P\{T > t, U > s\} = H(t, s) = \exp(-\lambda t - \mu s - \theta ts)$, where $0 \le \theta \le \lambda\mu$.

a. Find the marginal distribution function $H_1(t)$ and corresponding hazard $\lambda_1(t)$ for T; find the marginal distribution $H_2(t)$ and corresponding hazard $\lambda_2(t)$ for U.

b. Compute

$$\lambda_1^\#(t) = \frac{-\frac{\partial}{\partial u} P\{T \ge u, U \ge t\}|_{u=t}}{P\{T \ge t, U \ge t\}}$$

and

$$\lambda_2^\#(t) = \frac{-\frac{\partial}{\partial u} P\{T \ge t, U \ge u\}|_{u=t}}{P\{T \ge t, U \ge t\}}$$

c. Suppose $\lambda = 1, \mu = 2$ and $\theta = 2$. Compare $H_1(t)$ with $H_1^\#(t) \equiv \exp\{-\int_0^t \lambda_1^\#(s)ds\}$, and compare $H_2(t)$ with $H_2^\#(t) \equiv \exp\{-\int_0^t \lambda_2^\#(s)ds\}$. Why are such comparisons important?

d. Suppose one erroneously assumes T and U are independent, and takes their joint distribution to be

$$H^\#(t, s) \equiv H_1^\#(t) H_2^\#(s).$$

Find $H^\#(t, s)$ and observe that $H^\#(t, t) = H(t, t)$.

e. Suppose data consist of $\{[\min(T_i, U_i), I_{\{T_i \le U_i\}}], i = 1, \ldots, n\}$. The crude survival functions (which are important because they always can be consistently estimated whether or not independence holds) are defined by $Q_1(t) = P\{T > t, U > T\}$ and $Q_2(t) = P\{U > t, T > U\}$. Show that whether the joint distribution of (T, U) is given by $H(t, s) =$

exp($-\lambda t - \mu s - \theta st$) or by $H^{\#}(t, s)$ as calculated in part (d), one obtains the same $Q_1(t)$ and $Q_2(t)$. Thus the crude survival functions, even when fully known, cannot be used to establish whether or not independence holds. This example illustrates the general non-identifiability result proved by Tsiatis (1975): If data consist of $\{[\min(T_i, U_i), I_{\{T_i \le U_i\}}], i = 1, \ldots, n\}$, then the hypothesis $H : \lambda_1(t) = \lambda_1^{\#}(t)$ is not testable.

1.10 In Theorem 1.3.1, establish that M is adapted to $\{\mathcal{F}_t : t \ge 0\}$.

1.11 We can generalize Theorem 1.3.1. Suppose we have n independent and identically distributed pairs $(T_i, U_i), i = 1, \ldots, n$, where each pair satisfies Condition (3.1) from Chapter 1. Set $X_i = \min(T_i, U_i)$ and $\delta_i = I_{\{X_i = T_i\}}$, and let $N_i(t) = I_{\{X_i \le t, \delta_i = 1\}}$ and $N_i^U(t) = I_{\{X_i \le t, \delta_i = 0\}}$. Let λ denote the hazard function for T_i.

a. In Theorem 1.3.1, we showed $M_i = N_i - \int I_{\{X_i \ge u\}} \lambda(u) du$ is a martingale with respect to $\{\mathcal{F}_t^{(i)} : t \ge 0\}$, where $\mathcal{F}_t^{(i)} = \sigma\{N_i(u), N_i^U(u) : 0 \le u \le t\}$. Show that M_i is a martingale with respect to the richer σ-algebras $\{\mathcal{F}_t^n : t \ge 0\}$ where

$$\mathcal{F}_t^n = \sigma[\{N_i(u), N_i^U(u), 0 \le u \le t\}, i = 1, \ldots, n].$$

In turn, it immediately follows that $\{\overline{N}(t) - \int_0^t \overline{Y}(s)\lambda(s)ds : t \ge 0\}$ is a martingale with respect to $\{\mathcal{F}_t^n : t \ge 0\}$, where $\overline{N}(t) = \sum_{i=1}^n N_i(t)$ and $\overline{Y}(t) = \sum_{i=1}^n Y_i(t)$.
(Hint: Chung (1974, Section 9.2) shows that if $\mathcal{F}_1, \mathcal{F}_2$ and $\mathcal{F}_3$ are three σ-algebras such that $\sigma(\mathcal{F}_1, \mathcal{F}_2)$ is independent of $\mathcal{F}_3$, then for each integrable X which is measurable with respect to $\mathcal{F}_1$, we have

$$E\{X | \sigma(\mathcal{F}_2, \mathcal{F}_3)\} = E\{X | \mathcal{F}_2\}.$$

Note that σ-algebras $\mathcal{G}_1$ and $\mathcal{G}_2$ are defined to be independent if, for every $\Lambda_1 \in \mathcal{G}_1$ and $\Lambda_2 \in \mathcal{G}_2$, $EI_{\Lambda_1 \cap \Lambda_2} = EI_{\Lambda_1} EI_{\Lambda_2}$.)

b. Let $\{G_t : t \ge 0\}$ be any stochastic basis such that $\mathcal{F}_t^{(i)} \subset G_t$ for any t. Prove or disprove that $M_i = N_i - \int I_{\{X_i \ge u\}} \lambda(u) du$ is a martingale with respect to $\{G_t : t \ge 0\}$.

1.12 Prove Proposition 1.4.1 by establishing the following:

a. Let σ_p denote the predictable σ-algebra for a filtration $\{\mathcal{F}_t : t \ge 0\}$. Show that, for any constant k and any set $B \in \sigma_p$, $kI_B(t, \cdot)$ is $\mathcal{F}_{t-}$-measurable for each $t \ge 0$. (Hint: For $B \in \sigma_p$, define $B_t = \{\omega : (t, \omega) \in B\}$. Show that $\sigma_t \equiv \{B \in \sigma_p : B_t \in \mathcal{F}_{t-}\}$ is a σ-algebra and it contains the predictable rectangles).

b. Since any predictable process H can be written as

$$H(t, \omega) = \lim_{n \to \infty} \sum_{i=1}^{n} k_{i,n} I_{B_{i,n}}(t, \omega),$$

where the $k_{i,n}$ are constants and the sets $B_{i,n} \in \sigma_p$, use (a) to prove Proposition 1.4.1.

1.13 Let M be a process adapted to a filtration $\{\mathcal{F}_t : t \geq 0\}$ and of bounded variation on each finite interval $[0, t]$. Show that M is a martingale on $[0, \infty)$ if and only if for every predictable rectangle $B = (u, v] \times A, A \in \mathcal{F}_u$,

$$E \int_0^\infty I_B(s) dM(s) = 0.$$

(Hint: For $0 \leq u \leq v$, $E\{M(v)|\mathcal{F}_u\} = M(u)$ a.s. if and only if

$$\int_A M(u) dP = \int_A M(v) dP$$

for any $A \in \mathcal{F}_u$.)

1.14 Let $\{\Omega, \mathcal{F}, \{\mathcal{F}_t : t \geq 0\}, P\}$ denote a stochastic basis. Use the following steps to establish constructively that $\int_0^\cdot H dM$ is a martingale, whenever H is a bounded, adapted, left-continuous process, with right-hand limits, and M is a martingale having paths of bounded variation on each closed interval $[0, t]$. Let H^* denote a set of processes such that $H \in H^*$ if there exists a partition

$$0 = t_0 < t_1 < \ldots < t_{n-1} < t_n = \infty$$

on $[0, \infty]$ such that

$$H(t) = H_0 I_{[0]}(t) + \sum_{i=1}^{n} H_i I_{(t_{i-1}, t_i]}(t),$$

where H_0 is $\mathcal{F}_0$-measurable, H_i is $\mathcal{F}_{t_{i-1}}$-measurable, and $P\{|H_i| \leq \Gamma\} = 1, i = 0, \ldots, n$, for some $\Gamma < \infty$.

a. Show that if $H \in H^*$ and M has paths of bounded variation on each closed interval $[0, t]$, then $\int_0^\cdot H(s) dM(s)$ is a martingale. (Hint: begin with $H(t) = H_a I_{(a,b]}(t)$, where H_a is $\mathcal{F}_a$-measurable.)

b. Show that any bounded, adapted, left-continuous process with right-hand limits can be written as a limit of processes on H^*.

c. Use (a), (b) and the Dominated Convergence Theorem to show that $\int_0^\cdot H(s)\, dM(s)$ is a martingale, when M has paths of bounded variation on closed subintervals of R and H is a bounded, adapted, left-continuous process having right-hand limits.

2.1 Suppose $\{\mathcal{F}_t : t \geq 0\}$ is a filtration, and τ is a nonnegative random variable.

a. Prove $\{\tau \leq t\} \in \mathcal{F}_t$ for all $t \geq 0$ implies that $\{\tau < t\} \in \mathcal{F}_t$ for all $t \geq 0$.

b. Disprove $\{\tau < t\} \in \mathcal{F}_t$ for all $t \geq 0$ implies that $\{\tau \leq t\} \in \mathcal{F}_t$ for all $t \geq 0$.

c. If the filtration is right-continuous, prove $\{\tau < t\} \in \mathcal{F}_t$ for all $t \geq 0$ implies that $\{\tau \leq t\} \in \mathcal{F}_t$ for all $t \geq 0$.

2.2 If N is an arbitrary counting process adapted to a filtration $\{\mathcal{F}_t : t \geq 0\}$, prove N is a local submartingale with localizing sequence $\{\tau_n\}$ if, for any n,

$$\tau_n = n \wedge \sup\{t : N(t) < n\}.$$

2.3 Show that $\mathcal{F}_\tau$ in Definition 2.2.4 is a σ-algebra.

2.4 If τ is a stopping time with respect to a filtration $\{\mathcal{F}_t : t \geq 0\}$ and X is a left- or right-continuous $\mathcal{F}_t$- adapted process, prove $X(\cdot \wedge \tau)$ is adapted.

2.5 If τ_1 and τ_2 are $\mathcal{F}_t$-stopping times, show $\max(\tau_1, \tau_2)$ and $\min(\tau_1, \tau_2)$ are $\mathcal{F}_t$-stopping times.

2.6 **a.** Suppose the process H, with $H(0) = 0$, is adapted with respect to a filtration $\{\mathcal{F}_t : t \geq 0\}$. If $\tau_n = n \wedge \sup\{t : \sup_{0 \leq s \leq t} |H(s+)| < n\}$, prove $\{\tau_n\}$ is a localizing sequence if either

i. H is right-continuous with left-hand limits, and has bounded jump sizes; or

ii. $\{\mathcal{F}_t : t \geq 0\}$ is right-continuous, and H is left-continuous with finite right-hand limits.

b. Suppose H is left-continuous with finite right-hand limits, is adapted to a filtration $\{\mathcal{F}_t : t \geq 0\}$, and $H(0) = 0$. Define

$$\tau_n = n \wedge \sup\{t : \sup_{0 \leq s \leq t} |H(s)| \leq n\}.$$

Show $\{\tau_n\}$ need not be a localizing sequence in general, but must be when the filtration is right-continuous. (Hint: For any $t \leq n$, $\{\tau_n \geq t\} = \{\sup_{0 \leq s \leq t} |H(s)| \leq n\}$.)

2.7 Suppose $X_0 \equiv 0$ and $\{X_i, i = 1, 2, \ldots\}$ is a collection of independent and identically distributed random variables. Let $[t]$ denote the greatest integer $\leq t$. Set $\mathcal{F}_t = \sigma\{X_i, i \leq [t]\}$.

a. Suppose $X_i \sim N(0, 1)$, the standard normal distribution. Set

$$M(\cdot) = \sum_{i=0}^{[\cdot]} X_i.$$

i. Is M a martingale over $[0, \infty)$?

ii. Is M a square integrable martingale over $[0, \infty)$?

iii. If $\tau_n \equiv \sup\{t : \sup_{0 \leq s \leq t} |M(s)| < n\} \wedge n$, is M a local square integrable martingale with localizing sequence $\{\tau_n\}$?

b. Suppose X_i has a Cauchy distribution.

i. Is M a martingale?

ii. If $\tau_n \equiv \sup\{t : \sup_{0 \leq s \leq t} |M(s)| < n\} \wedge n$, is M a local martingale with localizing sequence $\{\tau_n\}$?

2.8 The two paragraphs preceding Theorem 1.4.2 provide a heuristic argument for the form of the processes $\langle M, M\rangle$ and $\langle M_1, M_2\rangle$ such that $M^2 - \langle M, M\rangle$ and $M_1 M_2 - \langle M_1, M_2\rangle$ are martingales.

Consider the setting in Theorem 2.4.2. Provide a simple heuristic argument for

$$\left\langle \int H_1, dM_1, \int H_2 dM_2 \right\rangle = \int H_1 H_2 d\langle M_1, M_2\rangle.$$

2.9 Let T_1 and T_2 be independent random variables, each exponentially distributed with rate λ. For any $t \geq 0$, define $N(t) = I_{\{T_1 \leq t\}}$, $\tilde{N}(t) = I_{\{T_2 \leq t\}}$, and $N^*(t) = N(t/2) + \tilde{N}(t)$, and set $\mathcal{F}_t = \sigma\{N(s), \tilde{N}(s) : 0 \leq s \leq t\}$.

a. Is $\{N(t), N^*(t) : t \geq 0\}$ a multivariate counting process?

b. Derive A and A^* such that $N - A$ and $N^* - A^*$ are martingales. Comment on whether you are surprised by the form of A or A^*.

c. Are $N(t)$ and $N^*(t)$ correlated? Are $\{N(t) - A(t)\}$ and $\{N^*(t) - A^*(t)\}$ correlated?

2.10 Consider the Cox partial likelihood score vector $\mathbf{U}(\beta_0)^{p \times 1}$ in Eq. (5.4) of Example 1.5.2, when $\mathbf{Z}_i$ is a bounded predictable process for $i = 1, \ldots, n$. For $k = 1, \ldots, p$, denote $\{\mathbf{U}(\beta_0)\}_k = U_k(\beta_0)$.

a. Show that $U_k(\beta_0)$ is a martingale over $[0, t]$, for any t.

b. Denote $U_k(\beta_0)$ at t by $U_k(\beta_0, t)$. Obtain $EU_k(\beta_0, t)$.

c. For $0 \leq s \leq t$ and $k, k' \in \{1, \ldots, p\}$, obtain $\operatorname{cov}\{U_k(\beta_0, s), U_{k'}(\beta_0, t)\}$. (Don't simplify.)

d. For the special case in which $p = 1, \beta_0 = 0$, and $Z_i \in \{0, 1\}$, (i.e., the logrank statistic), show the result in (c) simplifies to

$$E \int_0^s \frac{\overline{Y}_0(x)\overline{Y}_1(x)}{\overline{Y}_0(x) + \overline{Y}_1(x)} d\Lambda(x),$$

where $\overline{Y}_j(x) \equiv \sum_{i=1}^n Y_i(x) I_{\{Z_i = j\}}$ for $j = 0, 1$. (In the setting of Example 2.6.1, $d\Lambda(x)$ is replaced by $\{1 - \Delta\Lambda(x)\} d\Lambda(x)$.)

e. Obtain the result in (c) for the $(p + 1)$-sample logrank statistic testing $H : \beta = \beta_0 = \mathbf{0}$ where, for $i = 1, \ldots, n$ and for $k = 1, \ldots, p$, $\{\mathbf{Z}_i(t)\}_k \equiv Z_{ik} \in \{0, 1\}$ is an indicator for membership of individual i in sample k.

2.11 Verify Assumption 2.6.1 holds in Example 2.6.1.

2.12 **a.** Let $\{\mathcal{F}_t : t \geq 0\}$ be a filtration and X a random variable with $E|X| < \infty$. Show that the process $\{Y(t) : t \geq 0\}$ given by

$$Y(t) = E(X | \mathcal{F}_t)$$

is uniformly integrable.

b. Let M be a stochastic process and $\{\tau_n\}$ a sequence of stopping times with respect to $\{\mathcal{F}_t : t \geq 0\}$ such that $\{M(t \wedge \tau_n) : t \geq 0\}$ is an $\mathcal{F}_t$-martingale for each n. Show that the stopped process $M(\cdot \wedge \sigma_n), \sigma_n = \tau_n \wedge n$, is a uniformly integrable martingale.

3.1 Using Theorem 3.2.3 to formulate $\{\hat{S}(t) - S(t)\}$ in terms of martingales, motivate the large sample result

$$\sqrt{n}\,\{\hat{S}(t) - S(t)\} \xrightarrow{\mathcal{D}} N\left(0, S^2(t) \int_0^t \frac{dS(u)}{\pi(u) S(u)}\right) \quad \text{as } n \to \infty.$$

(Hint: See a similar argument for $\hat{\Lambda}$ given in section 3.2.)

3.2 Establish that the Gehan-Wilcoxon statistic (Gehan, 1965) arises for

$$K(s) = \left(\frac{n_1 + n_2}{n_1 n_2}\right)^{1/2} \frac{\overline{Y}_1(s)\overline{Y}_2(s)}{n_1 + n_2}.$$

3.3 Suppose K is the weight function in Eq. (3.3) of Chapter 3, which gives rise to the logrank statistic. Verify that $\hat{\sigma}^2$ in Eq. (3.7) of Chapter 3, is identical to the heuristically formulated estimator given by Mantel (1966), where the logrank variance was obtained through conditioning and use of the hypergeometric distribution.

3.4 Give an example in which Condition (4.2) of Chapter 3 holds while Condition (4.5) of Chapter 3 fails to hold.

4.1 Suppose $\{N(t) : t \geq 0\}$ is a counting process adapted to a right-continuous filtration $\{\mathcal{F}_t : t \geq 0\}$. For fixed t, let τ be the time of the first jump in N after time t. Prove τ is a stopping time, and the process $I_{(t,\tau]} = \{I_{\{t<s\leq\tau\}} : s \geq 0\}$ is adapted to $\{\mathcal{F}_s : s \geq 0\}$.

4.2 Prove Eq. (3.23) of Chapter 4:

$$\mathbf{V}(\beta, t) = \frac{n^{-1}\Sigma\{\mathbf{Z}_i(t) - \mathbf{E}(\beta, t)\}^{\otimes 2} Y_i(t) \exp\{\beta'\mathbf{Z}_i(t)\}}{S^{(0)}(\beta, t)}.$$

4.3 Prove

$$\mathcal{I}(\beta, t) \equiv -\frac{\partial}{\partial \beta}\mathbf{U}(\beta, t) = \sum_{i=1}^{n} \int_0^t \mathbf{V}(\beta, u) dN_i(u).$$

4.4 Find expressions for the first and second moments of $\hat{\Lambda}_0(t)$ defined in Eq. (3.29) of Chapter 4.

4.5 Consider the partial likelihood expressions given in Eq. (3.11) of Chapter 4:

$$L_e(\beta, \infty) = \prod_{i=1}^{L} \frac{\exp(\mathbf{S}_i'\beta)}{\sum_{l \in R_{D_i}(T_i^o)} \exp(\mathbf{S}_\ell'\beta)}$$

and

$$L_a(\beta, \infty) = \prod_{i=1}^{L} \frac{\exp(\mathbf{S}_i'\beta)}{\left\{\sum_{\ell \in R_i} \exp(\mathbf{Z}_\ell'\beta)\right\}^{D_i}}.$$

In the $(s + 1)$ sample problem (Groups $0, 1, \ldots, s$), $\mathbf{Z}_i^{s\times 1}$ denotes sample membership for the ith individual, where

$$(\mathbf{Z}_i)_j \equiv Z_{ij} = \begin{cases} 0 \text{ if not in Group } j \\ 1 \text{ if in Group } j, \end{cases}$$

$j = 1, \ldots, s$. At the L ordered observed failure times $\{T_i^o : i = 1, \ldots, L\}$, let N_{ji} denote the number at risk in Group j and D_{ji} denote the number failing in Group j, $j = 0, \ldots, s$. Set $N_i \equiv \sum_{j=0}^{s} N_{ji}$ and $D_i \equiv \sum_{j=0}^{s} D_{ji}$.

a. Using the likelihood $L_a(\boldsymbol{\beta}, \infty)$:

i. Derive $\mathbf{U}(\boldsymbol{\beta})$ in terms of N_{ji}, N_i, D_{ji}, D_i and $\boldsymbol{\beta}$. When $s = 1$ and $\boldsymbol{\beta} = 0$, $\mathbf{U}(\boldsymbol{\beta})$ reduces to the two-sample logrank score statistic $U(0)$.

ii. Derive $\widehat{\text{cov}}\{\mathbf{U}(\boldsymbol{\beta})\} = \mathcal{I}(\boldsymbol{\beta})^{s\times s} = -\frac{\partial}{\partial \boldsymbol{\beta}}\mathbf{U}(\boldsymbol{\beta})$. If $\delta_{jk} \equiv I_{\{j=k\}}$, show that

$$\{\mathcal{I}(\mathbf{0})\}_{jk} = \sum_{i=1}^{L} D_i \left\{ \frac{\delta_{jk} N_{ji}}{N_i} - \frac{N_{ji}}{N_i}\frac{N_{ki}}{N_i} \right\},$$

a binomial-type variance when $s + 1 = 2$.

iii. Formulate a score test of $H : \boldsymbol{\beta} = \boldsymbol{\beta}_0$ giving the distribution of the statistic under H.

b. Using the exact likelihood $L_e(\boldsymbol{\beta}, \infty)$:

i. Derive expressions for $\mathbf{U}(\boldsymbol{\beta})$ and $\{\mathcal{I}(\boldsymbol{\beta})\}_{jk}$.

ii. When $\boldsymbol{\beta} = \mathbf{0}$, show $\mathbf{U}(\boldsymbol{\beta})$ reduces to the same $\mathbf{U}(\boldsymbol{\beta})$ derived in (a) (i).

iii. Show that

$$\{\mathcal{I}(\mathbf{0})\}_{jk} = \sum_{i=1}^{L} \left\{ D_i \left(\frac{N_{ji}}{N_i}\delta_{jk} - \frac{N_{ki}N_{ji}}{N_i^2} \right) \left(\frac{N_i - D_i}{N_i - 1} \right) \right\}.$$

Note

$$\{\mathcal{I}(\mathbf{0})\}_{jj} = \sum_{i=1}^{L} \left\{ D_i \frac{N_{ji}}{N_i} \left(1 - \frac{N_{ji}}{N_i} \right) \left(\frac{N_i - D_i}{N_i - 1} \right) \right\},$$

a hypergeometric-type variance when $s + 1 = 2$.

4.6 Suppose one wishes to compare two treatments (R_x0 vs R_x1) with respect to length of survival following initiation of therapy, in a study with Q matched pairs. Within each pair, randomization is used to assign one member R_x0 and the other R_x1. Suppose one fits the stratified Cox model

$$\lambda_q(t|Z_{qi}) = \lambda_{0q}(t)\exp(\beta Z_{qi}),$$

$i = 1, 2, q = 1, \ldots, Q$. Without loss of generality, set $(Z_{q1}, Z_{q2}) = (0, 1)$ for all q.

a. Discuss modeling considerations which motivate choosing this model.

b. State the hypothesis of no treatment effect (H_0) in terms of this stratified model.

c. Making the standard assumption of independent and uninformative censoring, provide the Rao (i.e. score) test statistic for H_0 in the setting allowing for ties in failure times. Simplify the expression and discuss its relationship to the SIGN test. (In uncensored untied data, the SIGN test is based simply on the proportion of pairs in which the member on R_x1 fails first.)

d. Are you surprised by the simple form of the Rao test statistic in this setting? What information is "thrown away" and what motivates it being discarded?

4.7 (Sample Size Calculation) For the multiplicative intensity model in Definition 4.2.1,

$$l_i(t) = Y_i(t)\lambda_0(t)\exp\{\beta'\mathbf{Z}_i(t)\},$$

and using the corresponding partial likelihood given by Eq. (3.13) in Chapter 4, the score statistic for β from Eq. (3.24) in Chapter 4 is

$$\mathbf{U}(\beta) = \sum_{i=1}^{n} \int_0^\infty \left\{ \mathbf{Z}_i(u) - \frac{\mathbf{S}^{(1)}(\beta, u)}{S^{(0)}(\beta, u)} \right\} dN_i(u).$$

By Eq. (3.21) in Chapter 4 and Exercise 4.3, the observed information matrix is

$$\mathcal{I}(\beta) = \sum_{i=1}^{n} \int_0^\infty \left[\frac{\mathbf{S}^{(2)}(\beta, u)}{S^{(0)}(\beta, u)} - \left\{ \frac{\mathbf{S}^{(1)}(\beta, u)}{S^{(0)}(\beta, u)} \right\}^{\otimes 2} \right] dN_i(u). \quad (1)$$

In Chapter 8, it is established that, as $n \to \infty$,

$$\frac{1}{\sqrt{n}}\mathbf{U}(\beta) \xrightarrow{\mathcal{D}} N_p(\mathbf{0}, E\frac{1}{n}\mathcal{I}(\beta)) \quad (2)$$

and that $\hat{\beta}$ is consistent for β, where $\hat{\beta}$ is a solution for $\mathbf{U}(\hat{\beta}) = \mathbf{0}$.

a. By a first order Taylor expansion of $\mathbf{U}$ about β, show, for large n, we have the approximation

$$\hat{\beta} \sim N(\beta, \{\mathcal{I}(\hat{\beta})\}^{-1}) \quad (3)$$

b. Using (3), give a $(1 - \alpha)$-level two-sided confidence interval for $(\beta)_j \equiv \beta_j$.

In (c) through (f), consider the special case of the logrank statistic in Example 1.5.2, where $X_i = (T_i \wedge U_i)$, $N_i(t) = I_{\{X_i \le t, \delta_i = 1\}}$, $Y_i(t) = I_{\{X_i \ge t\}}$, and $\mathbf{Z}_i(\cdot) = Z_i \in \{0, 1\}$. Suppose T_i and U_i are independent, and U_i is independent of Z_i. Finally, assume $P\{Z_i = 0\} = 1/2 = P\{Z_i = 1\}$.

c. If $\{T_j^o, j = 1, \ldots, L\}$ denote the L ordered observed failure times, show $\mathcal{I}(\beta)$ in (1) reduces to

$$\mathcal{I}(\beta) = \sum_{j=1}^{L} \left[\frac{S^{(1)}(\beta, T_j^o)}{S^{(0)}(\beta, T_j^o)} \left\{ 1 - \frac{S^{(1)}(\beta, T_j^o)}{S^{(0)}(\beta, T_j^o)} \right\} \right]. \quad (4)$$

In turn, for large n, show

$$\frac{S^{(1)}(\beta, x)}{S^{(0)}(\beta, x)} \approx \frac{\exp\{-\Lambda_0(x)(e^\beta - 1)\}e^\beta}{1 + \exp\{-\Lambda_0(x)(e^\beta - 1)\}e^\beta}, \quad (5)$$

where $\Lambda_0 = \int \lambda_0$.

d. Using (4) and (5), it can be shown $\mathcal{I}(\beta) \approx \sum_{j=1}^{L}(1/4)$ for β in an interval containing zero. Using this approximation and (3), give a $(1-\alpha)$ level two-sided confidence interval for β.

e. Under the assumed model, $\lambda(t|Z=1) = \lambda(t|Z=0)e^{\beta}$. For numbers $\beta_0 < \beta_1$, suppose we wish to test $H_0 : \beta = \beta_0$ vs. $H_1 : \beta = \beta_1$. Establish that the number of events required for a two-sided level α test of H_0 to have power $(1-\beta)$ against H_1 is

$$L = \left\{ \frac{z_{1-\alpha/2} + z_{1-\beta}}{\frac{1}{2}(\beta_1 - \beta_0)} \right\}^2$$

f. Suppose $Z_i = 0$ for an experimental therapy and $Z_i = 1$ for standard therapy. For $\alpha = .05$ and $\beta = .95$, compute L in the active control setting where $(e^{\beta_0}, e^{\beta_1}) = (0.8, 1.2)$ and in a "superiority" setting where $(e^{\beta_0}, e^{\beta_1}) = (1.0, 1.5)$.

4.8 Verify that the martingale residuals satisfy, for $i \neq j$,

$$\text{cov}(\hat{M}_i, \hat{M}_j) \to 0 \text{ as } n \to \infty,$$

and $E\hat{M}_i \to 0$ as $n \to \infty$.

4.9 Use techniques developed in Chapter 4 to explore whether adding information in variables stage, SGOT and urine copper allows one to build a better model for survival of PBC patients.

4.10 **a.** Using covariate and follow-up data from Appendix D, evaluate the effect of gamma interferon in increasing the time to first serious infection and on reducing the overall rate of serious infections. Do exploratory analyses yield convincing evidence of any subset effects?

b. Let Y_i indicate whether the ith patient was followed up to t months post randomization. Assuming the ith patient's intensity for a new serious infection at t satisfies

$$l_i(t) = Y_i(t)\lambda_0(t)\exp(\beta' \mathbf{Z}_i), \tag{1}$$

build a "natural history" model by selecting an appropriate vector $\mathbf{Z}$ of predictor variables and by estimating the parameters β and $\Lambda_0 = \int \lambda_0(s)ds$. Assess whether λ_0 appears to be constant in time. Finally, use time-dependent covariates to assess whether the patient's intensity for a new serious infection at t, modeled by (1), is further altered by the pattern of previous infections.

5.1 In Lindeberg's Theorem (Theorem 5.2.1), prove Condition (2.3) of Chapter 5 is equivalent to

$$E\left(\sum_{j=1}^{r_n} X_{n,j}^2 I_{\{|X_{n,j}|>\epsilon\}}\right) \to 0 \text{ as } n \to \infty, \text{ for any } \epsilon > 0,$$

when C_n^2 is bounded away from 0 and ∞.

5.2 Prove Theorem 5.3.5 using Lemma C.3.1 in Appendix C.

5.3 Suppose one wishes to compare the effect of two interventions and one has a set of r outcome events, $\{E_\ell, \ell = 1, \ldots, r\}$. For the ith individual, $i = 1, \ldots, n$, let $Z_i \in \{0, 1\}$ denote which intervention is randomly assigned and let $T_{\ell i}$ denote the time to event E_ℓ. For $i = 1, \ldots, n$ and $\ell = 1, \ldots, r$, the counting process $N_{\ell i}$ indicates whether the ith individual has had an observed event E_ℓ by time t. Let the filtration be obtained by setting

$$\mathcal{F}_t = \sigma\{Z_i, N_{\ell i}(s) : 0 \le s \le t, i = 1, \ldots, n, \ell = 1, \ldots, r\}$$

for each $t \ge 0$. Suppose the left-continuous adapted process $Y_{\ell i}$ at t indicates whether the ith individual is "at risk" to have an observed event E_ℓ at time t. The logrank statistic to test the hypothesis, H_ℓ, that treatment is not associated with time to event E_ℓ can be written

$$V_\ell^{(n)} = \sum_{i=1}^{n} \int_0^\infty \left\{ Z_i - \frac{\sum_{j=1}^n Y_{\ell j}(s) Z_j}{\sum_{j=1}^n Y_{\ell j}(s)} \right\} dN_{\ell i}.$$

a. For any $\ell, \ell' \in \{1, \ldots, r\}$, verify that $\text{cov}(V_\ell^{(n)}, V_{\ell'}^{(n)}) = 0$, whenever:

i. the compensator for $N_{\ell i}$ satisfies $A_{\ell i} = \int_0^\cdot Y_{\ell i}(s)\lambda_\ell(s)ds$ for some hazard functions $\{\lambda_\ell : \ell = 1, \ldots, r\}$, and

ii. $\{N_{\ell i} : i = 1, \ldots, n; \ell = 1, \ldots, r\}$ is a multivariate counting process.

Hereafter, suppose $r = 2$ and (T_{1i}, T_{2i}) have joint distribution specified by

$$P(T_{1i} > t, T_{2i} > s) = \exp\{-t\mu_{1Z_i} - s\mu_{2Z_i} - \theta st\}$$

for $s, t \ge 0$, and $\theta \in [0, (\mu_{10}\mu_{20}) \wedge (\mu_{11}\mu_{21})]$.

b. Even though T_{1i} and T_{2i} are dependent when $\theta > 0$, verify that conditions (i) and (ii) hold in the competing risks setting (i.e. when $N_{\ell i}(t) = I_{\{T_{\ell i} \le (t \wedge T_{\ell' i})\}}$, $\ell' \ne \ell$, and $Y_{\ell i}(t) = I_{\{T_{1i} \wedge T_{2i} \ge t\}}$), as long as H_ℓ holds (i.e., $\mu_{\ell Z_i} = \mu_\ell$) for $\ell = 1, 2$. Find an expression for $\lambda_\ell, \ell = 1, 2$.

c. Suppose $N_{\ell i}(t) = I_{\{T_{\ell i} \le t\}}$ and $Y_{\ell i}(t) = I_{\{T_{\ell i} \ge t\}}$. Verify that, for $0 \le s < t$,

$$P\{T_{\ell i} \in [t, t + dt) | Z_i, T_{\ell i} \ge t, T_{\ell' i} \ge t\} = (\mu_{\ell Z_i} + \theta t)dt,$$

while

$$P\{T_{\ell i} \in [t, t + dt) | Z_i, T_{\ell i} \ge t, T_{\ell' i} \in [s, s + ds)\}$$
$$= \left(\mu_{\ell Z_i} + \theta s - \frac{\theta}{\mu_{\ell' Z_i} + \theta t} \right) dt.$$

In turn, deduce that for any $t \ge 0$,

$$A_{\ell i}(t | Z_i) = \int_0^{t \wedge T_{\ell' i}} Y_{\ell i}(x)(\mu_{\ell Z_i} + \theta x)dx$$
$$+ \int_{t \wedge T_{\ell' i}}^{t} Y_{\ell i}(x) \left(\mu_{\ell Z_i} + \theta T_{\ell' i} - \frac{\theta}{\mu_{\ell' Z_i} + \theta x} \right) dx.$$

Suppose $H_\ell : \mu_{\ell Z_i} = \mu_\ell$ holds for $\ell = 1, 2$. Observe that (i) holds if and only if $\theta = 0$, and so $\text{cov}(V_\ell^{(n)}, V_{\ell'}^{(n)})$ need not be zero when $\theta > 0$.

6.1 In Chapter 6, the Breslow-Crowley Theorem (i.e. Theorem 6.3.1 (1)) is proved using Part (1) of Theorem 6.2.1, which in turn is established using Theorem 5.3.4. Instead, develop a proof of the Breslow-Crowley Theorem using Theorem 5.3.2. Note that one can drop Condition (1.4) in Theorem 5.3.2 by using the argument in the proof of Theorem 5.3.3.

6.2 Using Eq. (5.12) of Chapter 7 and the convergence result in Eq. (3.4) of Chapter 6, develop expressions for one-sided Gill bands over $[0, t]$ when $\pi(t) > 0$.

6.3 Check the validity of Chapter 6 Conditions (3.7), (3.8), and (3.15) in Case 3 of Figure 6.3.4.

6.4 **a.** If $v(t) = \int_0^t \pi^{-1}(s)d\Lambda(s)$ and $K(t) \equiv v(t)/\{1+v(t)\}$, show that the Hall–Wellner confidence band will be at its narrowest relative to the estimated standard error of $\hat{F}$ when $K \approx \frac{1}{2}$.

b. Show that the Gill bands over an interval $[0, t_0]$ are narrowest, relative to the estimated standard error of $\hat{F}$, near t_0.

6.5 Consider the special case of the general Random Censorship Model (Definition 3.1.1) for $r = 1$ sample, in which T_j and U_j are independent, F is continuous and $L_j \equiv L$. Since $\pi(t) = S(t)C(t-), u = \sup\{t : S(t)C(t-) > 0\}$. Define $U \equiv \{U(t) : t \geq 0\}$ for each $t \geq 0$ by

$$U(t) = \hat{S}(t) \int_0^t \{n\hat{C}(s-)\}^{1/2} I_{\{\bar{Y}(s)>0\}} d\{\Lambda(s) - \hat{\Lambda}(s)\},$$

where $\hat{S}$ and $\hat{C}$ are Kaplan-Meier estimators and $\hat{\Lambda}$ is Nelson's estimator.

a. Establish

$$U \Longrightarrow W^o(F) \text{ over } D[0, t] \text{ as } n \to \infty,$$

for any $t \in \mathcal{I} \equiv \{t : S(t)C(t-) > 0\}$, where W^o denotes a Brownian bridge process. (Note, under our assumptions, $\bar{Y} = n\hat{S}^-\hat{C}^-$ a.s.).

b. Establish

$$U \Longrightarrow W^o(F) \text{ over } D[0, u] \text{ as } n \to \infty,$$

whenever

$$S(u) > 0.$$

Note that $S(u) > 0$ holds in a prospective clinical trial as illustrated in Case 2 of Figure 6.3.4.

c. Establish, whenever $S(u) > 0$,

$$\sup_{0 \leq t \leq u} |U(t)| \xrightarrow{\mathcal{D}} \sup_{0 \leq t \leq u} |W^o(F(t))|,$$

which can be used to formulate a two-sided goodness-of-fit test for a hypothesized F_0.

d. Suppose one rejects

$$H_0 : F = F_0 \text{ over } [0, u]$$

whenever $\sup_{0 \le t \le u} |U(t)|$ exceeds the upper $(1-\alpha)$ quantile of the distribution of $\sup_{0 \le t \le u} |W^o(F(t))|$, where $U(t)$ is defined by replacing $\Lambda(s)$ by $\{-\ln F_0(t)\}$. What types of departures of F_0 from the true underlying F should such a test be sensitive in detecting?

7.1 In Theorem 7.2.1, the G^ρ statistic arises when the weight function is $W(t) = \{\hat{S}(t-)\}^\rho$ for a constant $\rho \ge 0$. Assume $L_{ij}^n = L_i \equiv 1 - C_i$ and $F_i^n = F_i$ for all i, j, n. Also assume independence between censoring and survival distributions, so $\pi_i(t) = S_i(t-)C_i(t-)$. The theorem indicates, as $n \to \infty$, under $H_0 : F_1 = F_2$,

$$\frac{G^\rho(T)}{\{\hat{\sigma}^2(T)\}^{1/2}} \xrightarrow{\mathcal{D}} N(0, 1)$$

where $\hat{\sigma}^2$ is a consistent estimator of σ^2.

a. Obtain an expression for σ^2 and verify it is dependent upon the censoring distribution.

b. A statistic which is closely related to G^ρ is

$$\tilde{G}^\rho(\cdot) = \int_0^{\cdot} \{\hat{S}(t-)\}^{\rho+1/2} \left\{ \frac{\overline{Y}_1(t)\overline{Y}_2(t)}{\overline{Y}_1(t) + \overline{Y}_2(t)} \right\}^{1/2} \left\{ \frac{d\overline{N}_1(t)}{\overline{Y}_1(t)} - \frac{d\overline{N}_2(t)}{\overline{Y}_2(t)} \right\}.$$

Verify that, as $n \to \infty$, under $H_0 : F_1 = F_2$,

$$\tilde{G}^\rho(u) \xrightarrow{\mathcal{D}} N(0, \sigma_0^2(u)),$$

where $u = \sup\{t : \pi_1(t) \wedge \pi_2(t) > 0\}$.

Give the expression for $\sigma_0^2(u)$ and observe that it is independent of the censoring distribution, unlike $\sigma^2(u)$. Hence $\tilde{G}^\rho$, unlike G^ρ, is in Leurgans' (1983) "approximately distribution-free" class.

(Hint: Assume or prove for the Kaplan-Meier estimator $\hat{S}$ that

$$\lim_{t \uparrow u} \limsup_{n \to \infty} P\left\{ \int_t^T I_{\{t \le T\}} \hat{S} d\Lambda > \epsilon \right\} = 0$$

for any $\epsilon > 0$.)

7.2 Verify Condition (2b) of Corollary 7.2.1 fails to hold for the statistic of the class $\mathcal{K}$ in Example 7.2.1.

(Hint: For any $t_0 \in [0, u)$,

$$E \int_{t_0}^u I_{\{\overline{Y}_2(t)=1\}} I_{\{\overline{Y}_1(t)>1\}} \overline{Y}_2(t) d\Lambda(t) = E \int_{t_0}^u I_{\{\overline{Y}_2(t)=1\}} I_{\{\overline{Y}_1(t)>1\}} d\overline{N}_2(t).$$

The right-hand side, which is the expectation of the indicator of the event that the last failure in sample 2 occurs after t_0 and occurs with at least 2 items at risk in sample 1, monotonically increases to $\frac{1}{4}$ as $n_0 \to \infty$.)

7.3 Construct an example in which a one-sided statistic of the class $\mathcal{K}$ is not consistent against a stochastic ordering alternative.

7.4 Verify the G^ρ statistic is efficient against the alternative specified by Eq. (4.24) in Chapter 7.

7.5 The statistics of the class $\mathcal{K}$ were derived in Section 7.4 as providing efficient tests relative to a sequence $(\lambda_{\theta_1^n}, \lambda_{\theta_2^n})$, for θ_i^n converging to θ_0 as specified by Eq. (4.14) in Chapter 7. They can also be obtained as partial likelihood score statistics. Specifically, by writing the partial likelihood

$$L(\theta) = \prod_{t\geq 0}\left\{\frac{\overline{Y}_1(t)\lambda_\theta(t)}{\overline{Y}_1(t)\lambda_\theta(t)+\overline{Y}_2(t)\lambda_{\theta_0}(t)}\right\}^{\Delta\overline{N}_1(t)}\left\{\frac{\overline{Y}_2(t)\lambda_{\theta_0}(t)}{\overline{Y}_1(t)\lambda_\theta(t)+\overline{Y}_2(t)\lambda_{\theta_0}(t)}\right\}^{\Delta\overline{N}_2(t)},$$

show that the partial likelihood score statistic at $\theta = \theta_0$ is given by

$$\frac{\partial}{\partial\theta}\log L(\theta)\bigg|_{\theta=\theta_0} = \int\left(\frac{\partial}{\partial\theta}\log\lambda_\theta\bigg|_{\theta=\theta_0}\right)\frac{\overline{Y}_1\overline{Y}_2}{\overline{Y}_1+\overline{Y}_2}\left(\frac{d\overline{N}_1}{\overline{Y}_1}-\frac{d\overline{N}_2}{\overline{Y}_2}\right).$$

8.1 To complete the proof of Theorem 8.3.1, establish that $X_n(\beta,\tau)$ is a concave function of β with a unique maximum and that $A(\beta,\tau)$ has a unique maximum at $\beta = \beta_0$.

8.2 Through the following steps, establish that the convergence in Eq. (5.22) of Chapter 4 does follow from Eq. (4.8) of Chapter 8 and the consistency of $n^{-1}\mathcal{I}(\hat{\beta},\infty)$.

a. Fix $j \in \{1,\ldots,p\}$, and set $\sigma_{jk} \equiv \{\Sigma(\beta_0,t)\}_{jk}$. Recall $\Sigma(\beta_0,\cdot) = \int_0^{\cdot} \mathbf{v}(\beta_0,x)s^{(0)}(\beta_0,x)\lambda_0(x)dx$. Show that, if $\{\mathbf{v}(\beta,\cdot)\}_{jk}$ is proportional to $\{\mathbf{v}(\beta,\cdot)\}_{jj}$ for every k, (over the region where $s^{(0)}(\beta_0,\cdot)\lambda_0(\cdot) > 0$), then $\{\mathbf{v}(\beta,\cdot)\}_{jk} = \{\mathbf{v}(\beta,\cdot)\}_{jj}\{\sigma_{jk}(\infty)/\sigma_{jj}(\infty)\}$ and $\sigma_{jk}(\cdot) = \sigma_{jk}(\infty)\ \{\sigma_{jj}(\cdot)/\sigma_{jj}(\infty)\}$ for every k.

b. Show, for $0 \leq s \leq t$,

$$\begin{aligned}&\{\sigma_{jj}(\infty)\}\text{cov}[W^o\{\sigma_{jj}(s)/\sigma_{jj}(\infty)\}, W^o\{\sigma_{jj}(t)/\sigma_{jj}(\infty)\}]\\ &= \sigma_{jj}(s)\left\{1-\frac{\sigma_{jj}(t)}{\sigma_{jj}(\infty)}\right\}\\ &= [\Sigma(\beta_0,s)-\Sigma(\beta_0,s)\{\Sigma(\beta_0,\infty)\}^{-1}\Sigma(\beta_0,t)]_{jj}\end{aligned}$$

whenever $\sigma_{jk}(\cdot) = \sigma_{jk}(\infty)\{\sigma_{jj}(\cdot)/\sigma_{jj}(\infty)\}$ for every k.

Bibliography

Aalen, O. O. 1975. Statistical inference for a family of counting processes. Ph.D. dissertation, University of California, Berkeley.

Aalen, O. O. 1976. Nonparametric inference in connection with multiple decrement models. *Scand. J. Statist.* **3**: 15–27.

Aalen, O. O. 1977. Weak convergence of stochastic integrals related to counting processes. *Z. Wahrsch. Verw. Gebiete.* **38**: 261–77.

Aalen, O. O. 1978a. Nonparametric estimation of partial transition probabilities in multiple decrement models. *Ann. Statist.* **6**: 534–45.

Aalen, O. O. 1978b. Nonparametric inference for a family of counting processes. *Ann. Statist.* **6**: 701–26.

Aalen, O. O. and S. Johansen. 1978. An empirical transition matrix for nonhomogeneous Markov chains based on censored observations. *Scand. J. Statist.* **5**: 141–50.

Altschuler, B. 1970. Theory for the measurement of competing risks in animal experiments. *Math. Biosci.* **6**: 1–11.

Andersen, P. K. 1982. Testing goodness-of-fit for Cox's regression and life model. *Biometrics* **38**: 67–77.

Andersen, P. K. 1985. Statistical models for longitudinal labor market data based on counting processes, in *Longitudinal Analysis of Labor Market Data*, J. J. Heckman and B. Singer, eds. London: Cambridge University Press.

Andersen, P. K. 1986. *Time-dependent covariates and Markov processes*, in *Modern Statistical Methods in Chronic Disease Epidemology*. S. Moolgavkar and R. Prentice, eds. New York: Wiley.

Andersen, P. K., and R. D. Gill. 1982. Cox's regression model for counting processes: A large sample study. *Ann. Statist.* **10**: 1100–20.

Andersen, P. K., O. Borgan, R. D. Gill, N. Keiding. 1982. Linear nonparametric tests for comparison of counting processes with application to censored survival data (with discussion). *Int. Statist. Rev.* **50**: 219–58.

Andersen, P. K., and O. Borgan. 1985. Counting process models for life history data: A review. *Scand. J. Statist.* **12**: 97–158.

Apostol, T. M., 1974. *Mathematical Analysis*, 2nd Ed., Reading, MA: Addison-Wesley.

Aranda-Ordaz, R. J., 1983. An extension of the proportional hazards model for grouped data. *Biometrics* **39**: 109–17.

Arjas, E., and P. Haara. 1984. A marked point process approach to censored failure data with complicated covariates. *Scand. J. Statist.* **11**: 193–209.

Armitage, P. 1959. The comparison of survival curves. *J. Roy. Statist. Soc. A.* **122**: 279–92.

Ash, R. B. 1972. *Real Analysis and Probability*, New York: Academic Press.

Bailey, K. R. 1979. The general ML approach to the Cox regression model. Ph.D. Dissertation, University of Chicago.

Bailey, K. R. 1980. A comparison of the Cox estimator and the general ML estimators of β in the Cox model. Unpublished manuscript.

Bailey, K. R. 1983. The asymptotic joint distribution of regression and survival parameter estimates in the Cox regression model. *Ann. of Statist.* **11**: 39–58.

Barlow, W. E., and R. L. Prentice. 1988. Residuals for relative risk regression. *Biometrika* **75**: 65–74.

Barlow, R. E., and F. Proschan. 1975. *Statistical Theory of Reliability and Life Testing*, New York: Holt, Rinehart and Winston.

Belsley, D. A., E. Kuh, and R. E. Welsch. 1980. *Regression Diagnostics*, New York: Wiley.

Berkson, J., and R. Gage. 1950. Calculation of survival rates for cancer. *Proceedings of the Staff Meetings of the Mayo Clinic* **25**: 270–86.

Bickel, P. J., C. A. Klaassen, Y. Ritov, and J. A. Wellner. 1990. Efficient and adaptive estimation in semiparametric models. Forthcoming monograph, Baltimore: Johns Hopkins University Press.

Billingsley, P. 1968. *Convergence of Probability Measures*, New York: Wiley.

Billingsley, P. 1986. *Probability and Measure*, 2nd Ed., New York: Wiley.

Boel, R., P. Varaiya, and E. Wong. 1975a. Martingales on jump processes. Part I: Representation results. *SIAM. J. Control.* **13**: 99–1021.

Boel, R., P. Varaiya, and E. Wong. 1975b. Martingales on jump processes. Part II: Applications. *SIAM. J. Control.* **13**: 1022–61.

Borgan, O. 1983. Maximum likelihood estimation in a parametric counting process model, with applications to censored failure time data and multiplicative models. Technical Report No. 18, Department of Mathematics and Statistics, Agricultural University of Norway.

Borgan, O. 1984. Maximum likelihood estimation in parametric counting process models, with applications to censored failure time data. *Scand. J. Statist.* **11**: 1–16. Correction **11**: 275.

Breiman, L., J. Friedman, R. Olshen, and C. Stone. 1984. *Classification and Regression Trees*, Belmont: Wadsworth.

Brémaud, P. 1972. A martingale approach to point processes. Memorandum ERL–M345, University of California, Berkeley.

Brémaud, P. 1974. The martingale theory of point processes over the real half line admitting an intensity. *Proc. of the IRIA Coll. on Control Theory* Lect. Notes (grey) 107 Springer-Verlag. 519–42.

Brémaud, P. 1981. *Point Processes and Queues: Martingale Dynamics*, New York: Springer-Verlag.

Brémaud, P., and J. Jacod. 1977. Processus ponctuels et martingales: Résultats récents sur la modélisation et le filtrage. *Advances in Applied Probability* **9**: 362–416.

Breslow, N. E. 1970. A generalized Kruskal-Wallis test for comparing K samples subject to unequal patterns of censorship. *Biometrika* **57**: 579–94.

Breslow, N. E. 1972. Contribution to the discussion on the paper by DR Cox, Regression and life tables. *J. Roy. Statist. Soc. B.*, **34**: 216–7.

Breslow, N. E. 1974. Covariance analysis of censored survival data. *Biometrics* **30**: 89–99.

Breslow, N. E. 1975. Analysis of survival data under the proportional hazards model. *Int. Stat. Rev.* **43**: 45–58.

Breslow, N. E., and J. J. Crowley. 1974. A large sample study of the life table and product limit estimates under random censorship. *Ann. Statist.* **2**: 437–53.

Brown, B. M. 1971. Martingale central limit theorem. *Ann. Math. Statist.* **42**: 513–35.

Cain, K. C., and N. T. Lange. 1984. Approximate case influence for the proportional hazards regression model with censored data. *Biometrics* **40**: 493–9.

Chiang, C. L. 1968. *Introduction to Stochastic Processes in Biostatistics*, New York: John Wiley.

Chen, Y. Y., M. Hollander, and N. A. Langberg. 1982. Small sample results for the Kaplan-Meier estimator. *J. Am. Statist. Ass.* **77**: 141–4.

Chung, K. L. 1968. *A Course in Probability Theory*, New York: Harcourt, Brace and World.

Chung, K. L. 1974. *A Course in Probability Theory*, 2nd Ed., New York: Academic Press.

Clayton, D., and J. Cuzick. 1985. Multivariate generalizations of the proportional hazards model. *J. Roy. Statist. Soc. A.* **148**: Part 2, 82–117.

Cleveland, W. S. 1979. Robust locally weighted regression and smoothing scatterplots. *J. Am. Statist. Ass.* **74**, 829–36.

Conover, W. J., and R. L. Iman. 1981. Rank transformations as a bridge between parametric and nonparametric statistics. *The American Statistician* **35**: 124–9.

Cook, R. D., and S. Weisberg. 1982. *Residuals and Influence in Regression*, New York: Chapman and Hall.

Cox, D. R. 1959. The analysis of exponentially distributed life-times with two types of failure. *J. Roy. Statist. Soc. B.* **21**: 411–21.

Cox, D. R. 1972. Regression models and life tables (with discussion). *J. Roy. Statist. Soc. B.* **34**: 187–220.

Cox, D. R., 1975. Partial likelihood. *Biometrika* **62**: 269–76.

Cox, D. R., and D. Oakes. 1984. *Analysis of Survival Data*, London: Chapman and Hall.

Cox, D. R., and E. J. Snell. 1968. A general definition of residuals (with discussion) *J. Roy. Statist. Soc. B.* **30**: 248–75.

Crowley, J. J. 1970. A comparison of several life table estimates. Master's Thesis, University of Washington.

Crowley, J. J., and M. Hu. 1977. Covariance analysis of heart transplant survival data. *J. Am. Statist. Ass.* **72**: 27–36.

Crowley, J. J., and D. R. Thomas. 1975. Large sample theory for the logrank test. Technical Report No. 415, Department of Statistics, University of Wisconsin.

Csörgő, S., and L. Horváth. 1981. On the Koziol-Green model of random censorship. *Biometrika* **68**: 391–401.

Csörgő, S., and L. Horváth. 1986. Confidence bands from censored samples. *Canad. J. Statist.* **14**: 131–44.

Cuzick, J. 1985. Asymptotic properties of censored linear rank tests. *Ann. Statist.* **13**, 133–41.

Cuzick, J. 1988. Rank regression. *Ann. Statist.* **16**: 1369–89.

Dellacherie, C. 1972. *Capacités et Processus Stochastiques*, Berlin: Springer-Verlag.

Dellacherie, C. 1980. Un survol de la théorie de l'intégrale stochastique. *Stoch. Proc. Appl.* **10**: 115–44.

Dellacherie, C., and P. A. Meyer. 1975. *Probabilite et Potentiel* Vol 1., Paris: Hermann.

Dellacherie, C., and P. A. Meyer. 1980. *Probabilite et Potentiel* Vol 2., Paris: Hermann.

Dickson, E. R., T. R. Fleming, R. H. Wiesner, W. P. Baldus, C. R. Fleming, J. Ludwig, and J. T. McCall. 1985. Trial of penicillamine in advanced primary biliary cirrhosis. *N. Eng. J. Med.* **312**: 1011–5.

Dickson, E. R., P. M. Grambsch, T. R. Fleming, L. D. Fisher, and A. Langworthy. 1989. Prognosis in primary biliary cirrhosis: Model for decision making. *Hepatology* **10**: 1–7.

Doléans-Dade, C. 1969. Variation quadratique des martingales continues à droite. *Ann. Math. Statist.* **40**: 284–9.

Doléans-Dade, C., and P. A. Meyer. 1970. Intégrales stochastiques par rapport aux martingales locales. *Séminaire de Probabilités IV. Lecture Notes in Mathematics* **124**: 77–107. Berlin: Springer-Verlag.

Dolivo, F. 1974. Counting processes and integrated conditional rates: A martingale approach with application to detection. Technical Report, College of Engineering, University of Michigan.

Doob, J. L. 1949. Heuristic approach to the Kolmogorov-Smirnov theorems, *Ann. Math. Statistics*. **20**: 393–403.

Doob, J. L. 1953. *Stochastic Processes*, New York: Wiley.

Durrlemans, S., and R. Simon. 1989. Flexible regression methods with cubic splines. *Statistics in Medicine* **8**: 557–61.

Dvoretzky, A. 1972. Asymptotic normality for sums of dependent random variables. *Proc. Sixth Berkeley Symposium on Mathematical Statistics and Probability* 513–35.

Efron, B. 1967. The two-sample problem with censored data, *Proc. Fifth Berkeley Symposium on Mathematical Statistics and Probability* **4**, 831–53.

Efron, B. 1977. The efficiency of Cox's likelihood function for censored data. *J. Am. Statist. Ass.* **72**: 557–65.

Elliott, R. J. 1982. *Stochastic Calculus and Applications*, New York: Springer-Verlag.

Elveback, L. 1958. Actuarial estimation of survivorship in chronic disease. *J. Am. Statist. Ass.* **53**: 420–40.

Feigl, P., and M. Zelen. 1965. Estimation of exponential survival probabilities with concomitant information. *Biometrics* **21**: 826–38.

Feller, W. 1966. *An Introduction to Probability Theory and its Applications*, Vol 2. New York: Wiley.

Fleming, T. R. 1978. Nonparametric estimation for nonhomogeneous Markov processes in the problem of competing risks. *Ann. Statist.* **6**, 1057–79.

Fleming, T. R., and D. P. Harrington. 1978. Estimation for discrete time non-homogeneous Markov chains. *Stochastic Processes and Their Applications* **7**: 131–9.

Fleming, T. R., and D. P. Harrington. 1981. A class of hypothesis tests for one and two samples of censored survival data. *Comm. Statist.* **10**: 763–94.

Fleming, T. R., and D. P. Harrington. 1984a. Nonparametric estimation of the survival distribution in censored data. *Commun. Statist. - Theor. Meth.* **13**, 2469–86.

Fleming, T. R., and D. P. Harrington. 1984b. Evaluation of censored survival data test procedures based on single and multiple statistics. *Topics in Applied Statistics*, 97–123, New York: Marcel Dekker.

Fleming, T. R., D. P. Harrington, and M. O'Sullivan. 1987. Supremum versions of the logrank and generalized Wilcoxon statistics. *J. Am. Statist. Ass.* **82**: 312–20.

Fleming, T. R., J. R. O'Fallon, P. C. O'Brien, and D. P. Harrington. 1980. Modified Kolmogorov-Smirnov test procedures with application to arbitrarily right censored data. *Biometrics* **36**: 607–26.

Gastwirth, J. L. 1985. The use of maximum efficiency robust tests in combining contingency tables and survival analysis. *J. Am. Statis. Ass.* **80**: 380–84.

Gehan, E. A. 1965. A generalized Wilcoxon test for comparing arbitrarily single-censored samples. *Biometrika* **52**: 203–23.

Gill, R. D. 1980. *Censoring and Stochastic Integrals*, Mathematical Centre Tracts 124, Mathematisch Centrum, Amsterdam.

Gill, R. D. 1983. Large sample behavior of the product-limit estimator on the whole line. *Ann. Statist.* **11**: 49–58.

Gill, R. D. 1984. Understanding Cox's regression model: A martingale approach. *J. Am. Statist. Ass.* **79**: 441–7.

Gill, R. D. 1989. Non- and semi-parametric maximum likelihood estimators and the von Mises method (Part 1). *Scand. J. Statist.* **16**: 97–128.

Gill, R. D. and S. Johansen. 1990. A survey of product-integration with a view towards application in survival analysis. *Ann. Statist.* **18**: 1501–55.

Gill, R. D. and M. Schumacher. 1987. A simple test of the proportional hazards assumption. *Biometrika* **74**: 289–300.

Gillespie, M., and L. Fisher. 1979. Confidence bands for the Kaplan-Meier survival curve estimates. *Ann. Statist.* **7**: 920–4.

Gore, S. M., S. J. Pocock, and G. R. Kerr. 1984. Regression models and nonproportional hazards in the analysis of breast cancer survival data. *Appl. Statist.* **33**: 176–95.

Grambsch, P. M., E. R. Dickson, M. Kaplan, G. LeSage, T. R. Fleming, and A. L. Langworthy. 1989. Extramural cross-validation of the Mayo primary biliary cirrhosis survival model establishes its generalizability. *Hepatology* **10**: 846–50.

Gray, R. 1990. Some diagnostic methods for Cox regression models through hazard smoothing. *Biometrics* **46**: 93–102.

Greenwood, M. 1926. The natural duration of cancer. *Reports on Public Health and Medical Subjects* **33**: 1–26, London: Her Majesty's Stationery Office.

Greenwood, P. E., and W. Wefelmeyer. 1989a. Efficiency of estimators for partially specified filtered models. Preprints in Statistics, University of Cologne, Number 119.

Greenwood, P. E., and W. Wefelmeyer. 1989b. Efficient estimating equations for nonparametric filtered models. Preprints in Statistics, University of Cologne, Number 120.

Gross, A. J., and V. A. Clark. 1975. *Survival Distributions: Reliability Applications in the Biomedical Sciences*, New York: Wiley.

Hájek, J., and Z. Šidák. 1967. *Theory of Rank Tests*, New York: Academic Press.

Hall, P., and C. C. Heyde. 1980. *Martingale Limit Theory and Its Applications*, New York: Academic Press.

Hall, W. J., and J. A. Wellner. 1980. Confidence bands for a survival curve from censored data. *Biometrika* **67**: 133–43.

Harrell, F. E. 1986. The PHGLM procedure. *SAS Supplemental Library User's Guide*, Version 5. Cary, NC: SAS Institute, Inc.

Harrell, F. E., and K. L. Lee. 1986. Verifying assumptions of the Cox proportional hazards model. *Proceedings of the eleventh international conference of the SAS user's group*. Atlanta Georgia, February 9–12. 823–8.

Harrington, D. P. 1983. Review of *Statistical Analysis of Counting Processes* by M. Jacobsen. *Bull. Amer. Math. Soc.* **9**: 401–7.

Harrington, D. P., and T. R. Fleming. 1982. A class of rank test procedures for censored survival data. *Biometrika* **69**: 133–43.

Harrington, D. P., T. R. Fleming, and S. J. Green. 1982. Procedures for serial testing in censored survival data. In *Survival Analysis* (Crowley J, Johnson RA, eds.) Hayward CA: Institute of Mathematical Statistics, 287–301.

Hastie, T., and R. Tibshirani. 1986. Generalized additive models (with discussion). *Statistical Science* **1**: 297–319.

Helland, I. S. 1982. Central limit theorems for martingales with discrete or continuous time. *Scand. J. Statist.* **9**: 79–94.

Hettsmansperger, T. P. 1984. *Statistical Inference Based on Ranks*, New York: Wiley.

Hollander, M., and E. Peña. 1989. Families of confidence bands for the survival function under the general random censorship model and the Koziol-Green model. *Can. J. Statist.* **17**: 59–74.

Hollander, M., and D. A. Wolfe. 1973. *Nonparametric Statistical Methods*, New York: Wiley.

Hollander, M., F. Proschan, and J. Sconing. 1985. Efficiency loss with the Kaplan-Meier estimator. FSU Statistics Report M707, AFOSR Technical Report No. 85–181.

Holt, J. D., and R. L. Prentice. 1974. Survival analysis in twin studies and matched-pair experiments. *Biometrika* **61**: 17–30.

Hyde, J. 1977. Testing survival under right censoring and left truncation. *Biometrika* **64**: 225–30.

International Chronic Granulomatous Disease Cooperative Study Group. 1991. A phase III study establishing efficacy of recombinant human interferon gamma for infection prophylaxis in chronic granulomatous disease. *N. Eng. J. Med.* (in press).

Jacobsen, M. 1982. *Statistical Analysis of Counting Processes*, Lecture Notes in Statistics, Vol 12, New York: Springer-Verlag.

Jacobsen, M. 1984. Maximum likelihood estimation in the multiplicative intensity model. *Int. Statist. Review*. **52**: 193–207.

Jacod, J. 1973. On the stochastic intensity of a random point process over the half-line. Technical Report No. 15, Department of Statistics, Princeton University.

Jacod, J. 1975. Multivariate point processes: Predictable projection, Radon-Nikodym derivatives, representations of martingales. *Z. Wahrsch. Gebiete*. **31**: 235–53.

Jacod, J. 1979. Calcul stochastique et problèmes de martingales. *Lecture Notes in Mathematics*, Vol 714, Berlin: Springer-Verlag.

Jacod, J., and A. N. Shiryayev. 1987. *Limit Theorems for Stochastic Processes*, Berlin: Springer-Verlag.

Johansen, S. 1983. The product limit estimator as maximum likelihood estimator. *Scand. J. Statist.* **5**: 195–99.

Johansen, S. 1983. An extension of Cox's regression model. *Int. Statist. Rev.* **51**: 258–62.

Johnson, M. E., H. D. Tolley, M. C. Bryson, and A. S. Goldman. 1982. Covariate analysis of survival data: A small-sample study of Cox's model. *Biometrics* **38**: 685–98.

Kalbfleisch, J. D., and A. A. McIntosh. 1977. Efficiency in survival distributions with time-dependent covariates. *Biometrika* **64**: 47–50.

Kalbfleisch, J. D., and R. L. Prentice. 1973. Marginal likelihoods based on Cox's regression and life model. *Biometrika* **60**: 267–78.

Kalbfleisch, J. D., and R. L. Prentice. 1980 *The Statistical Analysis of Failure Time Data* New York: Wiley.

Kaplan, E. L., and P. Meier. 1958. Nonparametric estimator from incomplete observations. *J. Amer. Statist. Ass.* **53**, 457–81.

Karlin, S., and H. M. Taylor. 1975. *A First Course in Stochastic Processes*, 2nd Ed., New York: Academic Press.

Karlin, S., and H. M. Taylor. 1981. *A Second Course in Stochastic Processes*, New York: Academic Press.

Karr, A. F. 1986. *Point Processes and their Statistical Inference*, New York: Marcel Dekker.

Karr, A. F. 1987. Maximum likelihood estimation in the multiplicative intensity model via sieves. *Ann. Statist.* **15**: 473–90.

Kay, R. 1977. Proportional hazards regression models and the analysis of censored survival data. *Appl. Statist.* **26**: 227–37.

Kay, R. 1984. Goodness-of-fit methods for the proportional hazards model: A review. *Revue d'Épidémiologie et de Santé Publique* **32**: 185–98.

Kimball, A. W. 1960. Estimation of mortality intensities in animal experiments. *Biometrics* **16**: 505–21.

Koziol, J. A. 1978. A two-sample Cramer-von Mises test for randomly censored data. *Biometrical J.* **20**: 603–8.

Kunita, H., and S. Watanabe. 1967. On square integrable martingales. *Nagoya Mathematics Journal* **30**: 209–45.

Lagakos, S. W. 1980. The graphical evaluation of explanatory variables in proportional hazards regression models. *Biometrika* **68**: 93–8.

Lai, T. L., and Z. Ying. 1990. Linear rank statistics in regression analysis with censored or truncated data. *J. Mult. Anal.* (in press).

Lam, F. C., and M. T. Longnecker. 1983. A modified Wilcoxon rank sum test for paired data. *Biometrika* **70**: 510–3.

Lancaster, A., and S. J. Nickell. 1980. The analysis of re-employment probabilities for the unemployed. *J. Roy. Statist. Soc. A.* **143**: 141–65.

Lawless, J. F. 1982. *Statistical Models and Methods for Lifetime Data* , New York: Wiley.

Lehmann, E. L. 1959. *Testing Statistical Hypotheses*, New York: Wiley.

Lehmann, E. L. 1975. *Nonparametrics: Statistical Methods Based on Ranks*, San Francisco: Holden Day.

Lenglart, E. 1977. Relation de domination entre deux processus. *Ann. Inst. Henri Poincaré* **13**: 171–9.

Leurgans, S. 1983. Three classes of censored data rank tests: Strengths and weaknesses under censoring. *Biometrika* **70**: 651–8.

Lin, D. Y., and T. R. Fleming. 1991. Confidence Bands for survival curves under the proportional hazards model. *Biometrika* (in press).

Liptser, R. S., and A. N. Shiryayev. 1977. *Statistics of Random Processes I, General Theory*, New York: Springer-Verlag.

Liptser, R. S., and A. N. Shiryayev. 1978. *Statistics of Random Processes II, Applications*, New York: Springer-Verlag.

Liptser, R. S., and A. N. Shiryayev. 1980. A functional central limit theorem for semimartingales. *Theory of Probability and its Applications* **25**: 667–88.

Littel, A. S. 1952. Estimation of the T-year survival rate from follow-up studies over a limited period of time. *Human Biology* **24**: 87–116.

Liu, P. Y., and J. Crowley. 1978. Large sample theory of the mle based on Cox's regression model for survival data. Technical Report 1, Wisconsin Clinical Cancer Center, Biostatistics, University of Wisconsin-Madison.

Lustbader, E. D. 1980. Time dependent covariates in survival analysis. *Biometrika* **67**:, 697–8.

Mann, N. R., R. E. Schafer, and N. D. Singpurwalla. 1974. *Methods for Statistical Analysis of Reliability and Life Data* New York: Wiley.

Mansfield, E. R., and M. D. Conerly. 1987. Diagnostic value of residual and partial residual plots. *Am. Statist.* **41**: 107–16.

Mantel, N. 1966. Evaluation of survival data and two new rank order statistics arising in its consideration. *Cancer. Chemother. Rep.* **50**: 163–70.

Mantel, N., and J. Ciminera. 1979. Use of logrank scores in the analysis of litter-matched data on time to tumor appearance. *Cancer Research* **39**: 4308–15.

Markus, B. H., E. R. Dickson, P. M. Grambsch, T. R. Fleming, V. Mazzaferro, G. Klintmalm, R. H. Wiesner, D. H. Van Thiel, and T. E. Starzl. 1989. Efficacy of liver transplantation in patients with primary biliary cirrhosis. *N. Eng. J. Med.* **320**: 1709–13.

Mau, J. 1986. On a graphical method for the detection of time-dependent effects of covariates in survival data. *Appl. Statist.* **35**: 245–55.

McCullagh, P., and J. A. Nelder. 1989. *Generalized Linear Models*, 2nd Ed., London: Chapman and Hall.

McLeish, D. L. 1974. Dependent central limit theorems and invariance principles. *Ann Probability* **2**: 620–8.

McLeish, D. L. 1983. Martingales and estimating equations for censored and aggregate data. Laboratory for Statistics and Probability. Technical Report No. 12, University of Ottawa, Ottawa, Canada.

Mehrotra, K. G., J. E. Michalek, and D. Mihalko. 1982. A relationship between two forms of linear rank procedures for censored data. *Biometrika* **69**: 674–6.

Meier, P. 1975. Estimation of a distribution function from incomplete observations, in *Perspectives in Probability and Statistics*, J. Gani, ed., 67–87, Sheffield, England: Applied Probability Trust.

Meyer, P. A. 1966. *Probability and Potentials*, Waltham MA: Blaisdell.

Meyer, P. A. 1967. Intégrales stochastiques. *Séminaire de Probabilités I. Lecture Notes in Mathematics* **39**: 72–94. Berlin: Springer-Verlag.

Meyer, P. A. 1976. Un cours sur les intégrales stochastiques. *Séminaire de Probabilités X. Lecture Notes in Mathematics* **511**: 245–400. Berlin: Springer-Verlag.

Miller, R. G. 1981. *Survival Analysis*, New York: Wiley.

Miller, R. G. 1983. What price Kaplan-Meier? *Biometrics* **39**: 1077–82.

Næs, T. 1982. The asymptotic distribution of the estimator for the regression parameter in Cox's regression model. *Scandinavian J. Statist.* **9**: 107–15.

Nair, V. 1984. Confidence bands for survival functions with censored data: A comparative study. *Technometrics* **26**: 265–75.

Nelson, W. 1969. Hazard plotting for incomplete failure data. *J. Qual. Technol.* **1**: 27–52.

Nelson, W. 1972. Theory and applications of hazard plotting for censored failure data. *Technometrics* **14**: 945–65.

Oakes, D. 1977. The asymptotic information in censored survival data. *Biometrika* **64**: 441–8.

Oakes, D. 1981. Survival times: Aspects of partial likelihood. *Int. Statist. Review.* **49**: 235–64.

O'Brien, P. C., and T. R. Fleming. 1979. A multi-stage procedure for clinical trials. *Biometrics* **35**: 549–56.

O'Brien, P. C., and T. R. Fleming. 1987. A paired Prentice-Wilcoxon test for censored paired data. *Biometrics* **43**: 169–80.

O'Connell, M. J., D. P. Harrington, J. D. Earle, G. J. Johnson, J. H. Glick, P. P. Carbone, R. H. Creech, R. S. Neiman, R. B. Mann, M. N. Silverstein. 1987. Prospectively randomized clinical trial of three intensive chemotherapy regimens for the treatment of advanced unfavorable histology non-Hodgkin's lymphoma. *J. of Clinical Oncology* **5**: 1329–39.

O'Connell, M. J., D. P. Harrington, J. D. Earle, G. J. Johnson, G. H. Glick, R. S. Neiman, M. N. Silverstein. 1988. Chemotherapy followed by consolidation radiotherapy for treatment of clinical stage II aggressive histology non-Hodgkin's lymphoma. *Cancer* **61**: 1754–58.

Peduzzi, P., R. Hardy, and T. Holford. 1980. A stepwise variable selection procedure for nonlinear regression models *Biometrics* **36**: 511–6.

Pepe, M. S., and T. R. Fleming. 1989. Weighted Kaplan-Meier statistics: A class of distance tests for censored survival data. *Biometrics* **45**: 497–507.

Pepe, M. S., and T. R. Fleming. 1990. Weighted Kaplan-Meier statistics: Large sample and optimality considerations. *J. Roy. Statist. Soc.*, in press.

Peterson, A. V. 1976. Bounds for a joint distribution function with fixed sub-distribution functions: applications to competing risks. *Proc. Natl. Acad. Sci. (US)* **73**: 11–13.

Peterson, A. V. 1977. Expressing the Kaplan-Meier estimator as a function of empirical subsurvival functions. *J Am. Statist. Ass.* **72**: 854–8.

Peto, R. 1972. Contribution to the discussion of paper by DR Cox. *J. Roy. Statist. Soc. B*. **34**: 205–7.

Peto, R., and J. Peto. 1972. Asymptotically efficient rank invariant test procedures (with discussion). *J. Roy. Statist. Soc. A*. **135**: 185–206.

Pierce, D. A., W. H. Steward, and D. J. Kopecky. 1979. Distribution-free regression analysis of grouped survival data. *Biometrics* **34**: 57–67.

Pike, M. C. 1966. A method of analysis of a certain class of experiments in carcinogenesis. *Biometrics* **22**: 142–61.

Pollard, D. 1984. *Convergence of Stochastic Processes*, New York: Springer-Verlag.

Prentice, R. L. 1978. Linear rank tests with right censored data. *Biometrika* **65**: 167–79.

Prentice, R. L., and L. A. Gloeckler. 1978. Regression analysis of grouped survival data with application to breast cancer data. *Biometrics* **34**: 57–67.

Prentice, R. L., and P. Marek. 1979. A qualitative discrepancy between censored data rank tests. *Biometrics* **35**: 861–7.

Prentice, R. L., and S. G. Self. 1983. Asymptotic distribution theory for Cox-type regression models with general relative risk form. *Ann. of Statist.* **11**: 804–13.

Prentice, R. L., J. D. Kalbfleisch, A. V. Peterson, N. Flournoy, V. T. Farewell, and N. Breslow. 1978. The analysis of failure time data in the presence of competing risks. *Biometrics* **34**: 541–54.

Ramlau-Hansen, H. 1983. Smoothing counting process intensities by means of kernel functions. *Ann. Statist.* **11**: 453–66.

Randles, R. H., and D. A. Wolfe. 1979. *Introduction to the Theory of Nonparametric Statistics*. New York: Wiley.

Rao, C. R. 1973. *Linear Statistical Inference and Its Applications*, 2nd ed. New York: Wiley.

Rebolledo, R. 1977. Remarques sur la convergence en loi de martingales vers des martingales continues. *C. R. Acad. Sc. Paris* **285**: 517–20.

Rebolledo, R. 1978. Sur les applications de la théorie des martingales à l'étude statistique d'une famille de processus ponctuels. Springer Lecture Notes in Mathematics, **636**: 27–70. Berlin: Springer-Verlag.

Rebolledo, R. 1979. La méthode des martingales appliquée à l'étude de la convergence en loi de processus. *Mémoires de la Soc. Math. de France* 62.

Rebolledo, R. 1980. Central limit theorems for local martingales. *Z Wahrsch Verw Gebiete* **51**: 269–86.

Ross, S. M. 1983. *Stochastic Processes*, New York: Wiley.

Royden, H. L. 1968. *Real Analysis* 2nd Ed., London: Macmillan.

Schmidt, P., and A. D. Witte. 1988. *Prediction Recidivism using Survival Models* Berlin: Springer-Verlag.

Schoenfeld, D. 1980. Chi-squared goodness-of-fit tests for the proportional hazards regression model. *Biometrika* **67**: 145–53.

Schumacher, M. 1984. Two-sample tests of Cramer-von Mises and Kolmogorov-Smirnov type for randomly censored data. *Int. Statist. Rev.* **52**: 263–81.

Segall, A., and T. Kailath. 1975a. The modeling of randomly modulated jump processes. *IEEE Trans. Inform. Theory* IT-21, **2**: 135–43.

Segall, A., and T. Kailath. 1975b. Radon-Nikodym derivatives with respect to measures induced by discontinuous independent-increment processes. *Ann. Probability* **3**: 449–64.

Self, S. G., and R. L. Prentice. 1982. Commentary on Andersen and Gill's Cox regression model for counting processes: A large sample study. *Ann. Statist.* **10**: 1121–4.

Sellke, T., and D. Sigmund. 1983. Sequential analysis of the proportional hazards model. *Biometrika* **70**: 315–26.

Sen, P. K. 1981. The Cox regression model, invariance principles for some induced quantile processes, and some repeated significance tests. *Ann. Statist.* **9**: 109–21.

Shiryayev, A. N. 1981. Martingales: Recent developments, results and applications. *Int. Statist. Rev.* **49**: 199–233.

Shorack, G. R., and J. A. Wellner. 1986. *Empirical Processes*, New York: Wiley.

Shorack, G. R., and J. A. Wellner, 1989. Corrections and Changes for *Empirical Processes with Applications to Statistics*, University of Washington Statistics Department Technical Report # 167.

Sleeper, L. A., and D. P. Harrington. 1990. Regression splines in the Cox model, with application to covariate effects in liver disease. *J. Am. Statist. Ass.* **85**: 941-49.

Stone, C. 1963. Weak convergence of stochastic processes defined on semi-infinite time intervals. *Proc. Am. Math. Soc.* **14**: 694–6.

Storer, B. E., and J. J. Crowley. 1985. A diagnostic for Cox regression and general conditional likelihoods. *J. Am. Statist. Ass.* **80**: 139–47.

Struthers, C. A. 1984. Asymptotic properties of linear rank tests with censored data. University of Waterloo Statistics Department Ph.D. Dissertation.

Tanner, M. A., and W. H. Wong. 1983. The estimation of the hazard function from randomly censored data by the kernel method. *Ann. Statist.* **11**: 989–93.

Tanner, M. A., and W. H. Wong. 1984. Data-based nonparametric estimation of the hazard function with applications to model diagnostics and exploratory analysis. *J. Am. Statist. Ass.* **79**: 174–82.

Tarone, R. E. 1981. On the distribution of the maximum of the logrank statistic and the modified Wilcoxon statistic. *Biometrics* **37**: 79–85.

Tarone, R. E., and J. Ware. 1977. On distribution-free tests for equality of survival distributions. *Biometrika* **64**: 156–60.

Therneau, T. M., P. M. Grambsch, and T. R. Fleming. 1990. Martingale-based residuals for survival models. *Biometrika* **77**: 147–60.

Tsiatis, A. A. 1975. A nonidentifiability aspect of the problem of competing risks. *Proc. Nat. Acad. Sci. USA*. **72**: 20–2.

Tsiatis, A. A. 1981. A large sample study of Cox's regression model. *Ann. Statist.* **9**: 93–108.

Tsiatis, A. A. 1990. Estimating regression parameters using linear rank tests for censored data. *Ann. Statist.* **18**: 354–72.

Tuma, N. B., M. T. Hannan, and L. P. Groeneveld. 1979. Dynamic analysis of event histories. *Amer. J. Sociology* **84**: 820–54.

van Zuijlen, M. C. A. 1978. Properties of the empirical distribution function for independent nonidentically distributed random variables. *Ann. Probability* **6**: 250–66.

Voelkel, J. G., and J. J. Crowley, 1984. Nonparametric inference for a class of semi-Markov processes with censored observations. *Ann. Statist.* **12**: 142–60.

Wang, J. G. 1987. A note on the uniform consistency of the Kaplan-Meier estimator. *Ann. Statist.* **15**: 1313–16.

Wei, L. J. 1980. A generalized Gehan and Gilbert test for paired observations that are subject to arbitrary right censorship. *J. Am. Statist. Ass.* **75**: 634–7.

Wei, L. J. 1984. Testing goodness-of-fit for the proportional hazards model with censored observations. *J. Am. Statist. Ass.* **79**: 649–52.

Wei, L. J., Z. Ying, and D. Y. Lin. 1990. Linear regression analysis of censored survival data based on rank tests. *Biometrika* **77**: 845–51.

Wellner, J. A. 1978. Limit theorems for the ratio of the empirical distribution function to the true distribution function. *Z. Wahrsch. verw. Gebiete.* **45**: 73–88.

Wellner, J. A. 1982. Asymptotic optimality of the product limit estimator. *Ann. of Statist.* **10**: 595–602.

Wellner, J. A. 1985. A heavy censoring limit theorem for the product limit estimator. *Ann. of Statist.* **13**: 150–62.

Whittemore, A. S., and J. B. Keller. 1986. Survival estimation using splines. *Biometrics* **42**: 495–506.

Winkler, H. Z., L. M. Rainwater, R. P. Meyers, G. M. Farrow, T. M. Therneau, H. Zinke, and M. M. Lieber. 1988. Stage D1 prostatic adenocarcinoma: Significance of nuclear DNA ploidy patterns studied by flow cytometry. *Mayo Clinic Proceedings*. **63**: 103–12.

Wong, W. H. 1986. Theory of partial likelihood. *Ann Statist* **14**: 88–123.

Woolson, R. F., and P. A. Lachenbruch. 1980. Rank tests for censored matched pair. *Biometrika* **67**: 597–606.

Ying, Z. 1989. A note on the asymptotic properties of the product-limit estimator on the whole line. *Stat. Prob. Letters* **7**: 311–14

Notation

A partial listing of symbols and other notation used frequently.

Ω	an arbitrary space, typically a space of outcomes of an experiment
ω	an arbitrary element of Ω
$\emptyset$	the empty set
$\bar{A}$	for any set A in Ω, $\bar{A}$ denotes its complement
$\mathcal{F}$	a σ–algebra of sets from Ω
P	a probability measure on Ω
R	the real line, $(-\infty, \infty)$
R^+	the positive real line, $[0, \infty)$
$\mathcal{B}$	the Borel σ–algebra of the real line, that is, the smallest σ–algebra containing all open intervals of R
$E(Y\|\mathcal{G})$	the conditional expectation of an $\mathcal{F}$–measurable random variable Y with respect to a σ–algebra $\mathcal{G} \subset \mathcal{F}$
$I_{\{A\}}$	indicator of an event or a set A. It is a $0, 1$ random variable which is unity over the event or set A; sometimes written I_A when A is a set in Ω.
$P(A\|\mathcal{G})$	$E(I_A\|\mathcal{G})$
$\sigma\{\cdot\}$	the σ–algebra generated by random variables in $\{\cdot\}$; e.g. $\mathcal{F}_t \equiv \sigma\{X(s) : 0 \leq s \leq t\}$ is the smallest σ–algebra such that $X(s)$ measurable for every $s \in [0, t]$, and $\{\mathcal{F}_t : t \geq 0\}$ is called the history of the stochastic process X.
$E(Y\|X)$	$E\{Y\|\sigma(X)\}$, for random variables X, Y
$\mathcal{F}_{t+}$	the σ–algebra $\bigcap_{h>0} \mathcal{F}_{t+h}$, that is, $A \in \mathcal{F}_{t+}$ if and only if $A \in \mathcal{F}_{t+h}$ for every $h > 0$

$\mathcal{F}_{t-}$	the smallest σ–algebra containing all the sets in $\bigcup_{h>0}\mathcal{F}_{t-h}$; also written $\sigma(\bigcup_{h>0}\mathcal{F}_{t-h})$ or $\bigvee_{h>0}\mathcal{F}_{t-h}$
$X(t-)$	$\lim_{h\downarrow 0} X(t-h)$, when X is a stochastic process with sample paths $X(\cdot,\omega)$ which are real valued functions on R^+, and where the limit exists on a set of probability one
X^-	the process defined by $X^-(t) = X(t-)$; $X(t+)$ and X^+ are defined in the obvious manner
$X(\infty)$	$\lim_{t\to\infty} X(t)$, if the limit exists on a set of probability one
ΔX	$X - X^-$, if X is right continuous with left hand limits
$\vee$	$x \vee y = \max(x, y)$
$\wedge$	$x \wedge y = \min(x, y)$
$X^{\oplus}$	$\max(X, 0)$
$X^{\ominus}$	$\max(-X, 0)$
X^T	$X^T(t) = X(t \wedge T)$
$\mathbf{a}$	bold symbols, such as $\mathbf{a}$, denote column vectors and matrices
$\mathbf{a}'$	the transpose of $\mathbf{a}$
$\mathbf{a}^{\otimes 2}$	for any vector $\mathbf{a}$, $\mathbf{a}^{\otimes 2} = \mathbf{a}\mathbf{a}'$
$\Vert \mathbf{a} \Vert$	for any vector or matrix $\mathbf{a}$, $\Vert \mathbf{a} \Vert = \max_{ij} \vert(\mathbf{a})_{ij}\vert$
$\mathcal{L}$	for any likelihood L, $\mathcal{L}$ denotes the log–likelihood
$\mathbf{I}$, $\mathcal{I}$	frequently denotes Fisher's information matrix and the corresponding observed information matrix
$\langle M_1, M_2\rangle$	the predictable covariation process for local square integrable martingales M_1, M_2
$[M, M]$	the quadratic variation process for a local square integrable martingale M
$N(\mu, \sigma^2)$	the normal distribution with mean μ and variance σ^2
Φ	the cumulative distribution function for the $N(0, 1)$ distribution
$\equiv$	equal, by definition
$\sim$	$X \sim F$ means X has distribution function F. More commonly, $X \sim N(\mu, \sigma^2)$ means X is normally distributed with mean μ, variance σ^2.
$\approx$	$x \approx y$ means x is approximately equal to y.

The following four types of convergence are defined in Chapter 5 and in Appendix B.

$\xrightarrow{P}$	convergence in probability
$\xrightarrow{\mathcal{D}}$	convergence in distribution
$\xrightarrow{L^p}$	L^p convergence
$\Longrightarrow$	weak convergence

Throughout Chapter 0, the item or subject will be indexed by the subscript j. In other chapters requiring double subscripting for sample and subject (i.e., Chapters 3, 6 and 7), by convention i will denote sample and j will denote subject. In Chapters 2, 4, 5 and 8, subject will be indexed by i.

Throughout, we use the convention $0/0 = 0$.

Author Index

Subject Index